工业和信息化部"十二五"规划教材
国家级双语教学示范课程建设项目

Theoretical Mechanics

理论力学(双语)

WANG Kaifu
王开福 编著

北京航空航天大学出版社

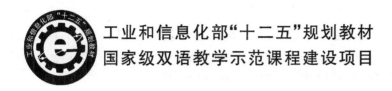

Synopsis

内容简介

This book consists of statics, kinematics, and kinetics. The main contents of the book include: statics of particle, reduction of force system, statics of rigid body, friction, kinematics of particle, kinematics of rigid body in plane motion, resultant motion of particle, kinetics of particle, kinetics of rigid body in plane motion, mechanical vibration, principle of virtual work, Lagrange's equations, and impact.

The book can be used as an English, Chinese, or bilingual textbook of theoretical mechanics for the student majoring in aeronautical, mechanical, civil, and hydraulic engineering.

本书由静力学、运动学和动力学组成。主要内容包括:质点静力学、力系简化、刚体静力学、摩擦、质点运动学、刚体平面运动学、质点合成运动、质点动力学、刚体平面动力学、机械振动、虚功原理、拉格朗日方程和碰撞。

本书可作为高等院校航空、机械、土木和水利等专业学生的英文、中文或双语理论力学教材。

图书在版编目(CIP)数据

理论力学 : 英汉对照 / 王开福编著. -- 北京 : 北京航空航天大学出版社, 2015.7

Theoretical Mechanics

ISBN 978-7-5124-1831-8

Ⅰ. ①理… Ⅱ. ①王… Ⅲ. ①理论力学—高等学校—双语教学—教材 Ⅳ. ①O31

中国版本图书馆 CIP 数据核字(2015)第 159519 号

Theoretical Mechanics

理论力学(双语)

WANG Kaifu 王开福 编著

责任编辑 宋淑娟 周方彦

*

北京航空航天大学出版社出版发行

北京市海淀区学院路 37 号(邮编 100191) http://www.buaapress.com.cn

发行部电话:(010)82317024 传真:(010)82328026

读者信箱:goodtextbook@126.com 邮购电话:(010)82316936

北京建宏印刷有限公司印刷 各地书店经销

*

开本:787×1 092 1/16 印张:19 字数:486 千字

2015 年 9 月第 1 版 2020 年 8 月第 4 次印刷 印数:1 601～2 100

ISBN 978-7-5124-1831-8 定价:59.00 元

Preface

Theoretical mechanics is a required subject for the student majoring in aeronautical, mechanical, civil, and hydraulic engineering and usually taught during the sophomore year. This book intends to provide the student with the theory and application of theoretical mechanics.

The book is written in English and Chinese, respectively; the English part of the book can be used as an English textbook of theoretical mechanics, while the Chinese part can be used as a Chinese textbook of theoretical mechanics.

Theoretical mechanics consists of statics, kinematics, and kinetics. Statics is the study of bodies at rest or in equilibrium, kinematics deals with the geometry of the motion without regard to the forces acting on bodies, and kinetics is with the relation between the motion of bodies and the forces acting on bodies.

The book is organized into fourteen chapters and two appendixes. Chapter 1 is an introduction to the fundamental concepts and general principles of theoretical mechanics. Chapter 2 discusses the resultant and equilibrium of concurrent forces acting on a particle. In Chapter 3 the reduction and equivalence of a force system acting on a rigid body are discussed, and in Chapter 4 the equilibrium of a rigid body, as well the internal force of a planar truss, is considered. The concepts of both sliding friction and rolling resistance are introduced in Chapter 5. The velocity and acceleration of a particle are analyzed in Chapter 6. Chapter 7 deals with the velocity and acceleration of a rigid body in translation, rotation, and general plane motion. The resultant motion of a particle is studied in Chapter 8. Chapter 9 and Chapter 10 are on the kinetics of a particle and of a rigid body in plane motion. Chapter 11 describes the mechanical vibrations of bodies. The principle of virtual work, and Lagrange's equations in analytical mechanics are, respectively, dealt with in Chapter 12, and 13. The impact is briefly introduced in Chapter 14.

The book can be used as an English, Chinese, or bilingual textbook of theoretical mechanics for the student majoring in aeronautical, mechanical, civil, and hydraulic engineering.

Kaifu Wang

Nanjing, December 2014

前　言

理论力学是航空、机械、土木和水利工程等专业学生的必修课，通常在大二学年讲授。本书旨在向学生传授理论力学的理论及其应用。

本书用英文和中文分别撰写：英文部分可作为理论力学的英文教材，中文部分可作为理论力学的中文教材。

理论力学由静力学、运动学和动力学组成。静力学研究物体的静止与平衡；运动学在不涉及作用力的情况下研究物体的运动；动力学研究物体的运动与作用力之间的关系。

全书由 14 章正文和 2 个附录组成。第 1 章介绍理论力学的基本概念与普遍原理。第 2 章讨论作用于质点上的汇交力系的合成与平衡。第 3 章讨论作用于刚体上的力系的简化与等效。第 4 章考虑刚体的平衡以及平面桁架的内力。第 5 章介绍滑动摩擦与滚动摩阻的概念。第 6 章分析质点的速度与加速度。第 7 章涉及平移、转动和一般平面运动刚体的速度与加速度。第 8 章研究质点合成运动。第 9 章和第 10 章分别研究质点动力学和刚体平面动力学。第 11 章描述物体的机械振动。第 12 章和第 13 章分别涉及分析力学的虚功原理和拉格朗日方程。第 14 章简要介绍碰撞。

本书可作为高等院校航空、机械、土木和水利工程等专业学生的英文、中文或双语理论力学教材。

王开福

2014 年 12 月于南京

Contents

Chapter 1 Basic Concepts and General Principles 1

1.1 Basic Concepts 1
1.2 General Principles 2

Chapter 2 Statics of Particle 5

2.1 Resultant of Coplanar Concurrent Forces 5
2.2 Equilibrium of Coplanar Concurrent Forces 10
2.3 Resultant of Spatial Concurrent Forces 12
2.4 Equilibrium of Spatial Concurrent Forces 14
Problems 15

Chapter 3 Reduction of Force System 19

3.1 Moment of Force about Point 19
3.2 Moment of Force about Given Axis 19
3.3 Principle of Moments 20
3.4 Components of Moment of Force about Point 20
3.5 Moment of Couple 22
3.6 Resultant of Couples 23
3.7 Equivalence of Force Acting on Rigid Body 24
3.8 Reduction of Force System 25
Problems 27

Chapter 4 Statics of Rigid Body 30

4.1 Equilibrium of Two-Dimensional Rigid Body 30
4.2 Two-Force and Three-Force Bodies 34
4.3 Planar Trusses 35
4.4 Equilibrium of Three-Dimensional Rigid Body 39
Problems 40

Chapter 5 Friction 44

5.1 Sliding Friction 44
5.2 Angles of Friction 45
5.3 Problems Involving Sliding Friction 46

5.4 Rolling Resistance 49
Problems 50

Chapter 6 Kinematics of Particle 53

6.1 Motion of Particle Represented by Vector 53
6.2 Motion of Particle Represented by Rectangular Coordinates 54
6.3 Motion of Particle Represented by Natural Coordinates 55
Problems 57

Chapter 7 Kinematics of Rigid Body in Plane Motion 59

7.1 Plane Motion of Rigid Body 59
7.2 Translation 60
7.3 Rotation about Fixed Axis 60
7.4 General Plane Motion 63
Problems 69

Chapter 8 Resultant Motion of Particle 71

8.1 Rates of Change of Vector 71
8.2 Resultant of Velocities 72
8.3 Resultant of Accelerations 73
Problems 76

Chapter 9 Kinetics of Particle 78

9.1 Equations of Motion of Particle 78
9.2 Method of Inertia Force for Particle in Motion 81
9.3 Method of Work and Energy for Particle in Motion 81
9.4 Method of Impulse and Momentum for Particle in Motion 84
Problems 86

Chapter 10 Kinetics of Rigid Body in Plane Motion 89

10.1 Motion for System of Particles 89
10.2 Motion of Mass Center of System of Particles 91
10.3 Motion of System of Particles about Its Mass Center 91
10.4 Equations of Motion for Rigid Body in Plane Motion 92
10.5 Method of Inertia Force for Rigid Body in Plane Motion 94
10.6 Method of Work and Energy for Rigid Body in Plane Motion 96
10.7 Method of Impulse and Momentum for Rigid Body in Plane Motion 98
Problems 100

Chapter 11 Mechanical Vibrations …… 104

11.1 Undamped Free Vibrations …… 104
11.2 Undamped Forced Vibrations …… 112
11.3 Damped Free Vibrations …… 115
11.4 Damped Forced Vibrations …… 119
Problems …… 121

Chapter 12 Principle of Virtual Work …… 127

12.1 Constraints and Virtual Work …… 127
12.2 Principle of Virtual Work …… 128
12.3 Generalized Coordinates and Generalized Forces …… 132
12.4 Conditions of Equilibrium Represented by Generalized Coordinates …… 133
Problems …… 133

Chapter 13 Lagrange's Equations …… 136

13.1 Lagrange's Equations …… 136
13.2 First Integrals of Lagrange's Equations …… 140
Problems …… 142

Chapter 14 Impact …… 144

14.1 Fundamental Principles Used for Impact …… 145
14.2 Coefficient of Restitution …… 146
Problems …… 150

Appendix Ⅰ Centers of Gravity and Centroids …… 153

Ⅰ.1 Center of Gravity and Centroid of Plate …… 153
Ⅰ.2 Center of Gravity and Centroid of Composite Plate …… 153
Ⅰ.3 Center of Gravity and Centroid of 3D Body …… 154
Ⅰ.4 Center of Gravityand Centroid of 3D Composite Body …… 155

Appendix Ⅱ Mass Moments of Inertia …… 157

Ⅱ.1 Moment of Inertia and Radius of Gyration …… 157
Ⅱ.2 Parallel-Axis Theorem …… 157

References …… 159

目　录

第 1 章　基本概念与普遍原理 …… 160

1.1　基本概念 …… 160
1.2　普遍原理 …… 161

第 2 章　质点静力学 …… 163

2.1　平面汇交力的合成 …… 163
2.2　平面汇交力的平衡 …… 167
2.3　空间汇交力的合成 …… 169
2.4　空间汇交力的平衡 …… 171
习　题 …… 172

第 3 章　力系简化 …… 175

3.1　力对点之矩 …… 175
3.2　力对轴之矩 …… 175
3.3　力矩定理 …… 175
3.4　力对点之矩的分量 …… 176
3.5　力偶矩 …… 178
3.6　力偶的合成 …… 179
3.7　作用于刚体上力的等效 …… 180
3.8　力系简化 …… 181
习　题 …… 182

第 4 章　刚体静力学 …… 185

4.1　二维刚体平衡 …… 185
4.2　二力和三力物体 …… 188
4.3　平面桁架 …… 189
4.4　三维刚体平衡 …… 192
习　题 …… 193

第 5 章　摩　擦 …… 197

5.1　滑动摩擦 …… 197
5.2　摩擦角 …… 198
5.3　含有滑动摩擦的问题 …… 198

5.4　滚动摩阻 …… 201
习　题 …… 202

第6章　质点运动学 …… 204

6.1　质点运动的矢量表示 …… 204
6.2　质点运动的直角坐标表示 …… 205
6.3　质点运动的自然坐标表示 …… 206
习　题 …… 207

第7章　刚体平面运动学 …… 209

7.1　刚体平面运动 …… 209
7.2　平　移 …… 210
7.3　定轴转动 …… 210
7.4　一般平面运动 …… 212
习　题 …… 218

第8章　质点合成运动 …… 219

8.1　矢量变化率 …… 219
8.2　速度合成 …… 219
8.3　加速度合成 …… 221
习　题 …… 223

第9章　质点动力学 …… 225

9.1　质点运动方程 …… 225
9.2　运动质点的惯性力法 …… 227
9.3　运动质点的功能法 …… 228
9.4　运动质点的冲量动量法 …… 230
习　题 …… 232

第10章　刚体平面动力学 …… 234

10.1　质点系的运动 …… 234
10.2　质点系质心的运动 …… 235
10.3　质点系相对质心的运动 …… 236
10.4　平面运动刚体的运动方程 …… 237
10.5　平面运动刚体的惯性力法 …… 239
10.6　平面运动刚体的功能法 …… 240
10.7　平面运动刚体的冲量动量法 …… 242
习　题 …… 243

第 11 章 机械振动 …… 247

11.1 无阻尼自由振动 …… 247
11.2 无阻尼受迫振动 …… 254
11.3 有阻尼自由振动 …… 256
11.4 有阻尼受迫振动 …… 259
习 题 …… 261

第 12 章 虚功原理 …… 266

12.1 约束与虚功 …… 266
12.2 虚功原理 …… 267
12.3 广义坐标和广义力 …… 270
12.4 平衡条件的广义坐标表示 …… 270
习 题 …… 271

第 13 章 拉格朗日方程 …… 273

13.1 拉格朗日方程 …… 273
13.2 拉格朗日方程的初积分 …… 277
习 题 …… 278

第 14 章 碰 撞 …… 280

14.1 用于碰撞的基本原理 …… 280
14.2 恢复系数 …… 281
习 题 …… 284

附录Ⅰ 重心与形心 …… 288

Ⅰ.1 薄板的重心与形心 …… 288
Ⅰ.2 组合薄板的重心与形心 …… 288
Ⅰ.3 三维物体的重心与形心 …… 289
Ⅰ.4 三维组合物体的重心与形心 …… 290

附录Ⅱ 转动惯量 …… 291

Ⅱ.1 转动惯量与回转半径 …… 291
Ⅱ.2 平行移轴定理 …… 291

参考文献 …… 293

Chapter 1 Basic Concepts and General Principles

Theoretical mechanics is the study of equilibrium or motion of bodies subjected to the action of forces, and consists of statics, kinematics and kinetics. Statics is the study of bodies at rest or in equilibrium; kinematics treats the geometry of the motion without regard to the forces acting on bodies; and kinetics deals with the relation between the motion of bodies and the forces acting on bodies.

In theoretical mechanics, bodies are assumed to be perfectly rigid. Though actual bodies are never absolutely rigid and deform under the action of forces, these deformations are usually small and do not affect the state of equilibrium or motion of bodies under consideration.

1.1 Basic Concepts

1. Length

Length is used to locate the position of a point in space. The position of a point can be defined by three lengths measured from a certain reference point in three given directions.

2. Time

Time is used to represent a nonspatial continuum in which events occur in irreversible succession from the past through the present to the future. To define an event, it is not sufficient to indicate its position in space. The time of the event should be given.

3. Mass

Mass is used to characterize the quantity of matter that a body contains. The mass of a body is not dependent on gravity and therefore is different from but proportional to its weight. Two bodies of the same mass, for example, will be attracted by the earth in the same manner; they will also offer the same resistance to a change in velocity.

4. Force

Force is used to represent the action of one body on another. A force tends to produce an acceleration of a body in the direction of its application. The effect of a force is completely characterized by its magnitude, direction, and point of application.

5. Particle

If the size and shape of a body do not affect the solution of the specific problem under consideration, then this body can be idealized as a particle, i.e., a particle has a mass, but its size and shape can be neglected. For example, the size and shape of the earth is insignificant compared to the size and shape of its orbit, and therefore the earth can be modeled as a particle when studying the orbital motion of the earth.

6. Rigid Body

A rigid body can be considered as a combination of a large number of particles in which all the particles occupy fixed positions with respect to each other within the body both before and after the action of forces, i.e., a rigid body is defined as one which does not deform when it is subjected to the action of forces.

7. Scalars

Scalars possess only magnitude, e. g. , length, time, mass, work, energy. Scalars are added by algebraic methods.

8. Vectors

Vectors possess both magnitude and direction (direction is understood to include both the inclination angle that the line of action makes with a given reference line and the sense of the vector along the line of action), e.g., force, displacement, impulse, momentum. Vectors are added by the parallelogram law.

9. Free Vectors

A free vector can be moved anywhere in space provided it remains the same magnitude and direction.

10. Sliding or Slip Vectors

A sliding or slip vector can be moved to any point along its line of action.

11. Fixed or Bound Vectors

A fixed or bound vector must remain at the same point of application.

1.2 General Principles

1. Parallelogram Law

This law states that two forces acting on a particle can be replaced by a single resultant force obtained by drawing the diagonal of the parallelogram which has sides equal to the given forces.

For example, two forces $\boldsymbol{F}_1$ and $\boldsymbol{F}_2$ acting on a particle O, as shown in Fig. 1.1(a), can be replaced by a single force $\boldsymbol{R}$, as shown in Fig. 1.1(b), which has the same effect on the particle O and is called the resultant force of the forces $\boldsymbol{F}_1$ and $\boldsymbol{F}_2$. The resultant force $\boldsymbol{R}$ can be obtained by drawing a parallelogram using $\boldsymbol{F}_1$ and $\boldsymbol{F}_2$ as two adjacent sides of the parallelogram. The diagonal that passes through O represents the resultant force $\boldsymbol{R}$, i.e., $\boldsymbol{R}=\boldsymbol{F}_1+\boldsymbol{F}_2$. This method for finding the resultant force of two forces is known as the parallelogram law.

From the parallelogram law, an alternative method for determining the resultant force of two forces, as shown in Fig. 1.2(a), by drawing a triangle, as shown in Fig. 1.2(b), can

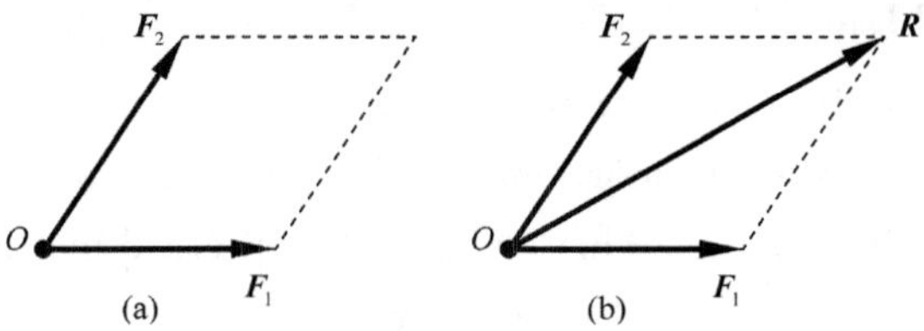

Fig. 1.1

be obtained. The resultant force $\boldsymbol{R}$ of the forces $\boldsymbol{F}_1$ and $\boldsymbol{F}_2$ can be found by arranging $\boldsymbol{F}_1$ and $\boldsymbol{F}_2$ in tip-to-tail fashion and then connecting the tail of $\boldsymbol{F}_1$ with the tip of $\boldsymbol{F}_2$, i.e., $\boldsymbol{R}=\boldsymbol{F}_1+\boldsymbol{F}_2$. This is known as the triangle rule.

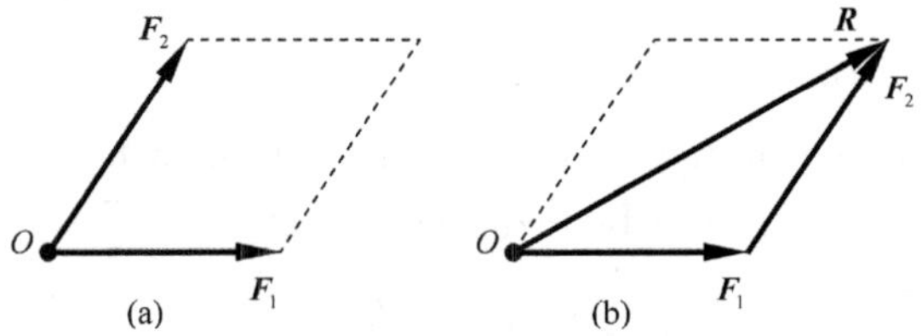

Fig. 1.2

2. Principle of Transmissibility

This principle states that the state of equilibrium or motion of a rigid body will remain unchanged if one force acting at a given point of the rigid body is replaced by another force of the same magnitude and same direction, but acting at a different point, provided that the two forces have the same line of action.

For example, a force $\boldsymbol{F}$, as shown in Fig. 1.3(a), acting at a given point O of a rigid body can be replaced by a force $\boldsymbol{F}'$, as shown in Fig. 1.3(b), of the same magnitude and same direction, but acting at a different point O' on the same line of action. The two forces $\boldsymbol{F}$ and $\boldsymbol{F}'$ have the same effect on the rigid body and are said to be equivalent. This principle shows that the effect of a force on a rigid body remains unchanged provided the force acting on the rigid body is moved along its line of action. Thus forces acting on a rigid body are sliding vectors.

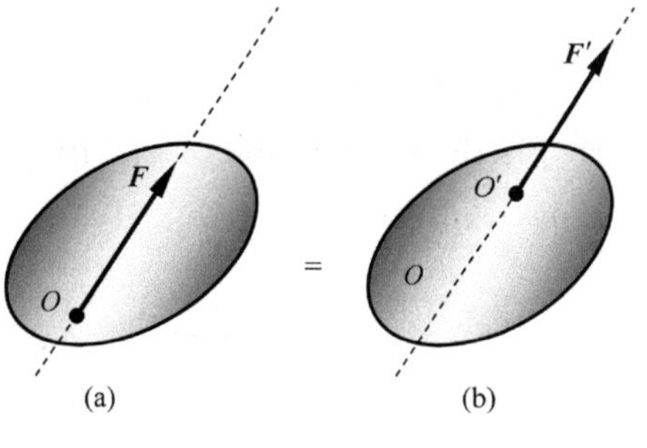

Fig. 1.3

3. Newton's First Law

This law states that if the resultant force acting on a particle is zero, then the particle will remain at rest (if originally at rest) or will move with constant velocity in a straight line (if originally in motion).

4. Newton's Second Law

This law states that if the resultant force acting on a particle is not zero, then the particle will have an acceleration proportional to the magnitude of the resultant force and in the direction of this resultant force. This law can be expressed mathematically as

$$\boldsymbol{F} = m\boldsymbol{a} \tag{1.1}$$

where $\boldsymbol{F}$, m, and $\boldsymbol{a}$ are, respectively, the resultant force acting on the particle, the mass of the particle, and the acceleration of the particle.

5. Newton's Third Law

This law states that the forces of action and reaction between two bodies in contact have the same magnitude (equal), same line of action (collinear), and opposite sense (direction).

6. Newton's Law of Gravitation

This law states that two particles are mutually attracted by equal and opposite forces. The magnitude of the two forces can be given by

$$F = G\frac{m_1 m_2}{r^2} \tag{1.2}$$

where F is the force of gravitation between the two particles, G is the universal constant of gravitation, m_1 and m_2 are, respectively, the mass of each of the two particles, and r is the distance between the two particles.

When a particle is located on or near the surface of the earth, the force exerted by the earth on the particle is defined as the weight of the particle. Taking m_1 equal to the mass M of the earth, m_2 equal to the mass m of the particle, and r equal to the radius R of the earth, and letting

$$g = G\frac{M}{R^2} \tag{1.3}$$

where g is the acceleration of gravity, then the magnitude of the weight of the particle can be given by

$$W = mg \tag{1.4}$$

The value of g is approximately equal to 9.81 m/s^2 in SI units, as long as the particle is located on or near the surface of the earth.

Chapter 2 Statics of Particle

A body under consideration can be idealized as a particle if its size and shape are able to be neglected. All the forces acting on this particle can be assumed to be applied at the same point and will thus form a system of concurrent forces.

2.1 Resultant of Coplanar Concurrent Forces

A coplanar system of concurrent forces consists of concurrent forces that lie in one plane.

1. Graphical Method for Resultant of Forces

The resultant force of a coplanar system of concurrent forces acting on a particle can be obtained by using the graphical method. If a particle is acted upon by three or more coplanar concurrent forces, the resultant force can be obtained by the repeated applications of the triangle rule.

Considering that a particle O is acted upon by coplanar concurrent forces $\boldsymbol{F}_1$, $\boldsymbol{F}_2$, and $\boldsymbol{F}_3$, as shown in Fig. 2.1(a), the resultant force $\boldsymbol{R}$ of these forces can be obtained graphically by arranging all the given forces in tip-to-tail fashion and connecting the tail of the first force with the tip of the last one, as shown in Fig. 2.1(b). This method is known as the polygon rule.

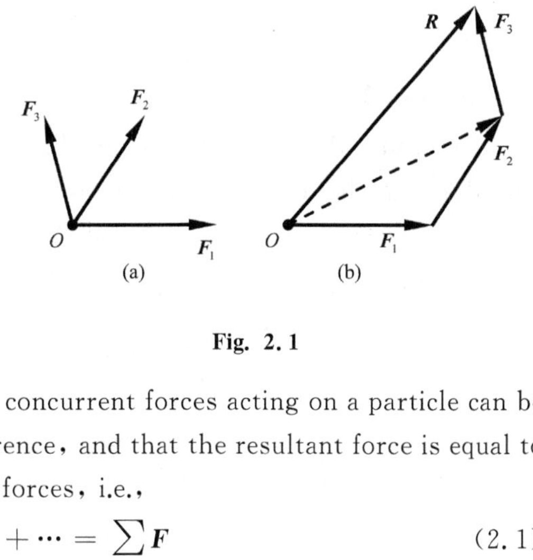

Fig. 2.1

We thus conclude that a coplanar system of concurrent forces acting on a particle can be replaced by a resultant force through the concurrence, and that the resultant force is equal to the vector sum of the given coplanar concurrent forces, i.e.,

$$\boldsymbol{R} = \boldsymbol{F}_1 + \boldsymbol{F}_2 + \boldsymbol{F}_3 + \cdots = \sum \boldsymbol{F} \tag{2.1}$$

Example 2.1

Two rods, AC and AD, are attached at A to column AB, as shown in Fig. E2.1(a). Knowing that the force in the left-hand rod is $F_1 = 150$ N, and that the inclination angles of the rods are $\theta_1 = 30°$ and $\theta_2 = 15°$, using the graphical method determine (a) the force F_2 in the right-hand rod if the resultant of the forces exerted by the rods on the column is to be vertical, (b) the corresponding magnitude of the resultant.

Solution

The forces $\boldsymbol{F}_1$ and $\boldsymbol{F}_2$ acting at A can be replaced by a resultant force $\boldsymbol{R}$ from the

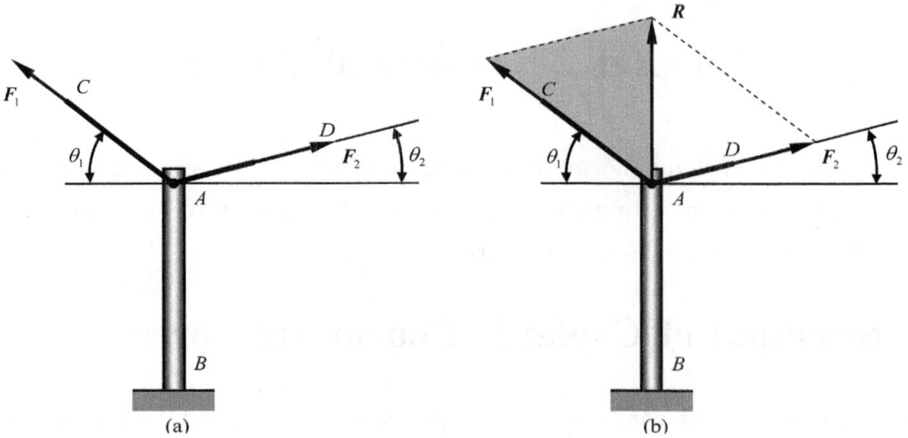

Fig. E2.1

parallelogram law, as shown in Fig. E2.1(b). Considering the shaded triangle shown in Fig. E2.1(b) and using the law of sines, we have

$$\frac{F_1}{\sin(90^\circ - \theta_2)} = \frac{F_2}{\sin(90^\circ - \theta_1)} = \frac{R}{\sin(\theta_1 + \theta_2)}$$

Using $F_1 = 150$ N, $\theta_1 = 30°$, and $\theta_2 = 15°$, we obtain

$$F_2 = \frac{\sin(90^\circ - \theta_1)}{\sin(90^\circ - \theta_2)} F_1 = 134.49 \text{ N}, \ R = \frac{\sin(\theta_1 + \theta_2)}{\sin(90^\circ - \theta_2)} F_1 = 109.81 \text{ N}$$

Example 2.2

Two rods, AC and AD, are attached at A to column AB, as shown in Fig. E2.2(a). Knowing that the forces in the rods are $F_1 = 120$ N and $F_2 = 100$ N, and that the inclination angles of the rods are $\theta_1 = 35°$ and $\theta_2 = 20°$, using the graphical method determine the resultant force.

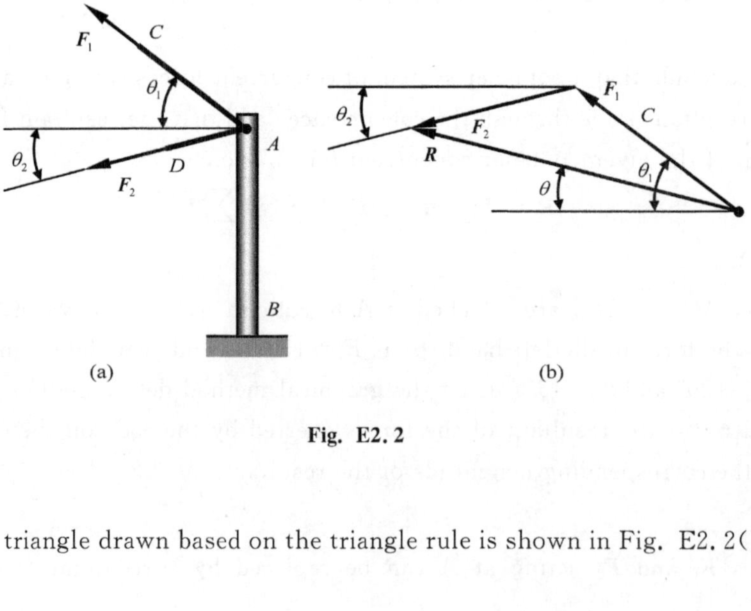

Fig. E2.2

Solution

The force triangle drawn based on the triangle rule is shown in Fig. E2.2(b). Using the law of cosines and the law of sines, we have

$$R^2 = F_1^2 + F_2^2 - 2F_1F_2\cos[180° - (\theta_1 + \theta_2)]$$

$$\frac{F_2}{\sin(\theta_1 - \theta)} = \frac{R}{\sin[180° - (\theta_1 + \theta_2)]}$$

Using $F_1=120$ N, $F_2=100$ N, $\theta_1=35°$, and $\theta_2=20°$, we can obtain

$$R = \sqrt{F_1^2 + F_2^2 + 2F_1F_2\cos(\theta_1 + \theta_2)} = 195.36 \text{ N}$$

$$\theta = \theta_1 - \arcsin\left[\frac{F_2}{R}\sin(\theta_1 + \theta_2)\right] = 17.93°$$

2. Components of Force

Two or more forces acting on a particle can be replaced by a single force which has the same effect on the particle. Conversely, one force acting on a particle can also be replaced by two or more forces which, together, have the same effect on the particle.

For example, a force $\boldsymbol{F}$ acting on a particle O, as shown in Fig. 2.2(a), can be replaced by $\boldsymbol{F}_1$ and $\boldsymbol{F}_2$. $\boldsymbol{F}_1$ and $\boldsymbol{F}_2$ are called the vector components of $\boldsymbol{F}$, and the process of substituting $\boldsymbol{F}_1$ and $\boldsymbol{F}_2$ for $\boldsymbol{F}$ is called the resolution of a force into components. Clearly, for a force $\boldsymbol{F}$ there exist an infinite number of possible sets of vector components, as shown in Fig. 2.2(b).

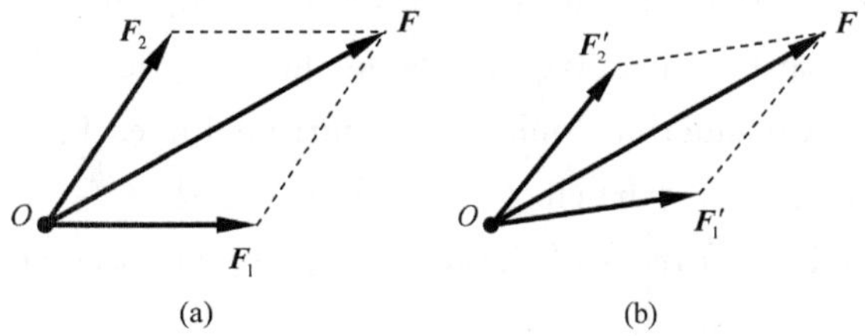

Fig. 2.2

3. Rectangular Components of Force

It is often convenient to resolve a force into components which are perpendicular to each other. For example, a force $\boldsymbol{F}$ acting on a particle O, as shown in Fig. 2.3, can be resolved into two vector components $\boldsymbol{F}_x$ and $\boldsymbol{F}_y$, respectively along the x and y axes, where $\boldsymbol{F}_x$ and $\boldsymbol{F}_y$ are called the rectangular components of the force $\boldsymbol{F}$. Thus we have

$$\boldsymbol{F} = \boldsymbol{F}_x + \boldsymbol{F}_y \tag{2.2}$$

By introducing two unit vectors $\boldsymbol{i}$, and $\boldsymbol{j}$, directed respectively along the positive x, and y axes, as shown in Fig. 2.4, $\boldsymbol{F}$ can also be expressed as

$$\boldsymbol{F} = F_x\boldsymbol{i} + F_y\boldsymbol{j} \tag{2.3}$$

where F_x and F_y are called the scalar components of the force $\boldsymbol{F}$ respectively along the x and y axes. F_x and F_y may be positive or negative, respectively depending upon the sense of $\boldsymbol{F}_x$ and $\boldsymbol{F}_y$. F_x is positive when $\boldsymbol{F}_x$ has the same sense as the positive x axis and is negative when $\boldsymbol{F}_x$ has the opposite sense. A similar conclusion can be drawn regarding the signs of F_y.

Denoting by F the magnitude of the force $\boldsymbol{F}$ and by θ the angle of $\boldsymbol{F}$ from the positive x axis, as shown in Fig. 2.5, we can express the scalar components F_x and F_y of $\boldsymbol{F}$ as follows:

$$F_x = F\cos\theta,\ F_y = F\sin\theta \tag{2.4}$$

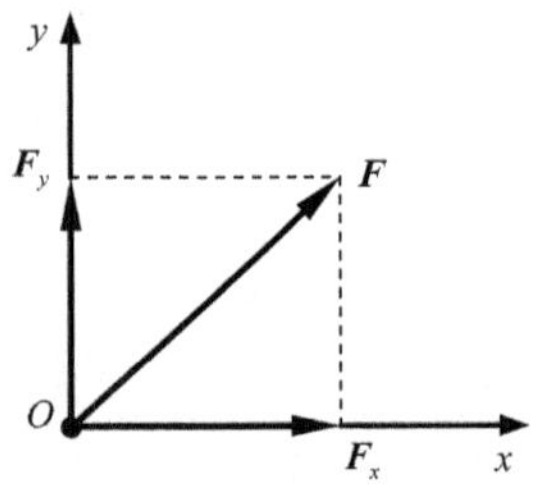

Fig. 2.3

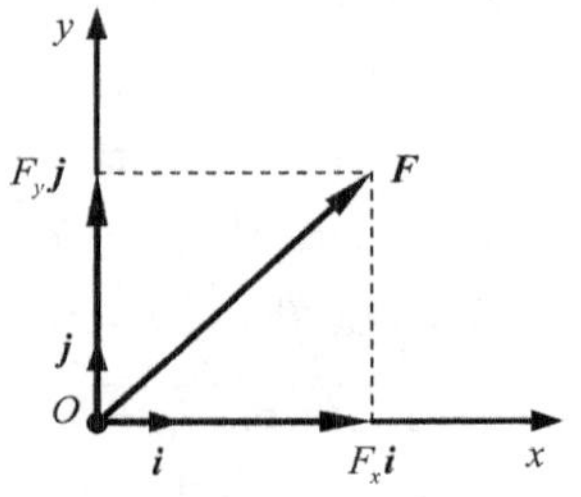

Fig. 2.4

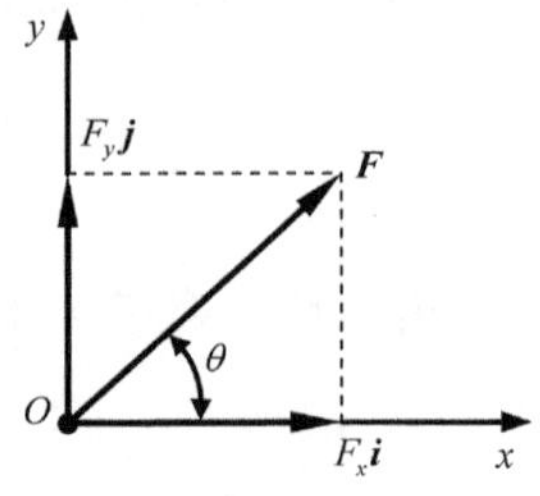

Fig. 2.5

4. Analytical Method for Resultant of Forces

Using the graphical method to determine the resultant force of coplanar concurrent forces often requires extensive geometric or trigonometric calculation, especially for finding the resultant force of three or more coplanar concurrent forces. Instead, problems of this type are easily solved by using the analytical method.

Considering coplanar concurrent forces $\boldsymbol{F}_1$, $\boldsymbol{F}_2$, and $\boldsymbol{F}_3$ acting on a particle O, as shown in Fig. 2.6, then the resultant force $\boldsymbol{R}$ of these forces can be expressed, using the graphical method, as

$$\boldsymbol{R} = \boldsymbol{F}_1 + \boldsymbol{F}_2 + \boldsymbol{F}_3 \tag{2.5}$$

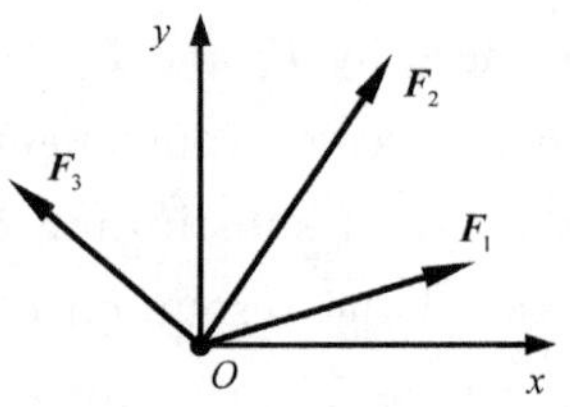

Fig. 2.6

Resolving each force, including the resultant force, into its rectangular components, we write

$$R_x\boldsymbol{i} + R_y\boldsymbol{j} = (F_{1x} + F_{2x} + F_{3x})\boldsymbol{i} + (F_{1y} + F_{2y} + F_{3y})\boldsymbol{j} \tag{2.6}$$

from which it follows that

$$R_x = F_{1x} + F_{2x} + F_{3x},\ R_y = F_{1y} + F_{2y} + F_{3y} \tag{2.7}$$

We thus conclude that the scalar components along the x and y axes of the resultant force of coplanar concurrent forces acting on a particle are respectively equal to the algebraic sums of the scalar components on the same axis of the given forces, i.e.,

$$R_x = \sum F_x,\ R_y = \sum F_y \tag{2.8}$$

The magnitude R of the resultant force and the angle θ that the resultant force forms with the positive x axis can be written as

$$R = \sqrt{R_x^2 + R_y^2},\ \theta = \arctan\frac{R_y}{R_x} \tag{2.9}$$

Example 2.3

Two rods, AC and AD, are attached at A to column AB, as shown in Fig. E2.3(a). Knowing that the force in the left-hand rod is $F_1 = 150$ N, and that the inclination angles of the rods are $\theta_1 = 30°$ and $\theta_2 = 15°$, using the analytical method determine (a) the force F_2 in the right-hand rod if the resultant of the forces exerted by the rods on the column is to be vertical, (b) the corresponding magnitude of the resultant.

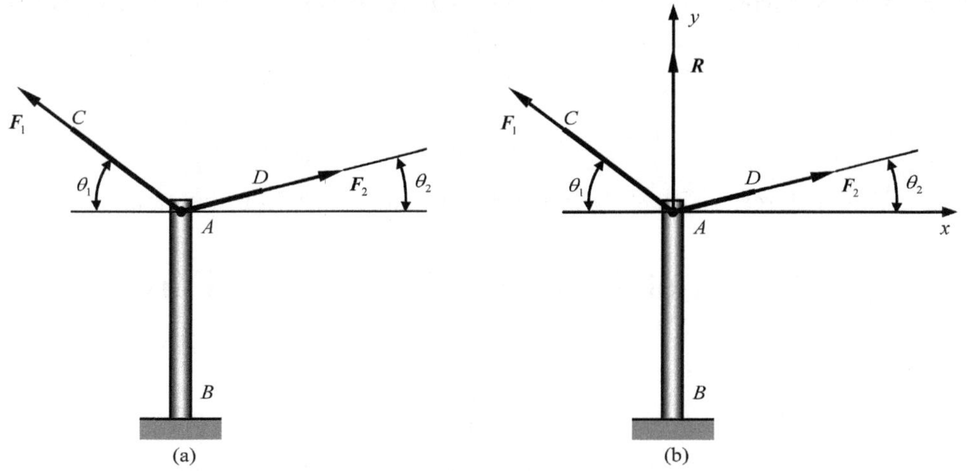

Fig. E2.3

Solution

Establishing the system of reference as shown in Fig. E2.3(b), then the scalar components of the resultant force can be expressed as

$$R_x = F_2\cos\theta_2 - F_1\cos\theta_1,\ R_y = F_2\sin\theta_2 + F_1\sin\theta_1$$

Since the resultant of the forces exerted by the rods on the column is vertical, i.e., $R_x=0$, we have

$$F_2 = \frac{\cos\theta_1}{\cos\theta_2}F_1 = 134.49\ \text{N}, R = R_y = F_2\sin\theta_2 + F_1\sin\theta_1 = 109.81\ \text{N}$$

Example 2.4

A block subjected to three forces is located on a surface of inclination angle $\alpha=25°$, as shown in Fig. E2.4(a). Assuming that $\theta=40°$, and that $F_1=150$ N, $F_2=250$ N, and $F_3=200$ N, using the analytical method determine the resultant of the forces acting on the block.

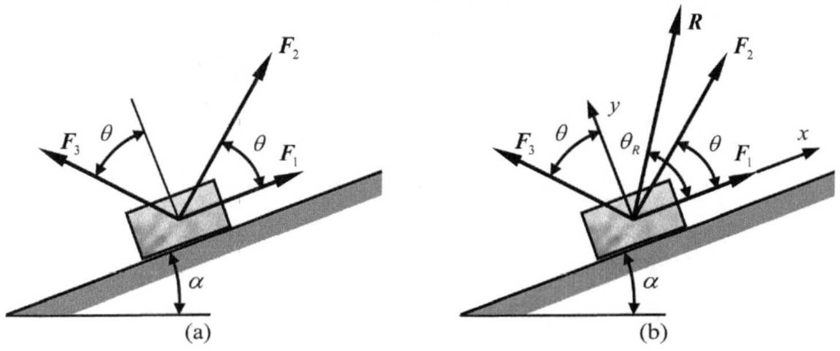

Fig. E2.4

Solution

Establishing the system of reference as shown in Fig. E2.4(b), then the scalar components of the resultant force can be expressed as

$$R_x = F_1 + F_2\cos\theta - F_3\sin\theta = 212.95\ \text{N},\ R_y = F_2\sin\theta + F_3\cos\theta = 313.91\ \text{N}$$

Using the scalar components above, we can obtain the magnitude and direction of the resultant as follows:

$$R = \sqrt{R_x^2 + R_y^2} = 379.32\ \text{N},\ \theta_R = \arctan\frac{R_y}{R_x} = 55.85\ °$$

Thus the resultant force is 379.32 N in magnitude and 80.85° in inclination.

2.2 Equilibrium of Coplanar Concurrent Forces

1. Free-Body Diagram

In solving a problem concerning the equilibrium of a particle, it is essential to consider all the forces acting on the particle. This can be done by choosing the particle under consideration and drawing a separate diagram to show this particle and all the forces acting on it. Such a diagram is called a free-body diagram.

2. Graphical Solution for Equilibrium of Forces

A particle is said to be in equilibrium if the resultant force of forces acting on the particle is zero. Thus the necessary and sufficient condition for equilibrium of a particle subjected to a coplanar system of concurrent forces can be expressed as

$$\sum \boldsymbol{F} = \boldsymbol{0} \tag{2.10}$$

It can be seen from Eq. (2.10) that a particle is in equilibrium if the given forces acting on the particle form a closed polygon.

Considering that a particle O is acted upon by forces $\boldsymbol{F}_1$, $\boldsymbol{F}_2$, and $\boldsymbol{F}_3$, as shown in Fig. 2.7(a), the resultant force of the given forces can be obtained by the polygon rule. Starting from point O with $\boldsymbol{F}_1$ and arranging the forces in tip-to-tail fashion, we find that the tip of $\boldsymbol{F}_3$ coincides with the starting point O, as shown in Fig. 2.7(b). Thus the resultant of the given forces is zero, and the particle is in equilibrium.

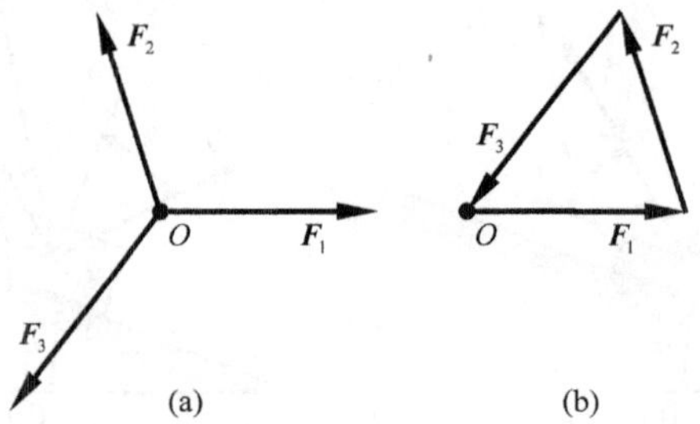

Fig. 2.7

Example 2.5

Three cables are tied together at A and are loaded as shown in Fig. E2.5(a). Using the graphical method determine the tension (a) in cable AB, (b) in cable AC.

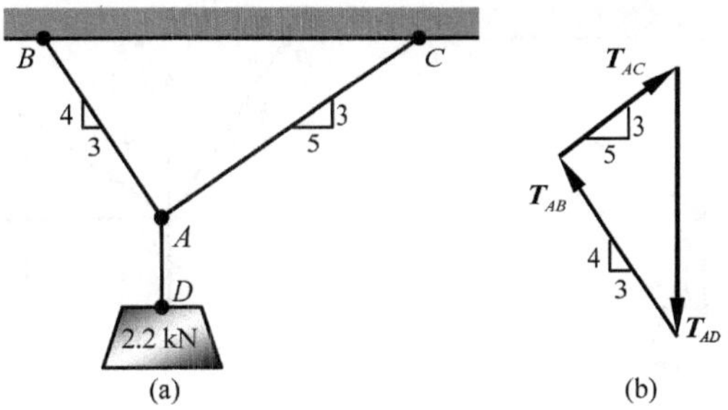

Fig. E2. 5

Solution

Since the joint A is in equilibrium, all the forces acting on A will form a closed triangle, as shown in Fig. E2. 5(b). Using the law of sines, we have

$$\frac{T_{AB}}{\sin[90° - \arctan(3/5)]} = \frac{T_{AC}}{\sin[90° - \arctan(4/3)]} = \frac{T_{AD}}{\sin[\arctan(4/3) + \arctan(3/5)]}$$

Using $T_{AD} = 2.2$ kN, we obtain

$$T_{AB} = 1.90 \text{ kN}, \quad T_{AC} = 1.33 \text{ kN}$$

3. Analytical Solution for Equilibrium of Forces

The necessary and sufficient condition for equilibrium of a particle subjected to a coplanar system of concurrent forces can be expressed as

$$\sum \boldsymbol{F} = \boldsymbol{0} \tag{2.11}$$

Resolving each force into its rectangular components, we have

$$(\sum F_x)\boldsymbol{i} + (\sum F_y)\boldsymbol{j} = \boldsymbol{0} \tag{2.12}$$

We thus conclude that necessary and sufficient condition for equilibrium of a particle subjected to a coplanar system of concurrent forces can be expressed as

$$\sum F_x = 0, \quad \sum F_y = 0 \tag{2.13}$$

which are called the equations of equilibrium for a coplanar system of concurrent forces.

Example 2. 6

Three cables are tied together at A and are loaded as shown in Fig. E2. 6(a). Using the analytical method determine the tension (a) in cable AB, (b) in cable AC.

Solution

Establishing the system of reference and considering the equilibrium of particle A as shown in Fig. E2. 6(b), then we have

$$\sum F_x = 0, \quad T_{AC} \times \frac{5}{\sqrt{5^2 + 3^2}} - T_{AB} \times \frac{3}{\sqrt{3^2 + 4^2}} = 0$$

$$\sum F_y = 0, \quad T_{AC} \times \frac{3}{\sqrt{5^2 + 3^2}} + T_{AB} \times \frac{4}{\sqrt{3^2 + 4^2}} - T_{AD} = 0$$

Solving the above equations for T_{AB} and T_{AC}, we obtain

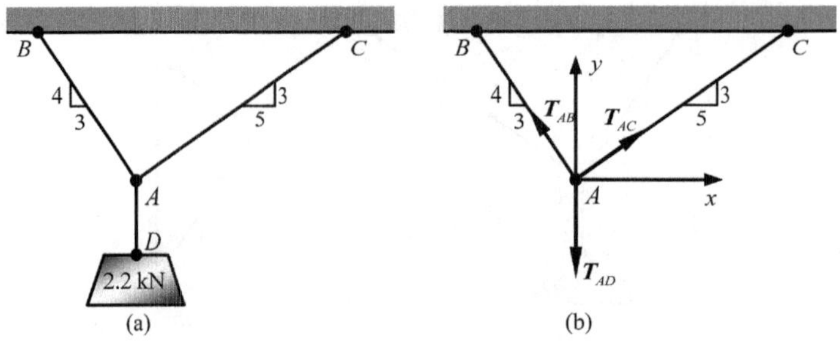

Fig. E2.6

$$T_{AB} = 1.90\ \text{kN},\ T_{AC} = 1.33\ \text{kN}$$

2.3 Resultant of Spatial Concurrent Forces

1. Components of Force in Space

A force $\boldsymbol{F}$ in space, as shown in Fig. 2.8, can be resolved into three vector components $\boldsymbol{F}_x$, $\boldsymbol{F}_y$, and $\boldsymbol{F}_z$ respectively along the x, y, and z axes, where $\boldsymbol{F}_x$, $\boldsymbol{F}_y$, and $\boldsymbol{F}_z$ are called the vector components of the force $\boldsymbol{F}$. Thus we have

$$\boldsymbol{F} = \boldsymbol{F}_x + \boldsymbol{F}_y + \boldsymbol{F}_z \tag{2.14}$$

By introducing three unit vectors $\boldsymbol{i}$, $\boldsymbol{j}$, and $\boldsymbol{k}$, directed respectively along the positive x, y, and z axes, as shown in Fig. 2.9, the force $\boldsymbol{F}$ can be expressed as

$$\boldsymbol{F} = F_x\boldsymbol{i} + F_y\boldsymbol{j} + F_z\boldsymbol{k} \tag{2.15}$$

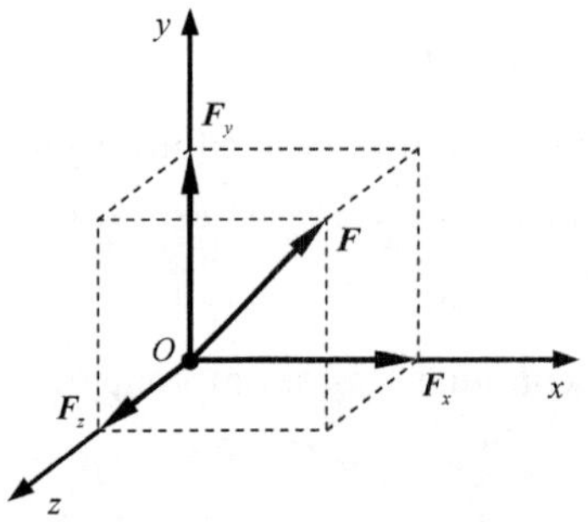

Fig. 2.8

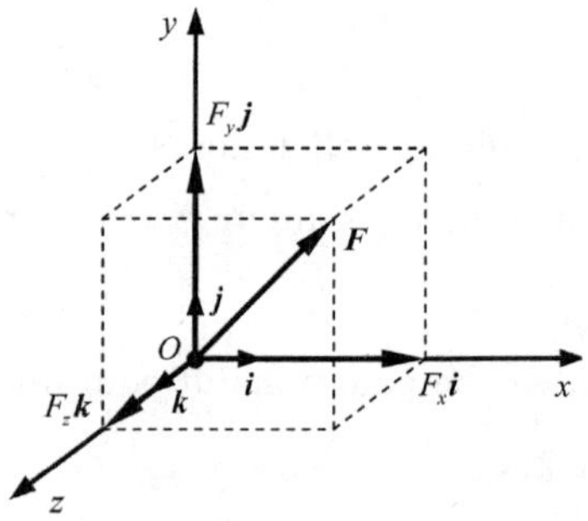

Fig. 2.9

where F_x, F_y, and F_z are called the scalar components of the force $\boldsymbol{F}$, respectively, along the x, y, and z axes. F_x, F_y, and F_y may be positive or negative, respectively depending upon the sense of $\boldsymbol{F}_x$, $\boldsymbol{F}_y$, and $\boldsymbol{F}_z$; i.e., F_x is positive when $\boldsymbol{F}_x$ has the same sense as the positive x axis and is negative when $\boldsymbol{F}_x$ has the opposite sense. A similar conclusion can be drawn regarding the sign of F_y or F_z.

Denoting by F the magnitude of the force $\boldsymbol{F}$ and by θ_x, θ_y, and θ_z the angles between $\boldsymbol{F}$ and the positive x, y, and z axes, as shown in Fig. 2.10, we can express the scalar components F_x, F_y, and F_z of $\boldsymbol{F}$ as follows:

$$F_x = F\cos\theta_x,\ F_y = F\cos\theta_y,\ F_z = F\cos\theta_z \tag{2.16}$$

where $\cos\theta_x$, $\cos\theta_y$, and $\cos\theta_z$ are the direction cosines of the force $\boldsymbol{F}$. These direction cosines satisfy the following relation:

$$\cos^2\theta_x + \cos^2\theta_y + \cos^2\theta_z = 1 \tag{2.17}$$

If we have known the angle γ between $\boldsymbol{F}$ and the positive y axis, and the angle φ between the positive z axis and the plane containing $\boldsymbol{F}$ and the y axis, as shown in Fig. 2.11, the corresponding scalar components F_x, F_y, and F_z of $\boldsymbol{F}$ can be expressed as

$$\left.\begin{aligned} F_x &= F_{xz}\sin\varphi = F\sin\gamma\sin\varphi \\ F_y &= F\cos\gamma \\ F_z &= F_{xz}\cos\varphi = F\sin\gamma\cos\varphi \end{aligned}\right\} \tag{2.18}$$

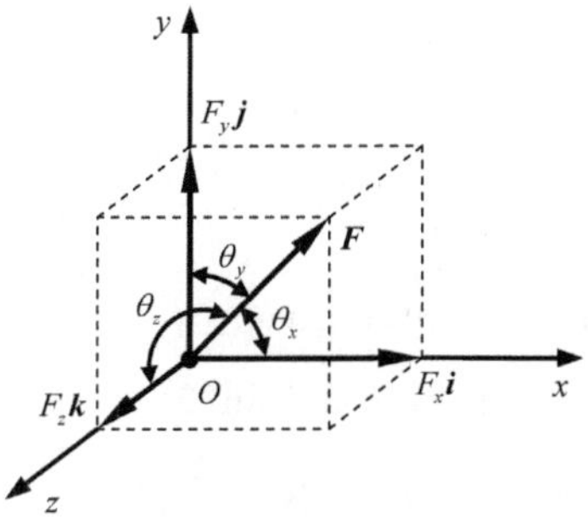

Fig. 2.10

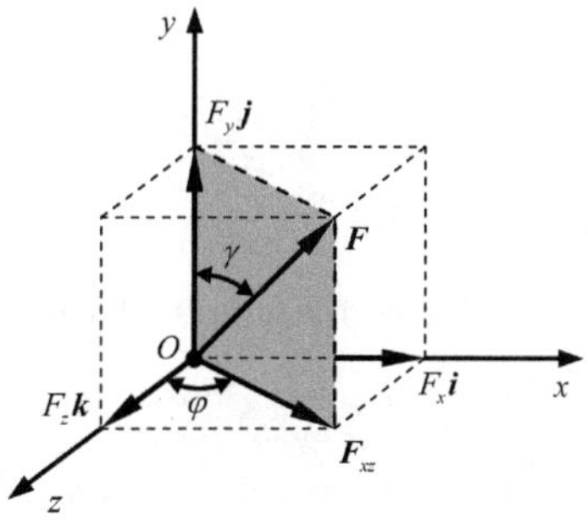

Fig. 2.11

2. Resultant of Forces in Space

Considering three forces $\boldsymbol{F}_1$, $\boldsymbol{F}_2$, and $\boldsymbol{F}_3$ acting on a particle O, as shown in Fig. 2.12, then, using the graphical method, the resultant force $\boldsymbol{R}$ of these forces can be expressed as

$$\boldsymbol{R} = \boldsymbol{F}_1 + \boldsymbol{F}_2 + \boldsymbol{F}_3 \tag{2.19}$$

Resolving each force, including the resultant force, into its rectangular components, we write

$$R_x\boldsymbol{i} + R_y\boldsymbol{j} + R_z\boldsymbol{k} = (F_{1x} + F_{2x} + F_{3x})\boldsymbol{i} + (F_{1y} + F_{2y} + F_{3y})\boldsymbol{j} + (F_{1z} + F_{2z} + F_{3z})\boldsymbol{k} \tag{2.20}$$

from which it follows that

$$R_x = F_{1x} + F_{2x} + F_{3x},\ R_y = F_{1y} + F_{2y} + F_{3y},\ R_z = F_{1z} + F_{2z} + F_{3z} \tag{2.21}$$

We therefore conclude that the scalar component along an arbitrary axis of the resultant force of the forces acting on a particle is equal to the algebraic sums of the scalar components on the same axis of the given forces, i.e.,

$$R_x = \sum F_x,\ R_y = \sum F_y,\ R_z = \sum F_z \tag{2.22}$$

The magnitude R of the resultant force and the angles θ_x, θ_y, and θ_z that the resultant force forms with the positive x, y, and z axes can be written as

$$R = \sqrt{R_x^2 + R_y^2 + R_z^2},\ \theta_x = \arccos\frac{R_x}{R},\ \theta_y = \arccos\frac{R_y}{R},\ \theta_z = \arccos\frac{R_z}{R} \tag{2.23}$$

Example 2.7

Determine the magnitude and direction of the resultant of the three forces, as shown in Fig. E2.7, knowing that $F_1=300$ N, $F_2=200$ N, and $F_3=100$ N.

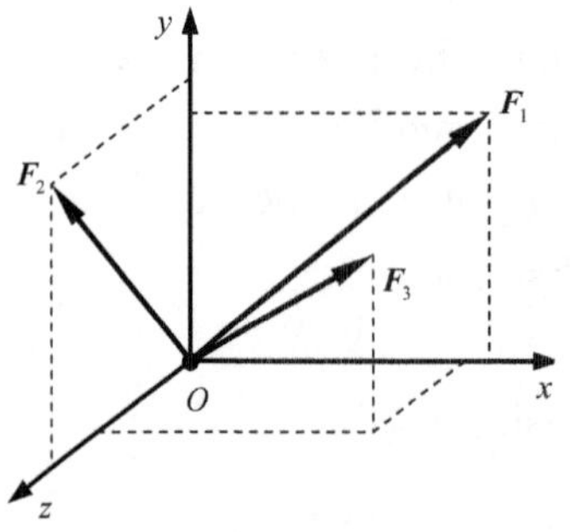

Fig. 2.12

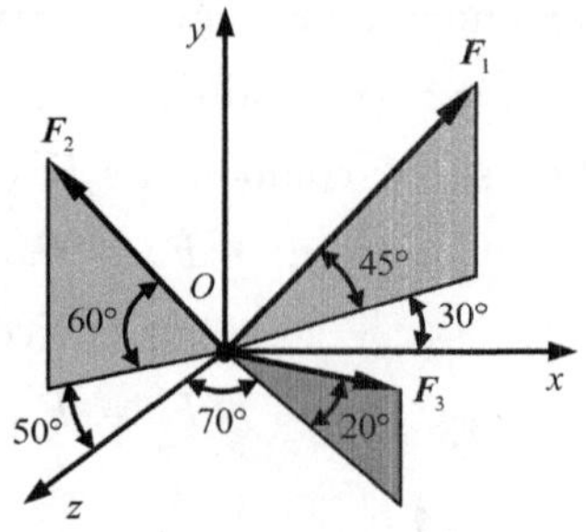

Fig. E2.7

Solution

The three scalar components of the resultant force can be given by

$$R_x = \sum F_x = F_1\cos45°\cos30° - F_2\cos60°\sin50° + F_3\cos20°\sin70° = 195.41\ \text{N}$$

$$R_y = \sum F_y = F_1\sin45° + F_2\sin60° + F_3\sin20° = 419.54\ \text{N}$$

$$R_z = \sum F_z = -F_1\cos45°\sin30° + F_2\cos60°\cos50° + F_3\cos20°\cos70° = -9.65\ \text{N}$$

Therefore, the magnitude and direction of the resultant force are

$$R = \sqrt{R_x^2 + R_y^2 + R_z^2} = 462.92\ \text{N}$$

$$\theta_x = \arccos\frac{R_x}{R} = 65.0°,\ \theta_y = \arccos\frac{R_y}{R} = 25.0°,\ \theta_z = \arccos\frac{R_z}{R} = 91.2°$$

2.4 Equilibrium of Spatial Concurrent Forces

The necessary and sufficient condition for equilibrium of a particle subjected to a space system of concurrent forces can be expressed as

$$\sum \boldsymbol{F} = \boldsymbol{0} \tag{2.24}$$

Resolving each force into its rectangular components, we have

$$(\sum F_x)\boldsymbol{i} + (\sum F_y)\boldsymbol{j} + (\sum F_z)\boldsymbol{k} = \boldsymbol{0} \tag{2.25}$$

We thus conclude that necessary and sufficient condition for equilibrium of a particle subjected to a space system of concurrent forces can be expressed as

$$\sum F_x = 0,\ \sum F_y = 0,\ \sum F_z = 0 \tag{2.26}$$

which are called the equations of equilibrium for a space system of concurrent forces.

Example 2.8

A horizontal homogeneous circular plate having a weight of $W = 300$ N is suspended from three wires which are attached to a support A and form $30°$ angles with the vertical, as shown in Fig. E2.8(a). Determine the tension in each wire.

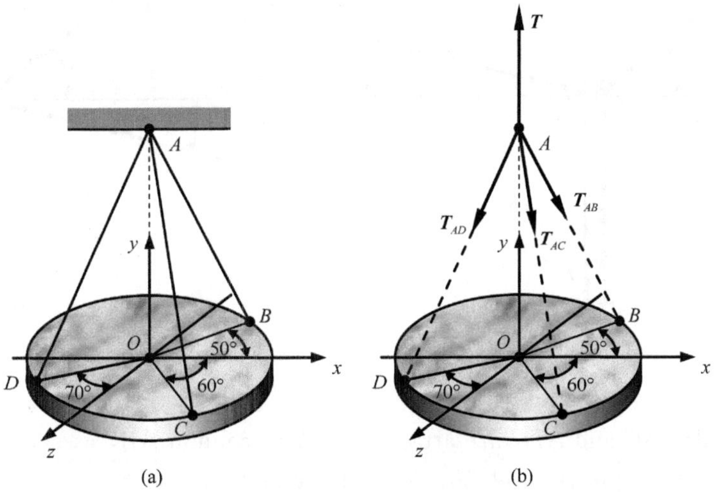

Fig. E2.8

Solution

Considering that the whole system is in equilibrium, we have

$$T = W = 300\ \text{N}$$

Taking particle A as a free body, and drawing its free-body diagram, as shown in Fig. E2.8(b), we have

$$\sum F_x = 0,\ T_{AB}\sin 30°\cos 50° + T_{AC}\sin 30°\cos 60° - T_{AD}\sin 30°\sin 70° = 0$$

$$\sum F_y = 0,\ T - T_{AB}\cos 30° - T_{AC}\cos 30° - T_{AD}\cos 30° = 0$$

$$\sum F_z = 0,\ -T_{AB}\sin 30°\sin 50° + T_{AC}\sin 30°\sin 60° + T_{AD}\sin 30°\cos 70° = 0$$

Solving the equations above for T_{AB}, T_{AC}, and T_{AD}, we obtain

$$T_{AB} = 140.71\ \text{N},\ T_{AC} = 71.44\ \text{N},\ T_{AD} = 134.26\ \text{N}$$

Problems

2.1 Two rods, AC and AD, are attached at A to column AB, as shown in Fig. P2.1. Knowing that the force in the right-hand rod is $F_2 = 100$ N, and that the inclination angles of the rods are $\theta_1 = 20°$ and $\theta_2 = 10°$, determine (a) the force F_1 in the left-hand rod if the resultant of the forces exerted by the rods on the column is vertical, (b) the corresponding magnitude of the resultant.

2.2 Two rods, AC and AD, are attached at A to column AB, as shown in Fig. P2.2. Knowing that the force in the upper rod is $F_1 = 150$ N, and that the inclination angles of the rods are $\theta_1 = 30°$ and $\theta_2 = 15°$, determine (a) the force F_2 in the lower rod if the resultant of the forces exerted by the rods on the column is horizontal, (b) the corresponding magnitude of the resultant.

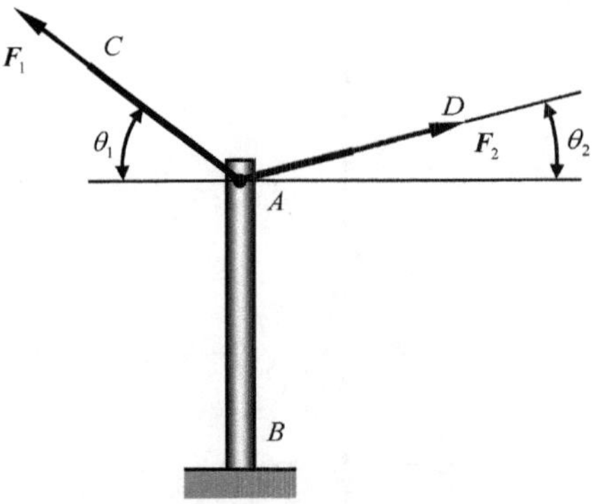

Fig. P2.1

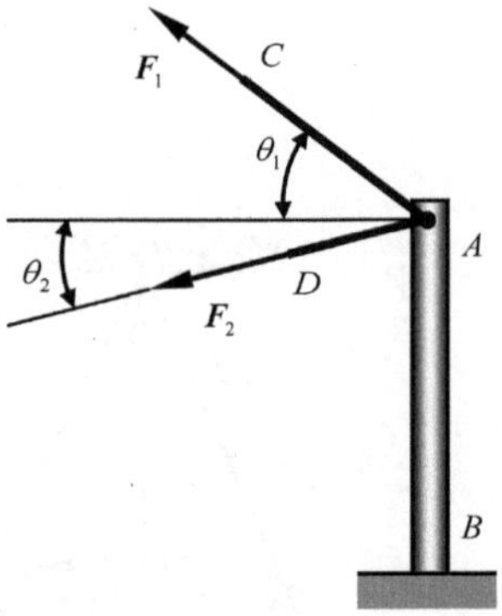

Fig. P2.2

2.3 Two rods, AC and AD, are attached at A to column AB, as shown in Fig. P2.3. Knowing that the forces in the rods are $F_1 = 110$ N and $F_2 = 90$ N, and that the inclination angles of the rods are $\theta_1 = 40°$ and $\theta_2 = 25°$, determine the resultant force.

2.4 Knowing that the tension in cable BC is 650 N, as shown in Fig. P2.4, determine the resultant of the three forces exerted at point B of beam AB.

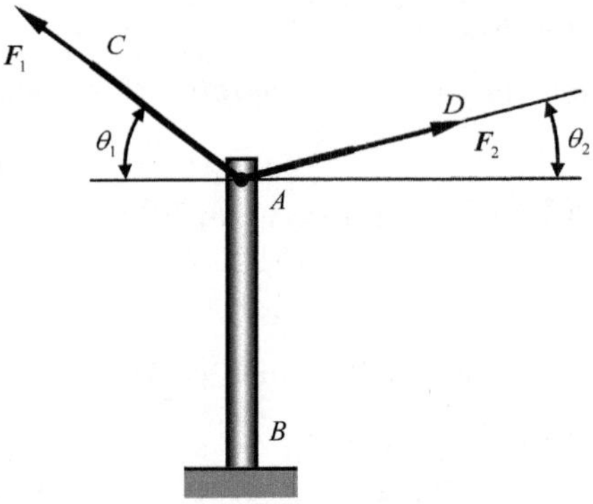

Fig. P2.3

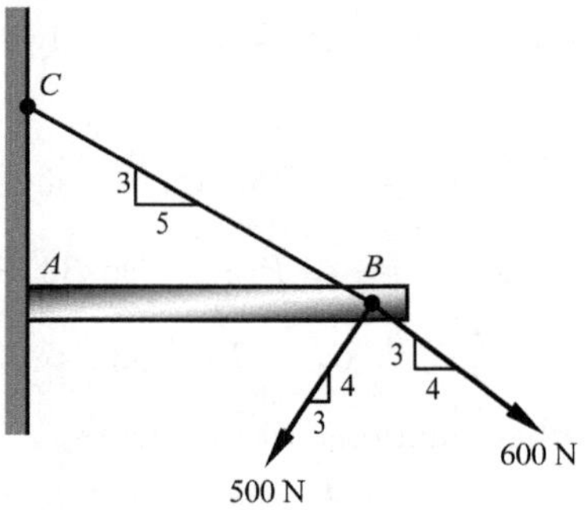

Fig. P2.4

2.5 Three cables are tied together at A and are loaded as shown in Fig. P2.5. Determine the tension (a) in cable AB, (b) in cable AC.

2.6 Two rods, AC and AD, are attached at A to lever AB, as shown in Fig. P2.6. Assuming the lever is in equilibrium, and knowing that $F_1 = 200$ N, $F_2 = 175$ N, and $\theta_1 = 30°$, determine (a) the inclination θ_2 of the right-hand rod, (b) the resultant of the forces exerted by the rods on the lever.

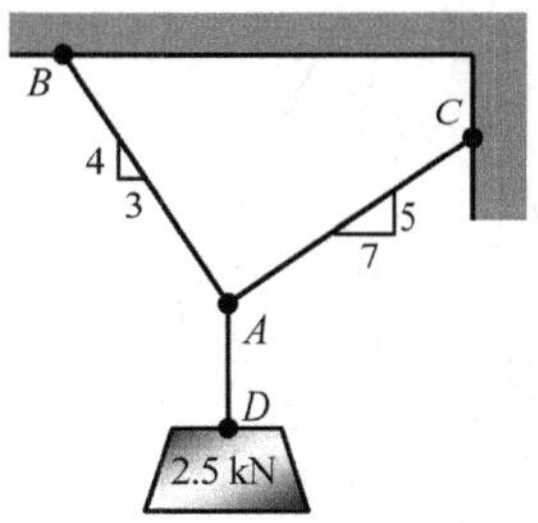

Fig. P2.5

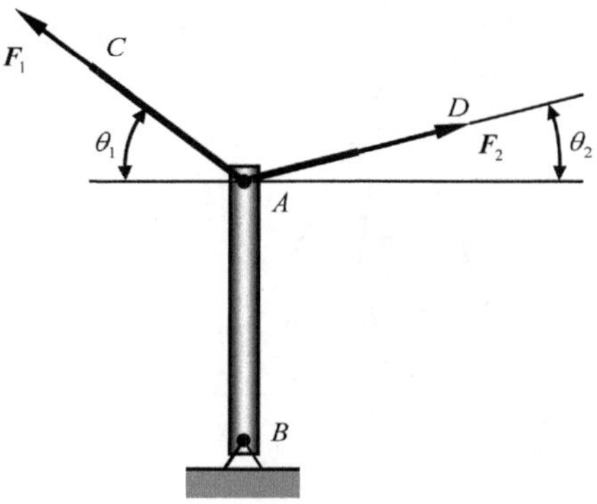

Fig. P2.6

2.7 Collar A can slide on a frictionless vertical rod and is attached to the spring C through a frictionless fixed pulley B, as shown in Fig. P2.7. Assuming that the spring is unstretched when $h=0.3$ m and that the collar is in equilibrium when $h=0.4$ m, and knowing that $a=0.4$ m and the spring stiffness $k=500$ N/m, determine the weight of the collar and the force exerted by the rod on the collar.

2.8 A collar B of weight W can move freely along the vertical rod, as shown in Fig. P2.8. Assuming that the spring stiffness is k and that the spring is unstretched when $\theta=0$, and knowing that $W=13.5$ N, $l=150$ mm, and $k=120$ N/m, determine the value of θ corresponding to equilibrium.

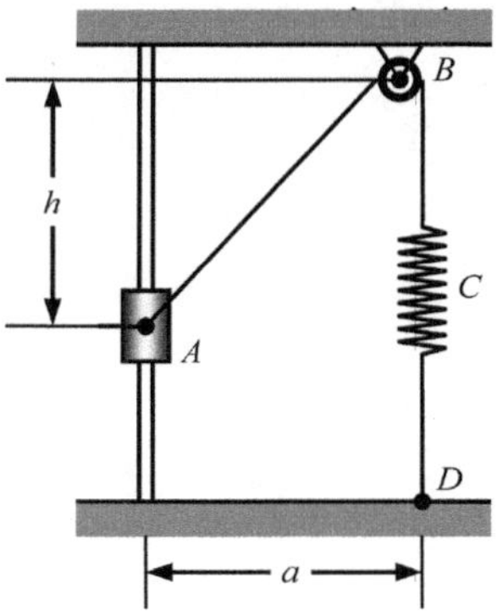

Fig. P2.7

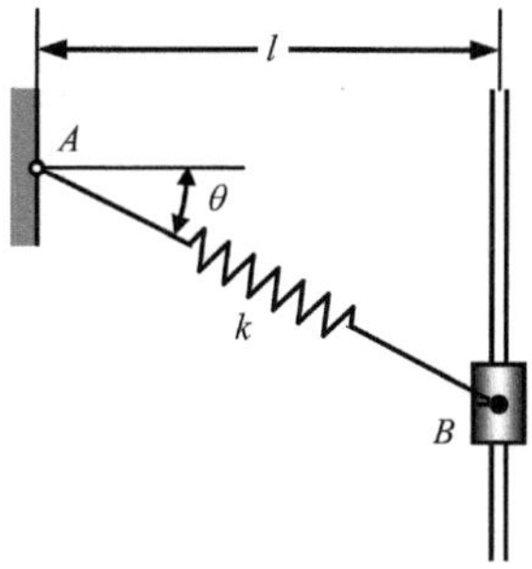

Fig. P2.8

2.9 A horizontal circular plate is suspended from three wires which are attached to a support at A and form 30° angles with the vertical, as shown in Fig. P2.9. Knowing that the x component of the force exerted by wire AB on the plate is 50 N, determine (a) the tension in wire AB, (b) the angles θ_x, θ_y, and θ_z that the force exerted at B forms with the coordinate axes.

2.10 A horizontal homogeneous circular plate having a weight of $W=200$ N is suspended from three wires which are attached to a support A and form 35° angles with the vertical, as shown in Fig. P2. 10. Determine the tension in each wire.

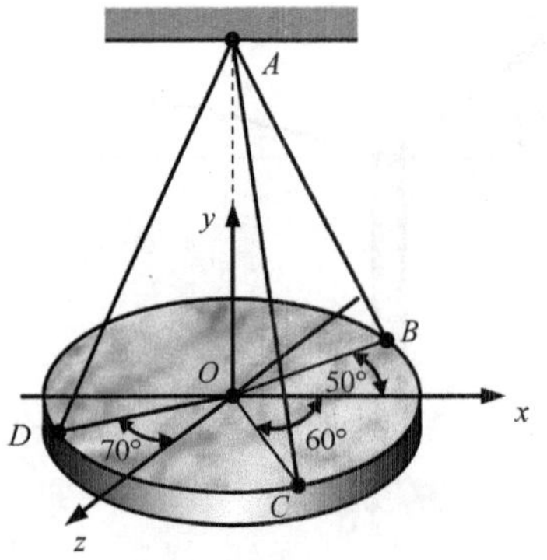

Fig. P2. 9

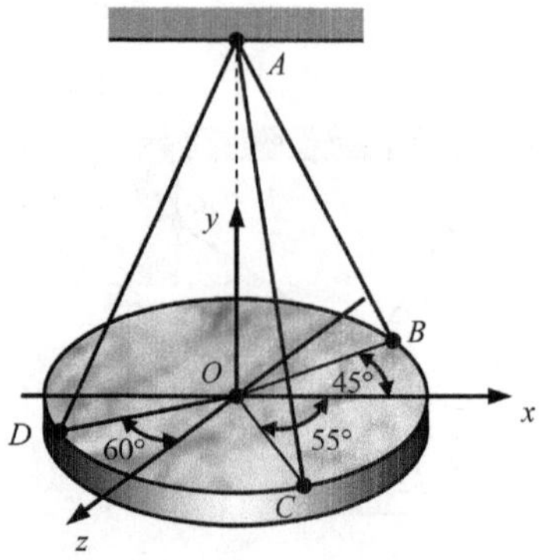

Fig. P2. 10

Chapter 3 Reduction of Force System

3.1 Moment of Force about Point

Consider a force $\boldsymbol{F}$ acting at a point A of a rigid body, as shown in Fig. 3.1, where A is the point of application of the force $\boldsymbol{F}$ and its position can be denoted by a vector $\boldsymbol{r}$ joining a fixed reference point O with A. $\boldsymbol{r}$ is called the position vector of A. The moment $\boldsymbol{M}_O(\boldsymbol{F})$ of $\boldsymbol{F}$ about O is defined as the vector product of $\boldsymbol{r}$ and $\boldsymbol{F}$, i.e.,

$$\boldsymbol{M}_O(\boldsymbol{F}) = \boldsymbol{r} \times \boldsymbol{F} \tag{3.1}$$

where $\boldsymbol{M}_O(\boldsymbol{F})$ satisfies:

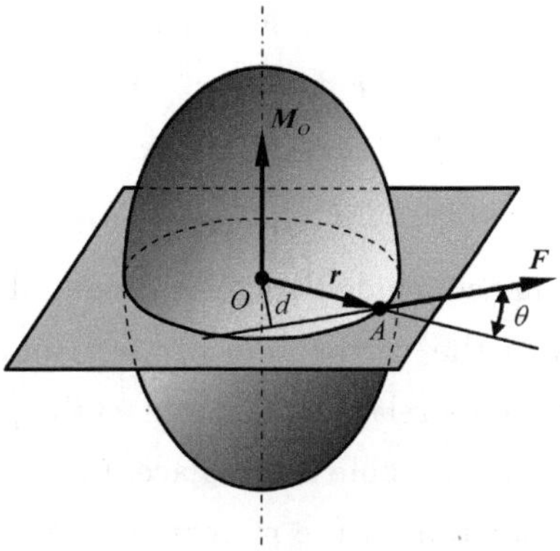

Fig. 3.1

(1) Its magnitude is equal to

$$M_O(\boldsymbol{F}) = rF\sin\theta = Fd \tag{3.2}$$

where θ is the included angle formed by $\boldsymbol{r}$ and $\boldsymbol{F}$ ($\theta \leqslant 180°$), and d is the perpendicular distance from O to the line of action of $\boldsymbol{F}$.

(2) Its line of action is perpendicular to the plane containing O and $\boldsymbol{F}$.

(3) Its direction is obtained from the right-hand rule, which can be stated as follows: if your right-hand fingers are curled from $\boldsymbol{r}$ to $\boldsymbol{F}$, then your right-hand thumb is pointed in the direction of $\boldsymbol{M}_O(\boldsymbol{F})$.

3.2 Moment of Force about Given Axis

Consider again a force $\boldsymbol{F}$ acting at a point A of a rigid body and the moment $\boldsymbol{M}_O(\boldsymbol{F})$ of $\boldsymbol{F}$ about O, as shown in Fig. 3.2. Assuming that OL is an axis through O, then the moment M_{OL} of $\boldsymbol{F}$ along OL is defined as the projection of the moment $\boldsymbol{M}_O(\boldsymbol{F})$ onto the axis OL, i.e.,

$$M_{OL} = \boldsymbol{\lambda} \cdot \boldsymbol{M}_O(\boldsymbol{F}) = \boldsymbol{\lambda} \cdot (\boldsymbol{r} \times \boldsymbol{F}) \tag{3.3}$$

where $\boldsymbol{\lambda}$ is the unit vector along OL.

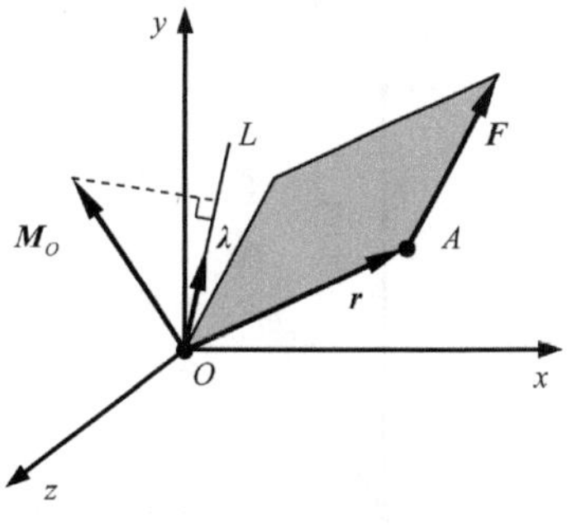

Fig. 3.2

3.3 Principle of Moments

If concurrent forces $\boldsymbol{F}_1$, $\boldsymbol{F}_2$, $\boldsymbol{F}_3$, $\cdots$ are applied at the same point A, and if the position vector of A with respect to a fixed reference point O is denoted by $\boldsymbol{r}$, as shown in Fig. 3.3, then we have

$$\boldsymbol{r}\times\boldsymbol{R}=\boldsymbol{r}\times(\boldsymbol{F}_1+\boldsymbol{F}_2+\boldsymbol{F}_3+\cdots)$$
$$=\boldsymbol{r}\times\boldsymbol{F}_1+\boldsymbol{r}\times\boldsymbol{F}_2+\boldsymbol{r}\times\boldsymbol{F}_3+\cdots \quad (3.4)$$

or

$$\boldsymbol{M}_O(\boldsymbol{R})=\sum\boldsymbol{M}_O(\boldsymbol{F}) \quad (3.5)$$

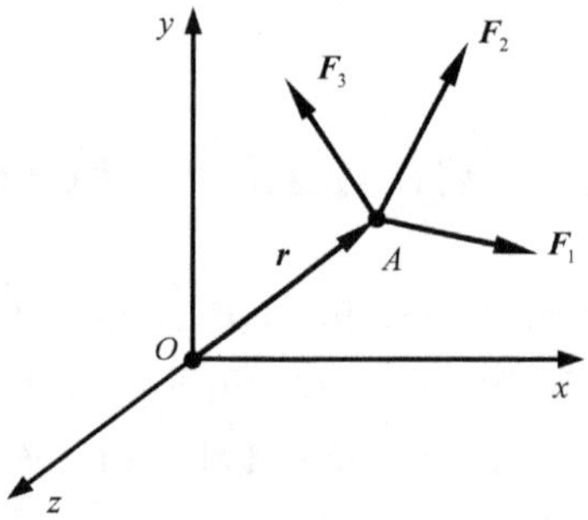

Fig. 3.3

We thus conclude that the moment about a given point of the resultant of concurrent forces is equal to the vector sum of the moments of the various concurrent forces about the same point. This relation is known as the principle of moments or Varignon's theorem. Eq. (3.4) makes it possible to replace the direct determination of the moment of a force by the determination of the moments of two or more component forces.

3.4 Components of Moment of Force about Point

Assume that a force $\boldsymbol{F}$ is applied at a point A, and that $\boldsymbol{r}$ represents the position vector of A with respect to the origin O of a fixed frame of reference $Oxyz$, as shown in Fig. 3.4.

Referring to Eq. (3.1), the moment $\boldsymbol{M}_O(\boldsymbol{F})$ of $\boldsymbol{F}$ about O can be expressed as

$$\boldsymbol{M}_O(\boldsymbol{F})=\begin{vmatrix}\boldsymbol{i} & \boldsymbol{j} & \boldsymbol{k}\\ x & y & z\\ F_x & F_y & F_z\end{vmatrix}=(yF_z-zF_y)\boldsymbol{i}+(zF_x-xF_z)\boldsymbol{j}+(xF_y-yF_x)\boldsymbol{k} \quad (3.6)$$

where x, y, z and F_x, F_y, F_z are the scalar components of the position vector $\boldsymbol{r}$ of A and of the force $\boldsymbol{F}$ acting on A, and $\boldsymbol{i}$, $\boldsymbol{j}$, $\boldsymbol{k}$ are the unit vectors. The scalar components of $\boldsymbol{M}_O$ can be expressed from Eq. (3.6) as

$$\left.\begin{aligned}[\boldsymbol{M}_O(\boldsymbol{F})]_x&=yF_z-zF_y\\ [\boldsymbol{M}_O(\boldsymbol{F})]_y&=zF_x-xF_z\\ [\boldsymbol{M}_O(\boldsymbol{F})]_z&=xF_y-yF_x\end{aligned}\right\} \quad (3.7)$$

Referring to Eq. (3.3) and using Eq. (3.6), the moments of $\boldsymbol{F}$ about the x, y, and z axes are expressed as

$$\left.\begin{aligned}M_x(\boldsymbol{F})&=\boldsymbol{i}\cdot\boldsymbol{M}_O=yF_z-zF_y\\ M_y(\boldsymbol{F})&=\boldsymbol{j}\cdot\boldsymbol{M}_O=zF_x-xF_z\\ M_z(\boldsymbol{F})&=\boldsymbol{k}\cdot\boldsymbol{M}_O=xF_y-yF_x\end{aligned}\right\} \quad (3.8)$$

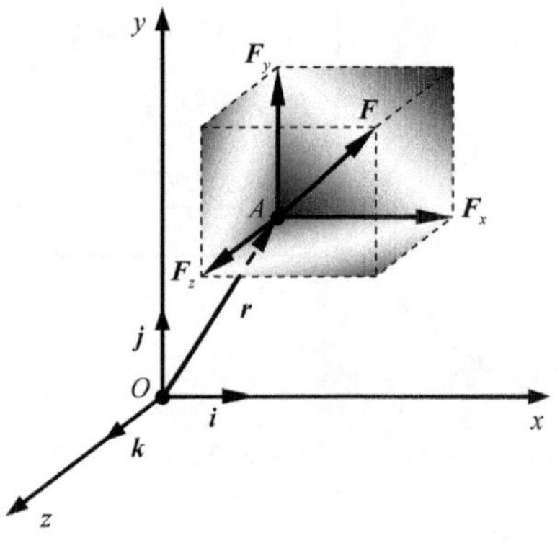

Fig. 3.4

It can be seen from Eqs. (3.7) and (3.8) that

$$M_x(\boldsymbol{F}) = [\boldsymbol{M}_O(\boldsymbol{F})]_x,\ M_y(\boldsymbol{F}) = [\boldsymbol{M}_O(\boldsymbol{F})]_y,\ M_z(\boldsymbol{F}) = [\boldsymbol{M}_O(\boldsymbol{F})]_z \tag{3.9}$$

Example 3.1

A plate is suspended from two chains AG and BH, as shown in Fig. E3.1(a). Knowing that the tension in BH is 200 N, determine (a) the moment about A of the force exerted by the chain BH, (b) the smallest force applied at E which creates the same moment about A.

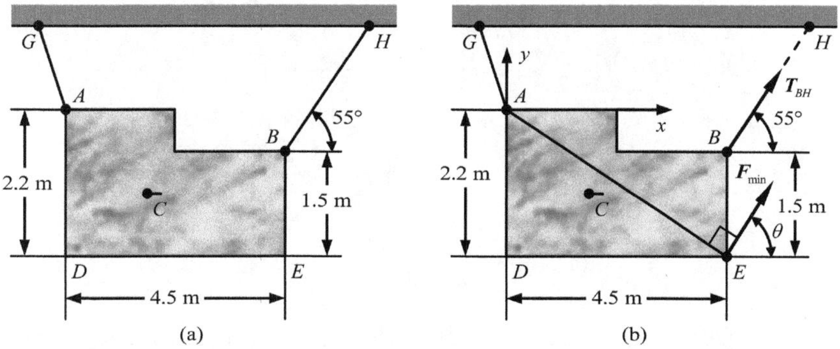

Fig. E3.1

Solution

(a) Establishing the coordinate system Axy as shown in Fig. E3.1(b), then we have

$$x_{B/A} = 4.5\ \text{m},\ y_{B/A} = -0.7\ \text{m};$$

$$(T_{BH})_x = T_{BH}\cos 55° = 114.72\ \text{N},\ (T_{BH})_y = T_{BH}\sin 55° = 163.83\ \text{N}$$

Thus we can obtain

$$M_A(\boldsymbol{T}_{BH}) = x_{B/A}(T_{BH})_y - y_{B/A}(T_{BH})_x = 817.54\ \text{N}\cdot\text{m}$$

(b) The force applied at E which creates the same moment about A is smallest when it is perpendicular to the line joining points A and E. Using $M_A(\boldsymbol{F}_{\min}) = r_{E/A}F_{\min} = M_A(\{\boldsymbol{T}_{BH})$, we obtain

$$F_{\min} = \frac{M_A(\boldsymbol{T}_{BH})}{r_{E/A}} = \frac{817.54}{\sqrt{2.2^2 + 4.5^2}} = 163.21\ \text{N},\ \theta = \arctan\frac{4.5}{2.2} = 63.95°$$

Example 3.2

Two cables AB and AC are attached to a concrete column, as shown in Fig. E3.2(a). Knowing that the tension in cables AB and AC are 800 N and 500 N, respectively, determine the moment about O of the resultant force exerted by the cables at A on the column.

Solution

Assuming that the unit vectors along x, y, and z axes are denoted by $\boldsymbol{i}$, $\boldsymbol{j}$, and $\boldsymbol{k}$, respectively, then we have

$$\boldsymbol{r}_{A/O} = 9\boldsymbol{j}\ \text{m},\ \boldsymbol{r}_{B/A} = 4\boldsymbol{i} - 9\boldsymbol{j} + \boldsymbol{k}\ \text{m},\ \boldsymbol{r}_{C/A} = -2\boldsymbol{i} - 9\boldsymbol{j} + 3\boldsymbol{k}\ \text{m}$$

$$\boldsymbol{T}_{AB} = T_{AB}\frac{\boldsymbol{r}_{B/A}}{r_{B/A}} = 800\frac{4\boldsymbol{i} - 9\boldsymbol{j} + \boldsymbol{k}}{\sqrt{4^2 + (-9)^2 + 1^2}} = 80.81(4\boldsymbol{i} - 9\boldsymbol{j} + \boldsymbol{k})\ \text{N}$$

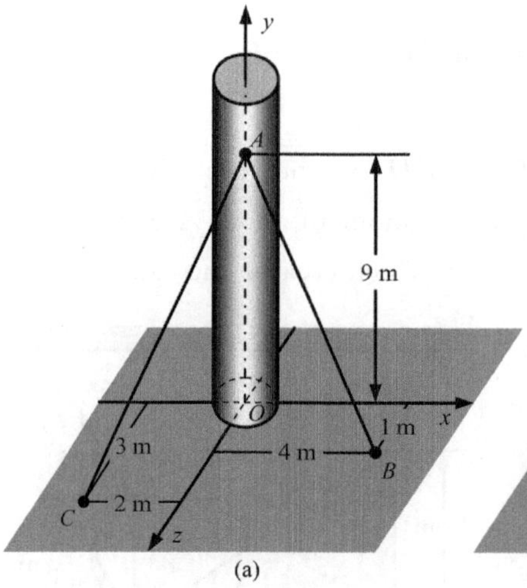

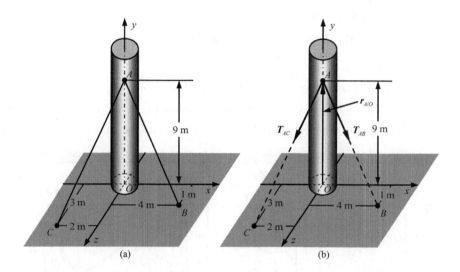

Fig. E3.2

$$\boldsymbol{T}_{AC} = T_{AC}\frac{\boldsymbol{r}_{C/A}}{r_{C/A}} = 500\frac{-2\boldsymbol{i}-9\boldsymbol{j}+3\boldsymbol{k}}{\sqrt{(-2)^2+(-9)^2+3^2}} = 51.57(-2\boldsymbol{i}-9\boldsymbol{j}+3\boldsymbol{k})\ \text{N}$$

Using $\boldsymbol{R}=\boldsymbol{T}_{AB}+\boldsymbol{T}_{AC}$, we have

$$\boldsymbol{R} = 220.1\boldsymbol{i} - 1\,191.4\boldsymbol{j} + 235.5\boldsymbol{k}\ \text{N}$$

Therefore, the moment about O of the resultant force exerted by the cables at A on the column is equal to

$$\boldsymbol{M}_O(\boldsymbol{R}) = \boldsymbol{r}_{A/O}\times\boldsymbol{R} = \begin{vmatrix} \boldsymbol{i} & \boldsymbol{j} & \boldsymbol{k} \\ 0 & 9 & 0 \\ 220.1 & -1\,191.4 & 235.5 \end{vmatrix}\ \text{N}\cdot\text{m} = 2\,119.5\boldsymbol{i} - 1\,980.9\boldsymbol{k}\ \text{N}\cdot\text{m}$$

3.5 Moment of Couple

Two forces $\boldsymbol{F}$ and $\boldsymbol{F}'$ having the same magnitude, parallel lines of action, and opposite sense are said to form a couple, as shown in Fig. 3.5. Clearly, the sum of the components of the two forces in any direction is zero. The sum of the moments of the two forces about a given point, however, is not zero. Therefore, the two forces do not translate the body, but tend to rotate it.

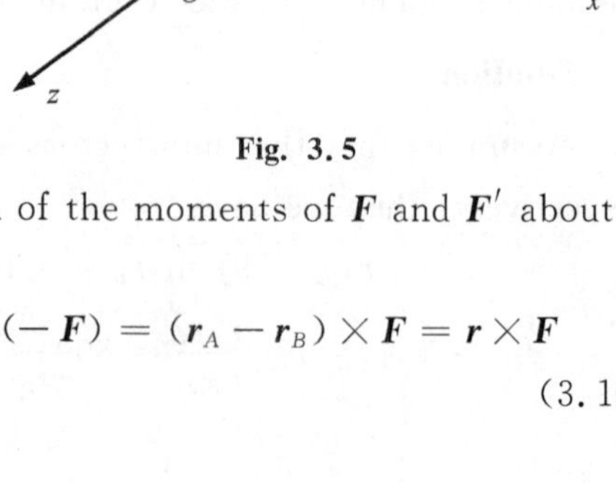

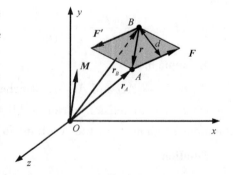

Fig. 3.5

Assuming that $\boldsymbol{r}_A$ and $\boldsymbol{r}_B$ are the position vectors of the points of application of $\boldsymbol{F}$ and $\boldsymbol{F}'$, then the sum of the moments of $\boldsymbol{F}$ and $\boldsymbol{F}'$ about O can be expressed as

$$\boldsymbol{M}_O(\boldsymbol{F},\boldsymbol{F}') = \boldsymbol{r}_A\times\boldsymbol{F}+\boldsymbol{r}_B\times\boldsymbol{F}' = \boldsymbol{r}_A\times\boldsymbol{F}+\boldsymbol{r}_B\times(-\boldsymbol{F}) = (\boldsymbol{r}_A-\boldsymbol{r}_B)\times\boldsymbol{F} = \boldsymbol{r}\times\boldsymbol{F} \tag{3.10}$$

where $\boldsymbol{r}=\boldsymbol{r}_A-\boldsymbol{r}_B$ is the position vector of the point of application A of $\boldsymbol{F}$ with respect to the point of application B of $\boldsymbol{F}'$. $\boldsymbol{M}_O(\boldsymbol{F},\boldsymbol{F}')=\boldsymbol{r}\times\boldsymbol{F}$ is not dependent on the choice of a reference point O, thus Eq. (3.10) can be rewritten as

$$\boldsymbol{M}=\boldsymbol{r}\times\boldsymbol{F} \tag{3.11}$$

where $\boldsymbol{M}$, called the moment of a couple, is a vector perpendicular to the plane containing the two forces, and its magnitude is equal to

$$M=rF\sin\theta=Fd \tag{3.12}$$

where d is the perpendicular distance between the lines of action of $\boldsymbol{F}$ and $\boldsymbol{F}'$. The sense of $\boldsymbol{M}$ is defined by the right-hand rule.

Since $\boldsymbol{M}$ is independent of the choice of a reference point, the moment $\boldsymbol{M}$ of a couple is a free vector which can be applied at any point of a rigid body. Thus two couples will have equal moments if the two couples lie in parallel planes (or in the same plane) and have the same magnitude and sense.

3.6 Resultant of Couples

Assume that P_1 and P_2 are two planes, and that AA' is the intersection line of P_1 and P_2, as shown in Fig. 3.6. Without any loss of generality, the couple $\boldsymbol{M}_1$ perpendicular to P_1 can be considered to consist of two forces $\boldsymbol{F}_1$ and $\boldsymbol{F}'_1$, which are contained in P_1, perpendicular to AA' and acting respectively at A and A'. Similarly, the couple $\boldsymbol{M}_2$ perpendicular to P_2 can be considered to consist of $\boldsymbol{F}_2$ and $\boldsymbol{F}'_2$ contained in P_2. It is clear that the resultant force $\boldsymbol{R}$ of $\boldsymbol{F}_1$ and $\boldsymbol{F}_2$ at point A and the resultant force $\boldsymbol{R}'$ of $\boldsymbol{F}'_1$ and $\boldsymbol{F}'_2$ at point A' form a couple. Denoting by $\boldsymbol{r}$ the position vector joining A' to A and using the principle of moments, we can express the moment $\boldsymbol{M}$ of the resulting couple as follows:

$$\boldsymbol{M}=\boldsymbol{r}\times\boldsymbol{R}=\boldsymbol{r}\times(\boldsymbol{F}_1+\boldsymbol{F}_2)=\boldsymbol{r}\times\boldsymbol{F}_1+\boldsymbol{r}\times\boldsymbol{F}_2=\boldsymbol{M}_1+\boldsymbol{M}_2 \tag{3.13}$$

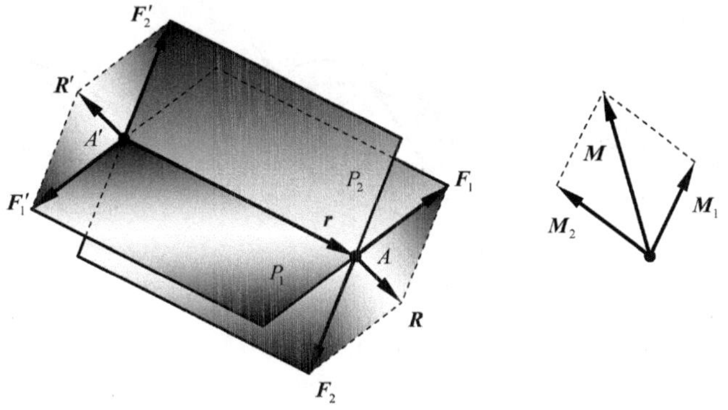

Fig. 3.6

where $\boldsymbol{M}_1=\boldsymbol{r}\times\boldsymbol{F}_1$ and $\boldsymbol{M}_2=\boldsymbol{r}\times\boldsymbol{F}_2$ are the couples acting in planes P_1 and P_2, respectively. We thus conclude that the moment $\boldsymbol{M}$ of the resulting couple is equal to the vector sum of the moments $\boldsymbol{M}_1$ and $\boldsymbol{M}_2$ of the two couples.

Example 3.3

A block is acted upon by three couples, as shown in Fig. E3.3. Knowing that $M_1 = 10$ N · m, $M_2 = 15$ N · m, and $M_3 = 8$ N · m, replace these couples with a single equivalent couple, and specify its magnitude and the direction of its axis.

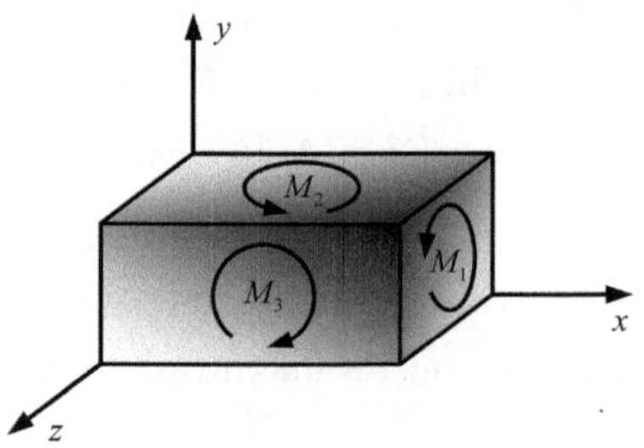

Fig. E3.3

Solution

Assuming that the unit vectors along x, y, and z axes are denoted by $\boldsymbol{i}$, $\boldsymbol{j}$, and $\boldsymbol{k}$, respectively, then the moments for the three couples can be written as

$$\boldsymbol{M}_1 = 10\boldsymbol{i}\ \text{N} \cdot \text{m},\ \boldsymbol{M}_2 = 15\boldsymbol{j}\ \text{N} \cdot \text{m},\ \boldsymbol{M}_3 = -8\boldsymbol{k}\ \text{N} \cdot \text{m}$$

Thus the resulting couple can be expressed as

$$\boldsymbol{M} = \boldsymbol{M}_1 + \boldsymbol{M}_2 + \boldsymbol{M}_3 = 10\boldsymbol{i} + 15\boldsymbol{j} - 8\boldsymbol{k}\ \text{N} \cdot \text{m}$$

And the magnitude and direction of this resulting couple are, respectively, given by

$$M = \sqrt{10^2 + 15^2 + (-8)^2} = 19.72\ \text{N} \cdot \text{m}$$

$$\theta_x = \arccos(10/19.72) = 59.53^\circ$$

$$\theta_y = \arccos(15/19.72) = 40.48^\circ$$

$$\theta_z = \arccos(-8/19.72) = 113.93^\circ$$

3.7 Equivalence of Force Acting on Rigid Body

Considera force $\boldsymbol{F}$ acting on a rigid body at a given point A defined by a position vector $\boldsymbol{r}$, as shown in Fig. 3.7(a). From the principle of transmissibility, $\boldsymbol{F}$ can be moved along its line of action without modifying the action of $\boldsymbol{F}$ on the rigid body.

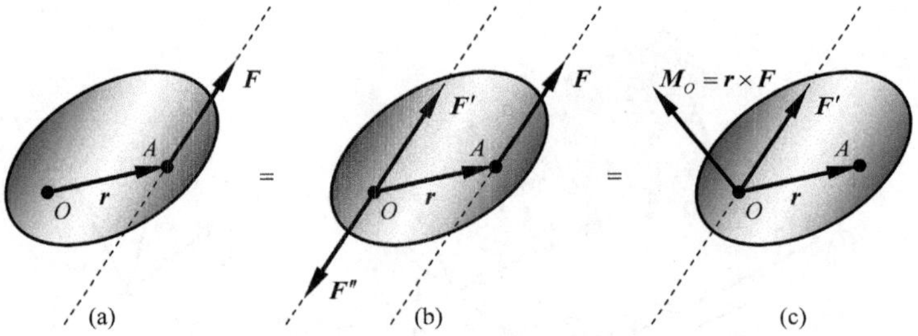

Fig. 3.7

For the force $\boldsymbol{F}$ acting on the rigid body, we cannot move it to a point O which does not lie on the original line of action without modifying the action of $\boldsymbol{F}$ on the body. We can, however, attach two forces $\boldsymbol{F}'$ and $\boldsymbol{F}''$ at point O, $\boldsymbol{F}' = \boldsymbol{F}$ and $\boldsymbol{F}'' = -\boldsymbol{F}$, without modifying the action of original force on the rigid body, as shown in Fig. 3.7(b). As a result of this transformation, a force $\boldsymbol{F}'$ is now applied at O; the other two forces $\boldsymbol{F}$ and $\boldsymbol{F}''$ will form a couple of moment $\boldsymbol{M}_O = \boldsymbol{r} \times \boldsymbol{F}$, as shown in Fig. 3.7(c).

We thus conclude that any force $\boldsymbol{F}$ acting on a rigid body can be moved to an arbitrary point O provided that a couple is added whose moment is equal to the moment $\boldsymbol{M}_O$ of $\boldsymbol{F}$ about O. The moment $\boldsymbol{M}_O$ of this couple is perpendicular to the plane containing $\boldsymbol{r}$ and $\boldsymbol{F}$. Since $\boldsymbol{M}_O$ is a free vector, it may be applied anywhere; for convenience, however, the couple vector is usually attached at O, together with $\boldsymbol{F}$, and combination obtained is referred to as a force-couple system.

Example 3.4

A vertical force $\boldsymbol{F}$ acts at C of a planar truss, as shown in Fig. E3.4(a). Knowing that $F = 80$ N, replace $\boldsymbol{F}$ with an equivalent force-couple system at G.

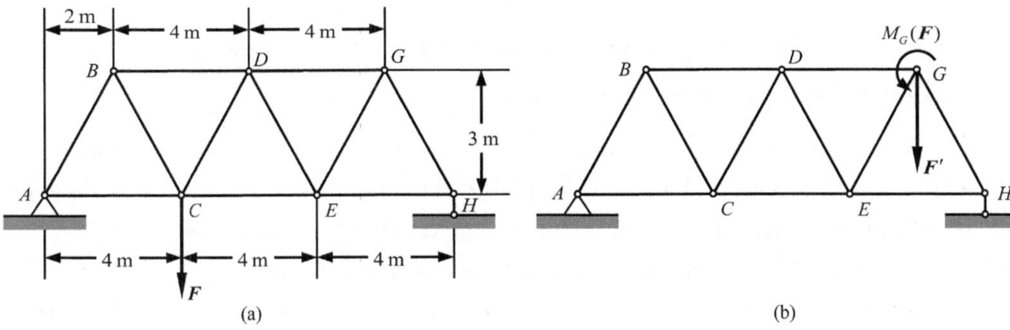

Fig. E3.4

Solution

The equivalent force-couple system, as shown in Fig. E3.4(b), can be expressed as

$$F' = F = 80 \text{ N}$$

$$M_G(\boldsymbol{F}) = 80 \times 6 = 480 \text{ N} \cdot \text{m}$$

3.8 Reduction of Force System

1. Reduction of Force System to Force-Couple System

Consider a system of forces $\boldsymbol{F}_1$, $\boldsymbol{F}_2$, $\boldsymbol{F}_3$, $\cdots$, acting on a rigid body at the points A_1, A_2, A_3, $\cdots$, defined by the position vectors $\boldsymbol{r}_1, \boldsymbol{r}_2, \boldsymbol{r}_3, \cdots$, as shown in Fig. 3.8(a).

$\boldsymbol{F}_1$ can be moved from A_1 to a given point O if a couple $\boldsymbol{M}_O(\boldsymbol{F}_1) = \boldsymbol{r}_1 \times \boldsymbol{F}_1$ is added to the system of forces. Repeating this procedure with $\boldsymbol{F}_2$, $\boldsymbol{F}_3$, $\cdots$, we obtain a new system of forces which consists of the forces $\boldsymbol{F}'_1$, $\boldsymbol{F}'_2$, $\boldsymbol{F}'_3$, $\cdots$, acting at O and the couples $\boldsymbol{M}_O(\boldsymbol{F}_1) = \boldsymbol{r}_1 \times \boldsymbol{F}_1$, $\boldsymbol{M}_O(\boldsymbol{F}_2) = \boldsymbol{r}_2 \times \boldsymbol{F}_2$, $\boldsymbol{M}_O(\boldsymbol{F}_3) = \boldsymbol{r}_3 \times \boldsymbol{F}_3$, $\cdots$, as shown in Fig. 3.8(b).

Since the forces $\boldsymbol{F}'_1$, $\boldsymbol{F}'_2$, $\boldsymbol{F}'_3$, $\cdots$, are concurrent, they can be added vectorially and replaced by a force $\boldsymbol{R}'$. Similarly, the couples $\boldsymbol{M}_O(\boldsymbol{F}_1) = \boldsymbol{r}_1 \times \boldsymbol{F}_1$, $\boldsymbol{M}_O(\boldsymbol{F}_2) = \boldsymbol{r}_2 \times \boldsymbol{F}_2$, $\boldsymbol{M}_O(\boldsymbol{F}_3) = \boldsymbol{r}_3 \times \boldsymbol{F}_3$, $\cdots$, can be added vectorially and replaced by a couple $\boldsymbol{M}_O$, as shown in Fig. 3.8(c). $\boldsymbol{R}'$ and $\boldsymbol{M}_O$ can be written as

$$R' = \sum F' = \sum F,\ M_O = \sum M_O(F) \tag{3.14}$$

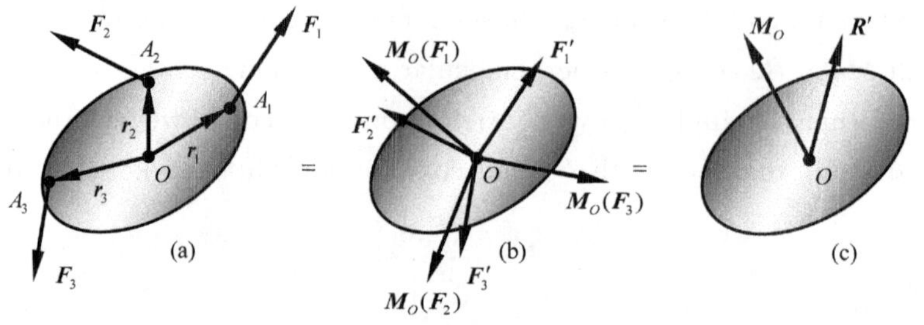

Fig. 3.8

which shows that the force $\boldsymbol{R}'$ is obtained by adding all the forces of the system, and that the couple $\boldsymbol{M}_O$ is obtained by adding the moments about O of all the forces of the system.

2. Further Reduction of Force-Couple System

Any given system of forces acting on a rigid body can be reduced to an equivalent force-couple system at O consisting of a force $\boldsymbol{R}'$ equal to the sum of the forces of the system and a couple $\boldsymbol{M}_O$ of moment equal to the sum of the moments about O of the system.

If $\boldsymbol{M}_O=0$ and $\boldsymbol{R}'=0$, the given system of forces is in equilibrium.

If $\boldsymbol{R}'\neq 0$ but $\boldsymbol{M}_O=0$, the given system of forces can be reduced to a single force $\boldsymbol{R}=\boldsymbol{R}'$, called the resultant force of the system.

If $\boldsymbol{M}_O\neq 0$ but $\boldsymbol{R}'=0$, the given system of forces can be reduced to a single couple $\boldsymbol{M}=\boldsymbol{M}_O$, called the resultant couple of the system.

If $\boldsymbol{R}'\neq 0$ and $\boldsymbol{M}_O\neq 0$, where $\boldsymbol{R}'$ and $\boldsymbol{M}_O$ are mutually perpendicular, the given system of forces can further be reduced to a single force $\boldsymbol{R}=\boldsymbol{R}'$, i.e., the resultant force of the system. $\boldsymbol{R}'$ and $\boldsymbol{M}_O$ are always mutually perpendicular for a system of (1) concurrent forces, (2) coplanar forces, or (3) parallel forces.

If $\boldsymbol{R}'\neq 0$ and $\boldsymbol{M}_O\neq 0$, where $\boldsymbol{R}'$ and $\boldsymbol{M}_O$ are not perpendicular, the given system of forces can be reduced to a force screw or wrench.

Example 3.5

A truss supports the loadings, as shown in Fig. E3.5(a). Knowing that $F_1=160$ N, $F_2=150$ N, and $F_3=80$ N, determine the equivalent force acting on the truss and the point of intersection of its line of action with a line through points A and H.

Solution

(1) Reduction of the original system of forces to an equivalent force-couple system at A is shown in Fig. E3.5(b). Assuming that the unit vectors along x and y axes are denoted by $\boldsymbol{i}$ and $\boldsymbol{j}$, respectively, and using $\boldsymbol{R}' = \sum \boldsymbol{F}$ and $M_A = \sum M_A(\boldsymbol{F})$, we have

$$\boldsymbol{R}' = \boldsymbol{F}_1 + \boldsymbol{F}_2 + \boldsymbol{F}_3 = 80\boldsymbol{i} - 310\boldsymbol{j}\ \text{N}$$

$$M_A = M_A(\boldsymbol{F}_1) + M_A(\boldsymbol{F}_2) + M_A(\boldsymbol{F}_3) = -2\ 080\ \text{N}\cdot\text{m}$$

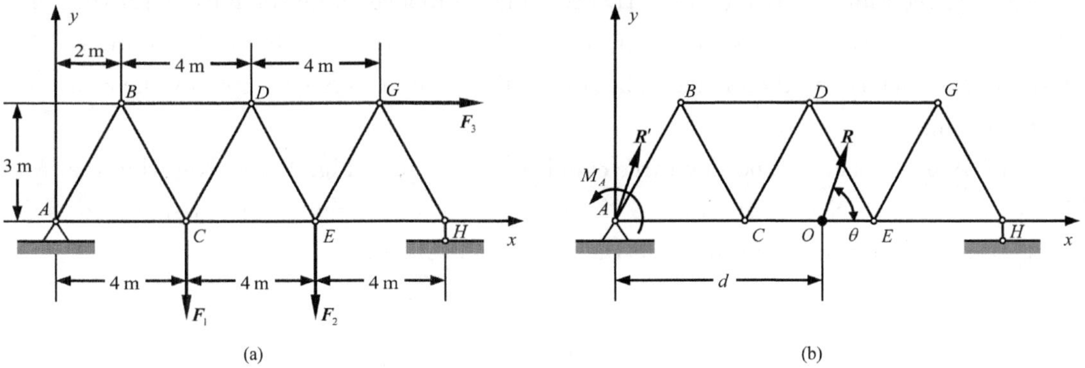

Fig. E3.5

(2) Reduction of the above equivalent force-couple system at A to an equivalent force at O is shown in Fig. E3.5(b). Using $\boldsymbol{R}=\boldsymbol{R}'$ and $M_A = R'_y d$, we have,

$$\boldsymbol{R} = \boldsymbol{R}' = 80\boldsymbol{i} - 310\boldsymbol{j}\ \text{N},\ d = \frac{M_A}{R'_y} = 6.71\ \text{m}$$

or

$$R = \sqrt{R_x^2 + R_y^2} = 320.16\ \text{N},\ \theta = \arctan\frac{R_y}{R_x} = -75.53°,\ d = \frac{M_A}{R'_y} = 6.71\ \text{m}$$

Problems

3.1 A plate is suspended from two chains AG and BH, as shown in Fig. P3.1. Knowing that the tension in AG is 300 N, determine (a) the moment about B of the force exerted by the chain AG, (b) the smallest force applied at D which creates the same moment about B.

3.2 A cable AB is attached to a concrete column, as shown in Fig. P3.2. Knowing that the tension in cable AB is 500 N, determine the moment about O of the force exerted by the cable at A on the column.

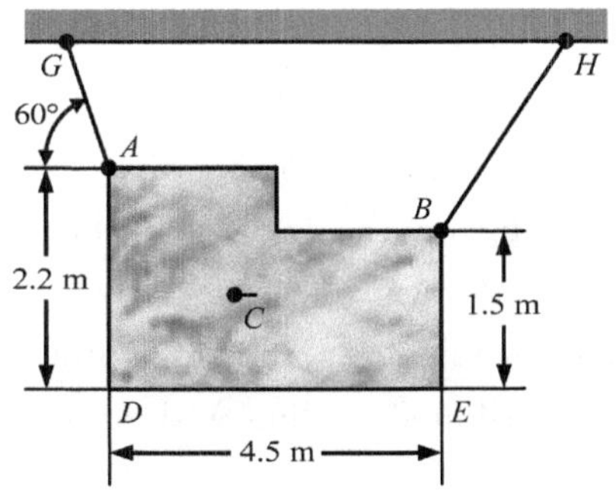

Fig. P3.1

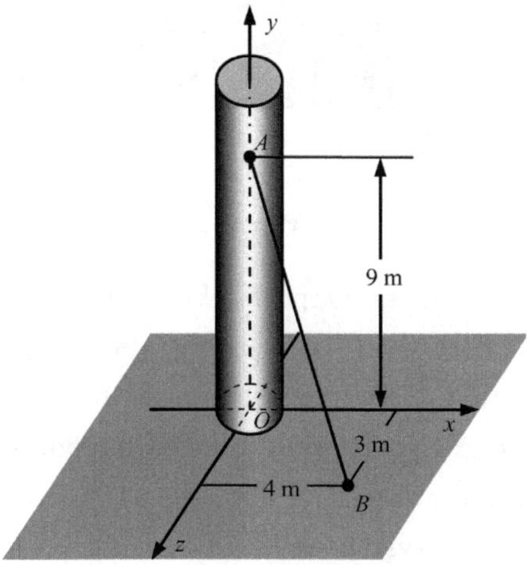

Fig. P3.2

3.3 Three cables AB, AC, and AD are attached to a concrete column, as shown in Fig. P3.3. Knowing that the tension in cables AB, AC, and AD are 800 N, 700 N, and 500 N, respectively, determine the moment about O of the resultant force exerted by the cables at A on the column.

3.4 A block is acted upon by three couples, as shown in Fig. P3.4. Knowing that $M_1=10$ N·m, $M_2=15$ N·m, and $M_3=8$ N·m, replace these couples with a single equivalent couple, and specify its magnitude and the direction of its axis.

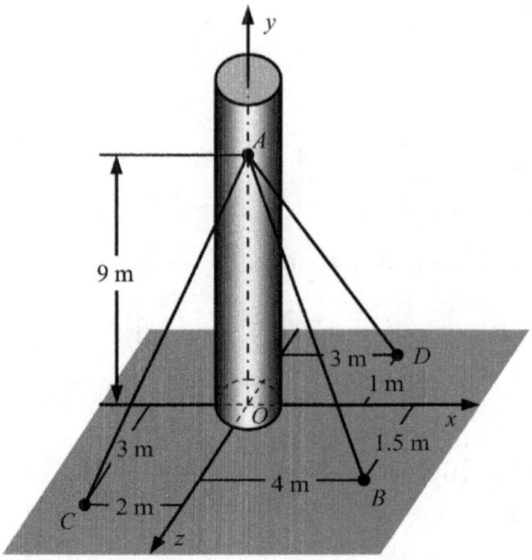

Fig. P3.3

3.5 A vertical force $\boldsymbol{F}$ acts at C of a planar truss, as shown in Fig. P3.5. Knowing that $F=100$ N, replace $\boldsymbol{F}$ with an equivalent force-couple system at B.

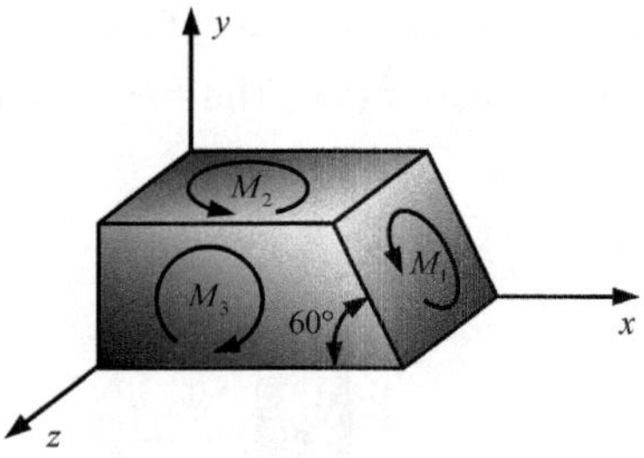

Fig. P3.4

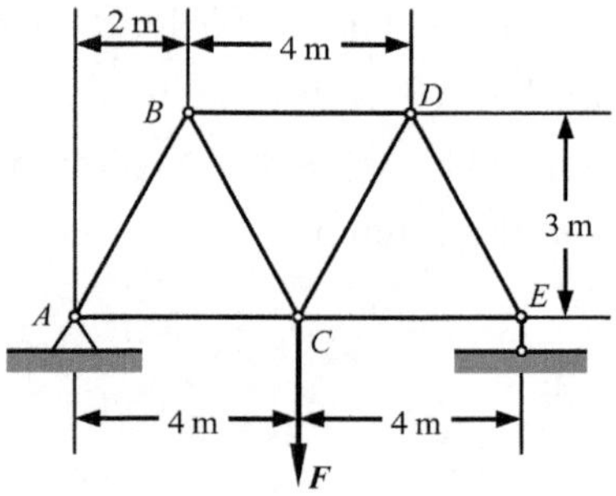

Fig. P3.5

3.6 A truss supports the loading, as shown in Fig. P3.6. Knowing that $F_1=F_2=100$ N, $F_3=90$ N, determine the equivalent force acting on the truss and the point of intersection of its line of action with a line through points A and H.

3.7 A cantilever beam AB supports a distributed loading, as shown in Fig. P3.7. Knowing that $a=2$ m and $q=100$ N/m, replace the distributed loading with an equivalent force-couple system at B.

3.8 A cantilever beam AB supports the loadings, as shown in Fig. P3.8. Knowing that

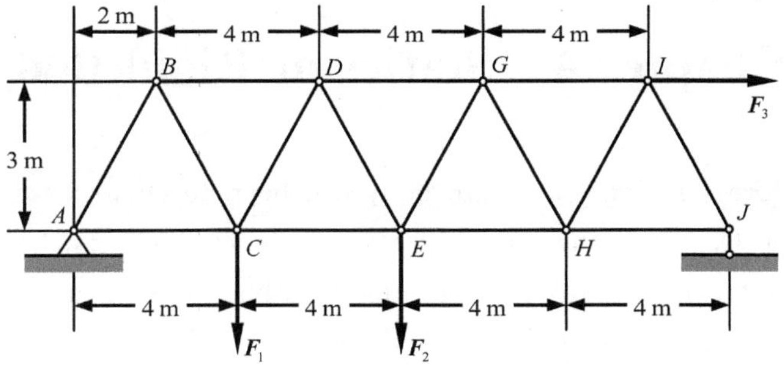

Fig. P3.6

$l=8$ m and $q=80$ N/m, determine the equivalent force acting on the beam and the point of intersection of its line of action with the beam.

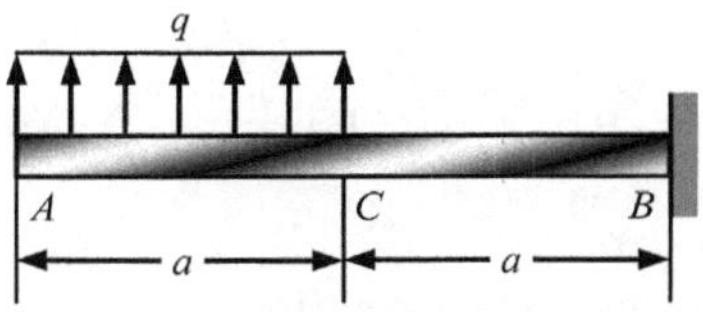

Fig. P3.7

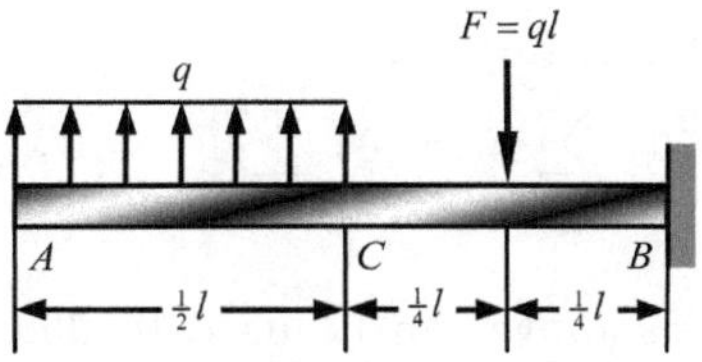

Fig. P3.8

Chapter 4 Statics of Rigid Body

The external forces acting on a rigid body can be reduced to a force-couple system at some arbitrary point O. When the force and the couple are both equal to zero, the external forces will form a system equal to zero, and the rigid body is said to be in equilibrium.

The necessary and sufficient conditions for the equilibrium of a rigid body can be expressed as

$$\sum \boldsymbol{F} = \boldsymbol{0},\ \sum \boldsymbol{M}_O(\boldsymbol{F}) = \boldsymbol{0} \tag{4.1}$$

Resolving each vector equation into its rectangular components, we can express the necessary and sufficient conditions for the equilibrium of a rigid body with the following six scalar equations of equilibrium:

$$\sum F_x = 0,\ \sum F_y = 0,\ \sum F_z = 0,\ \sum M_x(\boldsymbol{F}) = 0,\ \sum M_y(\boldsymbol{F}) = 0,\ \sum M_z(\boldsymbol{F}) = 0 \tag{4.2}$$

The equations above can be used to determine unknown forces acting on the rigid body or unknown reactions exerted by its supports.

In solving a problem concerning the equilibrium of a rigid body, we must consider all the forces acting on the body and exclude any force which is not directly applied to the body. Omitting a force or adding an extraneous one would destroy the conditions of equilibrium for the body considered. We should choose this rigid body as a free body and draw a separate free-body diagram showing this rigid body and all the forces acting on it.

4.1 Equilibrium of Two-Dimensional Rigid Body

1. Equations of Equilibrium

The equations of equilibrium for a 2D rigid body can be expressed, from Eq. (4.2), as

$$\sum F_x = 0,\ \sum F_y = 0,\ \sum M_A(\boldsymbol{F}) = 0 \tag{4.3}$$

where A is any point in the plane of the 2D rigid body. The three equations above are independent and can be solved for no more than three unknowns.

Although the three equations of equilibrium cannot be augmented by additional equation, any of them can be replaced by another equation. Therefore, an alternative form of equations of equilibrium can be given by

$$\sum F_x = 0,\ \sum M_A(\boldsymbol{F}) = 0,\ \sum M_B(\boldsymbol{F}) = 0 \tag{4.4}$$

where the line connecting points A and B is not perpendicular to the x axis.

A third possible set of equations of equilibrium is

$$\sum M_A(\boldsymbol{F}) = 0,\ \sum M_B(\boldsymbol{F}) = 0,\ \sum M_C(\boldsymbol{F}) = 0 \tag{4.5}$$

where the points A, B, and C do not lie in a straight line.

2. Reactions at Constraints for 2D Rigid Body

Supports or connections attached to a rigid body to restrict or limit its motion are called constraints. The reactions exerted by a constraint on a 2D rigid body can be divided into three groups corresponding to three types of constraints:

(1) Reactions equivalent to a force with known line of action. Constraints causing reactions of this type include rollers, rockers, cables, links, frictionless surfaces, frictionless collars on a rod, and frictionless pins in a slot.

(2) Reactions equivalent to a force with unknown direction and magnitude. Constraints causing reactions of this type include frictionless pins in a hole, frictionless hinges, and rough surfaces.

(3) Reactions equivalent to a force and a couple. Constraints causing reactions of this type include fixed supports and fixed connections.

Example 4.1

A T-shaped member $ABCD$ is supported at C by a pin and connected by a cable AED which passes over a fixed pulley at E, as shown in Fig. E4.1(a). Knowing that $F=150$ N and neglecting friction, determine the tension in the cable and the reaction at C.

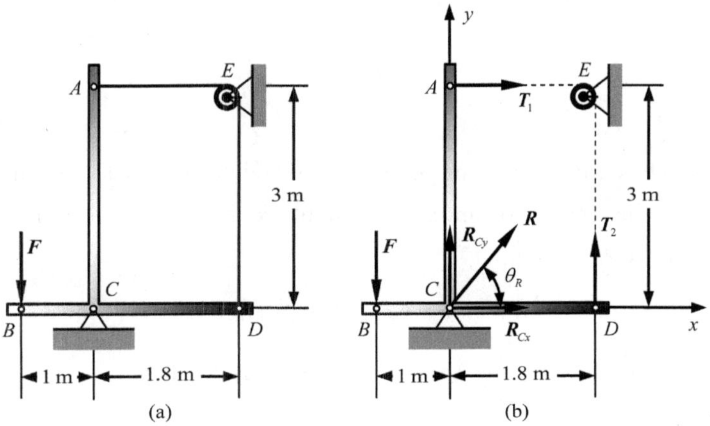

Fig. E4.1

Solution

Taking the T-shaped member as a free body and drawing the free-body diagram, as shown in Fig. E4.1(b), then from the equilibrium of the T-shaped member we have

$$\sum F_x = 0,\ R_{Cx} + T_1 = 0$$

$$\sum F_y = 0,\ R_{Cy} + T_2 - F = 0$$

$$\sum M_C = 0,\ F \times 1 - T_1 \times 3 + T_2 \times 1.8 = 0$$

Using $F=150$ N and $T_1=T_2$, and solving the above equations, we can obtain

$$R_{Cx} = -125\ \text{N},\ R_{Cy} = 25\ \text{N},\ T_1 = T_2 = 125\ \text{N}$$

Thus the tension in the cable is 125 N, and the reaction at C is

$$R = \sqrt{R_{Cx}^2 + R_{Cy}^2} = 127.48\ \text{N},\ \theta_R = 180° + \arctan\frac{R_{Cy}}{R_{Cx}} = 168.69°$$

Example 4.2

A homogeneous rod AB, of weight W, is attached to blocks A and B which move freely along the smooth surfaces, as shown in Fig. E4.2(a). The stiffness of the spring connected to block A is k, and the spring is unstretched when the rod is horizontal. (a) Neglecting the weight of the blocks, derive an equation in W, k, a, and θ which must be satisfied when the rod is in equilibrium. (b) Determine the value of θ when W=45 N, a=1 m, and k=50 N/m.

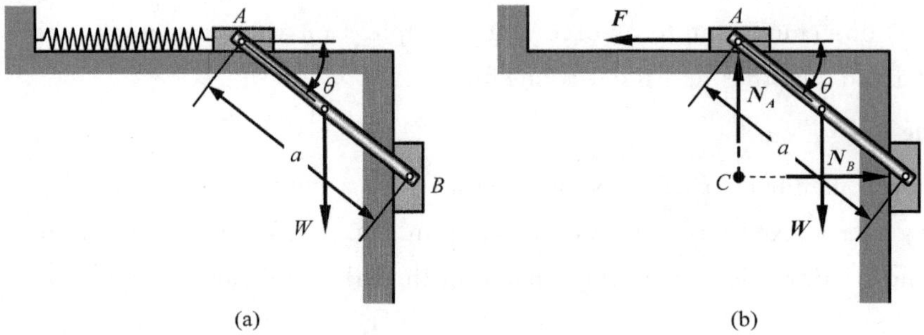

Fig. E4.2

Solution

(a) Take the rod as a free body and draw its free-body diagram, as shown in Fig. E4.2(b). Taking point C as the center of moment, then from the equilibrium of force moment about C of the rod, we can obtain

$$\sum M_C = 0,\ Fa\sin\theta - \frac{1}{2}Wa\cos\theta = 0$$

Using $F = ka(1-\cos\theta)$, and solving the above equation, we have

$$\tan\theta - \sin\theta = \frac{W}{2ka}$$

(b) Using W=45 N, a=1 m, and k=50 N/m, thus we obtain

$$\tan\theta - \sin\theta = 0.45$$

Solving the above equation, we can obtain

$$\theta = 50.7639°$$

Example 4.3

For the frame shown in Fig. E4.3(a), determine the components of the forces acting on rod EBG at E and B.

Solution

(1) Taking the entire frame as a free body and point A as the center of moment, as

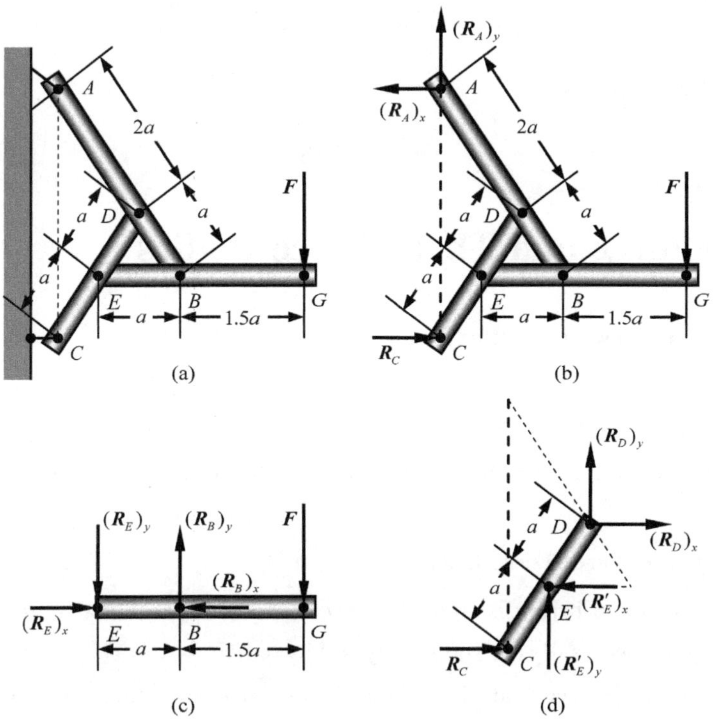

Fig. E4.3

shown in Fig. E4.3(b), then from the equilibrium of moment about A of the frame, we can obtain

$$\sum M_A = 0,\ R_C \cdot 2\sqrt{3}a - F \cdot 3a = 0,\ \text{i.e.},\ R_C = \frac{\sqrt{3}}{2}F$$

(2) Taking rod EBG as a free body, as shown in Fig. E4.3(c), then from the equilibrium of rod EBG, we can obtain

$$\sum M_B = 0,\ (R_E)y \cdot a - F \cdot 1.5a = 0$$

$$\sum M_E = 0,\ (R_B)_y \cdot a - F \cdot 2.5a = 0$$

$$\sum F_x = 0,\ (R_E)_x - (R_B)_x = 0$$

Solving the above equation, we have

$$(R_E)_y = \frac{3}{2}F,\ (R_B)_y = \frac{5}{2}F,\ (R_E)_x = (R_B)_x$$

(3) Taking rod CED as a free body and point D as the center of moment, as shown in Fig. E4.3(d), then from the equilibrium of moment about D of rod CED, we can obtain

$$\sum M_D = 0,\ R_C \cdot \sqrt{3}a - (R'_E)_x \cdot \frac{\sqrt{3}}{2}a - (R'_E)_y \cdot \frac{1}{2}a = 0$$

where $(R'_E)_y = (R_E)_y = \frac{3}{2}F$ and $R_C = \frac{\sqrt{3}}{2}F$. Solving the above equation, we have

$$(R'_E)_x = \frac{\sqrt{3}}{2}F, \text{ i.e. }, (R_E)_x = \frac{\sqrt{3}}{2}F$$

Using $(R_E)_x = (R_B)_x$, we obtain

$$(R_B)_x = \frac{\sqrt{3}}{2}F$$

4.2 Two-Force and Three-Force Bodies

1. Equilibrium of Two-Force Body

A particular case of equilibrium is that of a rigid body subjected to two forces. Such a body is called a two-force body. It can be shown that if a two-force body is in equilibrium, the two forces acting on it must have the same magnitude, the same line of action, and opposite sense.

2. Equilibrium of Three-Force Body

Another case of equilibrium is that of a three-force body, i.e., a rigid body subjected to three forces or, more generally, a rigid body subjected to forces at only three points. It can be shown that if a three-force body is in equilibrium, the lines of action of the three forces must be either concurrent or parallel.

Example 4.4

For the frame and loading, as shown in Fig. E4.4(a), determine the reactions at A and B knowing that $F=100$ N, $a=1$ m.

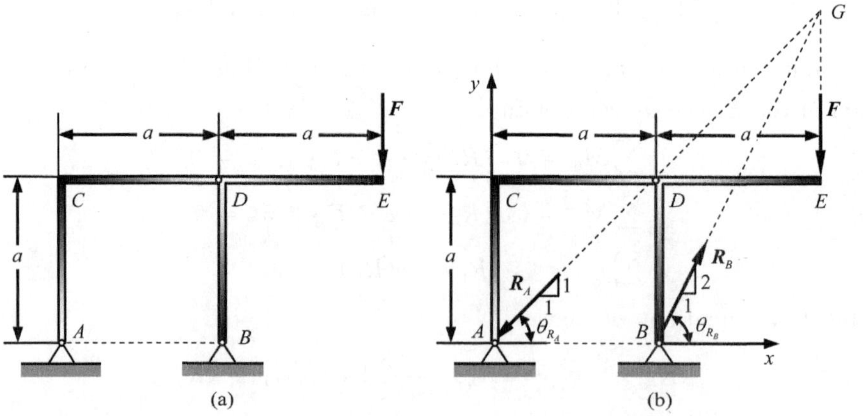

Fig. E4.4

Solution

Since the L-shaped bar ACD is a two-force member, the line of action of the reaction $\boldsymbol{R}_A$ at A must pass through A and D, as shown in Fig. E4.4(b). Since the entire structure is in equilibrium under the action of the three forces, i.e., $\boldsymbol{R}_A$ at A, $\boldsymbol{R}_B$ at B, and $\boldsymbol{F}$ at E, the line of action of $\boldsymbol{R}_B$ must pass through the point of intersection G of the lines of action of $\boldsymbol{R}_A$ and $\boldsymbol{F}$, as shown in Fig. E4.4(b), i.e., the entire structure can be considered to be subjected to three coplanar concurrent forces.

From the fact that the entire structure is in equilibrium, we have

$$\sum F_x = 0, \ -R_A\cos\theta_{R_A} + R_B\cos\theta_{R_B} = 0$$

$$\sum F_y = 0, \ -R_A\sin\theta_{R_A} + R_B\sin\theta_{R_B} - F = 0$$

Using $\tan\theta_{R_A}=1$, $\tan\theta_{R_B}=2$, and $F=100$ N, we have

$$R_A = 141.4 \text{ N}, \ \theta_{R_A} = 45^\circ$$

$$R_B = 223.6 \text{ N}, \ \theta_{R_B} = 63.43^\circ$$

4.3 Planar Trusses

A planar truss is a two-dimensional structure consisting of straight members connected together at their extremities by pins. Although the members of a planar truss are actually joined together by bolted or welded connections, it is customary to assume that the members are connected together by smooth pins. A planar truss is designed to carry loadings which act in the plane of the structure, it is, however, often to assume that all loadings are applied at the joints of the planar truss and that the weights of the members are also applied to the joints, half of the weight of each member being applied to each of the two joints of the member.

1. Simple Trusses

A truss is designed to carry loadings, thus it must be stable under the action of loadings. Consider the truss of Fig. 4.1(a), which is made of four members connected by four pins at A, B, C, and D. If a load is applied at C, the truss will be unstable. In contrast, the truss of Fig. 4.1(b), which is made of three members connected by three pins at A, B, and C, will be stable under the action of a load applied at C.

A large truss can be obtained by adding two members BD and CD to the basic triangular truss ABC, as shown in Fig. 4.2. This procedure can be repeated as many times as desired, and the resulting truss will be stable if each time two members are attached to two existing joints and connected at a new joint by a pin. A truss which can be constructed from a single triangular truss is called a simple truss.

Assuming that the total number of members and the total number of joints are respectively denoted by m and n in a simple truss, we have

$$(m-3) = 2(n-3), \text{ or } m = 2n-3 \qquad (4.6)$$

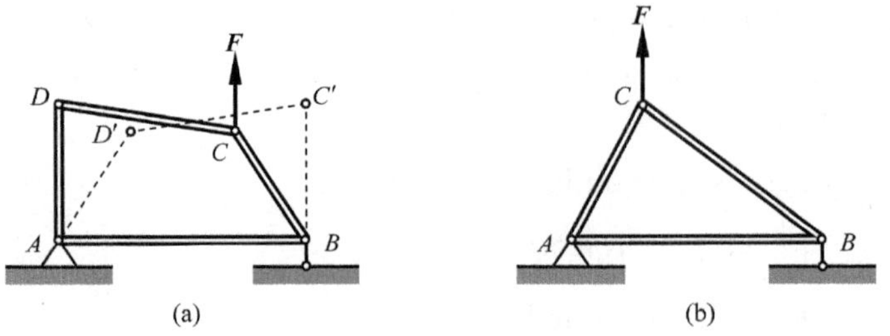

Fig. 4.1

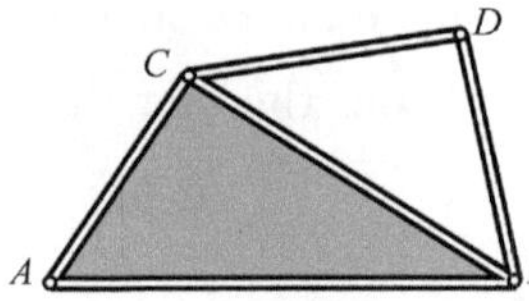

Fig. 4.2

2. Internal Forces of Trusses

Analysis of a truss requires the determination not only of the external forces acting on it but also of the internal forces which hold together the various parts of the truss. A truss can be considered as a combination of members and pins. Since the entire truss is in equilibrium, each member and each pin must also be in equilibrium. There are two methods, the method of joints and the method of sections, which can be used to determine the internal forces of a truss.

3. Method of Joints

Since the entire truss is in equilibrium, each pin of the truss must also be in equilibrium. The internal forces of the truss can be determined by drawing the free-body diagram of each pin and by using two equations of equilibrium for each pin.

Example 4.5

Knowing $F=80$ N, using the method of joints to determine the force in each member of the truss shown in Fig. E4.5(a) and state whether each member is in tension or compression.

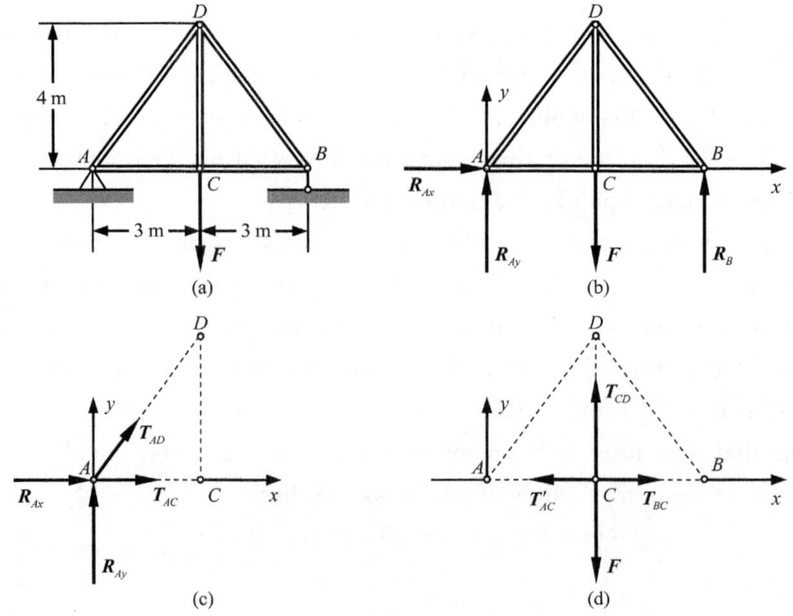

Fig. E4.5

Solution

(1) Taking the entire truss as free body and considering its equilibrium, as shown in Fig. E4.5(b), we have

$$\sum F_x = 0,\ R_{Ax} = 0$$

$$\sum M_A = 0,\ R_B \times 6 - F \times 3 = 0$$

$$\sum M_B = 0,\ -R_{Ay} \times 6 + F \times 3 = 0$$

Solving the above equations, we obtain

$$R_{Ax} = 0,\ R_{Ay} = 40\ \text{N},\ R_B = 40\ \text{N}$$

(2) Taking the joint A as free body and considering its equilibrium, as shown in Fig. E4.5(c), we have

$$\sum F_x = 0,\ R_{Ax} + T_{AC} + T_{AD} \times \frac{3}{5} = 0$$

$$\sum F_y = 0,\ R_{Ay} + T_{AD} \times \frac{4}{5} = 0$$

Solving the above equations, we obtain

$$T_{AC} = 30\ \text{N(tension)},\ T_{AD} = -50\ \text{N (compression)}$$

(3) Taking the joint C as free body and considering its equilibrium, as shown in Fig. E4.5(d), we have

$$\sum F_x = 0,\ T_{BC} - T'_{AC} = 0$$

$$\sum F_y = 0,\ T_{CD} - F = 0$$

Solving the above equations, we obtain

$$T_{BC} = 30\ \text{N (tension)},\ T_{CD} = 80\ \text{N (tension)}$$

(4) Similarly, taking the joint B or D as free body and considering its equilibrium, we can obtain

$$T_{BD} = -50\ \text{N (compression)}$$

4. Zero-Force Members

If the internal force in a member is zero, this member is said to be a zero-force member. A zero-force member is used to increase the stability of a truss and to provide support if the loading applied to the truss is changed.

Example 4.6

For the truss and loading shown in Fig. E4.6, determine the zero-force members.

Solution

Considering the joint G shown in Fig. E4.6(b), and using $\sum F_y = 0$, we can find

$$T_{CG} = 0$$

Similarly, by inspection of each joint of the truss, we can find

$$T_{CH} = T_{CK} = T_{DL} = T_{EI} = T_{EJ} = T_{EM} = 0$$

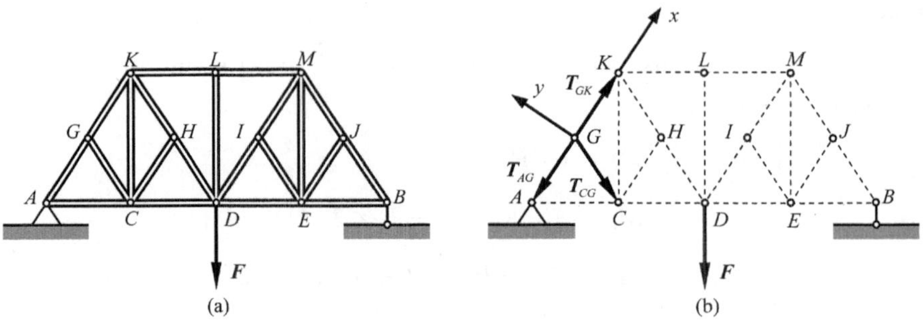

Fig. E4.6

5. Method of Sections

The method of joints is most effective when the forces in all the members of a truss are to be determined. If, however, the force in only one member or the forces in very few members are desired, the method of sections will be more efficient.

Since the entire truss is in equilibrium, any part of the truss must also be in equilibrium. The internal forces of any part of the truss can be determined by drawing the free-body diagram of this part and by using three equations of equilibrium.

Example 4.7

A truss is loaded as shown in Fig. E4.7(a). Knowing $F_1 = 80$ N and $F_2 = 40$ N, using the method of sections determine the force in members CD, CH, and GH.

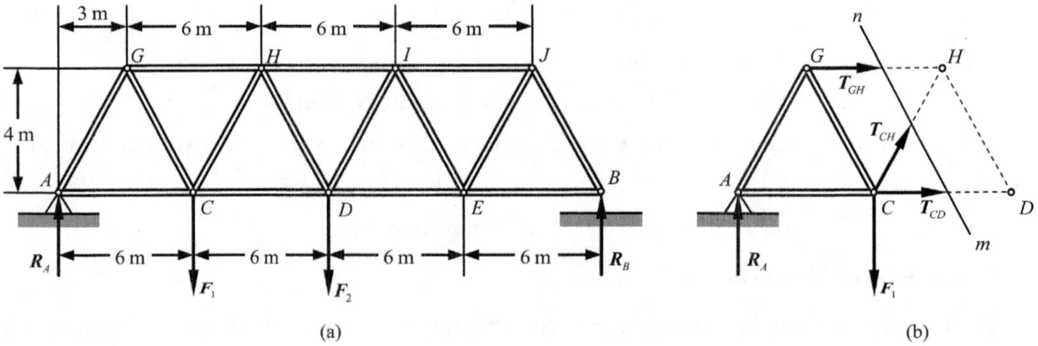

Fig. E4.7

Solution

Taking the entire truss as free body and considering its equilibrium, as shown in Fig. E4.7(a), we have

$$R_A = 80 \text{ N},\ R_B = 40 \text{ N}$$

To determine the forces in members CD, CH, and GH by using the method of sections, these three members should be sectioned, as shown in Fig. E4.7(b). Considering the equilibrium of the left part of the sectioned trusses, we can obtain

$$\sum M_H = 0,\ -R_A \times 9 + F_1 \times 3 + T_{CD} \times 4 = 0$$

$$\sum M_C = 0, \ -R_A \times 6 - T_{GH} \times 4 = 0$$

$$\sum F_y = 0, \ R_A - F_1 + T_{CH} \times \frac{4}{5} = 0$$

Solving the above equations, we obtain

$$T_{CD} = 120 \text{ N}, \ T_{CH} = 0, \ T_{GH} = -120 \text{ N}$$

4.4 Equilibrium of Three-Dimensional Rigid Body

1. Equations of Equilibrium

The equations of equilibrium for a 3D rigid body can be expressed, from Eq. (4.2), as

$$\sum F_x = 0, \ \sum F_y = 0, \ \sum F_z = 0, \ \sum M_x(\boldsymbol{F}) = 0, \ \sum M_y(\boldsymbol{F}) = 0, \ \sum M_z(\boldsymbol{F}) = 0 \quad (4.7)$$

These equations above are independent and can be solved for no more than six unknowns.

2. Reactions at Constraints for 3D rigid body

The reactions exerted on a 3D rigid body range from the single force of known direction exerted by a frictionless surface to the force-couple system exerted by a fixed support.

Example 4.8

A homogeneous square plate has a weight of 150 N and is supported by three vertical cables, as shown in Fig. E4.8(a). Determine the tension in each cable.

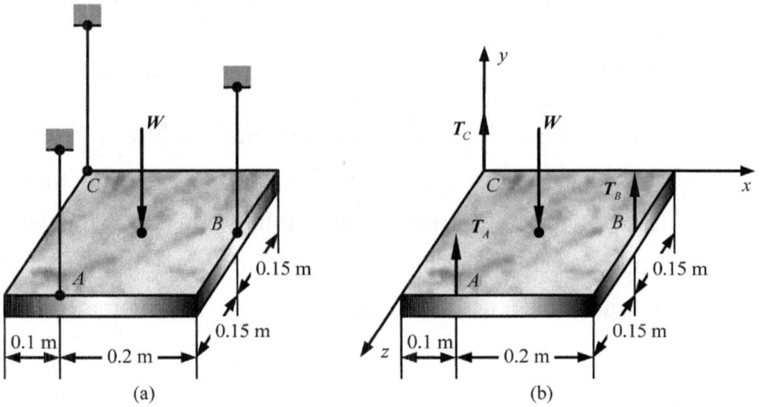

Fig. E4.8

Solution

Take the plate as a free body, as shown in Fig. E4.8(b). From the equilibrium of the plate, we have

$$\sum F_y = 0, \ T_A + T_B + T_C - W = 0$$

$$\sum M_x = 0, \ -T_A \times 0.3 - T_B \times 0.15 + W \times 0.15 = 0$$

$$\sum M_z = 0, \ T_A \times 0.1 + T_B \times 0.3 - W \times 0.15 = 0$$

Solving the above equations, we obtain

$$T_A = 45 \text{ N}, \ T_B = 60 \text{ N}, \ T_C = 45 \text{ N}$$

Problems

4.1 A T-shaped member $ABCD$ is supported at C by a pin and connected by a cable AED which passes over a fixed pulley at E, as shown in Fig. P4.1. Knowing that $q=250$ N/m and neglecting friction, determine the tension in the cable and the reaction at C.

4.2 A T-shaped member $ABCD$ is fixed at A and connected by a cable DE, as shown in Fig. P4.2. Knowing that $F=250$ N and the tension in the cable is $T=50$ N, determine the reaction at the fixed support A.

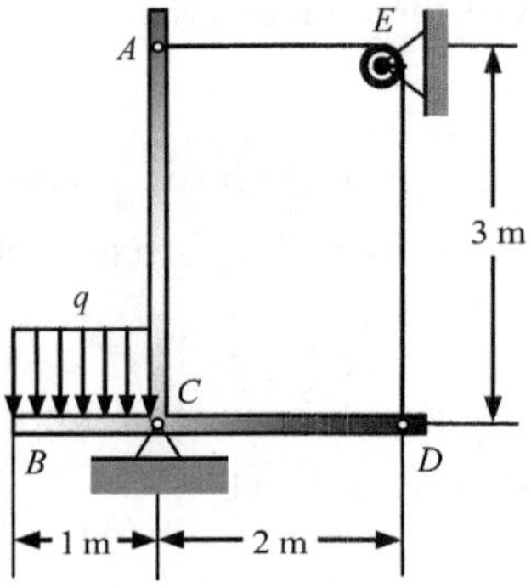

Fig. P4.1

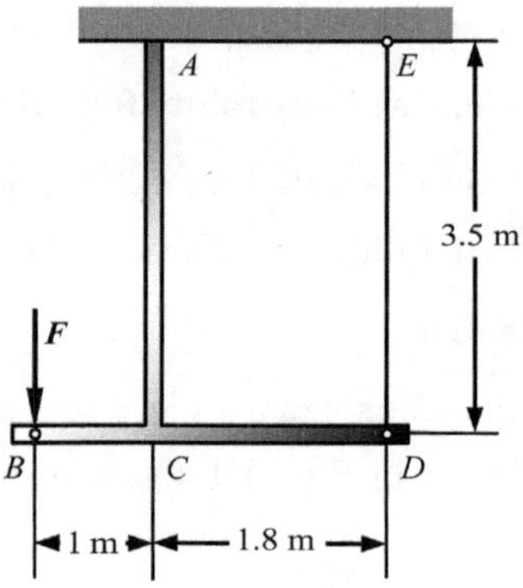

Fig. P4.2

4.3 A homogeneous rod AB, of weight W, is attached to blocks A and B which move freely along the smooth surfaces, as shown in Fig. P4.3. The stiffness of the spring connected to block A is k, and the spring is unstretched when the rod is horizontal. (a) Assuming that the weight of each of the blocks is W, derive an equation in W, k, a, and θ which must be satisfied when the rod is in equilibrium. (b) Determine the value of θ when $W=15$ N, $a=1$ m, and $k=50$ N/m.

4.4 A homogeneous rod AB, of weight W, is attached to blocks A and B, which move freely along the smooth surfaces, as shown in Fig. P4.4. The blocks are connected by a cable which passes over a pulley at C. (a) Neglecting the weight of the blocks, express the tension in the cable in terms of W and θ when the rod is in equilibrium. (b) Determine the value of θ for which the tension in the cable is equal to W.

4.5 The maximum allowable value of each of the reactions at A and B is 450 N, as shown in Fig. P4.5. Neglecting the weight of the beam, determine the range of values of the distance d for which the beam is safe.

4.6 A vertical force $\boldsymbol{F}$ acts at C of a planar truss, as shown in Fig. P4.6. Knowing that $F=80$ N, determine the reactions at A and H.

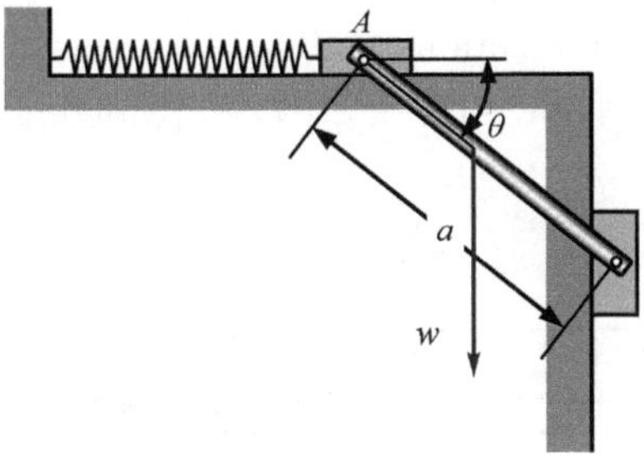

Fig. P4.3

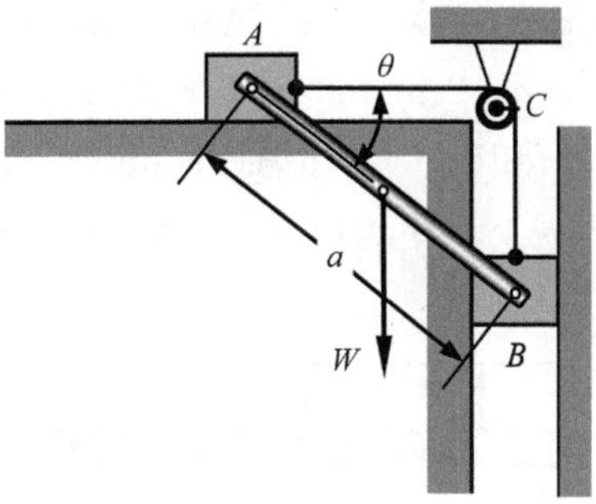

Fig. P4.4

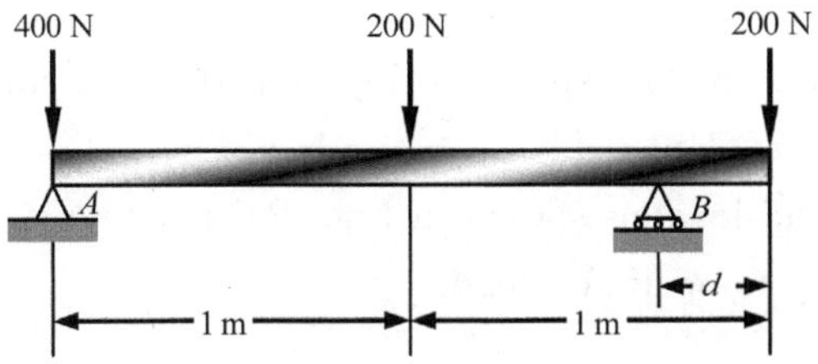

Fig. P4.5

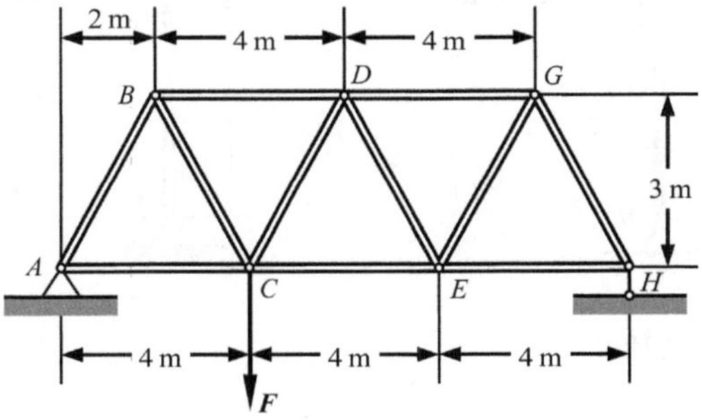

Fig. P4.6

4.7 A truss supports the loadings shown in Fig. P4.7. Knowing that $F_1 = 160$ N, $F_2 = 150$ N, and $F_3 = 80$ N, determine the reactions at A and H.

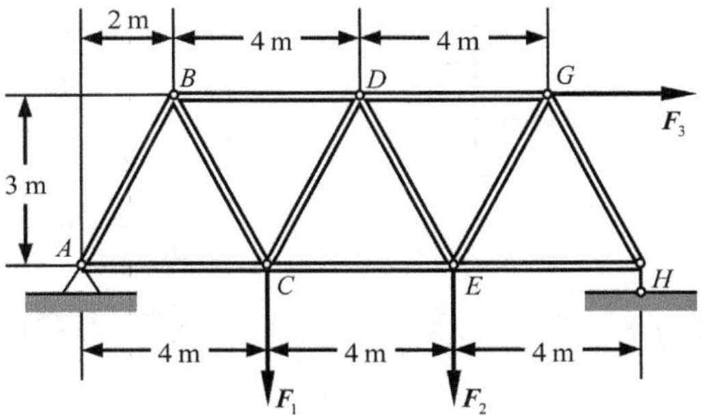

Fig. P4.7

4.8 A cantilever beam AB supports a distributed loading, as shown in Fig. P4.8. Knowing that $a=2$ m and $q=100$ N/m, determine the reactions at B.

4.9 A cantilever beam AB supports the loadings shown in Fig. P4.9. Knowing that $l=$ 8 m and $q=80$ N/m, determine the reactions at B.

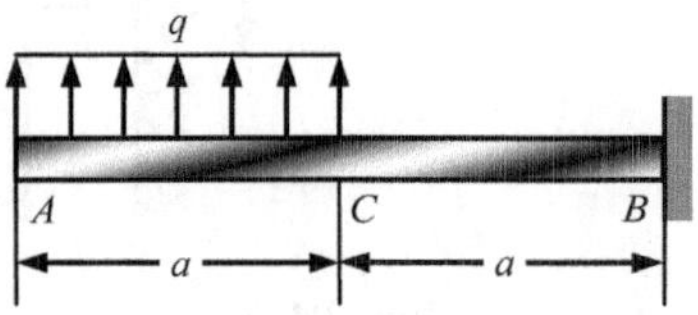

Fig. P4.8

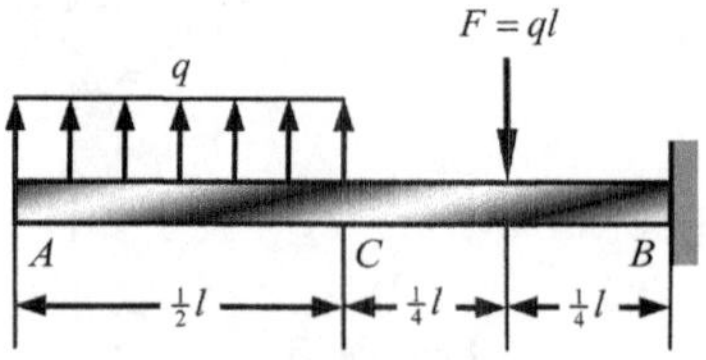

Fig. P4.9

4.10 For the frame and loading shown in Fig. P4.10, determine the reactions at A and B knowing $F=200$ N and $a=1$ m.

4.11 For the frame and loading shown in Fig. P4.11, knowing that $q=100$ N/m and $a=1$ m, determine the reactions at A and B.

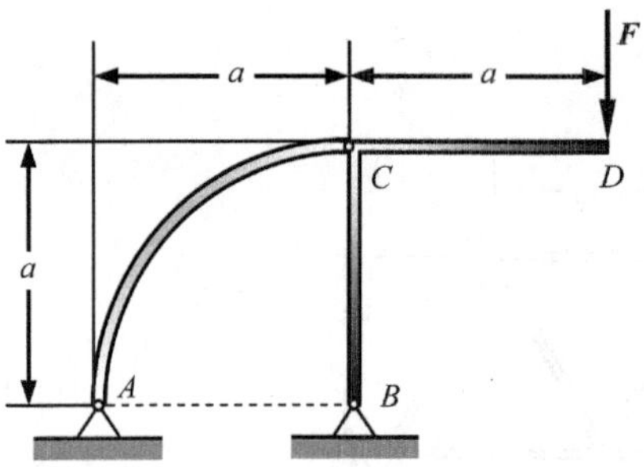

Fig. P4.10

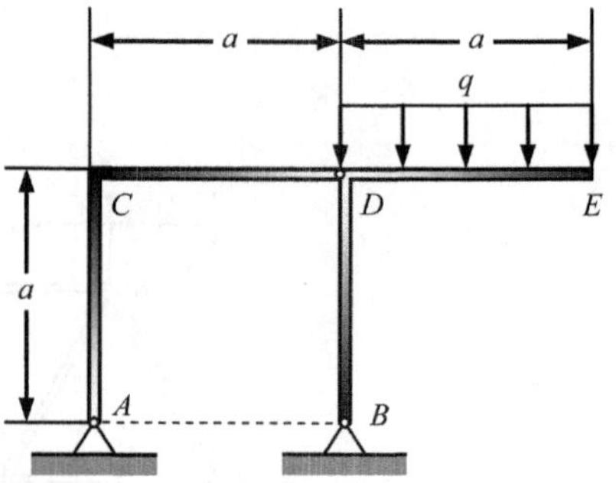

Fig. P4.11

4.12 Knowing $M_e=100$ N·m and $a=1$ m, determine the reactions at A and B, as shown in Fig. P4.12.

4.13 Using the method of joints to determine the force in each member of the truss shown in Fig. P4.13 and state whether each member is in tension or compression.

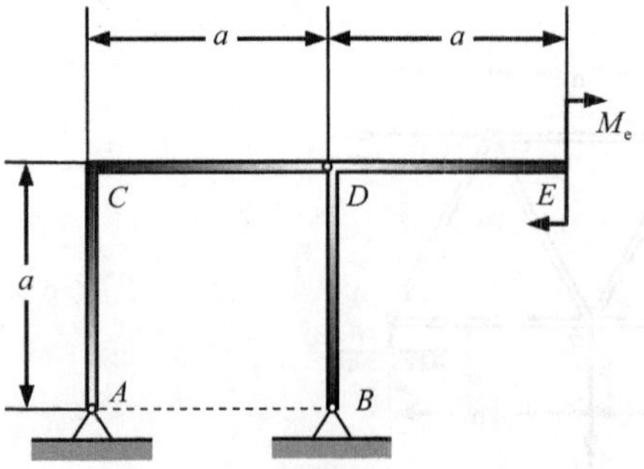

Fig. P4.12

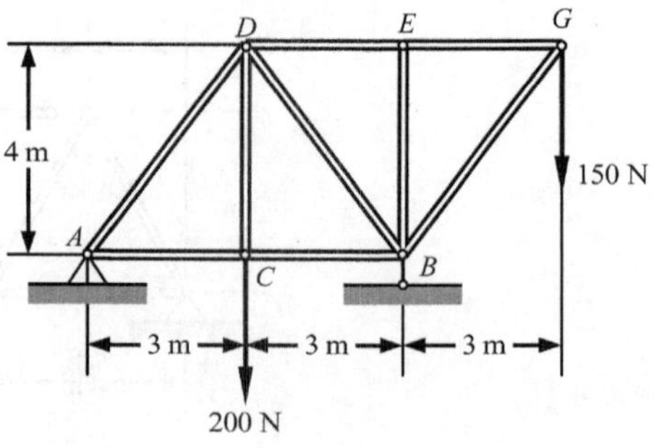

Fig. P4.13

4.14 For the truss and loading shown in Fig. P4.14, determine the zero-force members.

4.15 A truss is loaded as shown in Fig. P4.15. Knowing $F_1 = 90$ N and $F_2 = 60$ N, using the method of sections to determine the force in members CD, DG, and GH.

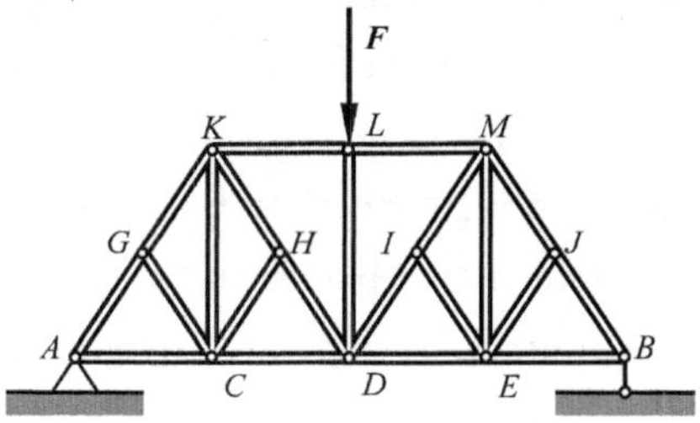

Fig. P4.14

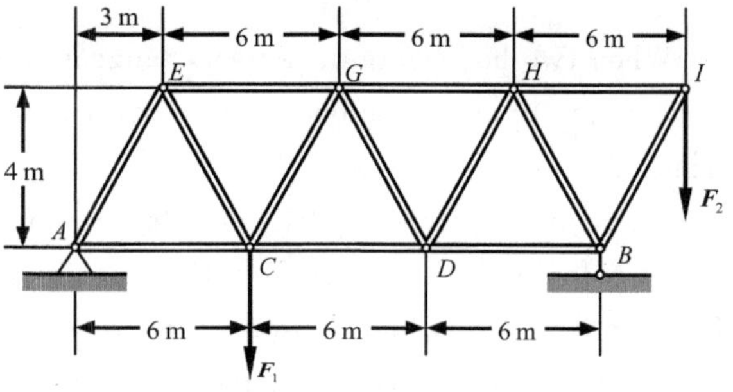

Fig. P4.15

4.16 A homogeneous composite plate has a weight of 250 N and is supported by three vertical cables, Fig. P4.16. Determine the tension in each cable.

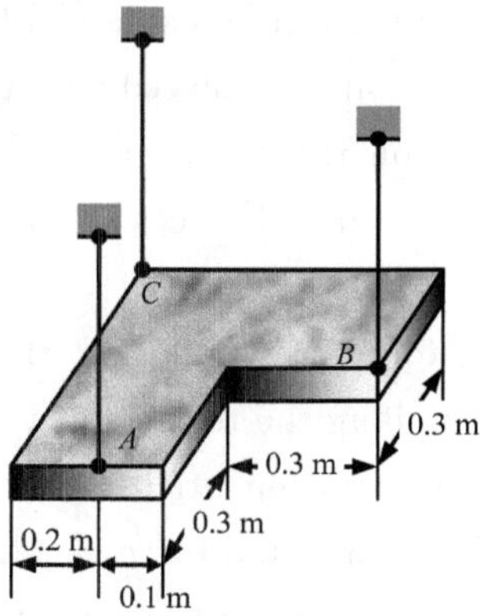

Fig. P4.16

Chapter 5 Friction

When two bodies are in contact, tangential forces, called friction forces, will develop if one attempts to move one body with respect to the other. These friction forces are limited in magnitude.

There are two types of friction: dry friction and fluid friction. Dry friction, sometimes called Coulomb friction, occurs on the surface of contact between two bodies in the absence of lubricating fluid. Fluid friction exists when two bodies in contact are separated by a film of fluid, and will be studied in the mechanics of fluids.

5.1 Sliding Friction

1. Analysis of Sliding Friction

The phenomenon of sliding friction can be explained by the following experiment:

(1) Consider that a block of weight $\boldsymbol{W}$ is placed on a rough horizontal surface, as shown in Fig. 5.1(a). The forces acting on the block are its weight $\boldsymbol{W}$ and the normal force $\boldsymbol{N}$ exerted by the rough surface. These vertical forces do not tend to move the block along the rough surface.

(2) Assume that a horizontal force $\boldsymbol{F}$ is applied to the block, as shown in Fig. 5.1(b). This force tends to move the block along the rough surface. However, when $\boldsymbol{F}$ is small, the block does not move. This shows that a tangential force exerted by the rough surface on the block must exist to balance $\boldsymbol{F}$. This tangential force is called the static friction force $\boldsymbol{F}_s$, and can be found by solving the equations of equilibrium for the block.

(3) If the horizontal force $\boldsymbol{F}$ is increased, the static friction force $\boldsymbol{F}_s$ also increases, continuing to balance $\boldsymbol{F}$, until it reaches the maximum static friction force $\boldsymbol{F}_{max}$, as shown in Fig. 5.1(c). When the maximum static friction force is reached, the block is on the verge of sliding, called impending motion.

(4) If the horizontal force $\boldsymbol{F}$ is further increased, the static friction force $\boldsymbol{F}_s$ cannot balance $\boldsymbol{F}$ any more and the block will begin to slide, as shown in Fig. 5.1(d). As soon as the block starts sliding, the friction force drops from the maximum static friction force $\boldsymbol{F}_{max}$ to a lower friction force $\boldsymbol{F}_k$, which is called the kinetic friction force. When the block is in motion, the kinetic friction force $\boldsymbol{F}_k$ remains approximately constant.

2. Law of Static Friction

Experimentally, it has been shown that the value F_{max} of the maximum static friction force is proportional to the value N of the normal force exerted by the rough surface on the block. We thus have

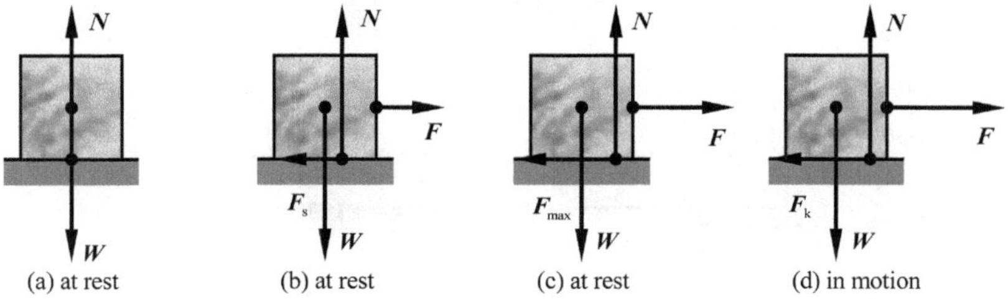

Fig. 5.1

$$F_{max} = \mu_s N \tag{5.1}$$

where μ_s is a constant called the coefficient of static friction. This relation is called the law of static friction.

3. Law of Kinetic Friction

Similarly the magnitude F_k of the kinetic friction force is also proportional to the magnitude N of the normal force acting on the block. Thus the law of kinetic friction can be expressed as

$$F_k = \mu_k N \tag{5.2}$$

where μ_k is the coefficient of kinetic friction.

The coefficients of friction μ_s and μ_k are not dependent on the contacting area of bodies in contact. They, however, will depend on both the material characteristics and the surface nature of bodies in contact.

5.2 Angles of Friction

1. Angle of Friction

Consider that a block of weight $\boldsymbol{W}$ is placed on a rough horizontal surface and a horizontal force $\boldsymbol{F}$ is applied to the block. When the static friction force reaches its maximum value F_{max}, the included angle between the normal force $\boldsymbol{N}$ and the resultant reaction $\boldsymbol{R}$ of $\boldsymbol{N}$ and $\boldsymbol{F}_{max}$ is defined as the angle of static friction, denoted by φ_s. From the geometry of Fig. 5.2(a), we have

$$\tan\varphi_s = \frac{F_{max}}{N} = \mu_s \tag{5.3}$$

If motion actually takes place, the included angle between the normal force $\boldsymbol{N}$ and the resultant reaction $\boldsymbol{R}$ drops from φ_s to a lower value φ_k, called the angle of kinetic friction. The angle of kinetic friction, from the geometry of Fig. 5.2(b), can be given by

$$\tan\varphi_k = \frac{F_k}{N} = \mu_k \tag{5.4}$$

2. Self-Locking

The angle of static friction φ_s is the largest included angle between the normal of the contacting surface and the resultant reaction $\boldsymbol{R}$, as shown in Fig. 5.3(a), thus if the included

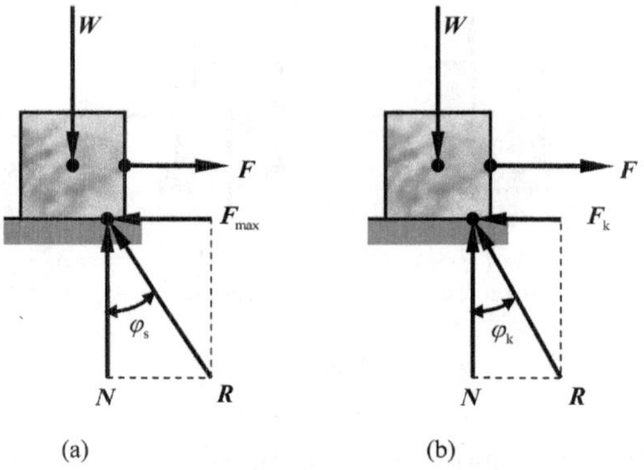

Fig. 5.2

angle φ between the normal line of the contacting surface and the resultant force $\boldsymbol{R}'$ of $\boldsymbol{W}$ and $\boldsymbol{F}$ is more than φ_s, i.e., $\varphi > \varphi_s$, then the block will be in motion due to the fact that $\boldsymbol{R}'$ and $\boldsymbol{R}$ do not have the same line of action, as shown in Fig. 5.3(a). However, if φ is equal to or less than φ_s, i.e., $\varphi \leqslant \varphi_s$, then the block will be at rest, i.e., the block is self-locking, Fig. 5.3(b) or Fig. 5.3(c).

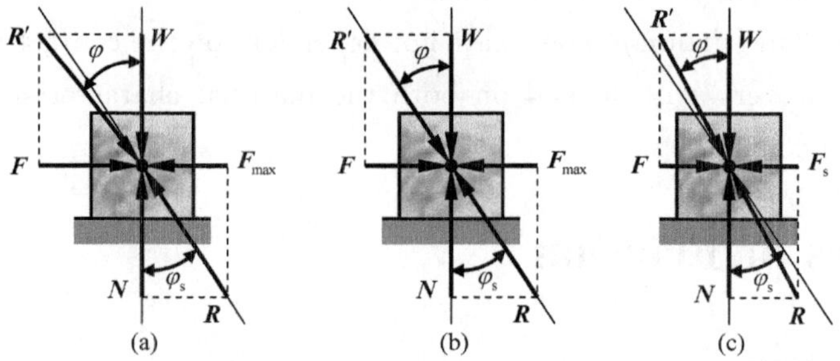

Fig. 5.3

5.3 Problems Involving Sliding Friction

Problems involving sliding friction fall into one of the following three types:

(1) The first type is that all applied forces are given and the coefficient of static friction is known, and that we need to determine whether the body under consideration will be at rest or in motion.

Example 5.1

A force of magnitude $F = 200$ N acts on a block of mass $m = 100$ kg placed on an inclined plane of inclination angle $\theta = 30°$, as shown in Fig. E5.1(a). Knowing that the coefficients of friction between the block and the plane are $\mu_s = 0.3$ and $\mu_k = 0.2$, determine whether the block is in equilibrium, and find the magnitude and direction of the friction force.

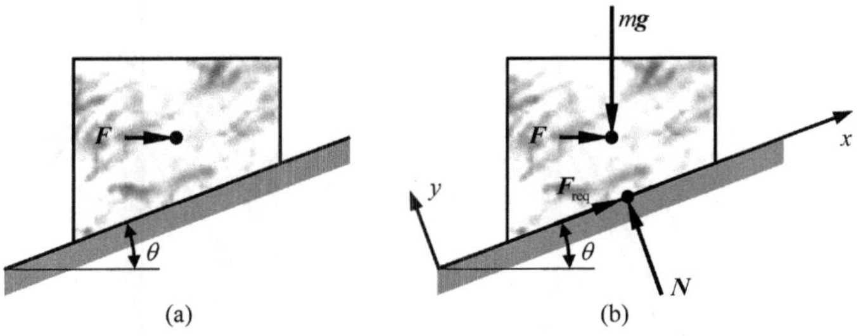

Fig. E5.1

Solution

Draw the free-body diagram of the block in Fig. E5.1(b). Assuming that the block is in equilibrium and that the required friction force along the plane is $\boldsymbol{F}_{req}$ directed upward and to the right, then the equations of equilibrium for the block can be expressed as

$$\sum F_x = 0, \ F_{req} + F\cos\theta - mg\sin\theta = 0$$

$$\sum F_y = 0, \ N - F\sin\theta - mg\cos\theta = 0$$

Solving the equations above for F_{req} and N, we obtain

$$F_{req} = 317.29 \text{ N}$$

$$N = 949.57 \text{ N}$$

The value of the maximum static friction force is equal to

$$F_{max} = \mu_s N = 284.87 \text{ N}$$

Since $F_{req} > F_{max}$, equilibrium will not be maintained and the block will slide down the plane. The magnitude of friction force can be given, according to the law of kinetic friction force, by

$$F_k = \mu_k N = 189.91 \text{ N}$$

and its direction is up the plane.

(2) The second type is that all applied forces are given and the motion is in impending, and that we need to determine the coefficient of static friction.

Example 5.2

A uniform ladder AB, having length l and mass m, leans against a wall, as shown in Fig. E5.2(a). Assuming the coefficient of static friction μ_s is the same at A and B, determine the smallest value of μ_s for which equilibrium is maintained when $\theta = 60°$.

Solution

Assume that motion is impending at both A and B, and draw the free-body diagram of the ladder, as shown in Fig. E5.2(b). Using the law of static friction, the supplementary equations can be given by

$$F_A = \mu_s N_A$$

$$F_B = \mu_s N_B$$

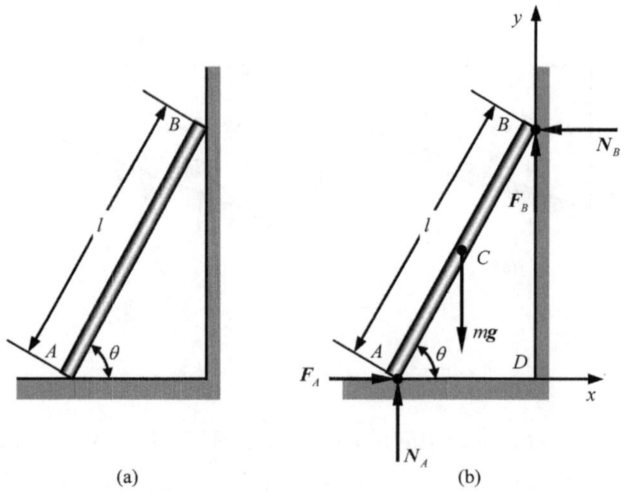

Fig. E5.2

Considering the equilibrium of the ladder, the equilibrium equations can be expressed as

$$\sum F_x = 0,\ F_A - N_B = 0$$

$$\sum F_y = 0,\ F_B + N_A - mg = 0$$

$$\sum M_D = 0,\ N_B l \sin\theta - N_A l \cos\theta + \frac{1}{2} mgl \cos\theta = 0$$

Solving the supplementary and equilibrium equations for μ_s, we have

$$\mu_s^2 + 2\mu_s \tan\theta - 1 = 0$$

Using $\theta = 60°$ and solving the above equation, we obtain

$$\mu_s = 0.27 \text{ or } \mu_s = -3.73$$

Physically, the positive root is possible, therefore the smallest value of μ_s is equal to 0.27.

(3) The third type is that the coefficient of static friction is known and the motion is impending, and that we need to determine the magnitude or the direction of the applied force.

Example 5.3

A collar B of weight $\boldsymbol{W}$ is attached to the spring AB and can move along the rod, as shown in Fig. E5.3(a). The spring stiffness is $k = 1.8$ kN/m and the spring is unstretched when $\theta = 0$. Knowing that the coefficient of static friction between the collar and the rod is $\mu_s = 0.25$, and that the horizontal distance between points A and B is $l = 0.6$ m, determine the range of values of $\boldsymbol{W}$ for which equilibrium is maintained when $\theta = 30°$.

Solution

Assume that the collar is in equilibrium and the static friction force $\boldsymbol{F}_s$ is directed upward, as shown in Fig. E5.3(b). From the free-body diagram of the collar, the equations of equilibrium for the collar can be expressed as

$$\sum F_x = 0,\ N - F_{spr} \cos\theta = 0$$

$$\sum F_y = 0,\ F_s + F_{spr} \sin\theta - W = 0$$

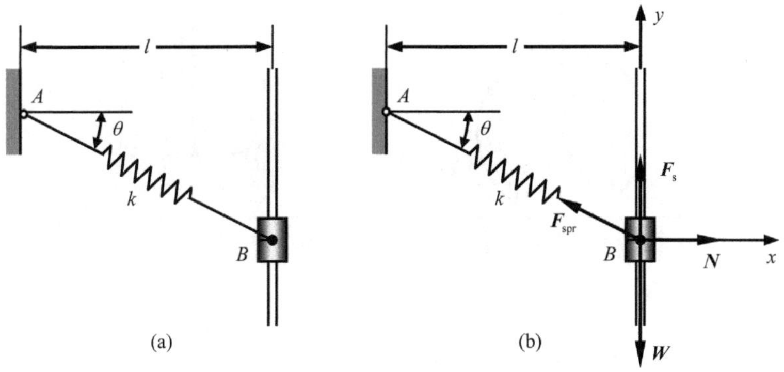

Fig. E5.3

Solving these two equations for N and F_s, we obtain

$$N = F_{spr}\cos\theta$$
$$F_s = W - F_{spr}\sin\theta$$

When the collar is in equilibrium, we have

$$|F_s| \leqslant \mu_s N$$

i.e.,

$$-\mu_s F_{spr}\cos\theta \leqslant W - F_{spr}\sin\theta \leqslant \mu_s F_{spr}\cos\theta$$

or,

$$F_{spr}(\sin\theta - \mu_s\cos\theta) \leqslant W \leqslant F_{spr}(\sin\theta + \mu_s\cos\theta)$$

Using $F_{spr} = kl(1/\cos\theta - 1)$ and $\theta = 30°$, then we obtain

$$47.37\ \text{N} \leqslant W \leqslant 119.71\ \text{N}$$

5.4 Rolling Resistance

Consider that a wheel of radius r rolls freely without sliding along a horizontal surface. If the wheel and surface are both rigid, then the normal force exerted by the surface on the wheel acts at the tangent point A of the wheel, as shown in Fig. 5.4(a), i.e., in this case, the wheel will be subjected to only two forces: its own weight $\boldsymbol{W}$ and the normal reaction $\boldsymbol{N}$ of the surface. Regardless of the value of the coefficient of friction between wheel and surface no friction force will act on the wheel. Thus the wheel rolling freely without sliding on a horizontal surface should keep rolling indefinitely.

However, experimental evidence shows that the wheel rolling without sliding will slow down and eventually come to rest. This is due to the fact that both the wheel and the surface are not perfectly rigid, and that the contact between wheel and surface takes always place over a certain area, rather than a single point. The resultant reaction exerted by the surface on the wheel over this area is a force $\boldsymbol{R}$ applied at a point B with a horizontal distance δ from A, as shown in Fig. 5.4(b). To balance the moment of $\boldsymbol{W}$ about point B and to keep the wheel rolling at constant velocity, it is necessary to apply a horizontal force $\boldsymbol{F}$ at the center of

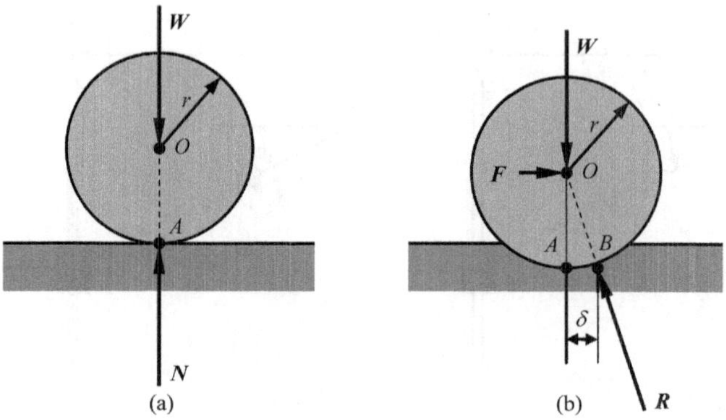

Fig. 5.4

the wheel.

Using $\sum M_B = 0$, we obtain

$$F = \frac{\delta}{r} W \tag{5.5}$$

where δ is called the coefficient of rolling resistance, which is usually expressed in mm. The value of the coefficient of rolling resistance δ is not dependent on the radius of the wheel. However, it will depend on both the material characteristics and the surface nature of the wheel and surface. Values of the coefficient of rolling resistance vary from about 0.25 mm for a steel wheel on a steel surface to 125 mm for the same wheel on a soft surface.

Since $\frac{\delta}{r} \ll \mu_s$ (or μ_k), i.e., $F_{\text{rolling}} = \frac{\delta}{r} W \ll F_{\text{sliding}} = \mu_s W$ (or $\mu_k W$), it is much easier to make a body roll along a surface than to make it slide along the same surface.

Problems

5.1 A force $\boldsymbol{F}$ acts on a block of mass $m = 100$ kg placed on an inclined plane of inclination angle $\theta = 30°$, as shown in Fig. P5.1. Knowing that the coefficients of friction between the block and the plane are $\mu_s = 0.3$ and $\mu_k = 0.2$, determine whether the block is in equilibrium, and find the magnitude and direction of the friction force when (a) $F = 500$ N, and (b) $F = 1200$ N.

5.2 A uniform ladder AB, having length l and mass m, leans against a wall, as shown in Fig. P5.2. Assuming that the coefficient of static friction is μ_s at A and zero at B, determine the smallest value of μ_s for which equilibrium is maintained when $\theta = 60°$.

5.3 A collar B of weight W is attached to the spring AB and can move along the rod, as shown in Fig. P5.3. The spring stiffness is $k = 1.8$ kN/m and the spring is unstretched when $\theta = 0$. Knowing that the coefficient of static friction between the collar and the rod is $\mu_s = 0.25$, and that the horizontal distance between points A and B is $l = 0.6$ m, determine the range of values of W for which equilibrium is maintained when $\theta = 15°$.

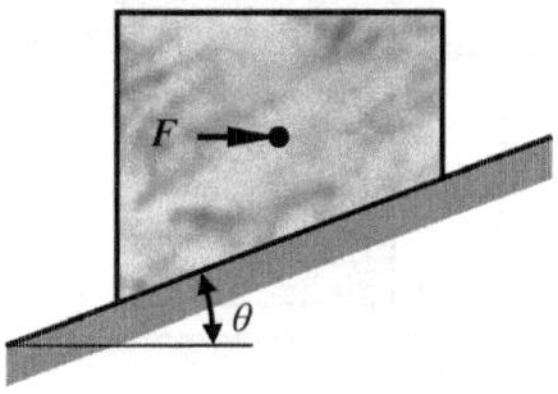

Fig. P5.1

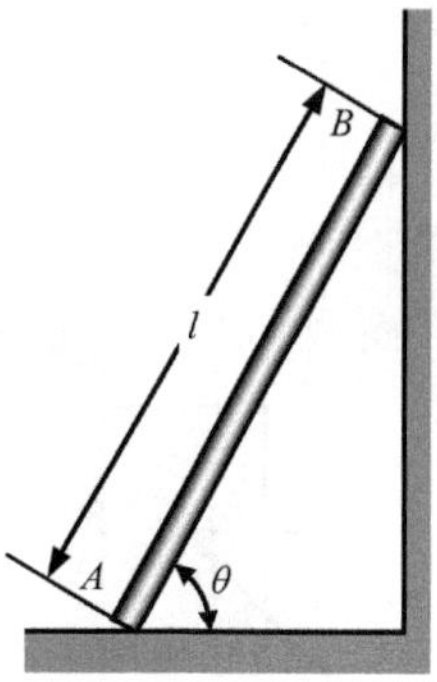

Fig. P5.2

5.4 A force of magnitude $F=100$ N acts on a block of mass m placed on a horizontal plane, as shown in Fig. P5.4. Knowing that the coefficients of friction between the block and the plane are $\mu_s=0.3$ and $\mu_k=0.2$, considering only values of θ less than or equal to 90°, determine the smallest value of θ for which motion of the block to the right is impending when (a) $m=15$ kg, (b) $m=30$ kg.

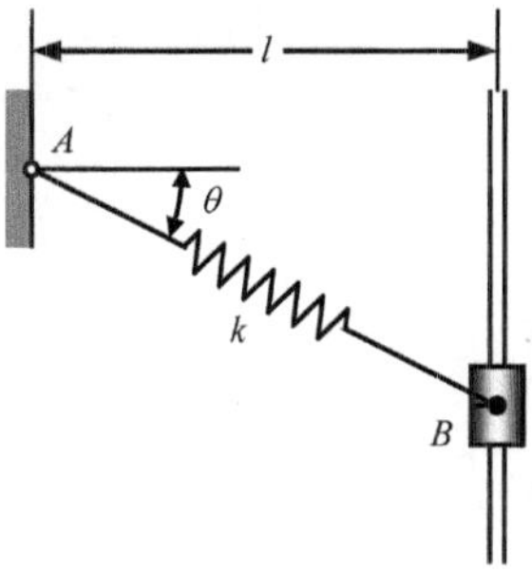

Fig. P5.3

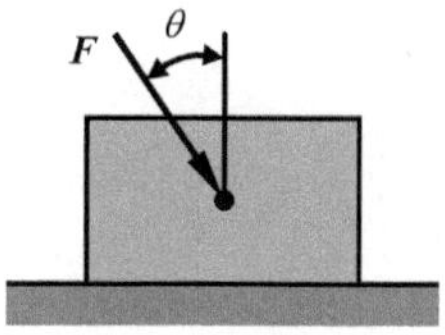

Fig. P5.4

5.5 A force of magnitude $F=150$ N acts on a block of mass $m=10$ kg, as shown in Fig. P5.5. Knowing that the coefficients of friction between the block and the plane are $\mu_s=0.3$ and $\mu_k=0.2$, determine the range of values of θ for which equilibrium of the block is maintained.

5.6 A cabinet of mass $m=50$ kg is mounted on casters which can be locked to prevent their rotation, as shown in Fig. P5.6. The coefficient of static friction between the floor and each caster is $\mu_s=0.3$. Knowing that $b=500$ mm and that $h=650$ mm, determine the magnitude of the force $\boldsymbol{F}$ required for impending motion of the cabinet to the right (a) if all casters are locked, (b) if the caster at B is locked and the caster at A is free to rotate, (c) if the caster at A is locked and the caster at B is free to rotate.

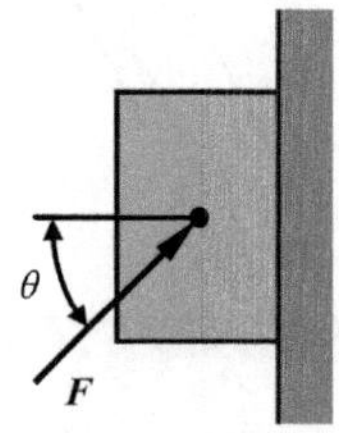

Fig. P5.5

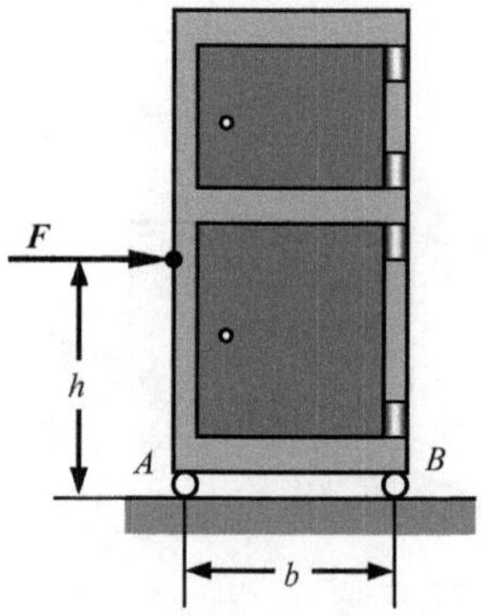

Fig. P5.6

Chapter 6 Kinematics of Particle

A particle moving along a straight line is said to be in rectilinear motion, whereas a particle moving along a curved line is said to be in curvilinear motion.

6.1 Motion of Particle Represented by Vector

Considering that a particle P moves along a curved line, as shown in Fig. 6.1, the position of the particle at time t can be represented by a position vector $\boldsymbol{r}$ joining O and P, where O is the reference point chosen in space. When the particle is in motion, the position vector $\boldsymbol{r}$ is a function of time t, that is,

$$\boldsymbol{r} = \boldsymbol{r}(t) \tag{6.1}$$

Assuming that the vector $\boldsymbol{r}'$ defines the position of the particle at a later time $t+\Delta t$, as shown in Fig. 6.2, then the vector $\Delta\boldsymbol{r}=\boldsymbol{r}'-\boldsymbol{r}$ represents the change of position vector $\boldsymbol{r}$ during the time interval Δt. Thus the velocity of the particle at time t can be expressed as

$$\boldsymbol{v} = \lim_{\Delta t \to 0} \frac{\Delta \boldsymbol{r}}{\Delta t} = \dot{\boldsymbol{r}} \tag{6.2}$$

The velocity is a vector, and its direction is always tangent to the path of motion, as shown in Fig. 6.3.

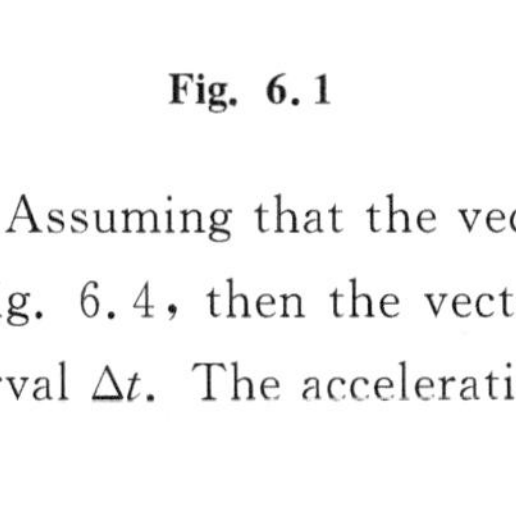

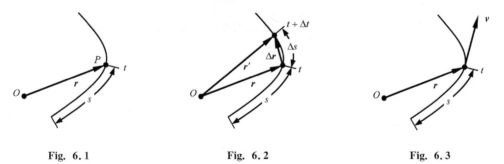

Fig. 6.1

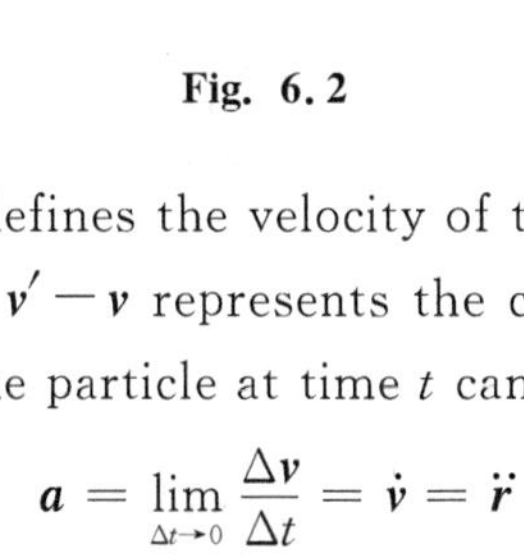

Fig. 6.2

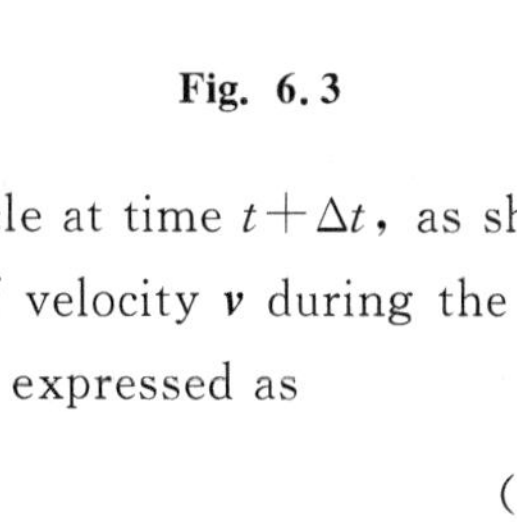

Fig. 6.3

Assuming that the vector $\boldsymbol{v}'$ defines the velocity of the particle at time $t+\Delta t$, as shown in Fig. 6.4, then the vector $\Delta\boldsymbol{v}=\boldsymbol{v}'-\boldsymbol{v}$ represents the change of velocity $\boldsymbol{v}$ during the time interval Δt. The acceleration of the particle at time t can thus be expressed as

$$\boldsymbol{a} = \lim_{\Delta t \to 0} \frac{\Delta \boldsymbol{v}}{\Delta t} = \dot{\boldsymbol{v}} = \ddot{\boldsymbol{r}} \tag{6.3}$$

The acceleration is a vector, and its direction is usually not tangent to the path of motion, as shown in Fig. 6.5.

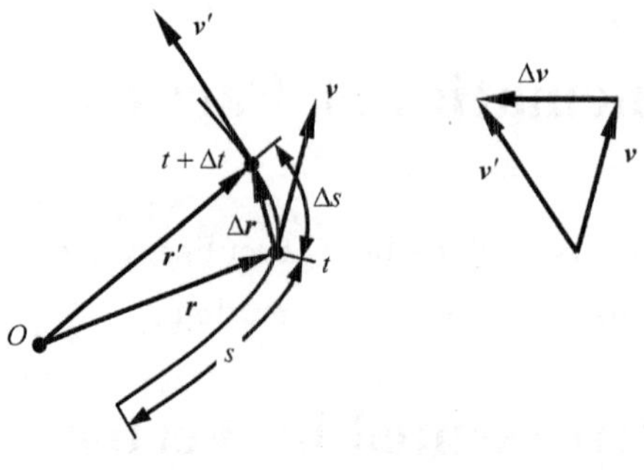

Fig. 6.4

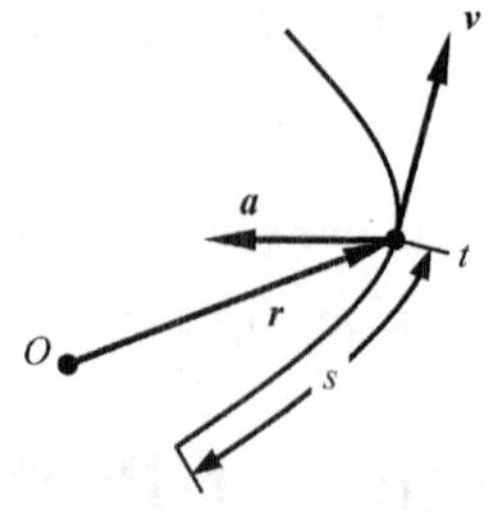

Fig. 6.5

6.2 Motion of Particle Represented by Rectangular Coordinates

Establishing a rectangular coordinate system $Oxyz$, as shown in Fig. 6.6, then the position vector $\boldsymbol{r}$ of the particle P at time t can be written as

$$\boldsymbol{r} = x\boldsymbol{i} + y\boldsymbol{j} + z\boldsymbol{k} \tag{6.4}$$

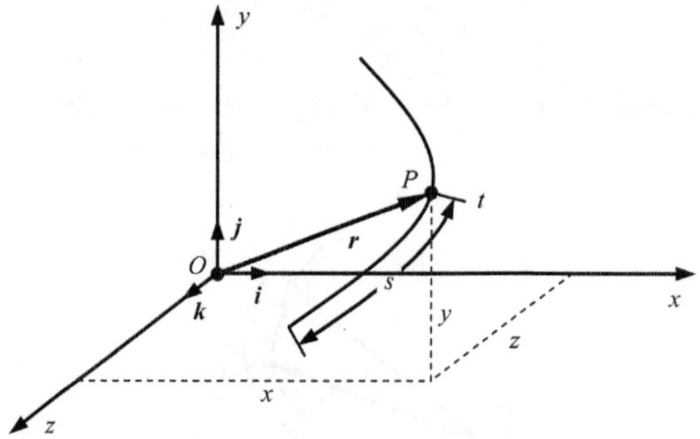

Fig. 6.6

where $\boldsymbol{i}$, $\boldsymbol{j}$, and $\boldsymbol{k}$ are the unit vectors along the positive coordinate axes respectively, $x=x(t)$, $y=y(t)$, and $z=z(t)$ are the corresponding scalar components of the position vector $\boldsymbol{r}$ on the three coordinate axes.

Since the magnitude and direction of $\boldsymbol{i}$, $\boldsymbol{j}$, and $\boldsymbol{k}$ are constants, then the velocity and acceleration of the particle P at time t can be expressed, respectively, as

$$\boldsymbol{v} = \dot{\boldsymbol{r}} = \dot{x}\boldsymbol{i} + \dot{y}\boldsymbol{j} + \dot{z}\boldsymbol{k} \tag{6.5}$$

$$\boldsymbol{a} = \ddot{\boldsymbol{r}} = \ddot{x}\boldsymbol{i} + \ddot{y}\boldsymbol{j} + \ddot{z}\boldsymbol{k} \tag{6.6}$$

where $\dot{x}$, $\dot{y}$, $\dot{z}$ and $\ddot{x}$, $\ddot{y}$, $\ddot{z}$ represent, respectively, the scalar components of the velocity $\boldsymbol{v}$ and the acceleration $\boldsymbol{a}$, i.e.,

$$v_x = \dot{x} \quad v_y = \dot{y} \quad v_z = \dot{z} \tag{6.7}$$

$$a_x = \ddot{x} \quad a_y = \ddot{y} \quad a_z = \ddot{z} \tag{6.8}$$

Example 6.1

The motion of a particle is defined by the position vector $\boldsymbol{r}=A(\cos t+t\sin t)\boldsymbol{i}+A(\sin t-t\cos t)\boldsymbol{j}$, where t is expressed in seconds. Determine the values of t for which the position vector and the acceleration vector are (a) perpendicular, (b) parallel.

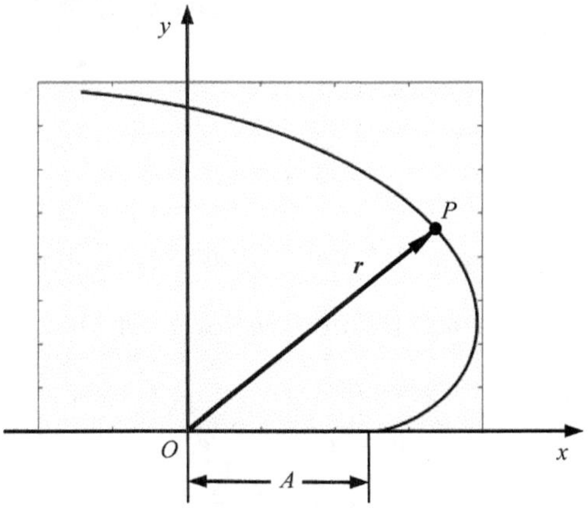

Fig. E6.1

Solution

Using $\boldsymbol{r}=A(\cos t+t\sin t)\boldsymbol{i}+A(\sin t-t\cos t)\boldsymbol{j}$, we have

$$\boldsymbol{v}=\dot{\boldsymbol{r}}=A(t\cos t)\boldsymbol{i}+A(t\sin t)\boldsymbol{j}$$

$$\boldsymbol{a}=\dot{\boldsymbol{v}}=A(\cos t-t\sin t)\boldsymbol{i}+A(\sin t+t\cos t)\boldsymbol{j}$$

(a) When the position vector and the acceleration vector are perpendicular, we have

$$\boldsymbol{r}\cdot\boldsymbol{a}=0$$

from which we obtain

$$t=1\ \text{s}$$

(b) When the position vector and the acceleration vector are parallel, we have

$$\boldsymbol{r}\times\boldsymbol{a}=0$$

from which we obtain

$$t=0$$

6.3 Motion of Particle Represented by Natural Coordinates

We establish a natural coordinate system attached to the particle P moving along a curved line, as shown in Fig. 6.7, and define three unit vectors $\boldsymbol{e}_t$, $\boldsymbol{e}_n$, and $\boldsymbol{e}_b$, where $\boldsymbol{e}_t$ is the tangential unit vector tangent to the path of motion pointed toward the direction of motion, $\boldsymbol{e}_n$ is the principal normal unit vector perpendicular to the path of motion pointed toward the center of curvature of the path of motion, and $\boldsymbol{e}_b$ is the binormal unit vector perpendicular to the plane containing $\boldsymbol{e}_t$ and $\boldsymbol{e}_n$ pointing in the direction of $\boldsymbol{e}_t\times\boldsymbol{e}_n$ indicated by the right-hand rule, i.e., $\boldsymbol{e}_b=\boldsymbol{e}_t\times\boldsymbol{e}_n$. The plane containing $\boldsymbol{e}_t$ and $\boldsymbol{e}_n$ is called the osculating plane.

In this natural coordinate system, the velocity vector $\boldsymbol{v}$ of the particle P at time t can be written as

$$\boldsymbol{v} = v\boldsymbol{e}_t \tag{6.9}$$

where $\boldsymbol{e}_t$ is the tangential unit vector. Using the relation above, the acceleration vector $\boldsymbol{a}$ of the particle P at time t can be given by

$$\boldsymbol{a} = \dot{\boldsymbol{v}} = \dot{v}\boldsymbol{e}_t + v\dot{\boldsymbol{e}}_t \tag{6.10}$$

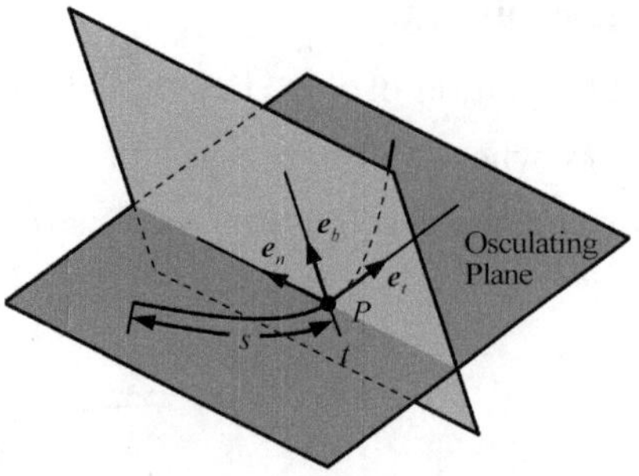

Fig. 6.7

Using $\dot{\boldsymbol{e}}_t = \frac{v}{\rho}\boldsymbol{e}_n$, where $\boldsymbol{e}_n$ is the principal normal unit vector and ρ is the radius of curvature of the path of motion, then the acceleration vector $\boldsymbol{a}$ of the particle P at time t can be rewritten as

$$\boldsymbol{a} = \dot{v}\boldsymbol{e}_t + \frac{v^2}{\rho}\boldsymbol{e}_n \tag{6.11}$$

where $\dot{v}$ and $\frac{v^2}{\rho}$ represent, respectively, the tangential and normal components of the acceleration $\boldsymbol{a}$, i.e.,

$$a_t = \dot{v},\ a_n = \frac{v^2}{\rho} \tag{6.12}$$

The relation above shows that the tangential component a_t of the acceleration is equal to the rate of the speed change of the particle, while the normal component a_n is equal to the square of the speed divided by the radius of curvature of the path of motion. If the speed of the particle increases, a_t is positive and $\boldsymbol{a}_t$ points in the direction of motion. If the speed of the particle decreases, a_t is negative and $\boldsymbol{a}_t$ points against direction of motion. However, a_n is always positive and $\boldsymbol{a}_n$ always points toward the center of curvature of the path of motion.

We thus conclude that the tangential component of the acceleration reflects a change in the magnitude of velocity of the particle, while its normal component reflects a change in direction of velocity of the particle.

Example 6.2

The motion of a particle is defined by the position vector $\boldsymbol{r} = (2\sin 4t)\boldsymbol{i} + (2\cos 4t)\boldsymbol{j} + (4t)\boldsymbol{k}$ m, where t is expressed in seconds. Determine the radius of curvature of the path along which the particle moves.

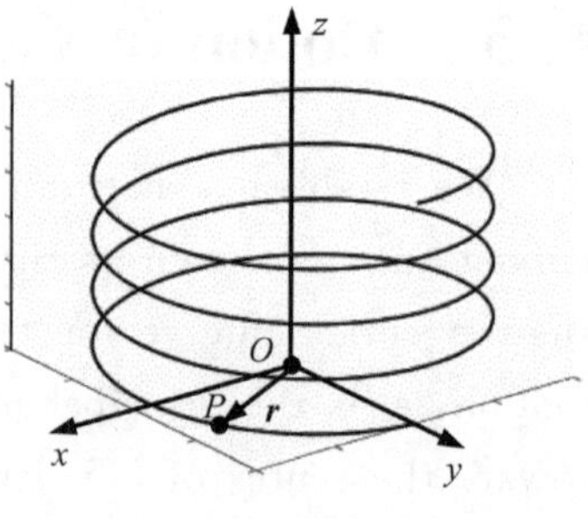

Fig. E6.2

Solution

From $\boldsymbol{r} = (2\sin 4t)\boldsymbol{i} + (2\cos 4t)\boldsymbol{j} + (4t)\boldsymbol{k}$ m, we have

$$\boldsymbol{v} = \dot{\boldsymbol{r}} = (8\cos 4t)\boldsymbol{i} + (-8\sin 4t)\boldsymbol{j} + (4)\boldsymbol{k} \text{ m/s}$$

$$\boldsymbol{a} = \dot{\boldsymbol{v}} = (-32\sin 4t)\boldsymbol{i} + (-32\cos 4t)\boldsymbol{j} \text{ m/s}^2$$

Thus we can obtain

$$v = \sqrt{(8\cos 4t)^2 + (-8\sin 4t)^2 + (4)^2} = 4\sqrt{5}\ \mathrm{m/s}$$

$$a = \sqrt{(-32\sin 4t)^2 + (-32\cos 4t)^2} = 32\ \mathrm{m/s^2}$$

Using $a^2=(a_t)^2+(a_n)^2$, where $a_t=\dot{v}=0$, $a_n=\frac{v^2}{\rho}$, we can find

$$\rho = \frac{v^2}{a_n} = \frac{v^2}{a} = 2.5\ \mathrm{m}$$

Problems

6.1 The motion of a particle is defined by the position vector $\boldsymbol{r}=(4t^2-3t)\boldsymbol{i}+t^3\boldsymbol{j}$, where r and t are expressed in meters and seconds, respectively, as shown in Fig. P6.1. Determine the velocity and acceleration of the particle when (a) $t=0.2$ s, (b) $t=1$ s.

6.2 A particle P moves along a parabolic path $y=\frac{1}{20}x^2$, where x and y are expressed in meters, as shown in Fig. P6.2. The particle has a speed of 6 m/s which is increasing at 2 m/s² when $x=6$ m. Determine the direction of the velocity and the magnitude and direction of the acceleration at this instant.

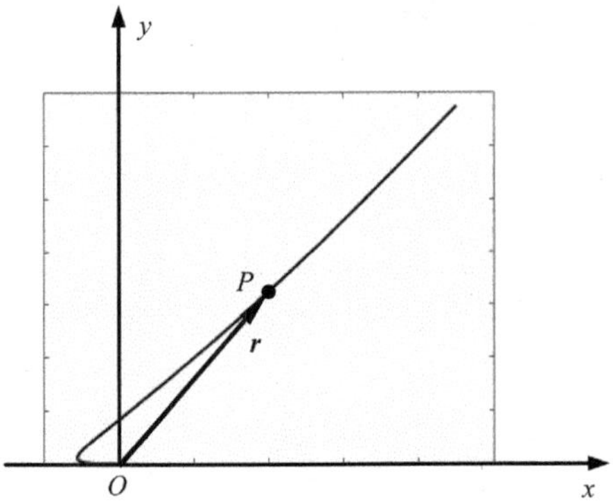

Fig. P6.1

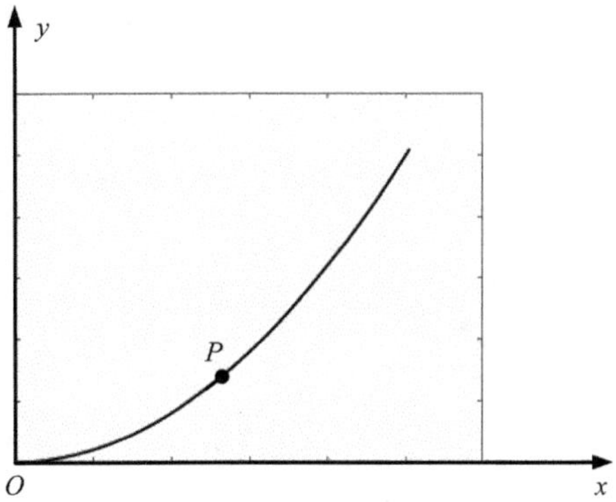

Fig. P6.2

6.3 A particle is defined by the position vector $\boldsymbol{r}=\frac{3}{4}[1-1/(t+1)]\boldsymbol{i}+\frac{1}{2}[\exp(-\pi t/2)\cos 2\pi t]\boldsymbol{j}$, where r and t are expressed in meters and seconds, respectively, as shown in Fig. P6.3. Determine the position, the velocity, and the acceleration of the particle when (a) $t=0.5$ s, (b) $t=1$ s.

6.4 A particle is defined by the position vector $\boldsymbol{r}=(\omega t-\sin\omega t)\boldsymbol{i}+(1-\cos\omega t)\boldsymbol{j}$ m, where ω and t are expressed in rad/s and s, respectively, as shown in Fig. P6.4. Determine the tangential and normal accelerations of the particle at time t.

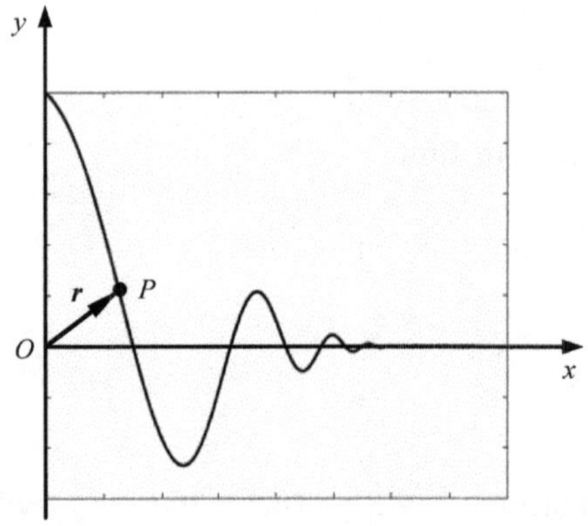

Fig. P6.3

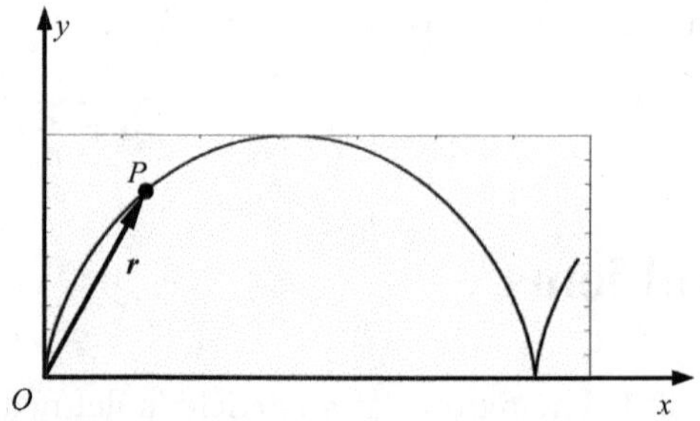

Fig. P6.4

Chapter 7 Kinematics of Rigid Body in Plane Motion

7.1 Plane Motion of Rigid Body

When all the particles of a rigid body move along paths which are equidistant from a fixed plane, this body is said to be in plane motion. There are three types of plane motion:

(1) Translation. A body is said to be in translation if any straight-line segment in the body remains the same direction during the motion. When a body is in translation, all the particles within the body will move along parallel paths of motion. If these paths of motion for a body in translation are straight lines the motion is called rectilinear translation, as shown in Fig. 7.1(a). However, if the paths of motion are curved lines the motion of a body in translation is called curvilinear translation, as shown in Fig. 7.1(b).

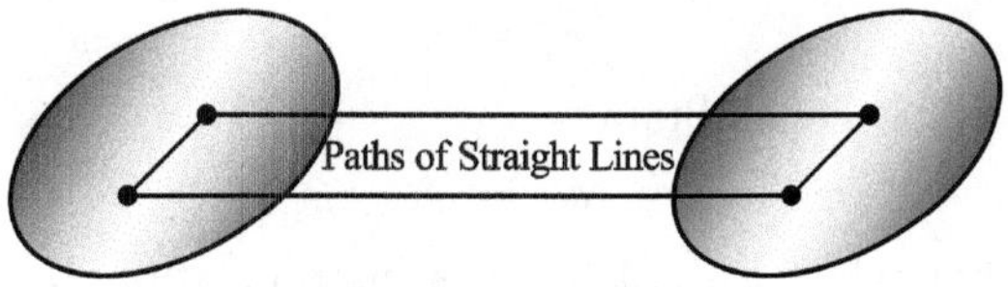

(a) Rectilinear Translation

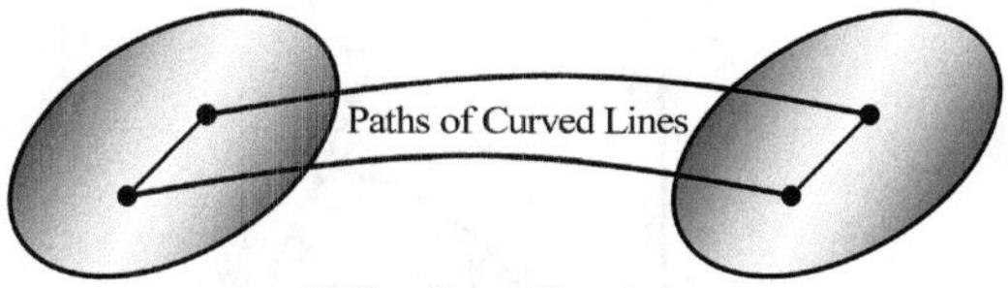

(b) Curvilinear Translation

Fig. 7.1

(2) Rotation about a fixed axis. When a body rotates about a fixed axis, all the particles forming the body, except for those which lie on the axis of rotation, move along circular paths, which are perpendicular to the axis of rotation, as shown in Fig. 7.2.

(3) General plane motion. When a body is subjected to general plane motion, it undergoes a combination of translation and rotation, as shown in Fig. 7.3. The translation takes place within a plane of reference, and the rotation takes place about an axis perpendicular to the plane of reference.

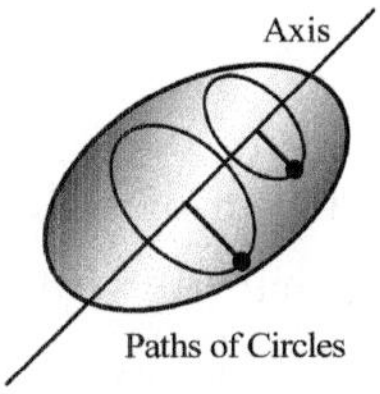

Fig. 7.2

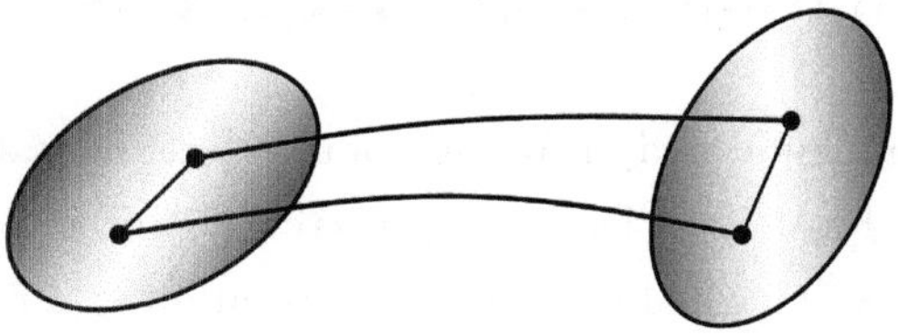
Fig. 7.3

7.2 Translation

Consider a rigid body in translation (either rectilinear translation or curvilinear translation), and let A and B be any two particles within the body, as shown in Fig. 7.4. Denote, respectively, by $\boldsymbol{r}_P$ and $\boldsymbol{r}_B$ the position vectors of particles A and B with respect to a fixed frame of reference $Oxyz$ and by $\boldsymbol{r}_{A/B}$ the position vector of particle A relative to particle B in a translating frame of reference $O'x'y'z'$ attached to the body at particle B. From Fig. 7.4, we have

$$\boldsymbol{r}_A = \boldsymbol{r}_B + \boldsymbol{r}_{A/B} \tag{7.1}$$

Differentiating Eq. (7.1) with respect to time t, we obtain

$$\boldsymbol{v}_A = \boldsymbol{v}_B + \boldsymbol{v}_{A/B} \tag{7.2}$$

where $\boldsymbol{v}_A = \dot{\boldsymbol{r}}_A$ and $\boldsymbol{v}_B = \dot{\boldsymbol{r}}_B$ are the velocities of particles A and B respectively, and $\boldsymbol{v}_{A/B} = \dot{\boldsymbol{r}}_{A/B}$ is the velocity of particle A relative to particle B. When the body is in translation, $\boldsymbol{r}_{A/B}$ is a constant vector, i.e., $\dot{\boldsymbol{r}}_{A/B} = 0$, thus we have

$$\boldsymbol{v}_A = \boldsymbol{v}_B \tag{7.3}$$

Differentiating Eq. (7.3), we obtain

$$\boldsymbol{a}_A = \boldsymbol{a}_B \tag{7.4}$$

Thus we conclude that when a rigid body is in translation all the particles of the body have the same velocity and the same acceleration at any given instant, as shown in Fig. 7.5.

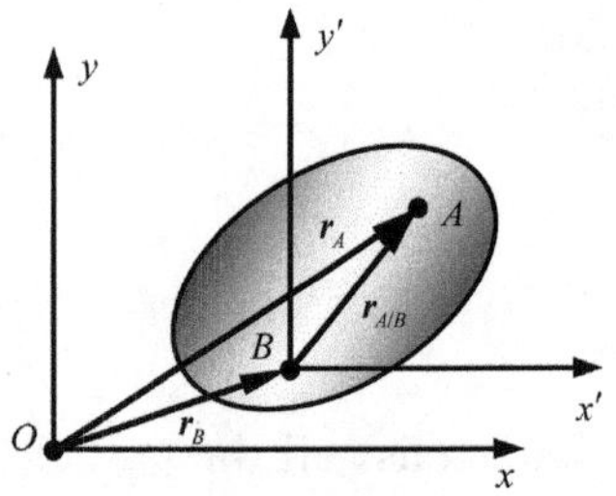

Fig. 7.4

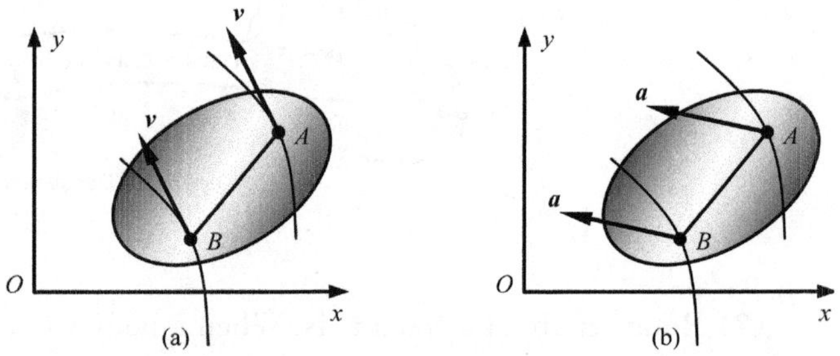

Fig. 7.5

In the case of curvilinear translation, the velocity and acceleration change in direction as well as in magnitude at every instant. However, in the case of rectilinear translation, the velocity and acceleration remain the same direction during the entire motion.

7.3 Rotation about Fixed Axis

Consider a rigid body rotating about a fixed axis λ, and let A be a particle on the body, as shown in Fig. 7.6. Denote by $\boldsymbol{r}_A$ the position vector of particle A with respect to an arbitrary point O located on the axis of rotation λ. From Fig. 7.6, we have

$$\boldsymbol{v}_A = \boldsymbol{\omega} \times \boldsymbol{r}_A \tag{7.5}$$

where $\boldsymbol{\omega}$ is the angular velocity of the body, which is directed along the axis of rotation, in the sense determined by the right-hand rule from the sense of rotation of the body, and equal in magnitude to the rate of change of the angular coordinate.

Differentiating Eq. (7.5) with respect to time t, we obtain

$$\boldsymbol{a}_A = \boldsymbol{\alpha} \times \boldsymbol{r}_A + \boldsymbol{\omega} \times \boldsymbol{v}_A \tag{7.6}$$

where $\boldsymbol{\alpha}=\dot{\boldsymbol{\omega}}$ is the angular acceleration of the body. In the case of a body rotating about a fixed axis, the angular acceleration $\boldsymbol{\alpha}$ is a vector directed along the axis of rotation, and is equal in magnitude to the rate of change of the angular velocity. Using Eq. (7.5), Eq. (7.6) can be rewritten as

$$\boldsymbol{a}_A = \boldsymbol{\alpha} \times \boldsymbol{r}_A + \boldsymbol{\omega} \times (\boldsymbol{\omega} \times \boldsymbol{r}_A) \tag{7.7}$$

The equation above can be also expressed as

$$\boldsymbol{a}_A = \boldsymbol{a}_t + \boldsymbol{a}_n \tag{7.8}$$

where $\boldsymbol{a}_t=\boldsymbol{\alpha}\times\boldsymbol{r}_A$ is the tangential component of the acceleration $\boldsymbol{a}_A$ of particle A tangent to the circle drawn by particle A, and $\boldsymbol{a}_n=\boldsymbol{\omega}\times\boldsymbol{v}_A=\boldsymbol{\omega}\times(\boldsymbol{\omega}\times\boldsymbol{r}_A)$ is the normal component of the acceleration $\boldsymbol{a}_A$ of particle A directed toward the center of the circle formed by particle A. Eq. (7.8) shows that the acceleration $\boldsymbol{a}_A$ of particle A is equal to the vector sum of the tangential component $\boldsymbol{a}_t$ and the normal component $\boldsymbol{a}_n$, as shown in Fig. 7.7.

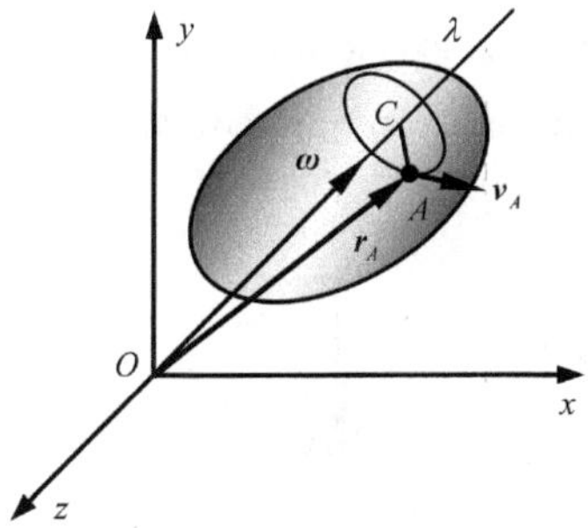

Fig. 7.6

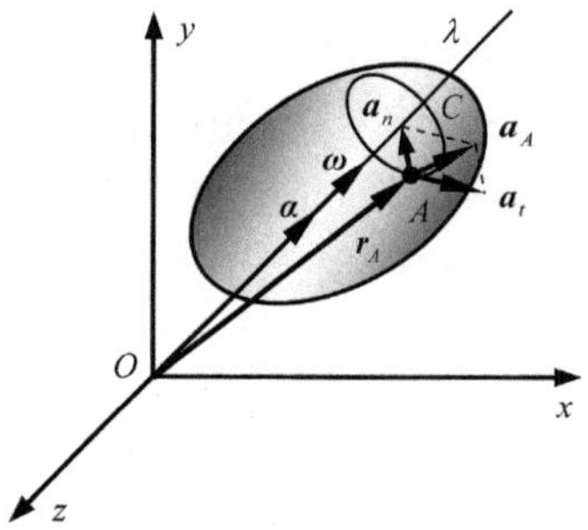

Fig. 7.7

The rotation of a rigid body about a fixed axis can be represented by the in-plane rotation of a plate perpendicular to the rotation axis of the body about the intersection of the plate and the rotation axis of the body, as shown in Fig. 7.8.

In the case of the in-plane rotation of a plate about a fixed point O, the angular velocity $\boldsymbol{\omega}$ can be expressed as

$$\boldsymbol{\omega} = \omega \boldsymbol{k} \tag{7.9}$$

where $\boldsymbol{k}$ is the unit vector, perpendicular to the plate and pointing out of the plate. ω is the magnitude of the angular velocity, which is positive if the plate rotates counterclockwise, and negative if the plate rotates clockwise.

Using Eq. (7.5), we can express the velocity of any given particle A on the plate, as shown in Fig. 7.9, as

$$\boldsymbol{v} = \boldsymbol{\omega} \times \boldsymbol{r} = r\omega\, \boldsymbol{e}_t \tag{7.10}$$

where $\boldsymbol{e}_t$ is the unit vector tangent to the circle described by particle A pointed toward the direction of motion. Thus the magnitude of the velocity is equal to

$$v = r\omega \tag{7.11}$$

and its direction can be obtained by rotating $\boldsymbol{r}$ through 90° in the sense of rotation of the plate.

Using Eq. (7.7) and (7.8), we can express the acceleration of any given particle A on the plate, as shown in Fig. 7.10, as

$$\boldsymbol{a} = \boldsymbol{a}_t + \boldsymbol{a}_n = \boldsymbol{\alpha} \times \boldsymbol{r} + \boldsymbol{\omega} \times (\boldsymbol{\omega} \times \boldsymbol{r}) = r\alpha\, \boldsymbol{e}_t + r\omega^2\, \boldsymbol{e}_n \tag{7.12}$$

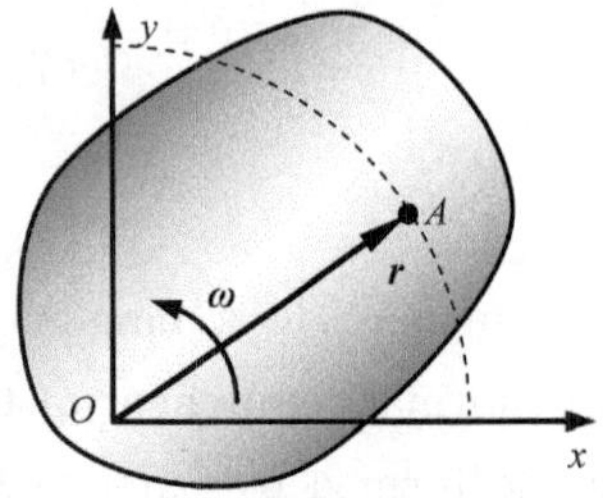

Fig. 7.8

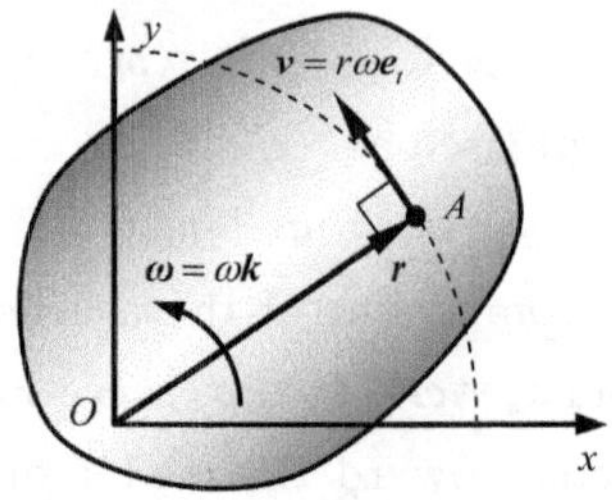

Fig. 7.9

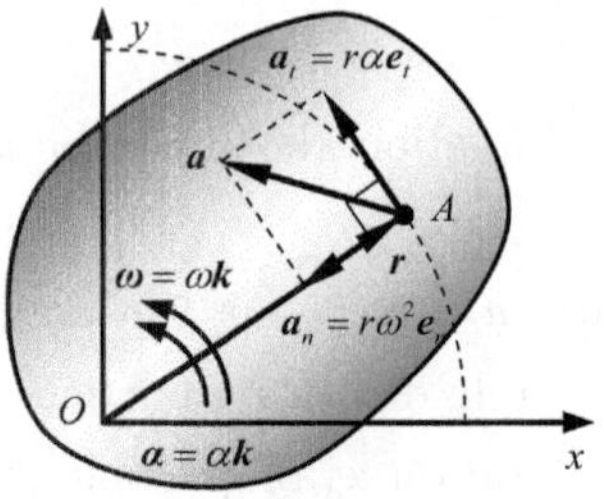

Fig. 7.10

where $\boldsymbol{e}_n$ is the unit vector, pointed toward the center of the circle described by particle A. The tangential component $\boldsymbol{a}_t = r\alpha\, \boldsymbol{e}_t$ points in the counterclockwise direction if α is positive, and is in the clockwise direction if α is negative. The normal component $\boldsymbol{a}_n = r\omega^2\, \boldsymbol{e}_n$ always points toward the center of the circle.

Example 7.1

The structure shown in Fig. E7.1 consists of two rods AE and CE and a rectangular plate $ABCD$ which are welded together. The structure rotates about the axis AE with a constant angular velocity of $\omega = 5$ rad/s. Knowing that the rotation is counterclockwise as viewed from A, determine the velocity and acceleration of point B.

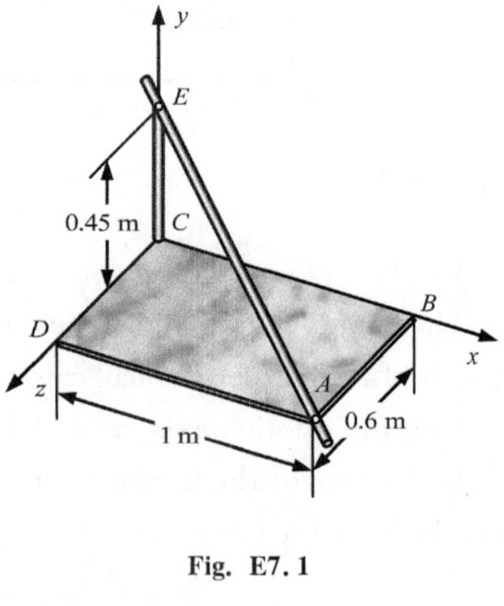

Fig. E7.1

Solution

Using $\boldsymbol{r}_{A/E} = \boldsymbol{i} - 0.45\boldsymbol{j} + 0.6\boldsymbol{k}$ m, we have

$$\boldsymbol{\omega} = \omega \frac{\boldsymbol{r}_{A/E}}{r_{A/E}} = 4\boldsymbol{i} - 1.8\boldsymbol{j} + 2.4\boldsymbol{k} \text{ rad/s}$$

Using $\boldsymbol{r}_{B/A} = -0.6\boldsymbol{k}$ m, we have

$$\boldsymbol{v}_B = \boldsymbol{\omega} \times \boldsymbol{r}_{B/A} = \begin{vmatrix} \boldsymbol{i} & \boldsymbol{j} & \boldsymbol{k} \\ 4 & -1.8 & 2.4 \\ 0 & 0 & -0.6 \end{vmatrix} = 1.08\boldsymbol{i} + 2.4\boldsymbol{j} \text{ m/s}$$

Using $\boldsymbol{\alpha}=0$, we have

$$\boldsymbol{a}_B=\boldsymbol{\alpha}\times\boldsymbol{r}_{B/A}+\boldsymbol{\omega}\times\boldsymbol{v}_B=\begin{vmatrix}\boldsymbol{i} & \boldsymbol{j} & \boldsymbol{k}\\ 4 & -1.8 & 2.4\\ 1.08 & 2.4 & 0\end{vmatrix}=-5.76\boldsymbol{i}+2.59\boldsymbol{j}+11.54\boldsymbol{k}\ \mathrm{m/s^2}$$

7.4 General Plane Motion

A general plane motion is neither a translation nor a rotation. However, it can always be considered as the combination of a translation and a rotation.

Consider, for example, a rod AB in general plane motion, as shown in Fig. 7.11(a), whose two ends slide along a horizontal and a vertical track, respectively. The motion of the rod AB from A_1B_1 to A_2B_2 can be replaced by a translation of the rod AB from A_1B_1 to A'_1B_2 with the base point B in a horizontal direction to the right, as shown in Fig. 7.11(b), and a rotation of the rod AB from A'_1B_2 to A_2B_2 about the base point B in a counterclockwise direction, as shown in Fig. 7.11(c).

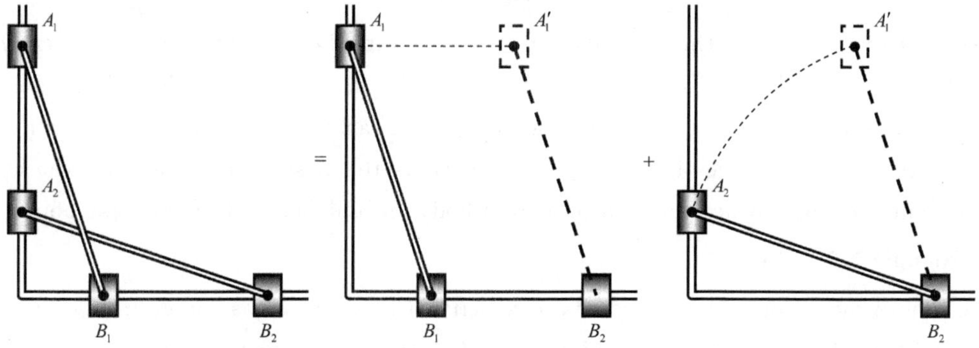

(a) General Plane Motion (b) Translation with Base Point (c) Rotation about Base Point

Fig. 7.11

Another example of general plane motion is a wheel rolling, without sliding, on a horizontal surface, as shown in Fig. 7.12(a). The motion of the wheel from A_1B_1 to A_2B_2 can be replaced by a translation of the wheel from A_1B_1 to A'_1B_2 with the base point B in a

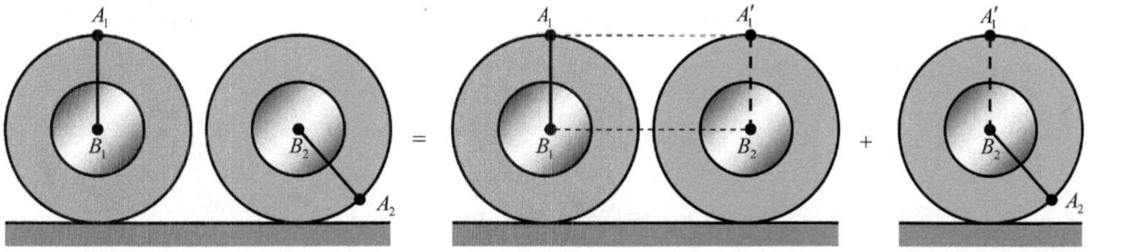

(a) General Plane Motion (b) Translation with Base Point (c) Rotation about Base Point

Fig. 7.12

horizontal direction to the right, as shown in Fig. 7.12(b), and a rotation of the wheel from $A_1'B_2$ to A_2B_2 about the base point B in a clockwise direction, as shown in Fig. 7.12(c).

1. Base-Point Method for Determining Velocity in General Plane Motion

Since any general plane motion of a rigid body can be replaced by a translation with the base point and a simultaneous rotation about the base point, the velocity $\boldsymbol{v}_A$ of a given particle A of the rigid body, Fig. 7.13(a), is can be expressed as

$$\boldsymbol{v}_A = \boldsymbol{v}_B + \boldsymbol{v}_{A/B} \tag{7.13}$$

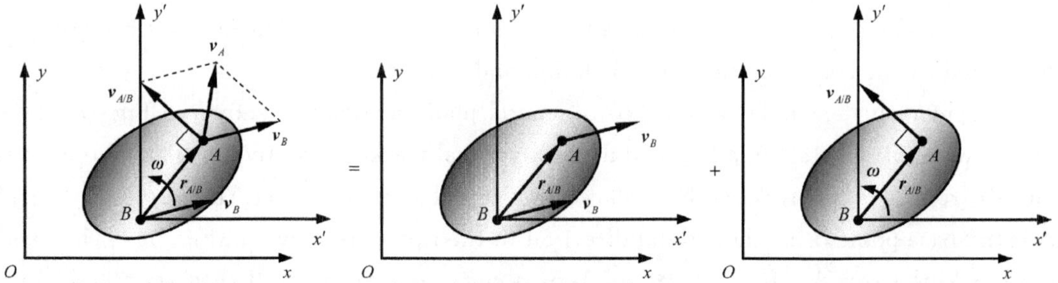

(a) General Plane Motion (b) Translation with the Base Point (c) Rotation about the Base Point

Fig. 7.13

where $\boldsymbol{v}_B$ is the velocity of the base point B, as shown in Fig. 7.13(b), and $\boldsymbol{v}_{A/B} = \boldsymbol{\omega} \times \boldsymbol{r}_{A/B}$ is the velocity of particle A with respect to the base point B, as shown in Fig. 7.13(c).

It should be noted that the choice of a base point is arbitrary, it is, however, convenient to select that point, where the motion is known, as the base point, and that the angular velocity and the angular acceleration of a rigid body are independent of the base point.

Example 7.2

Collar A moves down with a constant velocity of $v_A = 1$ m/s, as shown in Fig. E7.2(a). At the instant shown when $\theta = 30°$, determine (a) the angular velocity of rod AB, (b) the velocity of collar B.

Solution

Choosing point A as the base point, then the velocity of point B can be expressed as

$$\boldsymbol{v}_B = \boldsymbol{v}_A + \boldsymbol{v}_{B/A}$$

Using the fact that $\boldsymbol{v}_B$ is horizontal, and that $\boldsymbol{v}_{B/A}$ is perpendicular to the line joining A and B, we can obtain the velocity parallelogram shown in Fig. E7.2(b). From this parallelogram, then at the given instant we have

$$v_B = \frac{v_A}{\tan\theta} = 1.73 \text{ m/s}$$

$$v_{B/A} = \frac{v_A}{\sin\theta} = 2 \text{ m/s}$$

Using $v_{B/A} = AB \cdot \omega$, we have

$$\omega = \frac{v_{B/A}}{AB} = 4 \text{ rad/s (anticlockwise)}$$

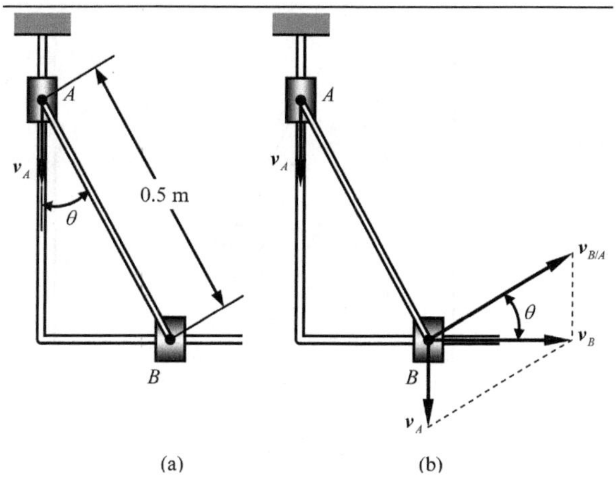

Fig. E7. 2

2. Instantaneous-Center Method for Determining Velocity in General Plane Motion

In velocity analysis of a rigid body in general plane motion, we often choose a point, which has zero velocity at the given instant, on the rigid body or outside the rigid body as the base point. Assuming that the velocity at point C is equal to zero at the instant considered, where C is called the instantaneous center of velocity, as shown in Fig. 7. 14, then the velocity $\boldsymbol{v}_A$ of a given particle A at the given instant can be given by

$$\boldsymbol{v}_A = \boldsymbol{\omega} \times \boldsymbol{r} \tag{7.14}$$

where $\boldsymbol{\omega}$ and $\boldsymbol{r}$ are, respectively, the angular velocity of the rigid body and the position vector of particle A with respect to the instantaneous center of velocity C. It should be noted that, in general, the instantaneous center of velocity does not have zero acceleration.

The instantaneous center of velocity at the given instant can be determined by using the following methods:

(1) Given the velocity $\boldsymbol{v}_A$ at point A of the body, and the angular velocity $\boldsymbol{\omega}$ of the body, as shown in Fig. 7. 15, the instantaneous center of velocity C is in the direction obtained by rotating $\boldsymbol{v}_A$ through 90° in the sense of rotation of $\boldsymbol{\omega}$, and the distance of C from A is equal to $r = v_A/\omega$.

(2) Given the lines of action of two nonparallel velocities $\boldsymbol{v}_A$ and $\boldsymbol{v}_B$, as shown in Fig. 7. 16, the instantaneous center of velocity C is located at the intersection of two perpendiculars to $\boldsymbol{v}_A$ and $\boldsymbol{v}_B$, respectively, drawn through A and B.

(3) Given the magnitude and direction of two parallel velocities $\boldsymbol{v}_A$ and $\boldsymbol{v}_B$, and the line connecting A and B perpendicular to $\boldsymbol{v}_A$ (or $\boldsymbol{v}_B$), as shown in Fig. 7. 17, the instantaneous center of velocity C is located at the intersection of line AB and the line joining the extremities of $\boldsymbol{v}_A$ and $\boldsymbol{v}_B$. A special case for Fig. 7. 17(a) should be noted that if $\boldsymbol{v}_A = \boldsymbol{v}_B$ at a given instant, the instantaneous center of velocity C will be located at infinity. This special case is called the instantaneous translation.

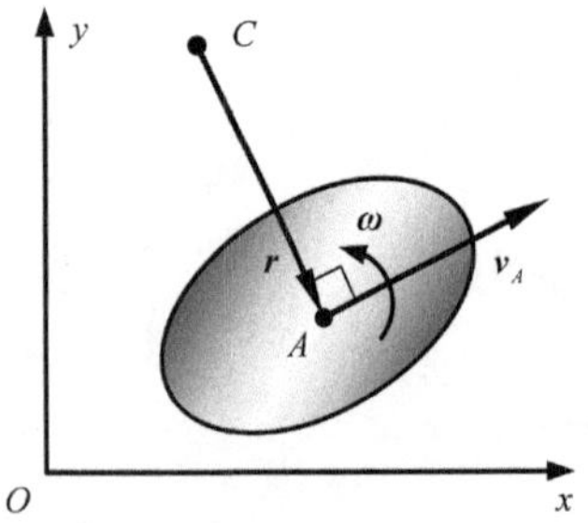

Fig. 7.14

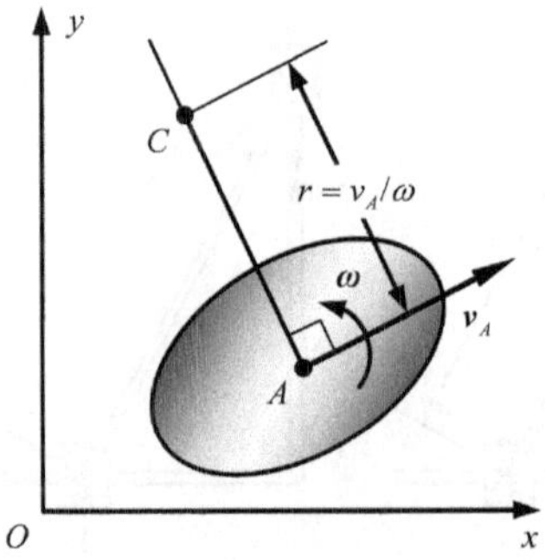

Fig. 7.15

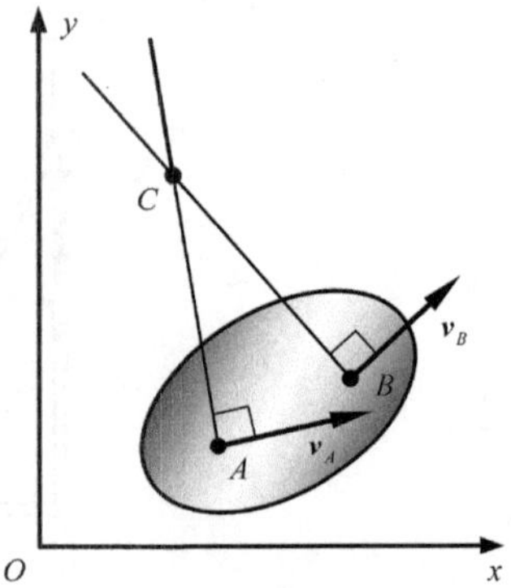

Fig. 7.16

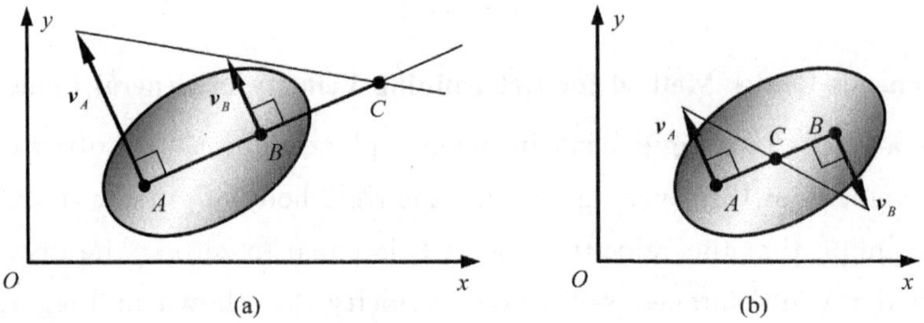

Fig. 7.17

The locus formed by points defining the positions of the instantaneous center of velocity is called the centrode. The locus in space is called the space centrode, whereas the locus on the body is called the body centrode. For example, the space centrode for a wheel rolling without sliding on a horizontal surface is a horizontal straight line coinciding with the horizontal surface, whereas the body centrode is a circle coinciding with the wheel rim.

Example 7.3

Collar A moves down with a constant velocity of $v_A = 1$ m/s, as shown in Fig. E7.3(a). At the instant shown when $\theta = 30°$, determine (a) the angular velocity of rod AB, (b) the velocity of collar B.

Solution

At the given instant the instantaneous center of velocity is located at C, as shown in Fig. E7.3(b), then we have

$$\omega = \frac{v_A}{AC} = \frac{v_A}{AB \cdot \sin\theta} = 4 \text{ rad/s}$$

$$v_B = BC \cdot \omega = (AB \cdot \cos\theta)\omega = 1.73 \text{ m/s}$$

3. Base-Point Method for Determining Acceleration in General Plane Motion

Since any general plane motion of a rigid body can be replaced by a translation with the base point, and a simultaneous rotation about the base point, the acceleration $\boldsymbol{a}_A$ of a given

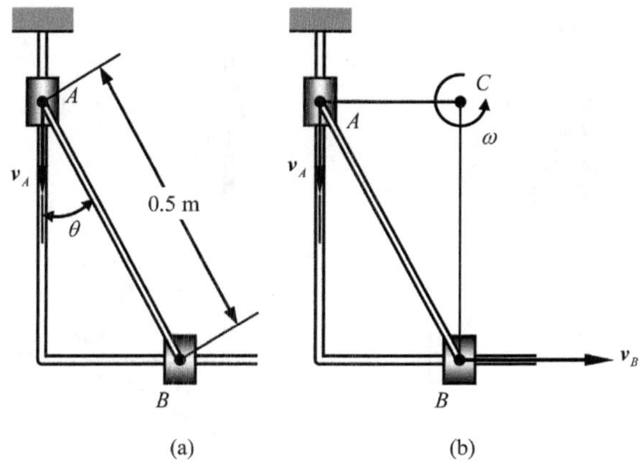

Fig. E7.3

particle A of the rigid body, as shown in Fig. 7.18(a), can be expressed as

$$\boldsymbol{a}_A = \boldsymbol{a}_B + (\boldsymbol{a}_{A/B})_t + (\boldsymbol{a}_{A/B})_n \tag{7.15}$$

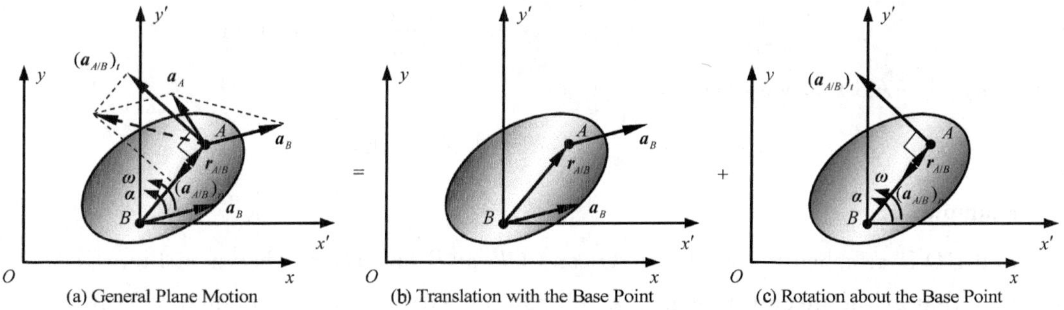

Fig. 7.18

where $\boldsymbol{a}_B$ is the acceleration of the base point, as shown in Fig. 7.18(b), $(\boldsymbol{a}_{A/B})_t = \boldsymbol{\alpha} \times \boldsymbol{r}_{A/B}$ and $(\boldsymbol{a}_{A/B})_n = \boldsymbol{\omega} \times (\boldsymbol{\omega} \times \boldsymbol{r}_{A/B})$ are the tangential and normal components of the acceleration of particle A with respect to the base point B, as shown in Fig. 7.18(c).

Example 7.4

Collar A moves down with a constant velocity of $v_A = 1$ m/s, as shown in Fig. E7.4(a). At the instant shown when $\theta = 30°$, determine (a) the angular acceleration of rod AB, (b) the acceleration of collar B.

Solution

Choosing point A as the base point, the acceleration of point B can be expressed as

$$\boldsymbol{a}_B = \boldsymbol{a}_A + (\boldsymbol{a}_{B/A})_t + (\boldsymbol{a}_{B/A})_n$$

Using the fact that $\boldsymbol{a}_B$ is horizontal and that $\boldsymbol{a}_A$ is identically equal to zero, we can obtain the acceleration parallelogram shown in Fig. E7.4(b). From this parallelogram, then at the given instant we have

$$a_B = \frac{(a_{B/A})_n}{\sin\theta}, \quad (a_{B/A})_t = \frac{(a_{B/A})_n}{\tan\theta}$$

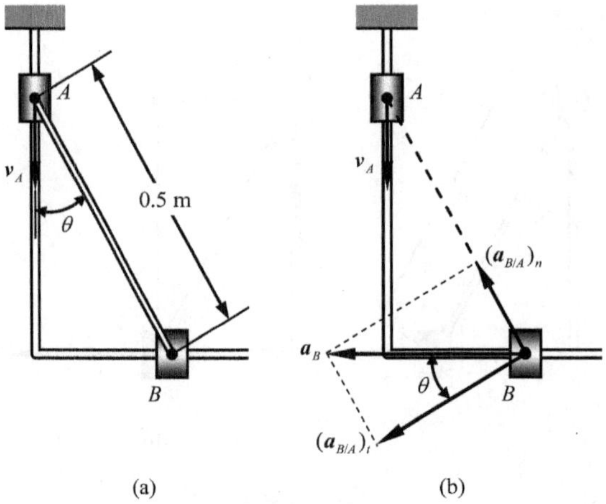

Fig. E7.4

Using $\omega=4$ m/s (referring to Example 7.2 or 7.3), i.e., $(a_{B/A})_n = AB \cdot \omega^2 = 8\ \text{m/s}^2$ (directed toward point A), we can obtain

$$a_B = \frac{(a_{B/A})_n}{\sin\theta} = 16\ \text{m/s}^2,\ (a_{B/A})_t = \frac{(a_{B/A})_n}{\tan\theta} = 13.86\ \text{m/s}^2$$

Using $(a_{B/A})_t = AB \cdot \alpha$, we have

$$\alpha = \frac{(a_{B/A})_t}{AB} = 27.72\ \text{rad/s}^2\ \text{(clockwise)}$$

Example 7.5

Rod BD is attached to two links AB and CD at B and D, as shown in Fig. E7.5(a). Knowing that at the instant shown link AB rotates with a constant angular velocity of $\omega=2$ rad/s clockwise, determine the angular velocity and angular acceleration (a) of rod BD, (b) of link CD.

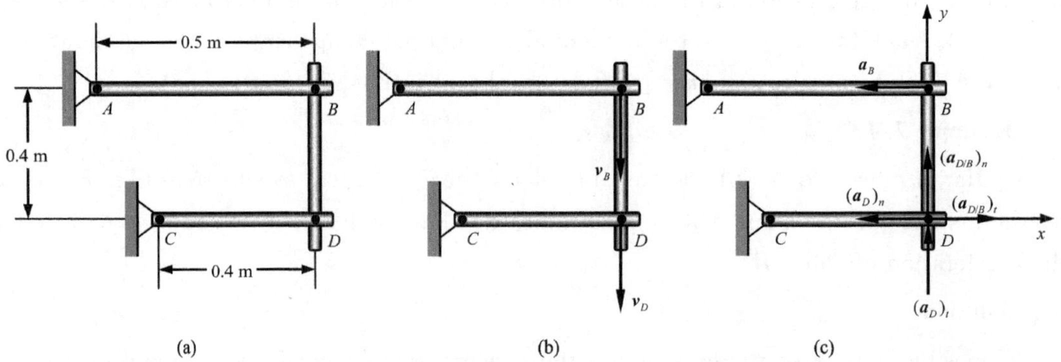

Fig. E7.5

Solution

Since link AB rotates about point A clockwise, the velocity of B is vertical down, as shown in Fig. E7.5(b). Similarly, the velocity of D is also vertical down due to CD rotating about point C clockwise. Therefore, the rod BD is in instantaneous translation due to the same velocity (including magnitude and direction) at B and D at the given instant, from which we have

$$v_D = v_B = AB \cdot \omega_{AB} = 1\ \mathrm{m/s},\ \omega_{BD} = 0,\ \omega_{CD} = \frac{v_D}{CD} = 2.5\ (\mathrm{rad/s})\ (\text{clockwise})$$

Choosing point B of rod BD in instantaneous translation at the given instant as the base point, then the acceleration of point D can be expressed as

$$(\boldsymbol{a}_D)_t + (\boldsymbol{a}_D)_n = \boldsymbol{a}_B + (\boldsymbol{a}_{D/B})_t + (\boldsymbol{a}_{D/B})_n$$

Establishing the reference frame xy, as shown in Fig. E7.5(c), and denoting by $\boldsymbol{i}$ and $\boldsymbol{j}$ the unit vectors respectively along the positive x and y axes, then we have

$$(a_D)_t\boldsymbol{j} - (a_D)_n\boldsymbol{i} = -a_B\boldsymbol{i} + (a_{D/B})_t\boldsymbol{i} + (a_{D/B})_n\boldsymbol{j}$$

or

$$-(a_D)_n = -a_B + (a_{D/B})_t,\ (a_D)_t = (a_{D/B})_n$$

Using $(a_D)_n = CD \cdot (\omega_{CD})^2 = 2.5\ \mathrm{m/s^2}$, $a_B = AB \cdot (\omega_{AB})^2 = 2\ \mathrm{m/s^2}$, and $(a_{D/B})_n = BD \cdot (\omega_{BD})^2 = 0$, we can obtain

$$(a_{D/B})_t = a_B - (a_D)_n = -0.5\ \mathrm{m/s^2}\,(\text{to the left}),\ (a_D)_t = 0$$

Using $(a_{D/B})_t = BD \cdot \alpha_{BD}$ and $(a_D)_t = CD \cdot \alpha_{CD}$, we have

$$\alpha_{BD} = -1.25\ \mathrm{rad/s^2}\,(\text{clockwise}),\ \alpha_{CD} = 0$$

Problems

7.1 The structure shown in Fig. P7.1 consists of two rods AE and CE and a rectangular plate $ABCD$ which are welded together. The structure rotates about the axis AE with an angular velocity of $\omega = 5$ rad/s and an angular acceleration of $\alpha = 10\ \mathrm{rad/s^2}$. Knowing that the angular velocity and angular acceleration are counterclockwise as viewed from A, determine the velocity and acceleration of point D.

7.2 Collar A moves up with a constant velocity of $v_A = 0.4$ m/s. At the instant shown in Fig. P7.2 determine (a) the angular velocity of rod AB and the velocity of collar B, (b) the angular acceleration of rod AB and the acceleration of collar B.

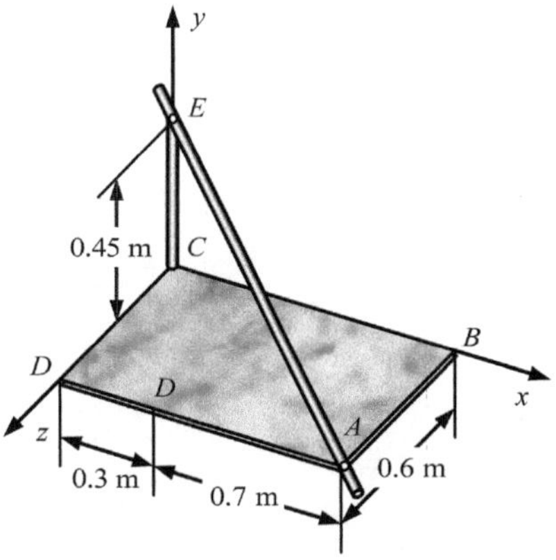

Fig. P7.1

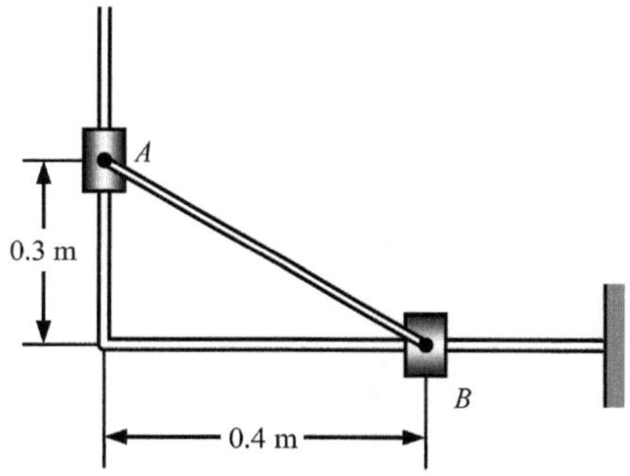

Fig. P7.2

7.3 Rod BDE is attached to two links AB and CD at B and D, as shown in Fig. P7.3. Knowing that at the instant shown link AB rotates with a constant angular velocity of $\omega=2$ rad/sclockwise, determine the velocity and acceleration of point E.

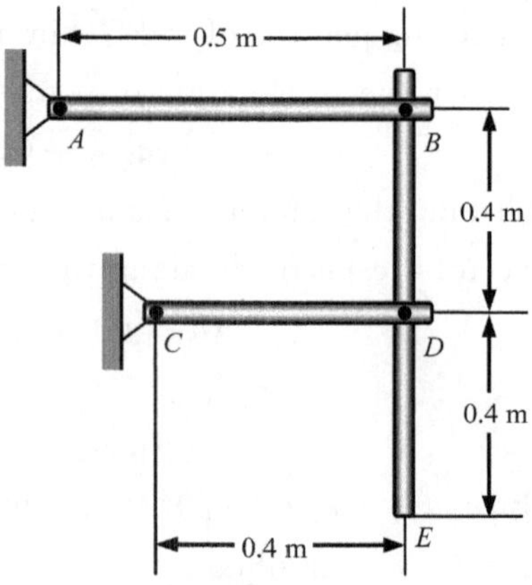

Fig. P7.3

7.4 The motion of a cylinder of radius r is controlled by a cable AG, as shown in Fig. P7.4. Knowing that end G of the cable has a velocity of $v_G=0.3$ m/s and an acceleration of $a_G=0.5$ m/s^2, both directed upward, and that $r=0.15$ m, determine the acceleration (a) of point B and of point D, (b) the velocity and acceleration of point E.

7.5 Rod AB moves over a small wheel at C while end A moves to the right with a constant velocity of 600 mm/s, as shown in Fig. P7.5. At the instant shown determine (a) the angular velocity of the rod and the velocity B of the rod, (b) the angular acceleration of the rod and the acceleration of end B of the rod.

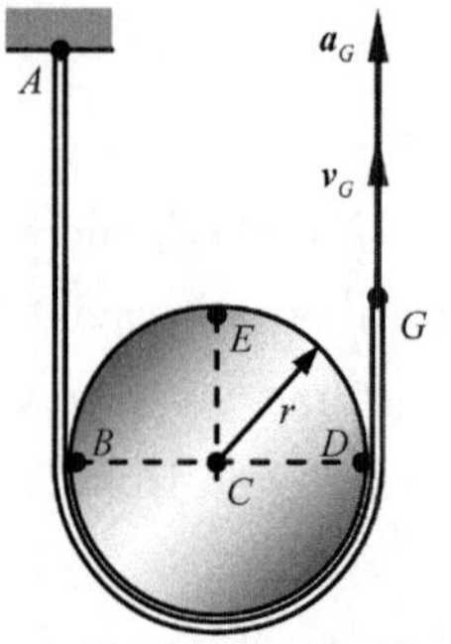

Fig. P7.4

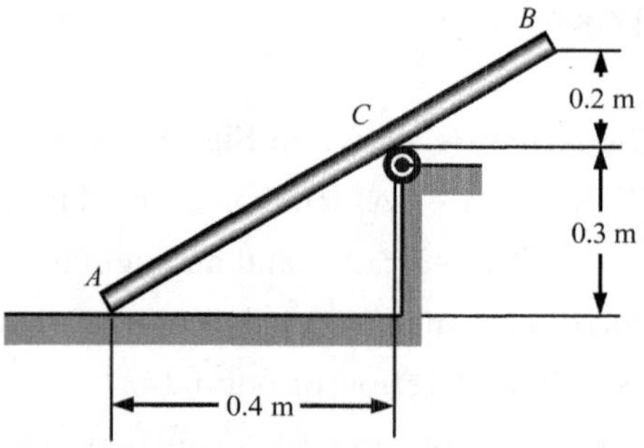

Fig. P7.5

Chapter 8 Resultant Motion of Particle

The motion of a particle has been analyzed by using a single frame of reference. There are many cases, however, where the path of motion for a particle is complicated, so that it may be feasible to analyze the motion of a particle by using two or more frames of reference. For example, the motion of a particle located at the tip of an airplane propeller, while the airplane is in flight, is more easily described if one observes first the motion of the airplane relative to the earth and then superposes, using the parallelogram law, the motion of the particle relative to the airplane.

One of the frames attached to the earth is called a fixed frame of reference, whereas the other frames moving relative to the earth are called moving frames of reference. It should be understood, however, that the selection of a fixed frame of reference is purely arbitrary.

The motion of a particle relative to a fixed frame of reference is referred to as absolute motion, whereas the motion of a particle relative to a moving frame of reference is referred to as relative motion.

8.1 Rates of Change of Vector

Assume that $Oxyz$ is a fixed frame of reference attached to the earth, and that $O'x'y'z'$ is a moving frame of reference with respect to $Oxyz$, as shown in Fig. 8.1.

If the moving frame of reference $O'x'y'z'$ is rotating about the axis λ with angular velocity $\boldsymbol{\omega}$ at a given instant, then a vector $\mathbf{V}$ can be expressed as

$$\mathbf{V} = V'_x\boldsymbol{i}' + V'_y\boldsymbol{j}' + V'_z\boldsymbol{k}' \qquad (8.1)$$

where V'_x, V'_y, V'_z and $\boldsymbol{i}'$, $\boldsymbol{j}'$, $\boldsymbol{k}'$ are, respectively, the rectangular components of the vector $\mathbf{V}$ in the rotating frame of reference $O'x'y'z'$ and the unit vectors used in the rotating frames of reference $O'x'y'z'$.

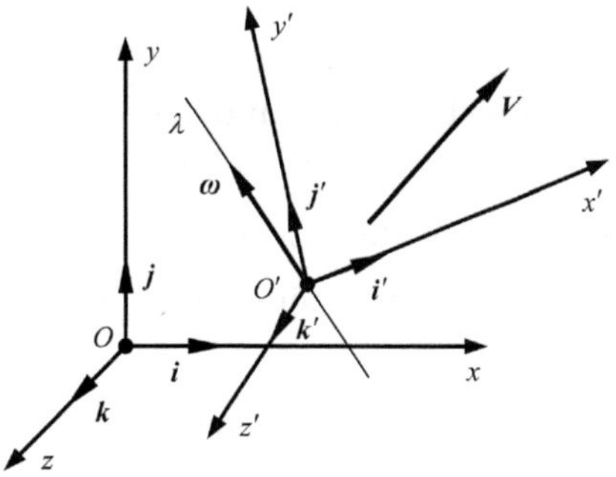

Fig. 8.1

In the rotating frame of reference $O'x'y'z'$, $\boldsymbol{i}'$, $\boldsymbol{j}'$, and $\boldsymbol{k}'$ remain unchanged in magnitude and direction, thus we have

$$\{\dot{\mathbf{V}}\}_{O'} = \dot{V}'_x\boldsymbol{i}' + \dot{V}'_y\boldsymbol{j}' + \dot{V}'_z\boldsymbol{k}' \qquad (8.2)$$

However, in the fixed frame of reference $Oxyz$, $\boldsymbol{i}'$, $\boldsymbol{j}'$, and $\boldsymbol{k}'$ are fixed in magnitude but variable in direction. We thus obtain

$$\{\dot{\mathbf{V}}\}_O = \dot{V}'_x\boldsymbol{i}' + \dot{V}'_y\boldsymbol{j}' + \dot{V}'_z\boldsymbol{k}' + V'_x\dot{\boldsymbol{i}}' + V'_y\dot{\boldsymbol{j}}' + V'_z\dot{\boldsymbol{k}}' \qquad (8.3)$$

Using $\dot{i}'=\omega\times i'$, $\dot{j}'=\omega\times j'$ and $\dot{k}'=\omega\times k'$, we have

$$\{\dot{V}\}_O = \dot{V}'_x i' + \dot{V}'_y j' + \dot{V}'_z k' + \omega\times(V'_x i' + V'_y j' + V'_z k') \tag{8.4}$$

Using Eqs. (8.1), and (8.2), we can obtain

$$\{\dot{V}\}_O = \{\dot{V}\}_{O'} + \omega\times V \tag{8.5}$$

8.2 Resultant of Velocities

Considering that a particle P moves along a curved line in space, and assuming that $Oxyz$ is a fixed frame of reference attached to the earth and that $O'x'y'z'$ is a moving frame of reference relative to $Oxyz$, as shown in Fig. 8.2, then we have

$$r_P = r_{O'} + r_{P/O'} \tag{8.6}$$

where r_P is the position vector of particle P, $r_{O'}$ is the position vector of the origin O' of the moving frame $O'x'y'z'$, and $r_{P/O'}$ is the position vector of particle P relative to the origin O' in the moving frame $O'x'y'z'$.

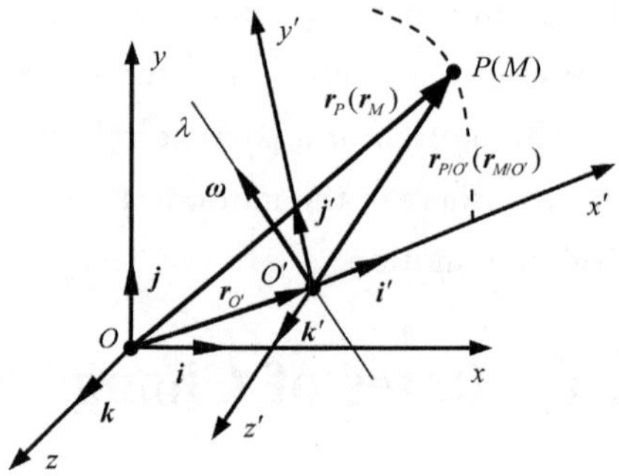

Fig. 8.2

Differentiating Eq. (8.6) with respect to time t within the fixed frame $Oxyz$, we obtain

$$\{\dot{r}_P\}_O = \{\dot{r}_{O'}\}_O + \{\dot{r}_{P/O'}\}_O \tag{8.7}$$

Assuming that the moving frame $O'x'y'z'$ is rotating about the axis λ with angular velocity ω at a given instant, and using Eq. (8.5), then we have

$$\{\dot{r}_{P/O'}\}_O = \{\dot{r}_{P/O'}\}_{O'} + \omega\times r_{P/O'} \tag{8.8}$$

Substituting Eq. (8.8) into Eq. (8.7), we obtain

$$\{\dot{r}_P\}_O = \{\dot{r}_{O'}\}_O + \{\dot{r}_{P/O'}\}_{O'} + \omega\times r_{P/O'} \tag{8.9}$$

Assuming that the convected point (i. e., the point located on the moving frame and coinciding with P at the instant considered) is denoted by M, then we have

$$r_M = r_{O'} + r_{M/O'} \tag{8.10}$$

Using a similar consideration, we can obtain

$$\{\dot{r}_M\}_O = \{\dot{r}_{O'}\}_O + \{\dot{r}_{M/O'}\}_{O'} + \omega\times r_{M/O'} \tag{8.11}$$

After Eq. (8.11) is subtracted from Eq. (8.9), we have

$$\{\dot{r}_P\}_O - \{\dot{r}_M\}_O = \{\dot{r}_{P/O'} - \dot{r}_{M/O'}\}_{O'} + \omega\times(r_{P/O'} - r_{M/O'}) \tag{8.12}$$

Using $\{\dot{r}_{P/O'} - \dot{r}_{M/O'}\}_{O'} = \{\dot{r}_{P/M}\}_{O'}$ and $r_{P/O'} - r_{M/O'} = r_{P/M} = 0$, we thus obtain

$$\{\dot{r}_P\} - \{\dot{r}_M\}_O = \{\dot{r}_{P/M}\}_{O'} \tag{8.13}$$

where $\{\dot{r}_P\}_O = v_P$ is the velocity of particle P, $\{\dot{r}_M\}_O = v_M$ is the velocity of the convected point M, and $\{\dot{r}_{P/M}\}_{O'} = v_{P/M}$ is the velocity of particle P relative to the convected point M. Therefore, Eq. (8.13) can be rewritten as

$$v_P = v_M + v_{P/M} \tag{8.14}$$

We thus conclude that, at any given instant, the absolute velocity $\boldsymbol{v}_P$ of particle P can be obtained by adding vectorially the convected velocity $\boldsymbol{v}_M$ of the convected point M and the relative velocity $\boldsymbol{v}_{P/M}$ of the particle P with respect to the convected point M.

Example 8.1

Two planes, A and B, are flying at the same altitude, as shown in Fig. E8.1(a). Assuming that plane A is flying south with a velocity of $v_A = 400$ km/h and that plane B is flying 30° north of east with a velocity of $v_B = 500$ km/h, determine the relative velocity of plane B with respect to plane A.

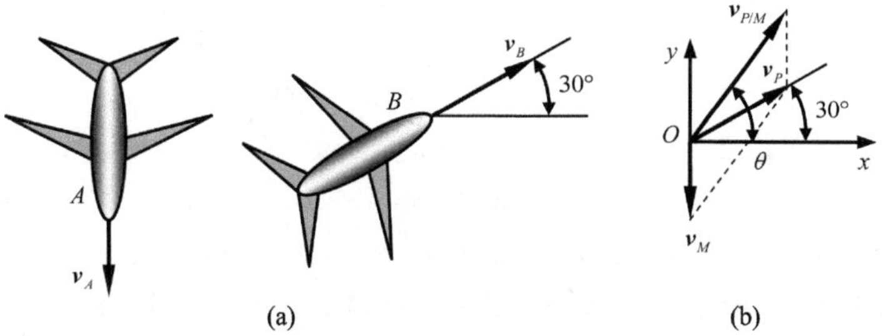

Fig. E8.1

Solution

Choosing the earth as a fixed frame of reference, plane A as a moving frame of reference, and plane B as a moving point, then we have

$$\boldsymbol{v}_P = \boldsymbol{v}_B \text{, and } \boldsymbol{v}_M = \boldsymbol{v}_A$$

Using $\boldsymbol{v}_P = \boldsymbol{v}_M + \boldsymbol{v}_{P/M}$, as shown in Fig. E8.1(b), we obtain

$$v_P \cos 30° \boldsymbol{i} + v_P \sin 30° \boldsymbol{j} = - v_M \boldsymbol{j} + v_{P/M} \cos\theta \boldsymbol{i} + v_{P/M} \sin\theta \boldsymbol{j}$$

where $\boldsymbol{i}$ and $\boldsymbol{j}$ are the unit vectors respectively corresponding to the positive x and y axes, and θ is the angle between $v_{P/M}$ and the positive x axis. Using $v_P = 500$ km/h, $v_M = 400$ km/h, and solving the above equation, we can obtain

$$v_{P/M} = 781.0 \text{ km/h}, \ \theta = 56.33°$$

Thus the relative velocity of plane B with respect to plane A is 781.0 km/h and 56.33° north of east.

8.3 Resultant of Accelerations

Differentiating Eq. (8.9) with respect to time t within the fixed frame $Oxyz$, as shown in Fig. 8.2, we have

$$\{\ddot{\boldsymbol{r}}_P\}_O = \{\ddot{\boldsymbol{r}}_{O'}\}_O + \left\{\frac{\mathrm{d}}{\mathrm{d}t}\{\dot{\boldsymbol{r}}_{P/O'}\}_{O'}\right\}_O + \dot{\boldsymbol{\omega}} \times \boldsymbol{r}_{P/O'} + \boldsymbol{\omega} \times \{\dot{\boldsymbol{r}}_{P/O'}\}_O \tag{8.15}$$

Using Eq. (8.5), we have

$$\left\{\frac{\mathrm{d}}{\mathrm{d}t}\{\dot{\boldsymbol{r}}_{P/O'}\}_{O'}\right\}_O = \{\ddot{\boldsymbol{r}}_{P/O'}\}_{O'} + \boldsymbol{\omega} \times \{\dot{\boldsymbol{r}}_{P/O'}\}_{O'} \tag{8.16}$$

and

$$\{\dot{r}_{P/O'}\}_O = \{\dot{r}_{P/O'}\}_{O'} + \boldsymbol{\omega} \times \boldsymbol{r}_{P/O'} \tag{8.17}$$

Substituting Eqs. (8.16) and (8.17) into Eq. (8.15), we have

$$\{\ddot{r}_P\}_O = \{\ddot{r}_{O'}\}_O + \{\ddot{r}_{P/O'}\}_{O'} + 2\boldsymbol{\omega} \times \{\dot{r}_{P/O'}\}_{O'} + \dot{\boldsymbol{\omega}} \times \boldsymbol{r}_{P/O'} + \boldsymbol{\omega} \times (\boldsymbol{\omega} \times \boldsymbol{r}_{P/O'}) \tag{8.18}$$

For the convected point M, using a similar consideration we can obtain

$$\{\ddot{r}_M\}_O = \{\ddot{r}_{O'}\}_O + \{\ddot{r}_{M/O'}\}_{O'} + 2\boldsymbol{\omega} \times \{\dot{r}_{M/O'}\}_{O'} + \dot{\boldsymbol{\omega}} \times \boldsymbol{r}_{M/O'} + \boldsymbol{\omega} \times (\boldsymbol{\omega} \times \boldsymbol{r}_{M/O'}) \tag{8.19}$$

After Eq. (8.19) is subtracted from Eq. (8.18), we have

$$\{\ddot{r}_P\}_O - \{\ddot{r}_M\}_O = \{\ddot{r}_{P/O'} - \ddot{r}_{M/O'}\}_{O'} + 2\boldsymbol{\omega} \times \{\dot{r}_{P/O'} - \dot{r}_{M/O'}\}_{O'} + \dot{\boldsymbol{\omega}} \times (\boldsymbol{r}_{P/O'} - \boldsymbol{r}_{M/O'}) + \boldsymbol{\omega} \times [\boldsymbol{\omega} \times (\boldsymbol{r}_{P/O'} - \boldsymbol{r}_{M/O'})] \tag{8.20}$$

Using $\{\ddot{r}_{P/O'} - \ddot{r}_{M/O'}\}_{O'} = \{\ddot{r}_{P/M}\}_{O'}$, $\{\dot{r}_{P/O'} - \dot{r}_{M/O'}\}_{O'} = \{\dot{r}_{P/M}\}_{O'}$ and $\boldsymbol{r}_{P/O'} - \boldsymbol{r}_{M/O'} = \boldsymbol{r}_{P/M} = 0$, then we obtain

$$\{\ddot{r}_P\}_O - \{\ddot{r}_M\}_O = \{\ddot{r}_{P/M}\}_{O'} + 2\boldsymbol{\omega} \times \{\dot{r}_{P/M}\}_{O'} \tag{8.21}$$

Using $\{\ddot{r}_P\}_O = \boldsymbol{a}_P$, $\{\ddot{r}_M\}_O = \boldsymbol{a}_M$, $\{\ddot{r}_{P/M}\}_{O'} = \boldsymbol{a}_{P/M}$, and $\{\dot{r}_{P/M}\}_{O'} = \boldsymbol{v}_{P/M}$, we have

$$\boldsymbol{a}_P = \boldsymbol{a}_M + \boldsymbol{a}_{P/M} + \boldsymbol{a}_C \tag{8.22}$$

where $\boldsymbol{a}_P$ is the absolute acceleration of particle P, $\boldsymbol{a}_M$ is the convected acceleration of the convected point M, $\boldsymbol{a}_{P/M}$ is the relative acceleration of particle P with respect to the convected point M, and $\boldsymbol{a}_C = 2\boldsymbol{\omega} \times \boldsymbol{v}_{P/M}$ is Coriolis acceleration.

We conclude that at any given instant the absolute acceleration $\boldsymbol{a}_P$ is a vector sum of the convected acceleration $\boldsymbol{a}_M$, the relative acceleration $\boldsymbol{a}_{P/M}$, and Coriolis acceleration $\boldsymbol{a}_C$.

Example 8.2

Knowing that at the instant shown in Fig. E8.2(a) rod BD rotates with a constant counterclockwise angular velocity $\omega_B = 2$ rad/s, determine (a) the angular velocity of rod AD and the relative velocity of collar D with respect to rod BD, (b) the angular acceleration of rod AD and the relative acceleration of collar D with respect to rod BD.

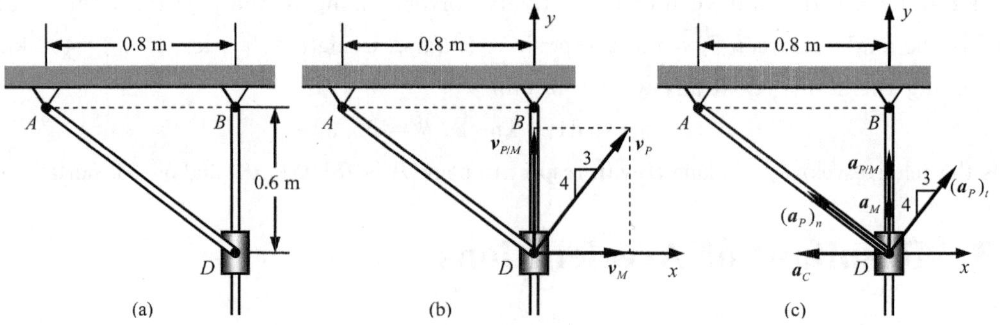

Fig. E8.2

Solution

Assume that a fixed frame of reference is attached to the earth, that a moving frame of reference is attached to rod BD, and that a moving point is attached to collar D.

(a) Using $\boldsymbol{v}_P = \boldsymbol{v}_M + \boldsymbol{v}_{P/M}$, as shown in Fig. E8.2(b), we have

$$\frac{3}{5}v_P\boldsymbol{i}+\frac{4}{5}v_P\boldsymbol{j}=v_M\boldsymbol{i}+v_{P/M}\boldsymbol{j}$$

where $\boldsymbol{i}$ and $\boldsymbol{j}$ are the unit vectors respectively corresponding to the positive x and y axes. Solving the above equation and using $v_M=BD\cdot\omega_B=1.2$ m/s, we obtain

$$v_P=\frac{5}{3}v_M=2\ \text{m/s},\ v_{P/M}=\frac{4}{3}v_M=1.6\ \text{m/s}$$

Using $v_P=AD\cdot\omega_A$, we can obtain

$$\omega_A=\frac{v_P}{AD}=2\ \text{rad/s}$$

(b) Using $(\boldsymbol{a}_P)_t+(\boldsymbol{a}_P)_n=\boldsymbol{a}_M+\boldsymbol{a}_{P/M}+\boldsymbol{a}_C$, as shown in Fig. E8.2(c), we have

$$\left[\frac{3}{5}(a_P)_t\boldsymbol{i}+\frac{4}{5}(a_P)_t\boldsymbol{j}\right]+\left[-\frac{4}{5}(a_P)_n\boldsymbol{i}+\frac{3}{5}(a_P)_n\boldsymbol{j}\right]=a_M\boldsymbol{j}+a_{P/M}\boldsymbol{j}-a_C\boldsymbol{i}$$

where $(a_P)_n=AD\cdot\omega_A^2=4\ \text{m/s}^2$, $a_M=BD\cdot\omega_B^2=2.4\ \text{m/s}^2$ and $a_C=2\omega_B v_{P/M}=6.4\ \text{m/s}^2$. Solving the above equation, we obtain

$$(a_P)_t=-5.33\ \text{m/s}^2,\ a_{P/M}=-4.27\ \text{m/s}^2$$

Using $(a_P)_t=AD\cdot\alpha_A$, we can obtain

$$\alpha_A=\frac{(a_P)_t}{AD}=-5.33\ \text{rad/s}^2$$

In the case of the moving frame of reference $O'x'y'z'$ in translation relative to the fixed frame of reference $Oxyz$, as shown in Fig. 8.3, i.e., $\boldsymbol{\omega}=0$, then Eq. (8.22) can be simplified as

$$\boldsymbol{a}_P=\boldsymbol{a}_M+\boldsymbol{a}_{P/M} \tag{8.23}$$

We thus conclude that for a translating frame of reference at any given instant the absolute acceleration $\boldsymbol{a}_P$ is a vector sum of the convected acceleration $\boldsymbol{a}_M$ and the relative acceleration $\boldsymbol{a}_{P/M}$.

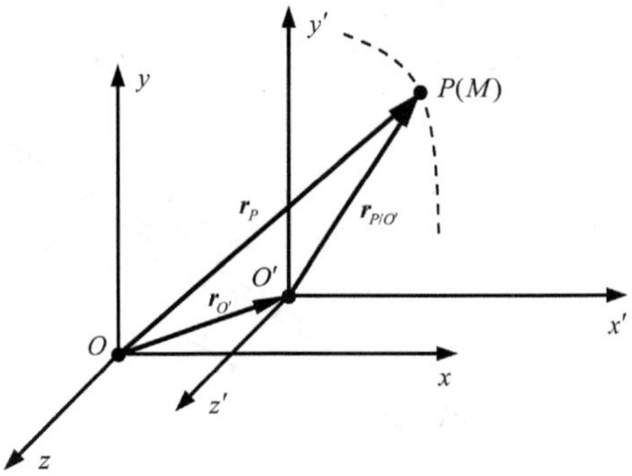

Fig. 8.3

Example 8.3

As the truck begins to move backward with an acceleration of 2 m/s^2, the outer section D of its boom starts to retract with an acceleration of 1 m/s^2 relative to the truck, as shown in Fig. E8.3(a). Determine the acceleration of section D.

Solution

Assume that a fixed frame of reference is attached to the earth, that a moving frame of reference is attached to the truck, and that a moving point is attached to section D.

Drawing the acceleration triangle based on $\boldsymbol{a}_P=\boldsymbol{a}_M+\boldsymbol{a}_{P/M}$, as shown in Fig. E8.3(b), we have

$$a_P=\sqrt{(a_M)^2+(a_{P/M})^2-2a_M a_{P/M}\cos45^\circ}=1.47\ \text{m/s}^2$$

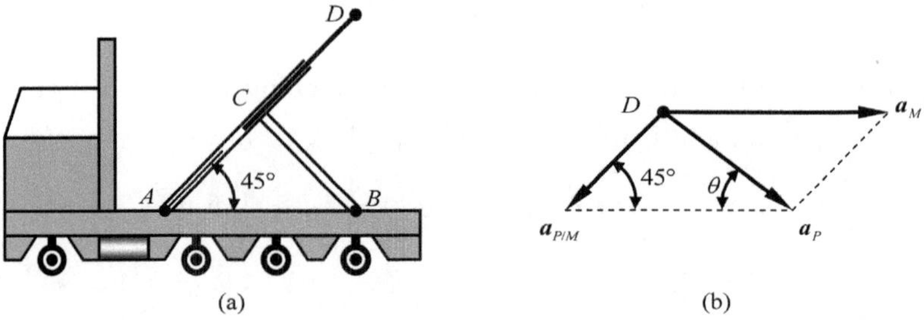

Fig. E8.3

$$\theta = \arcsin\left(\frac{a_{P/M}}{a_P}\sin 45^\circ\right) = 28.75^\circ$$

Thus the acceleration of section D is 1.47 m/s^2 and 28.75° south of east.

Problems

8.1 A collar slides outward at a constant relative speed v along the rod OAB, which rotates counterclockwise with a constant angular velocity of $\omega_B = 2$ rad/s, as shown in Fig. P8.1. Knowing that the collar is located at A when $\theta = 0$ and that the collar reaches B when $\theta = 90°$, determine the acceleration of the collar when $\theta = 30°$.

8.2 Knowing that at the instant rod BD rotates with a constant counterclockwise angular velocity $\omega_B = 2$ rad/s, as shown in Fig. P8.2, determine (a) the angular velocity of rod AD and the relative velocity of collar D with respect to rod AD, (b) the angular acceleration of rod AD and the relative acceleration of collar D with respect to rod AD.

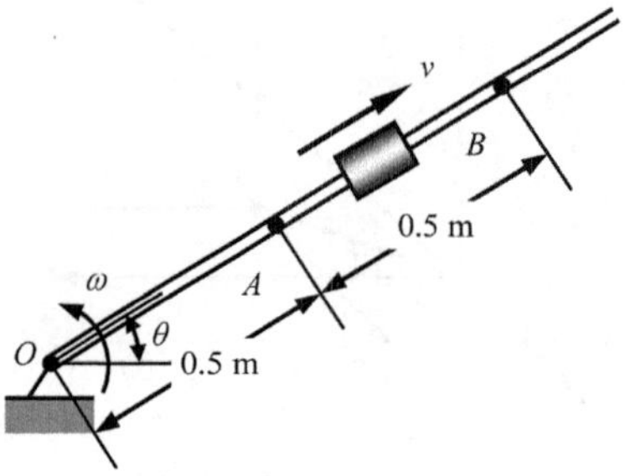

Fig. P8.1

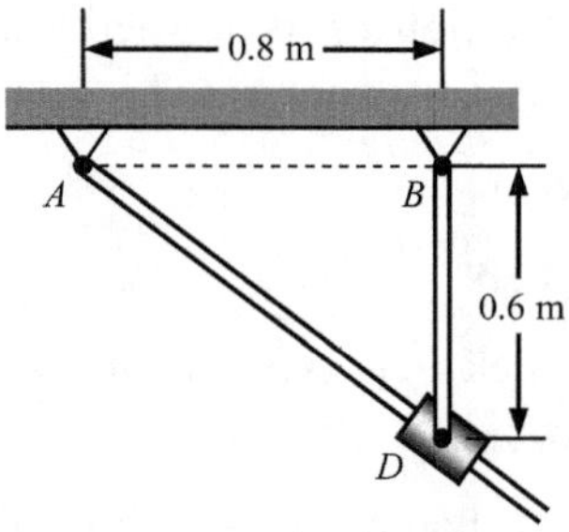

Fig. P8.2

8.3 As the truck begins to move forward with an acceleration of 3 m/s^2, the outer section D of its boom starts to retract with an acceleration of 2 m/s^2 relative to the truck, as shown in Fig. P8.3. Determine the acceleration of section D.

8.4 Collar E slides along rod BD and is attached to collar F that moves along a vertical rod AC, as shown in Fig. P8.4. Knowing that the angular velocity and angular acceleration of rod BD are $\omega = 6$ rad/s and $\alpha = 4$ rad/s^2, both clockwise, determine the velocity and acceleration of collar E.

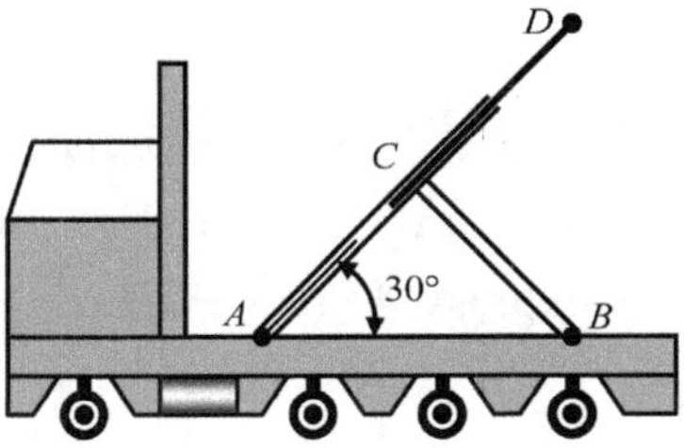

Fig. P8. 3

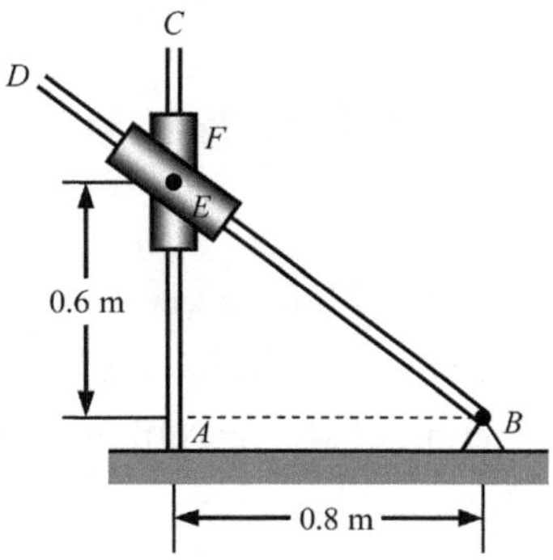

Fig. P8. 4

8. 5 Plane A flying horizontally in a straight line has a velocity of 300 km/h and an acceleration of 5 m/s^2 at the given instant. Plane B is flying at the same altitude as plane A and is following a circular path of radius 200 m, as shown in Fig. P8. 5. Knowing that at the given instant the velocity and deceleration of plane B are 400 km/h and 3 m/s^2 respectively, determine, at the instant shown, (a) the velocity of B relative to A, (b) the acceleration of B relative to A.

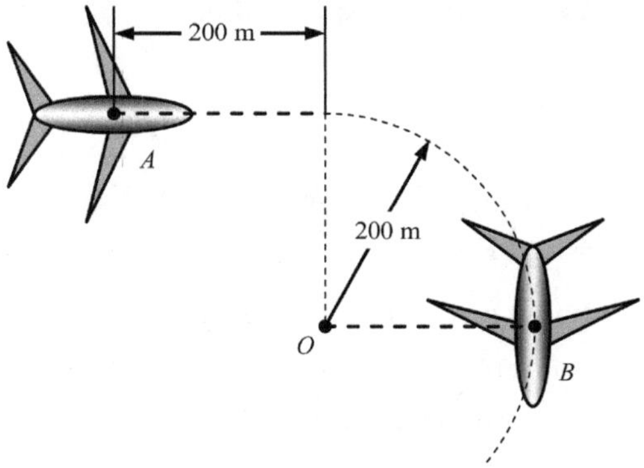

Fig. P8. 5

8. 6 A rod OA is always tangential to the surface of a semicylinder C of radius $r=$ 0. 4 m, as shown in Fig. P8. 6. Knowing that the velocity and acceleration of the semicylinder are $v=0.1$ m/s and $a=0.2$ m/s^2 respectively, both to the right, determine the angular velocity and angular acceleration of the rod when $\theta=30°$.

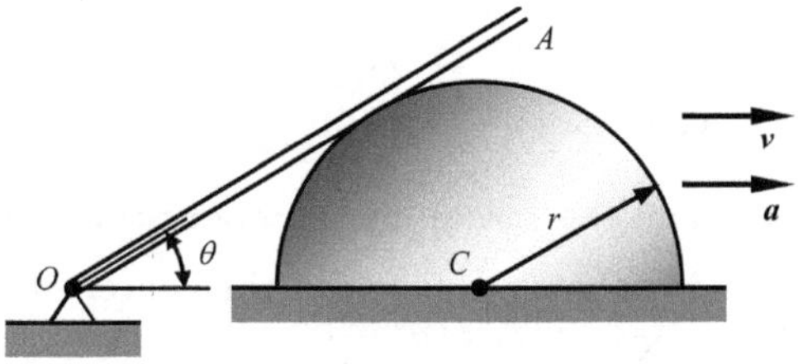

Fig. P8. 6

Chapter 9 Kinetics of Particle

Newton's second law can be stated as follows: if the resultant force acting on a particle is not zero, the particle will have acceleration proportional to the magnitude of the resultant and in the direction of this resultant. This law can be expressed mathematically as

$$m\boldsymbol{a} = \sum \boldsymbol{F} \tag{9.1}$$

where m is the mass of the particle, $\boldsymbol{a}$ is the acceleration of the particle under the action of the forces, and $\sum \boldsymbol{F}$ is the sum, or resultant, of all the forces acting on the particle.

It should be noted that Newton's second law of motion holds only with respect to a Newtonian frame of reference or an inertial frame of reference.

9.1 Equations of Motion of Particle

Considering a particle of mass m acted upon by several forces $\boldsymbol{F}_1$, $\boldsymbol{F}_2$, $\boldsymbol{F}_3$, ⋯, as shown in Fig. 9.1. Newton's second law can be expressed as

$$m\boldsymbol{a} = \sum \boldsymbol{F} \tag{9.2}$$

or

$$m\ddot{\boldsymbol{r}} = \sum \boldsymbol{F} \tag{9.3}$$

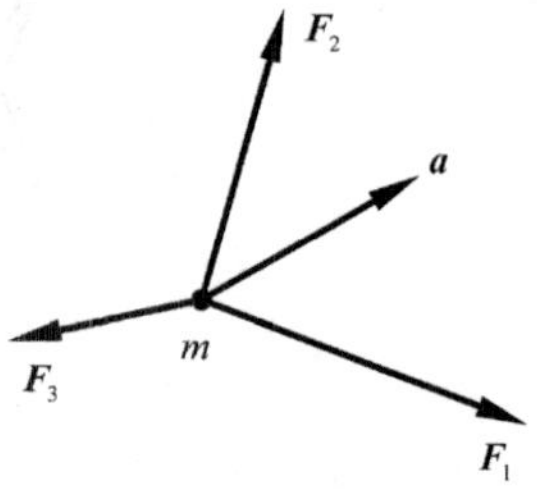

Fig. 9.1

In order to solve problems involving the motion of a particle, it will be more convenient to replace the vector equation above by equivalent scalar equations represented by rectangular or natural coordinates.

1. Rectangular Coordinate Method

Resolving the acceleration and each of the forces into rectangular components, we obtain

$$ma_x = \sum F_x,\ ma_y = \sum F_y,\ ma_z = \sum F_z \tag{9.4}$$

or

$$m\ddot{x} = \sum F_x,\ m\ddot{y} = \sum F_y,\ m\ddot{z} = \sum F_z \tag{9.5}$$

2. Natural Coordinate Method

Resolving the acceleration and the forces into components along the tangent to the path (in the direction of motion) and the normal (toward the center of curvature), we obtain

$$ma_t = \sum F_t,\ ma_n = \sum F_n,\ 0 = \sum F_b \tag{9.6}$$

or

$$m\dot{v} = \sum F_t, \ m\frac{v^2}{\rho} = \sum F_n, \ 0 = \sum F_b \tag{9.7}$$

Example 9.1

A spring AB of stiffness k is attached to a support A and to a collar B of mass m, as shown in Fig. E9.1(a). The unstretched length of the spring is l. Knowing that the collar is released from rest at $x=\sqrt{3}\,l$ and neglecting friction between the collar and the horizontal rod, determine the magnitude of the velocity of the collar as it passes through midpoint O.

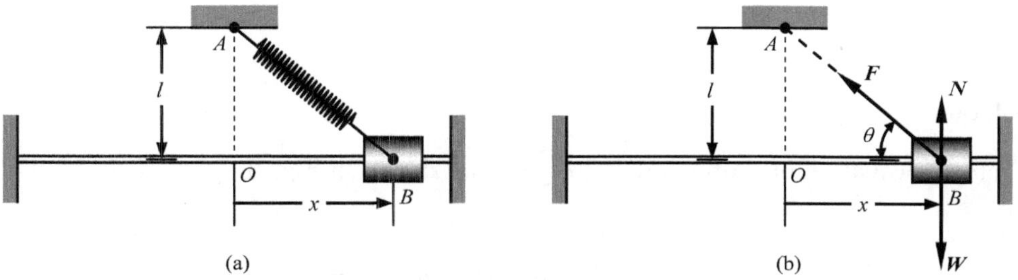

Fig. E9.1

Solution

When the collar is located at position x, then the elongation of the spring can be expressed as

$$\delta = \sqrt{l^2 + x^2} - l$$

Thus the corresponding spring force can be given by

$$F = k\delta = k(\sqrt{l^2 + x^2} - l)$$

From the equation of motion, $ma_x = \sum F_x$, we have

$$ma = -F\cos\theta$$

Using $a=\frac{dv}{dt}=\frac{dx}{dt}\frac{dv}{dx}=v\frac{dv}{dx}$ and $\cos\theta=\frac{x}{\sqrt{l^2+x^2}}$, we have

$$v\mathrm{d}v = -\frac{k}{m}\left(x - \frac{lx}{\sqrt{l^2 + x^2}}\right)\mathrm{d}x$$

Performing integration on the above equation, we obtain

$$\int_0^{v_O} v\mathrm{d}v = -\frac{k}{m}\int_{\sqrt{3}l}^{0}\left(x - \frac{lx}{\sqrt{l^2 + x^2}}\right)\mathrm{d}x$$

i.e.,

$$v_O = l\sqrt{k/m}$$

Example 9.2

The initial velocity of the block in position A is $v_A=8$ m/s, as shown in Fig. E9.2(a). Knowing that the coefficient of kinetic friction between the block and the plane of inclination is $\mu_k=0.25$, determine the distance it moves and the time it takes for the block to reach B with zero velocity if $\theta=10°$.

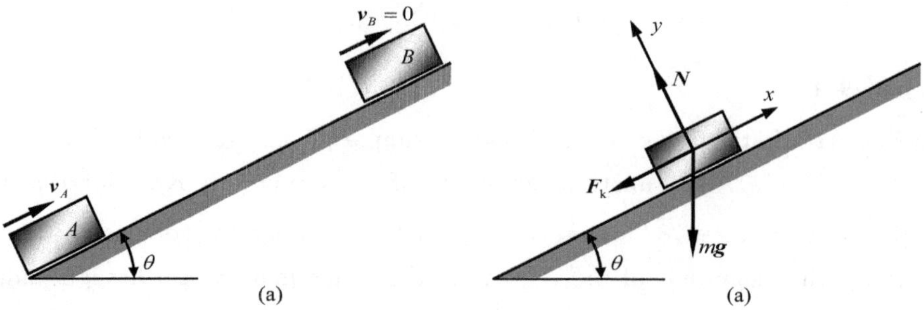

Fig. E9. 2

Solution

From the equation of motion $ma_x = \sum F_x$, as shown in Fig. E9. 2(b), we have

$$ma_x = \sum F_x,\ ma = -mg\sin\theta - F_k$$

$$ma_y = \sum F_y,\ 0 = N - mg\cos\theta$$

Using $F_k = \mu_k N$ and solving the above equations, we can obtain

$$a = -(\sin\theta + \mu_k\cos\theta)g$$

(1) Using $a = \frac{dv}{dt} = \frac{dx}{dt}\frac{dv}{dx} = v\frac{dv}{dx}$, we have

$$v\,dv = -(\sin\theta + \mu_k\cos\theta)g\,dx$$

Performing integration on the above equation, we obtain

$$\int_{v_A}^{v_B} v\,dv = \int_0^{x_B} -(\sin\theta + \mu_k\cos\theta)g\,dx$$

i.e.,

$$x_B = \frac{{v_A}^2 - {v_B}^2}{2(\sin\theta + \mu_k\cos\theta)g}$$

Using $v_A = 8$ m/s, $v_B = 0$, $\mu_k = 0.25$, $\theta = 10°$, and $g = 9.81$ m/s^2, we obtain

$$x_B = 7.77 \text{ m}$$

(2) Using $a = \frac{dv}{dt}$, we have

$$dv = -(\sin\theta + \mu_k\cos\theta)g\,dt$$

Performing integration on the above equation, we obtain

$$\int_{v_A}^{v_B} dv = \int_0^{t_B} -(\sin\theta + \mu_k\cos\theta)g\,dt$$

i.e.,

$$t_B = \frac{v_A - v_B}{(\sin\theta + \mu_k\cos\theta)g}$$

Using $v_A = 8$ m/s, $v_B = 0$, $\mu_k = 0.25$, $\theta = 10°$, and $g = 9.81$ m/s^2, we obtain

$$t_B = 1.94 \text{ s}$$

9.2 Method of Inertia Force for Particle in Motion

Returning to Eq. (9.1), we write Newton's second law of motion in the alternative form

$$\sum F - ma = 0 \tag{9.8}$$

Defining $\boldsymbol{F}_I = -m\boldsymbol{a}$, then Eq. (9.8) can be rewritten by

$$\sum \boldsymbol{F} + \boldsymbol{F}_I = 0 \tag{9.9}$$

where $\boldsymbol{F}_I = -m\boldsymbol{a}$ is called an inertia force, which is equal in magnitude to ma and opposite in direction to the acceleration $\boldsymbol{a}$. This method proposed by d'Alembert to analyze the motion of a particle is called the method of inertia force or d'Alembert's principle.

The method of inertia force can convert a dynamics problem into an equivalent problem in equilibrium, and this dynamics problem can be solved by the methods developed in statics. Although the particle is not in equilibrium, the equations of equilibrium obtained in statics can be applied if an inertia force is applied to the particle in motion.

It should be noted that an inertia force is an imaginary force, but not an actual force.

9.3 Method of Work and Energy for Particle in Motion

The method of work and energy can be used to solve problems involving force, mass, velocity and displacement.

1. Work of Force

Consider that a particle moves from a point A to a neighboring point A', as shown in Fig. 9.2. If the position vectors of the particle at points A and A' are denoted, respectively, by $\boldsymbol{r}$ and $\boldsymbol{r}'$, the differential $\mathrm{d}\boldsymbol{r} = \boldsymbol{r}' - \boldsymbol{r}$ joining A and A' is called the displacement of the particle. Assuming that a force $\boldsymbol{F}$ acts on the particle, then the elementary work of force $\boldsymbol{F}$ corresponding to the displacement $\mathrm{d}\boldsymbol{r}$ is defined as

$$\mathrm{d}W = \boldsymbol{F} \cdot \mathrm{d}\boldsymbol{r} \tag{9.10}$$

The work of force $\boldsymbol{F}$ during a finite displacement of the particle moving from A_1 to A_2 can be obtained by integrating the equation above along the path described by the particle. This work can be expressed as

$$W_{12} = \int_{A_1}^{A_2} \boldsymbol{F} \cdot \mathrm{d}\boldsymbol{r} \tag{9.11}$$

2. Principle of Work and Energy

Considering that a particle of mass m moves along a curved line under the action of forces $\boldsymbol{F}_1$, $\boldsymbol{F}_2$, $\boldsymbol{F}_3$, …, as shown in Fig. 9.3, then Newton's second law can be expressed as

$$m\dot{\boldsymbol{v}} = \sum \boldsymbol{F} \tag{9.12}$$

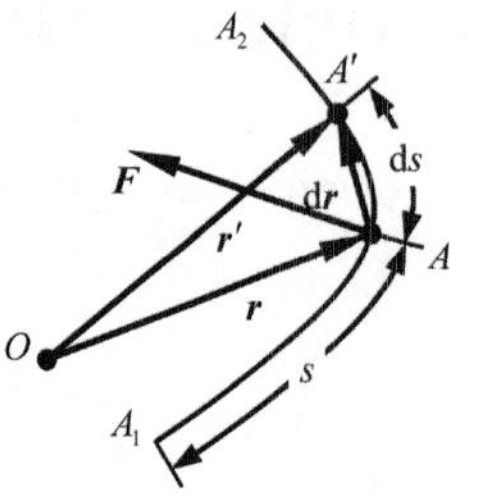

Fig. 9.2

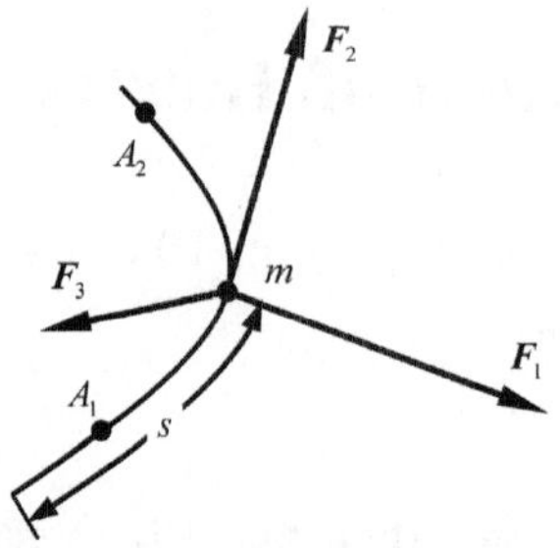

Fig. 9.3

Multiplying both sides of Eq. (9.12) by d$\boldsymbol{r}$ through a dot product, we obtain

$$m\dot{\boldsymbol{v}} \cdot \mathrm{d}\boldsymbol{r} = \sum \boldsymbol{F} \cdot \mathrm{d}\boldsymbol{r} \tag{9.13}$$

Using $\dot{\boldsymbol{v}} \cdot \mathrm{d}\boldsymbol{r} = \boldsymbol{v} \cdot \mathrm{d}\boldsymbol{v} = \frac{1}{2}\mathrm{d}(\boldsymbol{v} \cdot \boldsymbol{v}) = \frac{1}{2}\mathrm{d}(v^2)$, we have

$$\mathrm{d}T = \mathrm{d}W \tag{9.14}$$

where $\mathrm{d}T = \mathrm{d}\left(\frac{1}{2}mv^2\right)$ and $\mathrm{d}W = \sum \boldsymbol{F} \cdot \mathrm{d}\boldsymbol{r}$ are respectively the increment in kinetic energy of the particle and the elementary work done by the forces $\sum \boldsymbol{F}$ during a displacement d$\boldsymbol{r}$. Integrating from A_1 to A_2, we obtain

$$T_2 - T_1 = W_{12} \tag{9.15}$$

where $T_1 = \frac{1}{2}mv_1^2$ and $T_2 = \frac{1}{2}mv_2^2$ are the kinetic energy of the particle respectively at A_1 and A_2, $W_{12} = \sum \int_{A_1}^{A_2} \boldsymbol{F} \cdot \mathrm{d}\boldsymbol{r}$ is the work of the forces acting on the particle when the particle moves from A_1 to A_2. We thus conclude from Eq. (9.14) or (9.15) that, when a particle moves from one point to another, the change in kinetic energy of the particle is equal to the work of the forces acting on the particle. Eq. (9.15) can also expressed as an alternative form

$$T_1 + W_{12} = T_2 \tag{9.16}$$

The relation shows that the kinetic energy of the particle at A_2 can be obtained by adding to its kinetic energy at A_1 the work done during the displacement from A_1 to A_2 by the forces $\sum \boldsymbol{F}$ exerted on the particle.

The relation expressed by Eq. (9.14), (9.15) or (9.16) is called the principle of work and energy. It should be note that the principle of work and energy applies only with respect to a Newtonian's frame of reference.

Example 9.3

The initial velocity of the block in position A is $v_A = 8$ m/s, as shown in Fig. E9.3. Knowing that the coefficient of kinetic friction between the block and the plane of inclination is $\mu_k = 0.25$, determine the distance it moves for the block to reach B with zero velocity if $\theta = 10°$.

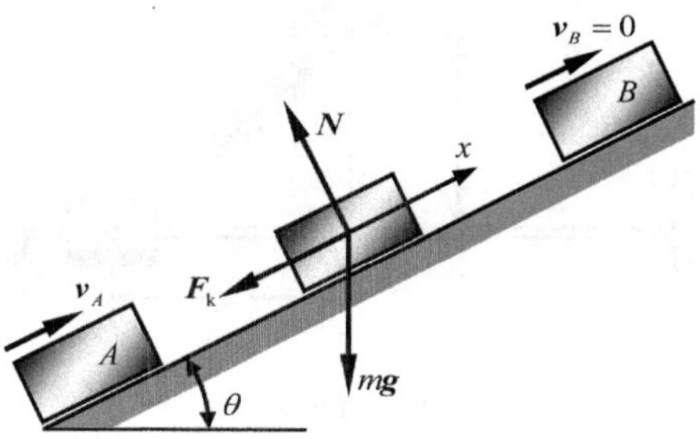

Fig. E9.3

Solution

From the principle of work and energy, $T_A + W_{AB} = T_B$, we have

$$\frac{1}{2}mv_A^2 + (-F_k - mg\sin\theta)x_B = \frac{1}{2}mv_B^2$$

where $F_k = \mu_k N = \mu_k mg\cos\theta$.

Using $v_A = 8$ m/s, $v_B = 0$, $\mu_k = 0.25$, $\theta = 10°$, and $g = 9.81$ m/s^2, we obtain

$$x_B = \frac{v_A^2 - v_B^2}{2(\sin\theta + \mu_k\cos\theta)g} = 7.77 \text{ m}$$

3. Conservation of Energy

A force acting on a particle is said to be conservative if its work is independent of the path followed by the particle as it moves from one point to another. The gravitational force and the elastic force are typical examples of conservative forces.

The work done by the conservative force acting on a particle moving from A_1 to A_2 is defined as the potential energy of the particle at A_1 with respect to A_2, and can be expressed as

$$V_1 - V_2 = W_{12} = \sum \int_{A_1}^{A_2} \boldsymbol{F} \cdot \mathrm{d}\boldsymbol{r} \tag{9.17}$$

where V_1 and V_2 respectively represent the potential energy of the particle at A_1 and A_2.

Using Eqs. (9.15) and (9.17), we can obtain

$$T_1 + V_1 = T_2 + V_2 \tag{9.18}$$

Eq. (9.18) shows that when a particle moves under the action of conservative forces, the sum of the kinetic energy and of the potential energy of the particle remains constant. This sum is called the mechanical energy of the particle.

Example 9.4

A spring AB of stiffness k is attached to a support A and to a collar B of mass m, as shown in Fig. E9.4. The unstretched length of the spring is l. Knowing that the collar is released from rest at $x = \sqrt{3}\,l$ and neglecting friction between the collar and the horizontal rod, determine the magnitude of the velocity of the collar as it passes through midpoint O.

Solution

When the collar is located at position $x = \sqrt{3}\,l$, we have

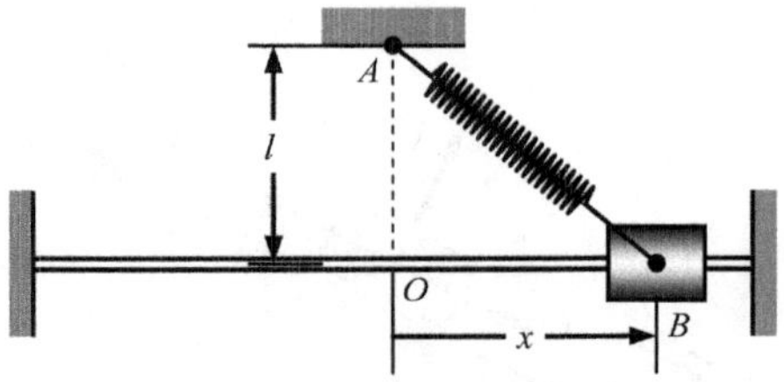

Fig. E9.4

$$T_1 = 0,\ V_1 = \frac{1}{2}k\left[\sqrt{l^2 + (\sqrt{3}l)^2} - l\right]^2 = \frac{1}{2}kl^2$$

And when at midpoint O, we have

$$T_2 = \frac{1}{2}mv_O^2,\ V_2 = 0$$

Using the conservation of mechanical energy, $T_1 + V_1 = T_2 + V_2$, we have

$$0 + \frac{1}{2}kl^2 = \frac{1}{2}mv_O{}^2 + 0,\ \text{or}\ v_O = l\sqrt{k/m}$$

9.4 Method of Impulse and Momentum for Particle in Motion

The method of impulse and momentum can be used to solve problems involving force, mass, velocity and time.

1. Principle of Impulse and Momentum

Considering that a particle of mass m moves along a curved line under the action of forces $\boldsymbol{F}_1$, $\boldsymbol{F}_2$, $\boldsymbol{F}_3$, …, as shown in Fig. 9.4, then Newton's second law can be expressed as

$$m\dot{\boldsymbol{v}} = \sum \boldsymbol{F} \tag{9.19}$$

Multiplying both sides of Eq. (9.19) by $\mathrm{d}t$, we have

$$\mathrm{d}\boldsymbol{L} = \mathrm{d}\boldsymbol{I} \tag{9.20}$$

where $\mathrm{d}\boldsymbol{L} = \mathrm{d}(m\boldsymbol{v})$ and $\mathrm{d}\boldsymbol{I} = \sum \boldsymbol{F}\mathrm{d}t$ are respectively the increment in momentum of the particle and the impulse of the forces $\sum \boldsymbol{F}$ acting on the particle during a time interval $\mathrm{d}t$. Integrating from t_1 to t_2, we obtain

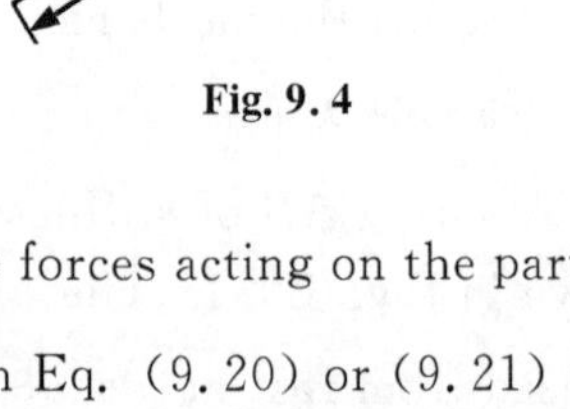

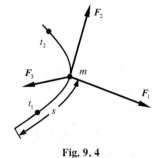

Fig. 9.4

$$\boldsymbol{L}_2 - \boldsymbol{L}_1 = \boldsymbol{I}_{12} \tag{9.21}$$

where $\boldsymbol{L}_1 = m\,\boldsymbol{v}_1$ and $\boldsymbol{L}_2 = m\,\boldsymbol{v}_2$ are respectively the momenta of the particle at t_1 and t_2, $\boldsymbol{I}_{12} = \sum \int_{t_1}^{t_2} \boldsymbol{F}\mathrm{d}t$ is the impulse of the forces acting on the particle during the interval of time from t_1 to t_2. We thus conclude from Eq. (9.20) or (9.21) that the change in momentum of the particle is equal to the impulse of the forces acting on the particle during the time interval considered. The above relation is called the principle of impulse and momentum, which can also be rewritten as

$$\boldsymbol{L}_1 + \boldsymbol{I}_{12} = \boldsymbol{L}_2 \tag{9.22}$$

This equation shows that, when a particle is acted upon by forces during a given time interval, the final momentum of the particle can be obtained by adding vectorially its initial momentum and the impulse of the forces during the time interval considered, as shown in Fig. 9.5.

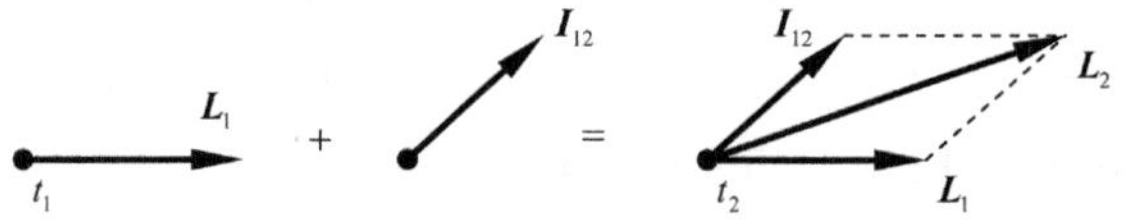

Fig. 9.5

The principle of impulse and momentum expressed by Eq. (9.20), (9.21), or Eq. (9.22) is a vector equation. When used to solve problems, Eq. (9.20), (9.21), or (9.22) is often resolved into rectangular coordinate equations, i.e.,

$$\begin{aligned}
(\mathrm{d}L)_x &= (\mathrm{d}I)_x \quad (L_2)_x - (L_1)_x = (I_{12})_x \quad\quad (L_1)_x + (I_{12})_x = (L_2)_x \\
(\mathrm{d}L)_y &= (\mathrm{d}I)_y, \ (L_2)_y - (L_1)_y = (I_{12})_y \ \text{or} \ (L_1)_y + (I_{12})_y = (L_2)_y \\
(\mathrm{d}L)_z &= (\mathrm{d}I)_z \quad (L_2)_z - (L_1)_z = (I_{12})_z \quad\quad (L_1)_z + (I_{12})_z = (L_2)_z
\end{aligned} \tag{9.23}$$

Example 9.5

The initial velocity of the block in position A is $v_A = 8$ m/s, as shown in Fig. E9.5. Knowing that the coefficient of kinetic friction between the block and the plane of inclination is $\mu_k = 0.25$, determine the time it takes for the block to reach B with zero velocity if $\theta = 10°$.

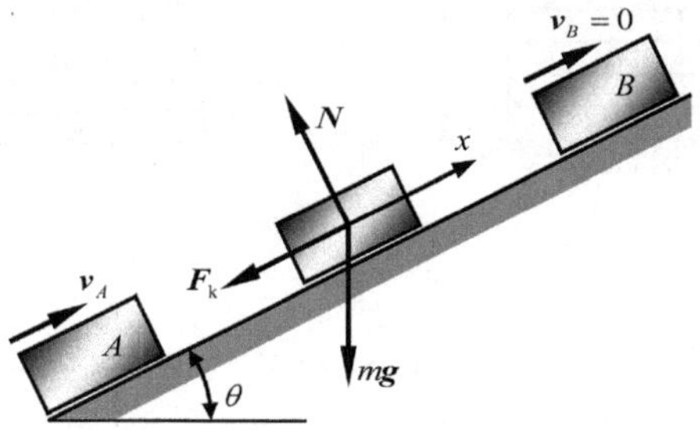

Fig. E9.5

Solution

From the principle of impulse and momentum, $mv_A + I_{AB} = mv_B$, we have

$$mv_A + (-F_k - mg\sin\theta)t_B = mv_B$$

where $F_k = \mu_k N = \mu_k mg\cos\theta$.

Using $v_A = 8$ m/s, $v_B = 0$, $\mu_k = 0.25$, $\theta = 10°$, and $g = 9.81$ m/s^2, we obtain

$$t_B = \frac{v_A - v_B}{(\sin\theta + \mu_k\cos\theta)g} = 1.94 \text{ s}$$

2. Conservation of Momentum

If the resultant of forces acting on a particle is zero, Eq. (9.21) or (9.22) reduces to

$$L_1 = L_2 \tag{9.24}$$

which expresses that the momentum of the particle is conserved.

Problems

9.1 To transport a block of mass m from the ground A to the roof B, a contractor uses a motor-driven lift consisting of a horizontal platform which can slide along a guide attached to a ladder, as shown in Fig. P9.1. The lift starts from rest at A and initially moves with a constant acceleration a_1. The lift then decelerates at a constant rate a_2 and comes to rest at B, near the top of the ladder. Knowing that the coefficient of static friction between the block and the horizontal platform is $\mu_s = 0.3$ and the inclination of the ladder is $\theta = 60°$, determine the largest allowable acceleration a_1 and the largest allowable deceleration a_2 if the block is not to slide on the platform.

9.2 To unload a stone block from a truck, the driver first tilts the bed of the truck and then accelerates from rest, as shown in Fig. P9.2. Knowing that the coefficients of friction between the block and the bed are $\mu_s = 0.4$ and $\mu_k = 0.3$, determine the smallest acceleration of the truck which will cause the block to slide.

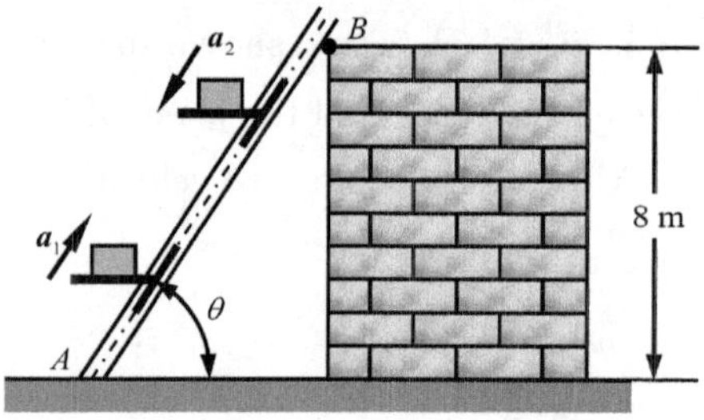

Fig. P9.1

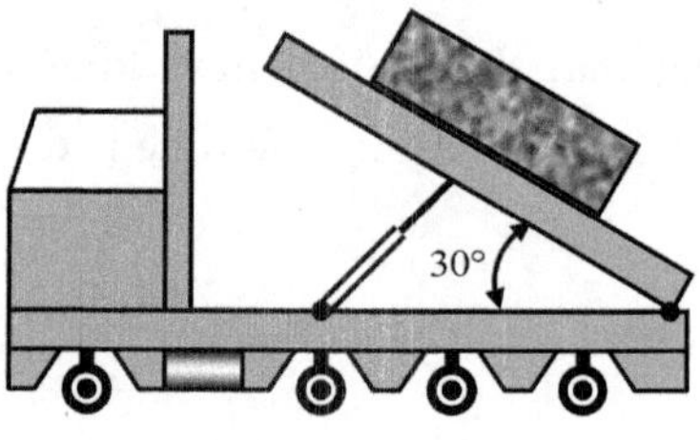

Fig. P9.2

9.3 The block B of weight $W_B = 100$ N is supported by the block A of weight $W_A = 300$ N and is attached to a cord to which a horizontal force of magnitude $F = 200$ N is applied, as shown in Fig. P9.3. Neglecting friction, determine (a) the acceleration of block A, (b) the acceleration of block B relative to A.

9.4 A collar of weight 30 N can slide without friction on a vertical rod and is held so it just touches an undeformed spring, as shown in Fig. P9.4. Knowing that the stiffness of spring is $k = 2$ kN/m, determine the maximum deflection of the spring (a) if the collar is slowly released until it reaches an equilibrium position, (b) if the collar is suddenly released.

9.5 A collar of mass 2 kg is attached to a spring and slides without friction along a circular rod in a vertical plane, as shown in Fig. P9.5. The spring has an undeformed length of 0.1 m and a stiffness k. The collar is at rest at point A and is given a slight push to get it moving to the right. Knowing that the maximum velocity of the collar is achieved as it passes through point B, determine (a) the spring stiffness k, (b) the maximum velocity of the collar.

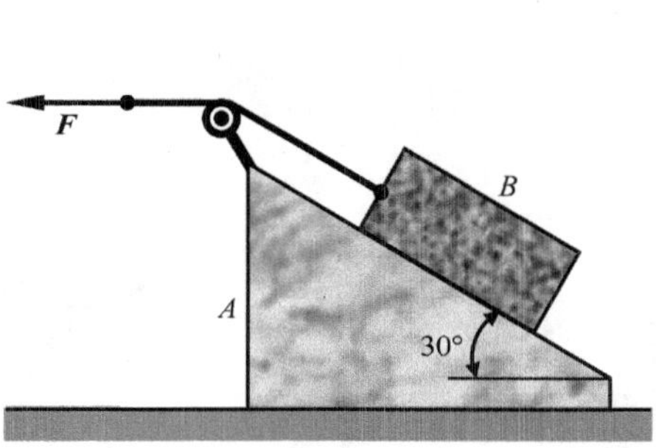

Fig. P9.3

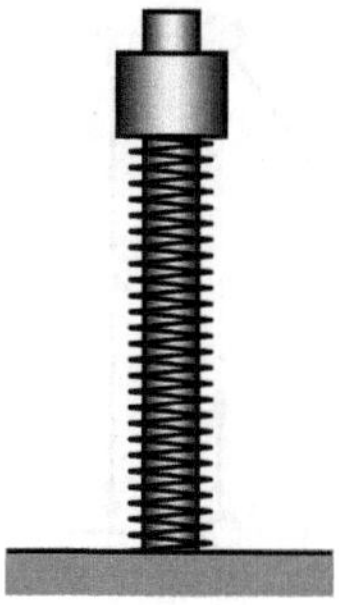

Fig. P9.4

9.6 The initial velocity of the block in position A shown in Fig. P9.6 is $v_A = 2$ m/s. Knowing that the coefficient of kinetic friction between the block and the horizontal plane is $\mu_k = 0.2$, determine the distance it moves and the time it takes for the block to reach B with zero velocity.

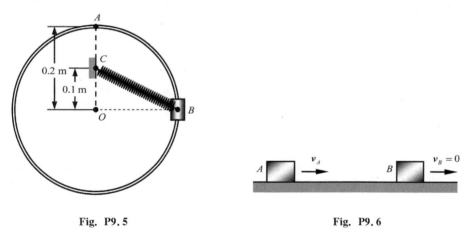

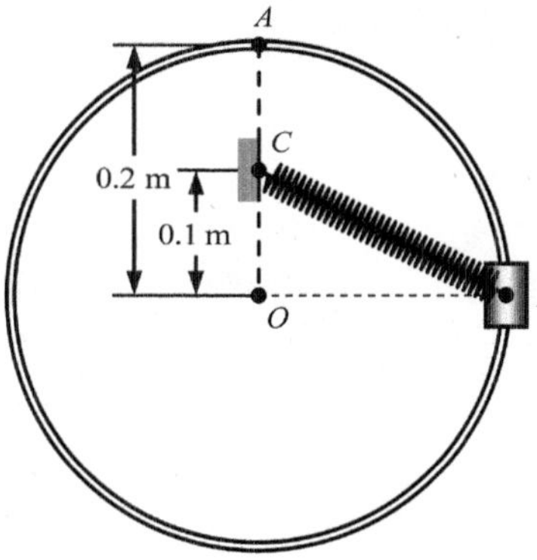

Fig. P9.5

Fig. P9.6

9.7 A 2 kg sphere is connected to a fixed point O by an inextensible cord of length 0.5 m, as shown in Fig. P9.7. The sphere is resting at A on a frictionless horizontal surface at a distance of 0.3 m from O when it is given a velocity $\boldsymbol{v}_0$ in a direction perpendicular to line OA. It moves freely until it reaches position B, when the cord becomes taut. Determine the maximum allowable velocity if the impulse of the force exerted on the cord is not to exceed $I = 5$ N · s.

9.8 The triple jump shown in Fig. P9.8 is a track-and-field event. Assuming that an athlete of weight 800 N approaches the takeoff line from the left with a horizontal velocity of 8 m/s, remains in contact with the ground for 0.2 s, and takes off at a 60° angle with a velocity of 10 m/s, determine the average impulsive force exerted by the ground on his foot.

9.9 A block of mass $m = 10$ kg slides from rest down a plane inclined $\theta = 30°$ with the horizontal, as shown in Fig. P9. 9. Knowing that the coefficient of kinetic friction between the block and the plane is $\mu_k = 0.3$, determine the velocity of the block at the end of $t = 5$ s.

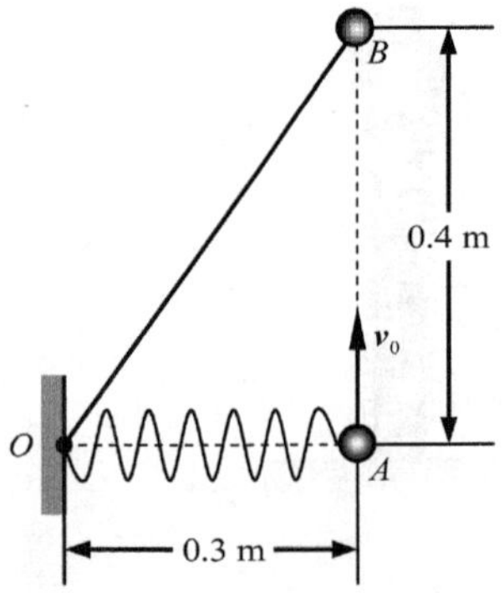

Fig. P9. 7

Fig. P9. 8

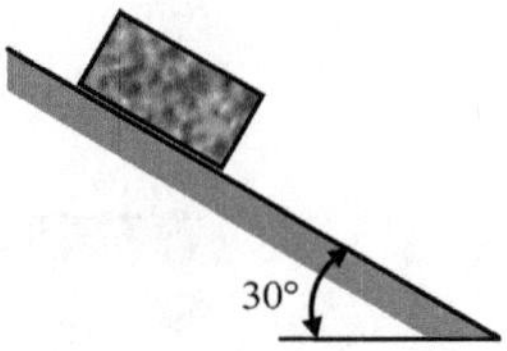

Fig. P9. 9

Chapter 10 Kinetics of Rigid Body in Plane Motion

10.1 Motion for System of Particles

In order to derive the equations of motion for a system of n particles, we apply Newton's second law to each individual particle of the system. Consider the particle P_i of mass m_i acted upon by the resultant $\boldsymbol{F}_i^{(e)}$ of the external forces acting on the particle P_i and by the resultant $\boldsymbol{F}_i^{(i)}$ of the internal forces exerted on the particle P_i by all the other particles $\sum_{j=1,j\neq i}^{n} P_j$ of the system. Newton's second law for the particle P_i can be expressed as

$$m_i \boldsymbol{a}_i = \boldsymbol{F}_i^{(e)} + \boldsymbol{F}_i^{(i)} \quad (i = 1, 2, \cdots, n) \tag{10.1}$$

where $\boldsymbol{a}_i$ is the acceleration of the particle P_i with respect to a Newtonian frame of reference. Taking the moments about a fixed point O of all the terms in Eq. (10.1), we have

$$\boldsymbol{r}_i \times m_i \boldsymbol{a}_i = \boldsymbol{r}_i \times \boldsymbol{F}_i^{(e)} + \boldsymbol{r}_i \times \boldsymbol{F}_i^{(i)} \quad (i = 1, 2, \cdots, n) \tag{10.2}$$

where $\boldsymbol{r}_i$ is the position vector of the particle P_i with respect to point O.

Considering all the particles in the system considered, then from Eqs. (10.1) and (10.2) we can obtain

$$\sum_{i=1}^{n} m_i \boldsymbol{a}_i = \sum_{i=1}^{n} \boldsymbol{F}_i^{(e)} + \sum_{i=1}^{n} \boldsymbol{F}_i^{(i)} \tag{10.3}$$

$$\sum_{i=1}^{n} (\boldsymbol{r}_i \times m_i \boldsymbol{a}_i) = \sum_{i=1}^{n} (\boldsymbol{r}_i \times \boldsymbol{F}_i^{(e)}) + \sum_{i=1}^{n} (\boldsymbol{r}_i \times \boldsymbol{F}_i^{(i)}) \tag{10.4}$$

Using $\sum_{i=1}^{n} \boldsymbol{F}_i^{(i)} = 0$ and $\sum_{i=1}^{n} (\boldsymbol{r}_i \times \boldsymbol{F}_i^{(i)}) = 0$, Eqs. (10.3) and (10.4) can be simplified as

$$\sum_{i=1}^{n} m_i \boldsymbol{a}_i = \sum_{i=1}^{n} \boldsymbol{F}_i^{(e)} \tag{10.5}$$

$$\sum_{i=1}^{n} (\boldsymbol{r}_i \times m_i \boldsymbol{a}_i) = \sum_{i=1}^{n} (\boldsymbol{r}_i \times \boldsymbol{F}_i^{(e)}) \tag{10.6}$$

The linear momentum of a system of particles is defined as the sum of the linear momenta of the various particles of the system, i.e.,

$$\boldsymbol{L} = \sum_{i=1}^{n} m_i \boldsymbol{v}_i \tag{10.7}$$

Differentiating Eq. (10.7) with respect to time t, we have

$$\dot{\boldsymbol{L}} = \sum_{i=1}^{n} m_i \boldsymbol{a}_i \tag{10.8}$$

The angular momentum about a fixed point O of a system of particles is defined as the sum of the angular momenta about the same point O of the various particles of the system, i.e.,

$$\boldsymbol{H}_O = \sum_{i=1}^{n} (\boldsymbol{r}_i \times m_i \boldsymbol{v}_i) \tag{10.9}$$

where $\boldsymbol{r}_i \times m_i \boldsymbol{v}_i$ is the angular momentum about the fixed point O of the particle P_i.

Differentiating Eq. (10.9) with respect to time t, we have

$$\dot{\boldsymbol{H}}_O = \sum_{i=1}^{n} (\boldsymbol{v}_i \times m_i \boldsymbol{v}_i) + \sum_{i=1}^{n} (\boldsymbol{r}_i \times m_i \boldsymbol{a}_i) \tag{10.10}$$

Using $\boldsymbol{v}_i \times \boldsymbol{v}_i = 0$, Eq. (10.10) can be simplified as

$$\dot{\boldsymbol{H}}_O = \sum_{i=1}^{n} (\boldsymbol{r}_i \times m_i \boldsymbol{a}_i) \tag{10.11}$$

Substituting Eqs. (10.8) and (10.11) into Eqs. (10.5) and (10.6), we obtain

$$\dot{\boldsymbol{L}} = \sum_{i=1}^{n} \boldsymbol{F}_i^{(e)} \tag{10.12}$$

$$\dot{\boldsymbol{H}}_O = \sum_{i=1}^{n} (\boldsymbol{r}_i \times \boldsymbol{F}_i^{(e)}) = \sum_{i=1}^{n} \boldsymbol{M}_O(\boldsymbol{F}_i^{(e)}) \tag{10.13}$$

We thus conclude that the rates of change of the linear momentum and of the angular momentum about a fixed point O of the system of particles are respectively equal to the resultant force and the resultant moment about O of the external forces acting on the particles of the system.

Example 10.1

A system consists of three particles, A, B, and C, as shown in Fig. E10.1. Knowing that $m_A = m_B = m_C = 1$ kg and that the velocities of the particles are $\boldsymbol{v}_A = 2\boldsymbol{i} + 3\boldsymbol{j}$ m/s, $\boldsymbol{v}_B = -2\boldsymbol{i} + 3\boldsymbol{j}$ m/s, and $\boldsymbol{v}_C = -3\boldsymbol{j} - 2\boldsymbol{k}$ m/s respectively, determine the linear momentum $\boldsymbol{L}$ of the system and the angular momentum $\boldsymbol{H}_O$ of the system about O.

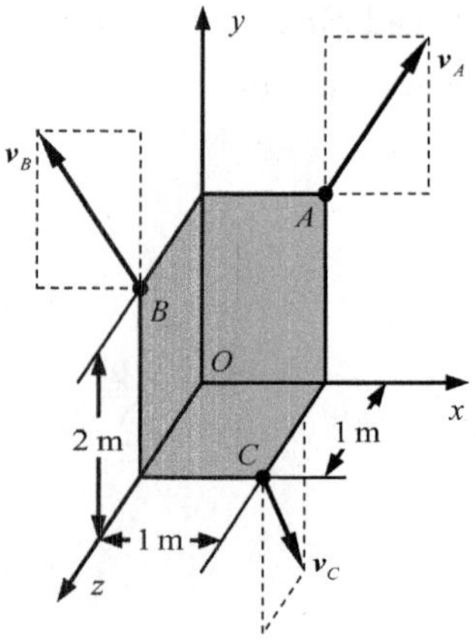

Fig. E10.1

Solution

The position vectors of particles A, B, and C can be expressed as

$$\boldsymbol{r}_A = \boldsymbol{i} + 2\boldsymbol{j} \text{ m}, \quad \boldsymbol{r}_B = 2\boldsymbol{j} + \boldsymbol{k} \text{ m}, \quad \boldsymbol{r}_C = \boldsymbol{i} + \boldsymbol{k} \text{ m}$$

Thus the linear momentum $\boldsymbol{L}$ of the system and the angular momentum $\boldsymbol{H}_O$ of the system about O can be given by

$$\begin{aligned} \boldsymbol{L} &= m_A \boldsymbol{v}_A + m_B \boldsymbol{v}_B + m_C \boldsymbol{v}_C \\ &= (2\boldsymbol{i} + 3\boldsymbol{j}) + (-2\boldsymbol{i} + 3\boldsymbol{j}) + (-3\boldsymbol{j} - 2\boldsymbol{k}) \\ &= 3\boldsymbol{j} - 2\boldsymbol{k} \text{ kg} \cdot \text{m/s} \end{aligned}$$

$$\boldsymbol{H}_O = \boldsymbol{r}_A \times m_A \boldsymbol{v}_A + \boldsymbol{r}_B \times m_B \boldsymbol{v}_B + \boldsymbol{r}_C \times m_C \boldsymbol{v}_C$$

$$= \begin{vmatrix} \boldsymbol{i} & \boldsymbol{j} & \boldsymbol{k} \\ 1 & 2 & 0 \\ 2 & 3 & 0 \end{vmatrix} + \begin{vmatrix} \boldsymbol{i} & \boldsymbol{j} & \boldsymbol{k} \\ 0 & 2 & 1 \\ -2 & 3 & 0 \end{vmatrix} + \begin{vmatrix} \boldsymbol{i} & \boldsymbol{j} & \boldsymbol{k} \\ 1 & 0 & 1 \\ 0 & -3 & -2 \end{vmatrix}$$

$$= 0$$

10.2 Motion of Mass Center of System of Particles

Assuming that $\boldsymbol{r}_C$ represents the position vector of the mass center of the system of particles, then we have

$$\boldsymbol{r}_C = \frac{\sum_{i=1}^{n} m_i \boldsymbol{r}_i}{m} \tag{10.14}$$

where $m = \sum_{i=1}^{n} m_i$ is the total mass of the system of particles.

Differentiating Eq. (10.14) twice with respect to time t, we can obtain

$$\boldsymbol{a}_C = \frac{\sum_{i=1}^{n} m_i \boldsymbol{a}_i}{m} \tag{10.15}$$

where $\boldsymbol{a}_C$ is the acceleration of the mass center of the system of particles.

Substituting $\dot{\boldsymbol{L}} = \sum_{i=1}^{n} m_i \boldsymbol{a}_i = m \boldsymbol{a}_C$ into Eq. (10.12), we obtain

$$m \boldsymbol{a}_C = \sum_{i=1}^{n} \boldsymbol{F}_i^{(e)} \tag{10.16}$$

Eq. (10.16) describes the motion of the mass center of the system of particles. We therefore conclude that the mass center of a system of particles moves as if the entire mass of the system and all the external forces were concentrated at that point.

10.3 Motion of System of Particles about Its Mass Center

It is often convenient to consider the motion of the particles of the system with respect to a centroidal frame of reference which translates with respect to the Newton's frame of reference. Denoting, respectively, by $\boldsymbol{r}'_i$ and $\boldsymbol{v}'_i$, the position vector and the velocity of the particle P_i relative to the moving frame of reference $O'x'y'z'$ as shown in Fig. 10.1, we define the angular momentum of the system of particles about the mass center as follows:

$$\boldsymbol{H}'_C = \sum_{i=1}^{n} (\boldsymbol{r}'_i \times m_i \boldsymbol{v}'_i) \tag{10.17}$$

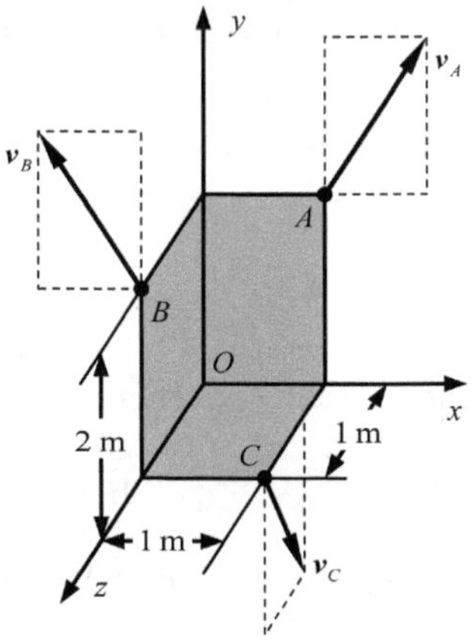

Fig. E10.1

In a similar manner, we can define

$$\boldsymbol{H}_C = \sum_{i=1}^{n}(\boldsymbol{r}'_i \times m_i \boldsymbol{v}_i) \tag{10.18}$$

where $\boldsymbol{v}_i$ is the absolute velocity observed from the Newtonian frame of reference $Oxyz$.

Using $\boldsymbol{v}_i - \boldsymbol{v}'_i = \boldsymbol{v}_C$, we have

$$\boldsymbol{H}_C - \boldsymbol{H}'_C = \sum_{i=1}^{n}(\boldsymbol{r}'_i \times m_i \boldsymbol{v}_i) - \sum_{i=1}^{n}(\boldsymbol{r}'_i \times m_i \boldsymbol{v}'_i) = \left(\sum_{i=1}^{n} m_i \boldsymbol{r}'_i\right) \times \boldsymbol{v}_C \tag{10.19}$$

Substituting $\sum_{i=1}^{n} m_i \boldsymbol{r}'_i = m\boldsymbol{r}'_C = 0$ into Eq. (10.19), we obtain

$$\boldsymbol{H}_C = \boldsymbol{H}'_C \tag{10.20}$$

Using $\boldsymbol{r}_i - \boldsymbol{r}'_i = \boldsymbol{r}_C$, we have

$$\boldsymbol{H}_O - \boldsymbol{H}_C = \sum_{i=1}^{n}(\boldsymbol{r}_i \times m_i \boldsymbol{v}_i) - \sum_{i=1}^{n}(\boldsymbol{r}'_i \times m_i \boldsymbol{v}_i) = \boldsymbol{r}_C \times \left(\sum_{i=1}^{n} m_i \boldsymbol{v}_i\right) \tag{10.21}$$

Substituting $\sum_{i=1}^{n} m_i \boldsymbol{v}_i = m\boldsymbol{v}_C$ into Eq. (10.21), we obtain

$$\boldsymbol{H}_O - \boldsymbol{H}_C = \boldsymbol{r}_C \times m\boldsymbol{v}_C \tag{10.22}$$

Differentiating both sides of Eq. (10.22) and using $\boldsymbol{v}_C \times \boldsymbol{v}_C = 0$, we have

$$\dot{\boldsymbol{H}}_O - \dot{\boldsymbol{H}}_C = \boldsymbol{r}_C \times m\boldsymbol{a}_C \tag{10.23}$$

Using Eqs. (10.13) and (10.16), Eq. (10.23) can be written as

$$\dot{\boldsymbol{H}}_C = \dot{\boldsymbol{H}}_O - \boldsymbol{r}_C \times m\boldsymbol{a}_C = \sum_{i=1}^{n}(\boldsymbol{r}_i \times \boldsymbol{F}_i^{(e)}) - \boldsymbol{r}_C \times \sum_{i=1}^{n} \boldsymbol{F}_i^{(e)} = \sum_{i=1}^{n}[(\boldsymbol{r}_i - \boldsymbol{r}_C) \times \boldsymbol{F}_i^{(e)}] \tag{10.24}$$

Using $\boldsymbol{r}_i - \boldsymbol{r}_C = \boldsymbol{r}'_i$, we have

$$\dot{\boldsymbol{H}}_C = \sum_{i=1}^{n}(\boldsymbol{r}'_i \times \boldsymbol{F}_i^{(e)}) = \sum_{i=1}^{n} \boldsymbol{M}_C(\boldsymbol{F}_i^{(e)}) \tag{10.25}$$

We therefore conclude that the rate of change of the angular momentum about the mass center C of the system of particles is equal to the resultant moment about C of the external forces acting on the particles of the system.

10.4 Equations of Motion for Rigid Body in Plane Motion

Referring to Eq. (10.17) and assuming that a rigid body is made of a large number n of particles P_i of mass m_i, then we obtain

$$\boldsymbol{H}_C = \boldsymbol{H}'_C = \sum_{i=1}^{n}(\boldsymbol{r}'_i \times m_i \boldsymbol{v}'_i) \tag{10.26}$$

Using $\boldsymbol{v}'_i = \boldsymbol{\omega} \times \boldsymbol{r}'_i$, where $\boldsymbol{\omega}$ is the angular velocity of the rigid body, we have

$$\boldsymbol{H}_C = \sum_{i=1}^{n}[\boldsymbol{r}'_i \times m_i(\boldsymbol{\omega} \times \boldsymbol{r}'_i)] = \left\{\sum_{i=1}^{n}[(r'_i)^2 m_i]\right\}\boldsymbol{\omega} \tag{10.27}$$

When $n \to \infty$, i.e., $m_i \to 0$, then Eq. (10.27) can be rewritten as

$$\boldsymbol{H}_C = \left(\int r'^2 \mathrm{d}m\right)\boldsymbol{\omega} \tag{10.28}$$

Defining $I_C = \int r'^2 \mathrm{d}m$, which is called the mass moment of inertia about the mass center of the rigid body, then we obtain

$$\boldsymbol{H}_C = I_C\boldsymbol{\omega} \tag{10.29}$$

Differentiating with respect to time t, we obtain

$$\dot{\boldsymbol{H}}_C = I_C\boldsymbol{\alpha} \tag{10.30}$$

Referring to Eqs. (10.16) and (10.25), and assuming that the rigid body with a plane of symmetry is subjected to a plane motion in the plane of symmetry under the action of coplanar forces $\boldsymbol{F}_1, \boldsymbol{F}_2, \boldsymbol{F}_3, \cdots$, contained in the plane of symmetry, we thus obtain the equations of motion:

$$m\boldsymbol{a}_C = \sum \boldsymbol{F},\ I_C\alpha = \sum M_C(\boldsymbol{F}) \tag{10.31}$$

Resolving the above equations into rectangular coordinate components, then we have

$$m(a_C)_x = \sum F_x,\ m(a_C)_y = \sum F_y,\ I_C\alpha = \sum M_C(\boldsymbol{F}) \tag{10.32}$$

Example 10.2

A 40 kg uniform thin panel is placed in a truck with end A supported by a smooth vertical surface and end B resting on a rough horizontal surface, as shown in Fig. E10.2(a). Knowing that the deceleration of the truck is 2 m/s^2 and that $l=2$ m and $\theta=60°$, determine (a) the reactions at ends A and B, (b) the minimum required coefficient of static friction at end B.

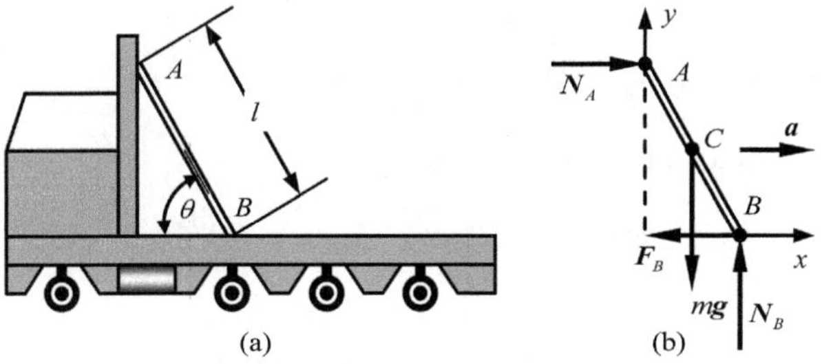

Fig. E10.2

Solution

Taking AB as a free body and drawing its free-body diagram, as shown in Fig. E10.2(b), then we have

$$m(a_C)_x = \sum F_x,\ ma = N_A - F_B$$

$$m(a_C)_y = \sum F_y,\ 0 = N_B - mg$$

$$I_C\alpha = \sum M_C(\boldsymbol{F}),\ 0 = -N_A\left(\frac{1}{2}l\sin\theta\right) + N_B\left(\frac{1}{2}l\cos\theta\right) - F_B\left(\frac{1}{2}l\sin\theta\right)$$

Solving the above equations and using $g=9.81$ m/s^2, we can obtain

$$N_A = 153.3\ \text{N},\ N_B = 392.4\ \text{N},\ F_B = 73.27\ \text{N}$$

Using $F_B \leqslant \mu_s N_B$, we have

$$\mu_s \geqslant 0.1867$$

Example 10.3

A uniform rod AB of length l and mass m is supported as shown in Fig. E10.3(a). If the cable attached at B suddenly breaks, determine (a) the angular acceleration of rod AB, (b) the reaction at the pin support A.

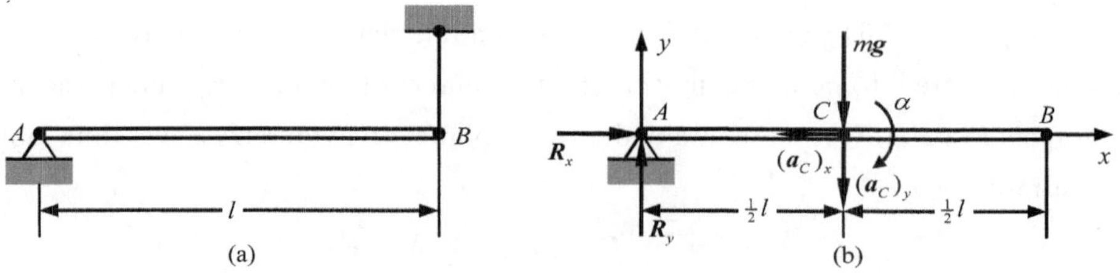

Fig. E10.3

Solution

Taking AB as a free body and drawing its free-body diagram, as shown in Fig. E10.3(b), then we have

$$m(a_C)_x = \sum F_x, \quad -m(a_C)_x = R_x$$

$$m(a_C)_y = \sum F_y, \quad -m(a_C)_y = R_y - mg$$

$$I_C\alpha = \sum M_C(\boldsymbol{F}), \ -I_C\alpha = -R_y\left(\frac{1}{2}l\right)$$

where $(a_C)_x = AC \cdot \omega^2 = 0$, $(a_C)_y = AC \cdot \alpha = \frac{1}{2}l\alpha$, and $I_C = \frac{1}{12}ml^2$.

Solving the above equations, we can obtain

$$R_x = 0, \ R_y = \frac{1}{4}mg, \ \alpha = \frac{3g}{2l}$$

10.5 Method of Inertia Force for Rigid Body in Plane Motion

The method of inertia force can be used to analyze the motion of a particle. Similarly, this method can also be applied to a rigid body in motion. In the case of rigid body in motion, not only an inertia force equal and opposite to $m\boldsymbol{a}_C$ must be applied to the body at the mass center but also a inertia couple equal and opposite to $I_C\boldsymbol{\alpha}$ must be applied to the body. Hence, the equations of motion for a rigid body can be expressed as

$$\sum \boldsymbol{F} + \boldsymbol{F}_I = 0, \ \sum \boldsymbol{M}_C(\boldsymbol{F}) + \boldsymbol{M}_{IC} = 0 \qquad (10.33)$$

where $\boldsymbol{F}_I = -m\,\boldsymbol{a}_C$ and $\boldsymbol{M}_{IC} = -I_C\boldsymbol{\alpha}$ are respectively called the inertia force and the inertia couple.

Although the rigid body considered is not in equilibrium, the equations of equilibrium can be used to analyze the motion of the rigid body if an inertia force and an inertia couple are applied to the rigid body in motion in the method of inertia force based on d'Alembert's principle.

For a rigid body with a plane of symmetry subjected to a plane motion in the plane of symmetry under the action of coplanar forces $\boldsymbol{F}_1$, $\boldsymbol{F}_2$, $\boldsymbol{F}_3$, $\cdots$, contained in the plane of symmetry, we thus obtain the equations of motion for the rigid body:

$$\sum \boldsymbol{F} + \boldsymbol{F}_{\mathrm{I}} = 0, \quad \sum M_C(\boldsymbol{F}) + M_{\mathrm{I}C} = 0 \tag{10.34}$$

Resolving the above equations into rectangular coordinate components, then we have

$$\sum F_x + (F_{\mathrm{I}})_x = 0, \quad \sum F_y + (F_{\mathrm{I}})_y = 0, \quad \sum M_C(\boldsymbol{F}) + M_{\mathrm{I}C} = 0 \tag{10.35}$$

where $(F_{\mathrm{I}})_x = m(a_C)_x$, $(F_{\mathrm{I}})_y = m(a_C)_y$, and $M_{\mathrm{I}C} = I_C\alpha$.

The advantage of the inertia-force method is that it converts the dynamics problem of a rigid body into an equivalent problem in equilibrium and that it allows moments to be conveniently taken about any axis and not only a centroidal axis. Therefore, the above equations can also be expressed as

$$\sum F_x + (F_{\mathrm{I}})_x = 0, \quad \sum F_y + (F_{\mathrm{I}})_y = 0, \quad \sum M_A(\boldsymbol{F}) + M_A(\boldsymbol{F}_{\mathrm{I}}) + M_{\mathrm{I}C} = 0 \tag{10.36}$$

where A is an arbitrary point of the rigid body in plane motion.

Two alternative forms of the above equations can be given respectively by

$$\sum F_x + (F_{\mathrm{I}})_x = 0, \quad \sum M_A(\boldsymbol{F}) + M_A(\boldsymbol{F}_{\mathrm{I}}) + M_{\mathrm{I}C} = 0, \quad \sum M_B(\boldsymbol{F}) + M_B(\boldsymbol{F}_{\mathrm{I}}) + M_{\mathrm{I}C} = 0 \tag{10.37}$$

where the line connecting points A and B are not perpendicular to the x axis, and

$$\begin{aligned} &\sum M_A(\boldsymbol{F}) + M_A(\boldsymbol{F}_{\mathrm{I}}) + M_{\mathrm{I}C} = 0 \\ &\sum M_B(\boldsymbol{F}) + M_B(F_{\mathrm{I}}) + M_{\mathrm{I}C} = 0 \\ &\sum M_C(\boldsymbol{F}) + M_C(\boldsymbol{F}_{\mathrm{I}}) + M_{\mathrm{I}C} = 0 \end{aligned} \tag{10.38}$$

where the points A, B, and C do not lie in a straight line.

Example 10.4

A uniform rod AB of length l and mass m is supported as shown in Fig. E10.4(a). If the cable attached at B suddenly breaks, determine (a) the angular acceleration of rod AB, (b) the reaction at the pin support A.

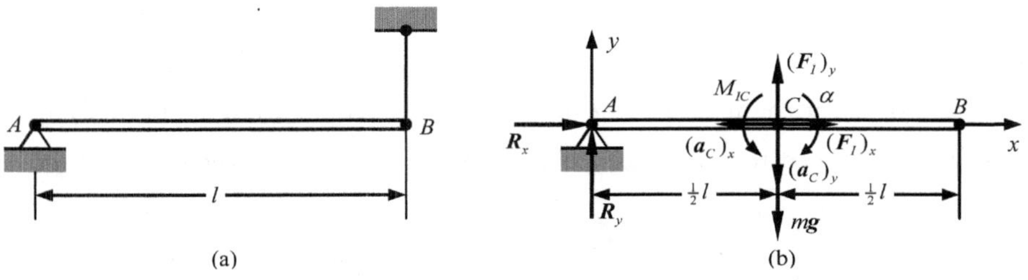

Fig. E10.4

Solution

Taking AB as a free body and drawing its free-body diagram, as shown in Fig. E10.4(b), then we have

$$\sum F_x + (F_I)_x = 0, \ R_x + m(a_C)_x = 0$$

$$\sum F_y + (F_I)_y = 0, \ R_y - mg + m(a_C)_y = 0$$

$$\sum M_A(\boldsymbol{F}) + M_A(\boldsymbol{F}_I) + M_{IC} = 0, \ -mg\left(\frac{1}{2}l\right) + m(a_C)_y\left(\frac{1}{2}l\right) + I_C\alpha = 0$$

Using $(a_C)_x = AC \cdot \omega^2 = 0$, $(a_C)_y = AC \cdot \alpha = \frac{1}{2}l\alpha$, and $I_C = \frac{1}{12}ml^2$, and solving the above equations, we can obtain

$$R_x = 0, \ R_y = \frac{1}{4}mg, \ \alpha = \frac{3g}{2l}$$

10.6 Method of Work and Energy for Rigid Body in Plane Motion

1. Works of Force and Couple Acting on Rigid Body

The work done by an external force $\boldsymbol{F}$ acting on a rigid body during a displacement of the point of application from A_1 to A_2 can be expressed as

$$W_{12} = \int_{A_1}^{A_2} \boldsymbol{F} \cdot \mathrm{d}\boldsymbol{r} \tag{10.39}$$

Similarly, the work done by an external couple of moment $\boldsymbol{M}$ acting on a rigid body during an angular displacement of the couple from θ_1 to θ_2 can be expressed as

$$W_{12} = \int_{\theta_1}^{\theta_2} M \mathrm{d}\theta \tag{10.40}$$

2. Kinetic Energy of Rigid Body in Plane Motion

The kinetic energy of a rigid body is defined as the sum of the kinetic energies of the various particles forming the body. We thus have

$$T = \frac{1}{2}\sum_{i=1}^{n} m_i v_i^2 \tag{10.41}$$

It is often convenient when computing the kinetic energy of a rigid body to consider separately the motion of the mass center of the body and the motion of the body relative to a translating frame attached to the mass center. Using $\boldsymbol{v}_i = \boldsymbol{v}_C + \boldsymbol{v}'_i$, then Eq. (10.41) can be expressed as

$$T = \frac{1}{2}\sum_{i=1}^{n} m_i(\boldsymbol{v}_i \cdot \boldsymbol{v}_i) = \frac{1}{2}\sum_{i=1}^{n} m_i v_C^2 + \boldsymbol{v}_C \cdot \left(\sum_{i=1}^{n} m_i \boldsymbol{v}'_i\right) + \frac{1}{2}\sum_{i=1}^{n} m_i v'^2_i \tag{10.42}$$

Using $\sum_{i=1}^{n} m_i \boldsymbol{v}'_i = m\boldsymbol{v}'_C = 0$, the we have

$$T = \frac{1}{2}mv_C^2 + \frac{1}{2}\sum_{i=1}^{n} m_i v'^2_i \tag{10.43}$$

Using $v'_i = r'_i\omega$, we obtain

$$T = \frac{1}{2}mv_{C^2} + \frac{1}{2}\left\{\sum_{i=1}^{n}[m_i (r'_i)^2]\right\}\omega^2 = \frac{1}{2}mv_{C^2} + \frac{1}{2}I_C\omega^2 \tag{10.44}$$

where $I_C = \sum_{i=1}^{n}[m_i (r'_i)^2]$ is the mass moment of inertia of the body.

This equation above shows that the kinetic energy of a rigid body in plane motion is equal to the sum of the translational kinetic energy of the body with the mass center and the rotational kinetic energy of the body relative to the translating frame attached to the mass center.

3. Principle of Work and Energy for Rigid Body in Plane Motion

The principle of work and energy for a particle can be applied to each of the particles forming the rigid body. Adding the kinetic energies of the various particles of the body and considering the work done by all the forces acting on the body, then we can obtain the principle of work and energy for a rigid body in plane motion:

$$\mathrm{d}T = \mathrm{d}W \tag{10.45}$$

where $\mathrm{d}T$ and $\mathrm{d}W$ are the increment in kinetic energy of the body and the elementary work done by all the forces acting on the body. Integrating from A_1 to A_2, we have

$$T_2 - T_1 = W_{12}, \text{ or } T_1 + W_{12} = T_2 \tag{10.46}$$

where T_1 and T_2 are the initial and final kinetic energies of the body, and W_{12} is the work of all the forces acting on the body.

4. Conservation of Energy

If all the forces acting on the rigid body are conservative, the principle of work and energy can be replaced by

$$T_1 + V_1 = T_2 + V_2 \tag{10.47}$$

where V is the potential energy of the body. This equation expresses the principle of conservation of mechanical energy for a rigid body.

Example 10.5

A uniform rod of length l and mass m is pivoted about a point O located at a distance d from its center C, as shown in Fig. E10.5(a). It is released from rest in a horizontal position and swings freely. Determine (a) the distance d for which the angular velocity of the rod as it passes through a vertical position is maximum, (b) the corresponding values of its angular velocity and of the reaction at O.

Solution

At position 1 in Fig. E10.5(a), we have

$$T_1 = V_1 = 0$$

And at position 2 in Fig. E10.5(b), we have

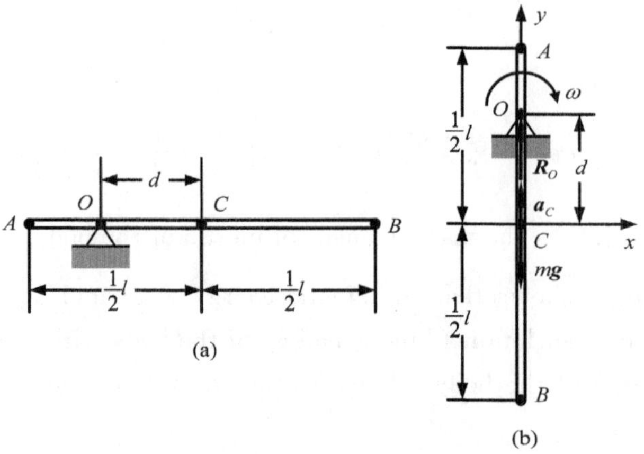

Fig. E10.5

$$T_2 = \frac{1}{2}I_O\omega^2 = \frac{1}{2}(I_C + md^2)\omega^2 = \frac{1}{2}(\frac{1}{12}ml^2 + md^2)\omega^2, \ V_2 = -mgd$$

Using $T_1+V_1=T_2+V_2$, we can obtain

$$\omega = \sqrt{\frac{2g}{d + \frac{1}{12}l^2/d}}$$

It is obvious that when $d=\frac{1}{12}l^2/d$, i.e., $d=\frac{\sqrt{3}}{6}l$, the angular velocity ω will have a maximum value, i.e.,

$$\omega_{\max} = \sqrt{\frac{2g}{2\sqrt{d(\frac{1}{12}l^2/d)}}} = \sqrt[4]{12}\sqrt{\frac{g}{l}} = 1.86\sqrt{\frac{g}{l}}$$

Using $ma_C = \sum F_y = R_O - mg$ and $a_C = d\,(\omega_{\max})^2 = g$ at position 2 in Fig. E10.5(b), we have

$$R_O = 2mg$$

10.7 Method of Impulse and Momentum for Rigid Body in Plane Motion

1. Principle of Impulse and Momentum for Rigid Body in Plane Motion

Considering that a rigid body of mass m in plane motion under the action of coplanar forces $\boldsymbol{F}_1$, $\boldsymbol{F}_2$, $\boldsymbol{F}_3$, …, we have

$$\mathrm{d}\boldsymbol{L} = \mathrm{d}\boldsymbol{I}, \ \mathrm{d}\boldsymbol{H}_C = \mathrm{d}\boldsymbol{G}_C \tag{10.48}$$

where $\mathrm{d}\boldsymbol{L}=\mathrm{d}(m\boldsymbol{v}_C)$ and $\mathrm{d}\boldsymbol{H}_C=\mathrm{d}(I_C\boldsymbol{\omega})$ are respectively the increments in linear momentum and the angular momentum about the mass center C of the body, and $\mathrm{d}\boldsymbol{I} = \sum \boldsymbol{F}\mathrm{d}t$ and $\mathrm{d}\boldsymbol{G}_C = \sum \boldsymbol{M}_C(\boldsymbol{F})\mathrm{d}t$ are the linear impulse and the angular impulse about the mass center C of the forces acting on the body during the interval of time dt. Integrating from t_1 to t_2, we write

$$\boldsymbol{L}_2 - \boldsymbol{L}_1 = \boldsymbol{I}_{12},\ (\boldsymbol{H}_C)_2 - (\boldsymbol{H}_C)_1 = (\boldsymbol{G}_C)_{12},\ \text{or } \boldsymbol{L}_1 + \boldsymbol{I}_{12} = \boldsymbol{L}_2,\ (\boldsymbol{H}_C)_1 + (\boldsymbol{G}_C)_{12} = (\boldsymbol{H}_C)_2 \tag{10.49}$$

where $\boldsymbol{L} = m\boldsymbol{v}_C$ and $\boldsymbol{H}_C = I_C\boldsymbol{\omega}$ are respectively the linear momentum and the angular momentum about the mass center C of the body, and $\boldsymbol{I}_{12} = \sum\int_{t_1}^{t_2}\boldsymbol{F}dt$ and $(\boldsymbol{G}_c)_{12} = \sum\int_{t_1}^{t_2}\boldsymbol{M}_C(\boldsymbol{F})dt$ are the linear impulse and the angular impulse about the mass center C of the forces acting on the body during the interval of time from t_1 to t_2. Resolving Eq. (10.48) or (10.49) into rectangular coordinate components, then we have

$$\begin{aligned}
&(d_L)_x = (dI)_x,\ (L_2)_x - (L_1)_x = (I_{12})_x, && (L_1)_x + (I_{12})_x = (L_2)_x \\
&(d_L)_y = (dI)_y,\ (L_2)_y - (L_1)_y = (I_{12})_y, && \text{or } (L_1)_y + (I_{12})_y = (L_2)_y \\
&dH_C = dG_C,\quad (H_C)_2 - (H_C)_1 = (G_C)_{12}, && (H_C)_1 + (G_C)_{12} = (H_C)_2
\end{aligned} \tag{10.50}$$

Example 10.6

A uniform cylinder of radius r and weight W with an initial anticlockwise angular velocity ω_0 is placed in the corner formed by the floor and a vertical wall, as shown in Fig. E10.6(a). Denoting by μ_k the coefficient of kinetic friction at A and B, derive an expression for the time required for the cylinder to come to rest.

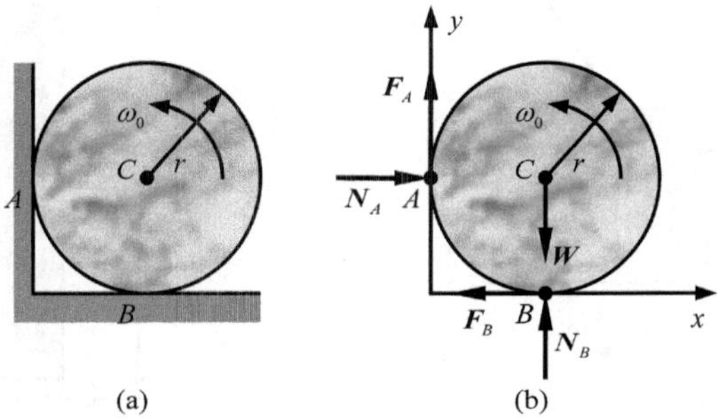

Fig. E10.6

Solution

Taking the cylinder as a free body and drawing its free-body diagram, Fig. E10.6(b), then we have

$$\begin{aligned}
&(L_1)_x + (I_{12})_x = (L_2)_x, && 0 + (N_A - F_B)t = 0 \\
&(L_1)_y + (I_{12})_y = (L_2)_y, && 0 + (N_B + F_A - W)t = 0 \\
&(H_C)_1 + (G_C)_{12} = (H_C)_2, && I_C\omega_0 - (F_A + F_B)rt = 0
\end{aligned}$$

Using $F_A = \mu_k N_A$, $F_B = \mu_k N_B$, and $I_C = \dfrac{1}{2}mr^2$, and solving the equations above, we can obtain

$$t = \frac{(1+\mu_k^2)I_C\omega_0}{\mu_k(1+\mu_k)Wr} = \frac{(1+\mu_k^2)r\omega_0}{2\mu_k(1+\mu_k)g}$$

where g is the acceleration of gravity.

2. Conservation of Momentum

If no external force acts on a rigid body, Eq. (10.49) reduces to

$$\boldsymbol{L}_1 = \boldsymbol{L}_2,\ (\boldsymbol{H}_C)_1 = (\boldsymbol{H}_C)_2 \tag{10.51}$$

which expresses that the linear momentum and the angular momentum about the mass center C of the body are conserved.

Problems

10.1 The system shown in Fig. P10.1 consists of two particles, A and B. Knowing that $m_A = 1$ kg and $m_B = 2$ kg, and that the velocities of the particles are $\boldsymbol{v}_A = 2\boldsymbol{i} + 3\boldsymbol{j}$ m/s and $\boldsymbol{v}_B = -2\boldsymbol{i} + 3\boldsymbol{j}$ m/s respectively, determine the angular momentum $\boldsymbol{H}_O$ of the system about O and the angular momentum $\boldsymbol{H}_C$ of the system about its mass center C.

10.2 A uniform cabinet of mass $m = 20$ kg is mounted on casters that allow it to move freely on the floor, as shown in Fig. P10.2. Knowing $F = 100$ N, $b = 500$ mm, and $h_C = 800$ mm, determine (a) the acceleration of the cabinet, (b) the range of values of h for which the cabinet will not tip.

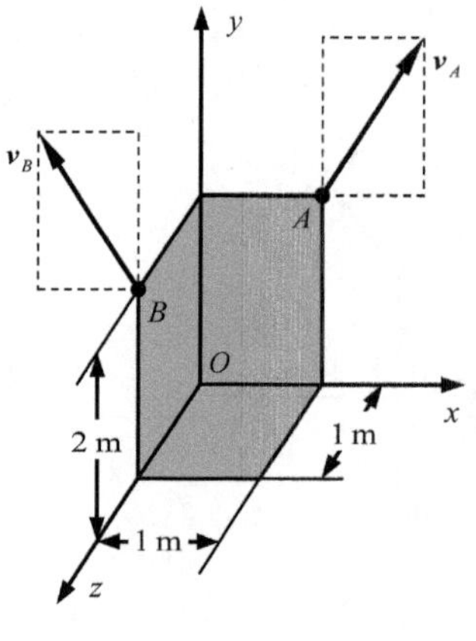

Fig. P10.1

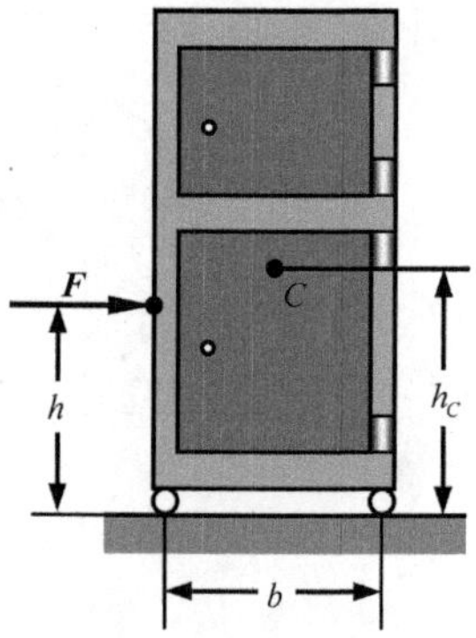

Fig. P10.2

10.3 A uniform slender rod AB of mass 2 kg is held in position by three ropes of the same length, as shown in Fig. P10.3. Determine, immediately after rope BE has been cut, (a) the acceleration of rod AB, (b) the tension in each rope.

10.4 A uniform circular plate of mass 5 kg is attached to three cables of the same length, as shown in Fig. P10.4. Knowing that the lines joining C to A and B are, respectively, horizontal and vertical, determine, immediately after cable BF has been cut, (a) the acceleration of the plate, (b) the tension in each cable.

10.5 A uniform cylinder of mass $m = 8$ kg and radius $r = 0.15$ m rolls without sliding down a plane inclined at an angle of $\theta = 30°$ with the horizontal, as shown in Fig. P10.5. Determine the friction force and the acceleration of the mass center.

10.6 The flywheel shown in Fig. P10.6. has a radius of 500 mm, a mass of 150 kg, and a radius of gyration of 400 mm. A block A of mass 20 kg is attached to a wire that is wrapped around the flywheel, and the system is released from rest. Neglecting the effect of friction, determine (a) the acceleration of block A, (b) the speed of block A after it has moved 2 m.

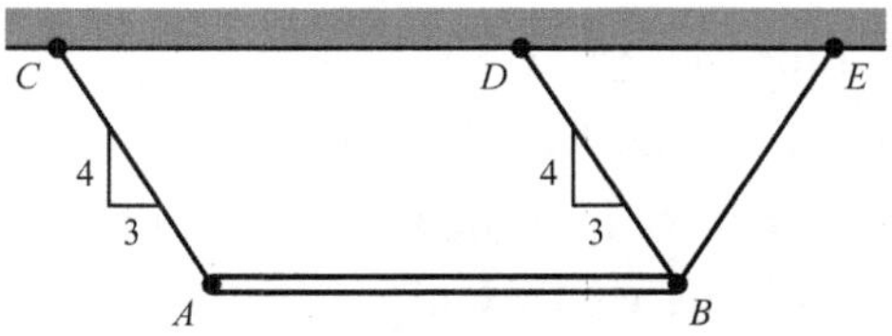

Fig. P10.3

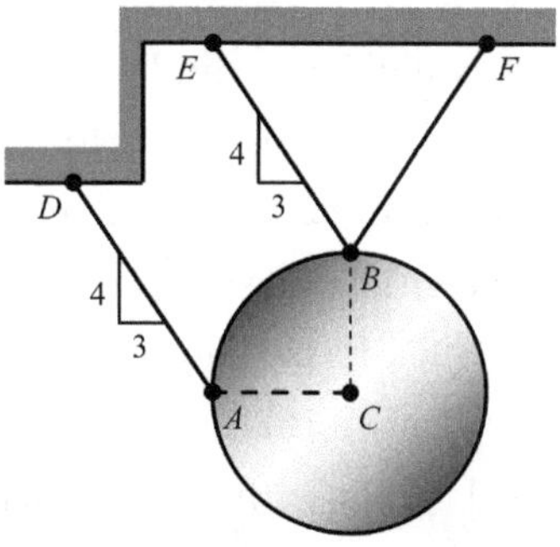

Fig. P10.4

10.7 A drum, as shown in Fig. P10.7, of 0.4 m radius is attached to a disk of radius 0.3 m. The disk and drum have a combined mass of 5 kg and a combined radius of gyration of 0.25 m and are suspended by two cords. Knowing that $T_A = 50$ N and $T_B = 30$ N, determine the accelerations of points A and B on the cords.

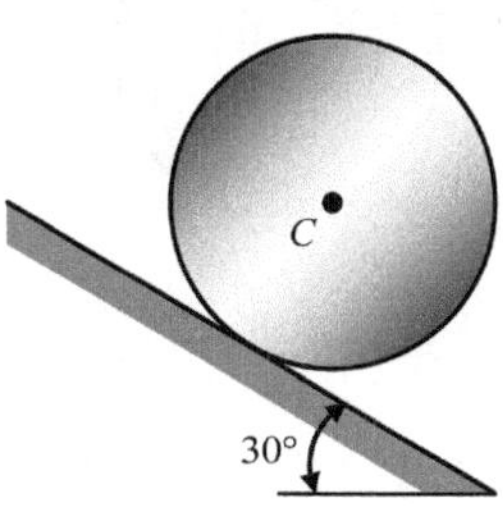

Fig. P10.5

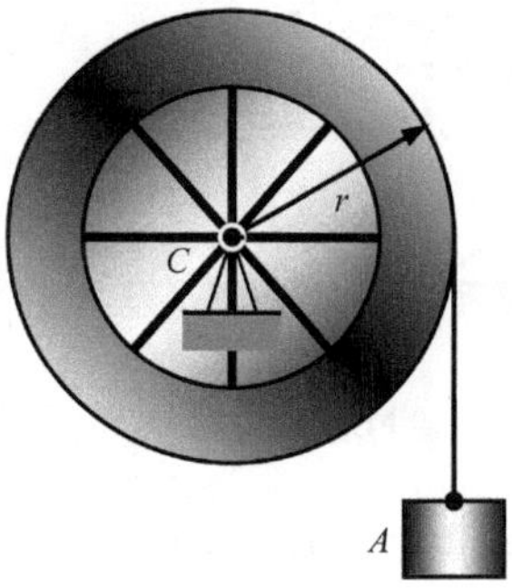

Fig. P10.6

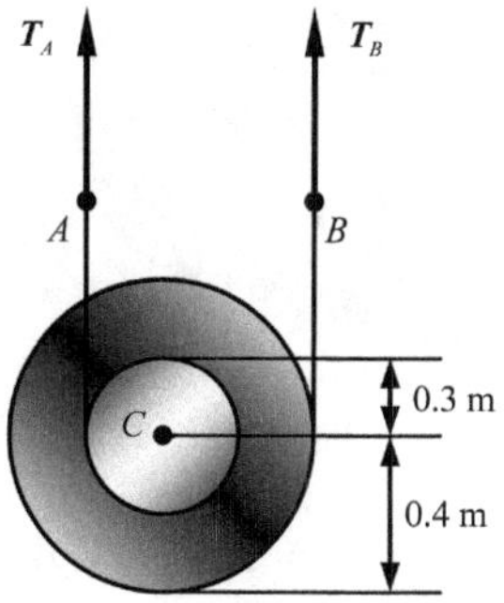

Fig. P10.7

10.8 A uniform rod AB of length l and mass m is supported from two springs, as shown in Fig. P10.8. If the spring attached at B suddenly breaks, determine, at this instant, (a) the angular acceleration of rod AB, (b) the acceleration of point A.

10.9 A uniform rod of length l and mass m is pivoted about a point O located at a distance d from its center C, as shown in Fig. P10.9. It is released from rest in a horizontal position and swings freely. Knowing $d=\frac{3}{8}l$, determine the angular velocity of rod AB and the reaction at point O after the rod has rotated through 90°.

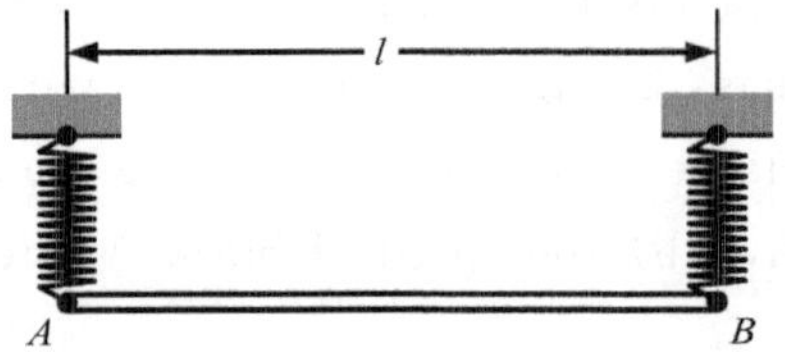

Fig. P10.8

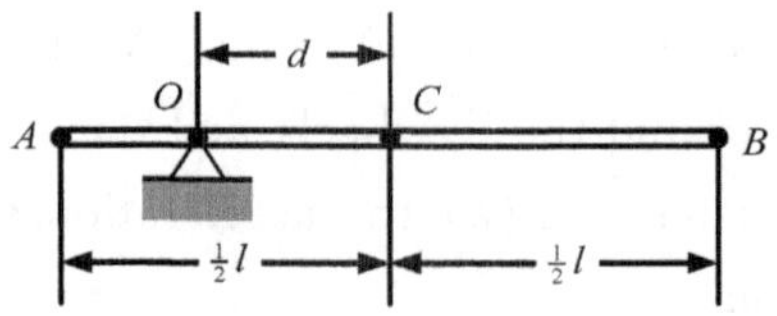

Fig. P10.9

10.10 A uniform cylinder, as shown in Fig. P10.10, of radius r and weight W with an initial anticlockwise angular velocity ω_0 is placed in the corner formed by the rough floor and a smooth wall. Denoting by μ_k the coefficient of kinetic friction between the cylinder and the floor, derive an expression for the time required for the cylinder to come to rest.

10.11 A uniform cylindrical roller, as shown in Fig. P10.11, of mass $m=25$ kg and radius $r=0.2$ m, initially at rest, is acted upon by a force of magnitude $F=120$ N. Knowing that the body rolls without slipping, determine (a) the velocity of its center C after it has moved 1.5 m, (b) the friction force required to prevent slipping.

10.12 A rope is wrapped around a uniform cylinder of radius r and mass m, as shown in Fig. P10.12. Knowing that the cylinder is released from rest, determine the velocity of the center C of the cylinder after it has moved downward a distance s.

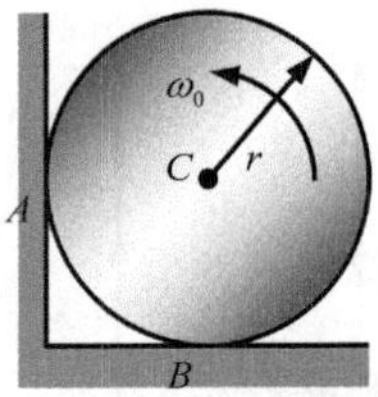

Fig. P10.10

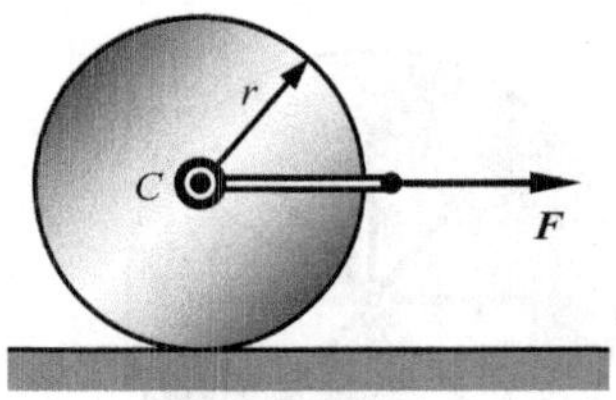

Fig. P10.11

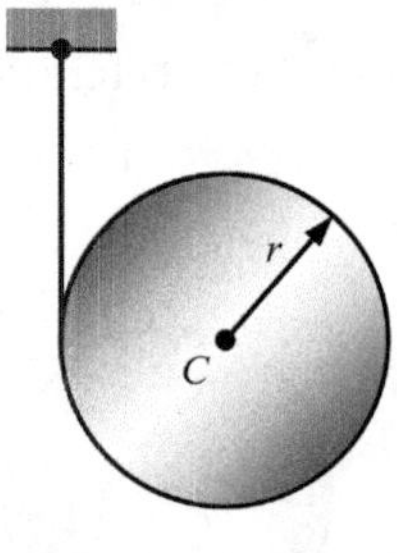

Fig. P10.12

10.13 A 10-kg cradle, subjected to a force of magnitude $F=30$ N, is supported by two uniform disks that roll without sliding at all surfaces of contact, as shown in Fig. P10.13. The mass of each disk is $m=5$ kg and the radius of each disk is $r=0.1$ m. Knowing that the system is initially at rest, determine the velocity of the cradle after it has moved 0.5 m.

10.14 A uniform cylinder, as shown in Fig. P10.14, of radius r and mass m is placed on a horizontal floor with no linear velocity but with an anticlockwise angular velocity ω_0. Denoting by μ_k the coefficient of kinetic friction between the cylinder and the floor, determine (a) the time at which the cylinder will start rolling without sliding, (b) the linear and angular velocities of the cylinder as the cylinder starts rolling without sliding.

10.15 Two blocks, A and B, are connected by a cable AB which is wrapped over the surface of a disk O, as shown in Fig. P10.15. Assume that there is no relative sliding between the cable and the disk and that the disk rotates in frictionless bearings. Knowing that at the instant shown block B is moving down 0.4 m/s and that the spring is compressed

0.2 m, determine the velocity of block B after it has dropped 0.5 m.

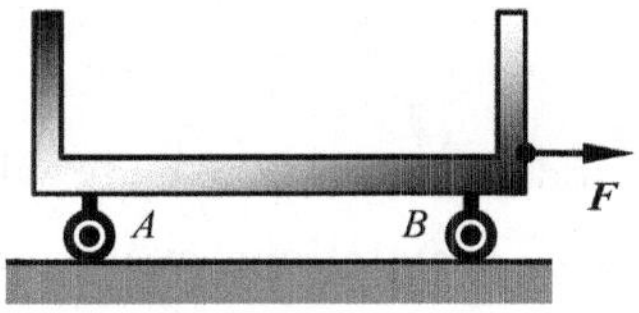

Fig. P10.13

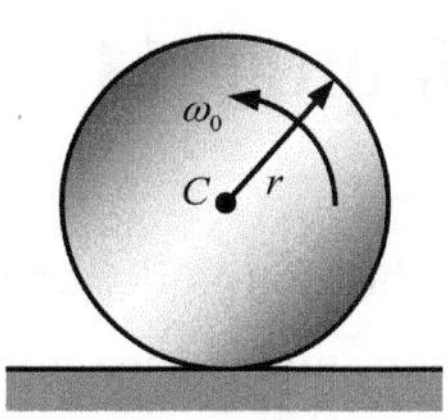

Fig. P10.14

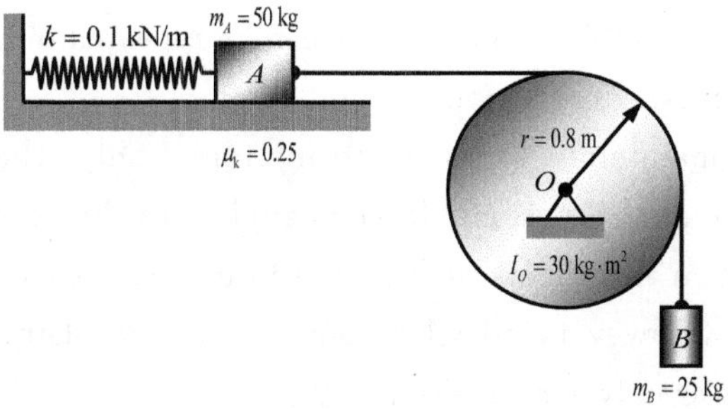

Fig. P10.15

10.16 Two blocks, A and B, are connected by a cable AB which is wrapped over the surface of a disk O, as shown in Fig. P10.16. Assuming that there is no relative sliding between the cable and the disk and that the disk rotates in frictionless bearings, determine the mass m_B of block B necessary to cause block A to change its velocity from 4 to 8 m/s in 6 s.

10.17 A wheel of mass 75 kg has a radius of gyration of 0.9 m with respect to its centroid, as shown in Fig. P10.17. Assume that the fixed pulley is massless and runs in frictionless bearings. Knowing that the wheel is rolling initially 10 rad/s anticlockwise, determine the time it will take until it is rolling 6 rad/s clockwise.

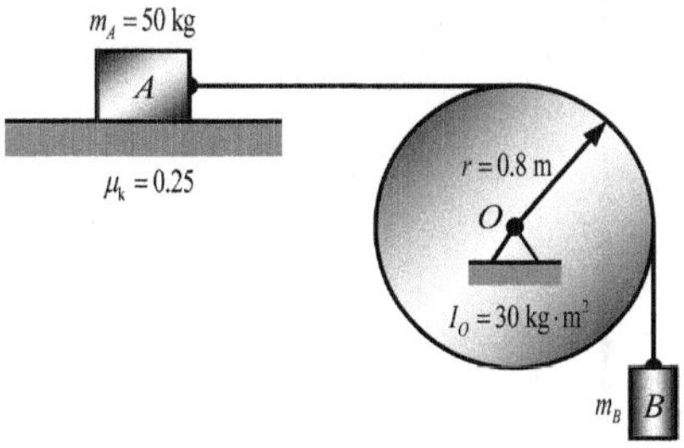

Fig. P10.16

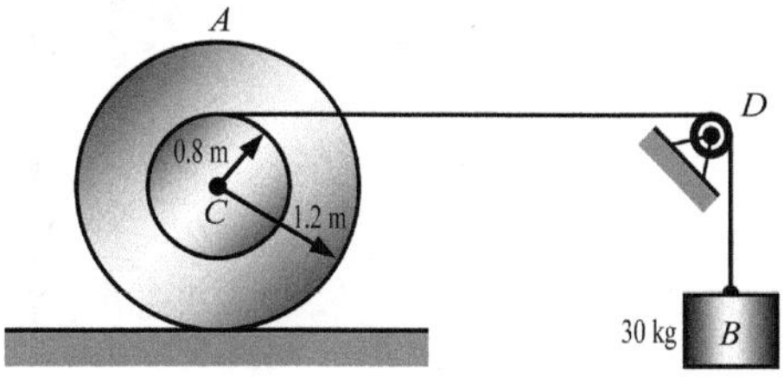

Fig. P10.17

Chapter 11 Mechanical Vibrations

A mechanical vibration is defined as the periodic motion of a body which oscillates at a position of equilibrium under the action of elastic or gravitational restoring forces.

A mechanical vibration will occur when a body is displaced from its position of equilibrium. The body tends to return to this position under the action of restoring forces. However, the body generally reaches its position of equilibrium with a certain velocity which carries it beyond that position. Since the process can be repeated indefinitely, the body keeps moving back and forth across its position of equilibrium.

When the motion is maintained by the restoring forces only, the vibration is defined as a free vibration. When an external periodic force is applied to the body, the vibration is called as a forced vibration. When the effects of friction can be neglected, the vibrations are said to be undamped vibrations. However, all vibrations are actually damped vibrations due to the fact that frictional forces are always present.

11.1 Undamped Free Vibrations

1. Undamped Free Vibrations of Particle

Consider a body of mass m attached to a spring of stiffness k. When the body is in static equilibrium, as shown in Fig. 11.1(a), the forces acting on it are its weight and the force exerted by the spring, of magnitude $k\delta_{st}$, where δ_{st} denotes the elongation of the spring in the position of equilibrium. We have, therefore, $mg=k\delta_{st}$.

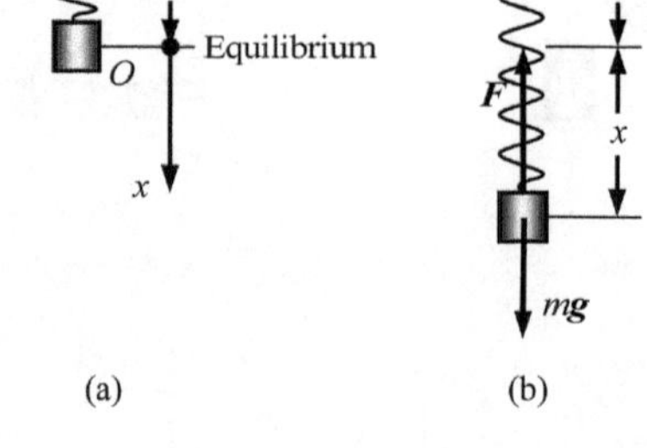

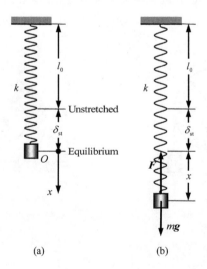

Fig. 11.1

Assuming that the body is displaced through an arbitrary displacement x from the position of equilibrium O, as shown in Fig. 11.1(b), then the forces acting on the body are its weight and the force exerted by the spring which, in this position, has a magnitude $F = k(\delta_{st} + x)$. Using $ma_x = \sum F_x$, we have

$$m\ddot{x} = mg - k(\delta_{st} + x) \tag{11.1}$$

Using $mg = k\delta_{st}$, Eq. (11.1) can be rewritten as

$$m\ddot{x} + kx = 0 \tag{11.2}$$

This is a homogeneous second-order differential equation. The motion defined by Eq. (11.2) is referred to as a simple harmonic motion. The general solution of Eq. (11.2) can be expressed as

$$x = C_1 \cos\omega_n t + C_2 \sin\omega_n t \tag{11.3}$$

where $\omega_n = \sqrt{k/m}$ is called the natural or circular frequency of the vibration, expressed in rad/s, and C_1 and C_2 are two constants which can be determined by using the initial conditions of the motion. The general solution of Eq. (11.2) can also be written, in a more compact form, as

$$x = A\sin(\omega_n t + \alpha) \tag{11.4}$$

where A is called the amplitude of the vibration, i.e., the maximum displacement of the body from its position of equilibrium, α is called the initial phase.

Differentiating Eq. (11.4) with respect to t, the velocity and the acceleration at time t can be expressed, respectively, as

$$\begin{aligned} v &= \dot{x} = \omega_n A\cos(\omega_n t + \alpha) \\ a &= \ddot{x} = -\omega_n^2 A\sin(\omega_n t + \alpha) \end{aligned} \tag{11.5}$$

The maximum values of the velocity and acceleration can be written as

$$\begin{aligned} v_{max} &= \omega_n A \\ a_{max} &= \omega_n^2 A \end{aligned} \tag{11.6}$$

The time interval required for the body to complete a full cycle of motion is called the period of the vibration, denoted by T_n. We note from Eq. (11.4) that a full cycle is described as the phase increases 2π rad. We have, therefore,

$$T_n = \frac{2\pi}{\omega_n} \tag{11.7}$$

The number of cycles per unit of time is denoted by f_n and is known as the frequency of the vibration. We write

$$f_n = \frac{\omega_n}{2\pi} \tag{11.8}$$

The unit of frequency is Hz in SI units.

Example 11.1

The motion of a particle is described by the equation $x = 0.4\sin 2t + 0.3\cos 2t$, where x is expressed in meters and t in seconds. Determine (a) the amplitude, (b) the period, and (c) the initial phase of the motion.

Solution

Using trigonometric identity, i.e., $\sin(A+B)=\sin A\cos B+\cos A\sin B$, the equation $x=0.4\sin 2t+0.3\cos 2t$ can be rewritten as

$$x=0.5\sin[2t+\arctan(3/4)]$$

(a) The amplitude is

$$A=0.5\ \text{m}$$

(b) The period is

$$T_{\text{n}}=\frac{2\pi}{\omega_{\text{n}}}=\frac{2\pi}{2}=3.14\ \text{s}$$

(c) The initial phase is

$$\alpha=\arctan(3/4)=0.75\ \text{rad}=36.9^{\circ}$$

Example 11.2

A 5-kg block is initially held so that the vertical spring attached as shown is undeformed. Knowing that the stiffness of the spring is 3 kN/m and that the block is suddenly released from rest, determine (a) the amplitude and frequency of the motion, (b) the maximum velocity and maximum acceleration of the block.

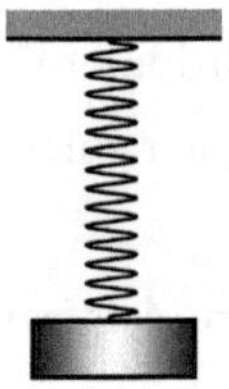

Fig. E11.2

Solution

(a) The amplitude is

$$A=\delta_{\text{st}}=\frac{mg}{k}=16.4\ \text{mm}$$

The frequency is

$$f_{\text{n}}=\frac{\omega_{\text{n}}}{2\pi}=\frac{1}{2\pi}\sqrt{\frac{k}{m}}=3.90\ \text{Hz}$$

(b) The maximum velocity is

$$v_{\max}=\omega_{\text{n}}A=\sqrt{\frac{k}{m}}\frac{mg}{k}=g\sqrt{\frac{m}{k}}=400\ \text{mm/s}$$

And the maximum acceleration is

$$a_{\max}=\omega_{\text{n}}^{2}A=\frac{k}{m}\frac{mg}{k}=g=9.81\ \text{m/s}^{2}$$

Example 11.3

A 2-kg block A rests on top of a 8-kg plate B which is attached to an unstretched spring of stiffness $k=400$ N/m. Plate B is slowly moved 65 mm to the right and released from rest. Assuming that block A does not slip on the plate, determine (a) the amplitude and period of the motion, (b) the corresponding smallest required value of the coefficient of static friction.

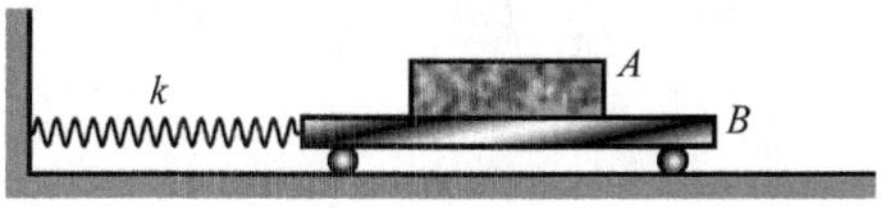

Fig. E11.3

Solution

(a) The amplitude is

$$A = 65\ \text{mm}$$

The period is

$$T_n = \frac{2\pi}{\omega_n} = 2\pi\sqrt{\frac{m_A + m_B}{k}} = 0.993\ \text{s}$$

(b) Using $a_{max} = \omega_n^2 A = \dfrac{k}{m_A + m_B} A = 2.60\ \text{m/s}^2$, we have

$$\mu_s = \frac{F_s}{N_A} = \frac{m_A a_{max}}{m_A g} = \frac{a_{max}}{g} = 0.265$$

Example 11.4

A block of mass m vibrates in vertical direction as shown. Knowing that the stiffnesses of two springs are respectively k_1 and k_2, determine the period of the vibration.

Solution

(a) The springs in parallel. Since for a deflection δ of the block the magnitudes of the forces exerted by the springs are, respectively, $F_1 = k_1\delta$ and $F_2 = k_2\delta$, we have

$$F = F_1 + F_2 = k_1\delta + k_2\delta = (k_1 + k_2)\delta$$

Therefore, the stiffness k_{eq} of the single equivalent spring can be expressed as

$$k_{eq} = \frac{F}{\delta} = k_1 + k_2$$

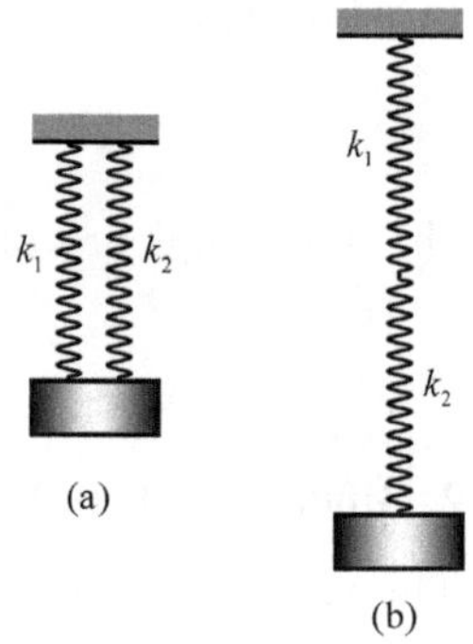

Fig. E11.4

Using $T_n = 2\pi/\omega_n$ and $\omega_n = \sqrt{k_{eq}/m}$, then the frequency of the vibration can be given by

$$T_n = \frac{2\pi}{\omega_n} = 2\pi\sqrt{\frac{m}{k_{eq}}} = 2\pi\sqrt{\frac{m}{k_1 + k_2}}$$

(b) The springs in series. Since for a force F exerted by the springs on the block the deflections of the springs are, respectively, $\delta_1 = F/k_1$ and $\delta_2 = F/k_2$, we have

$$\delta = \delta_1 + \delta_2 = \frac{F}{k_1} + \frac{F}{k_2} = F\left(\frac{1}{k_1} + \frac{1}{k_2}\right)$$

Therefore, the stiffness k_{eq} of the single equivalent spring is equal to

$$k_{eq} = \frac{F}{\delta} = \frac{k_1 k_2}{k_1 + k_2}$$

and the frequency of the vibration can be expressed as

$$T_n = \frac{2\pi}{\omega_n} = 2\pi\sqrt{\frac{m}{k_{eq}}} = 2\pi\sqrt{\frac{m(k_1+k_2)}{k_1k_2}}$$

Example 11.5

From mechanics of materials it is known that for a simply supported beam of uniform cross section a static load F applied at the center will cause a deflection $w_C = Fl^3/48EI$, where l is the length of the beam, E is the modulus of elasticity, and I is the moment of inertia of the cross sectional area of the beam. Knowing that $l = 5$ m, $E=200$ GPa, and $I=$ 2 370 cm^4, determine (a) the equivalent spring stiffness of the beam, (b) the frequency of vibration of a 250 kg block attached to the center of the beam. Neglect the mass of the beam and assume that the load remains in contact with the beam.

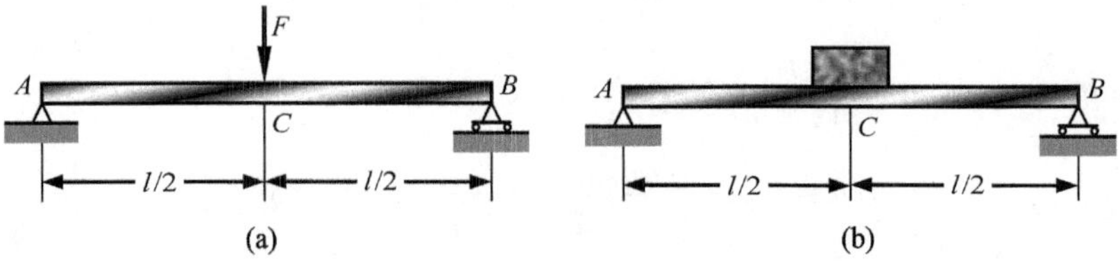

Fig. E11.5

Solution

(a) The equivalent spring stiffness of the beam is equal to

$$k_{eq} = \frac{48EI}{l^3} = 1820 \text{ kN/m}$$

(b) The frequency of vibration is

$$f_n = \frac{\omega_n}{2\pi} = \frac{1}{2\pi}\sqrt{\frac{k_{eq}}{m}} = 13.58 \text{ Hz}$$

2. Undamped Free Vibrations of Rigid Body

The vibration analysis of a rigid body is similar to that of a particle. An appropriate variable, such as a linear displacement x or an angular displacement φ, is chosen to define the position of the body, and an equation relating this variable and its second derivative with respect to t is written. If the equation obtained is of the same form as Eq. (11.2), i.e., if we have

$$\ddot{x} + \omega_n^2 x = 0 \text{ or } \ddot{\varphi} + \omega_n^2\varphi = 0 \tag{11.9}$$

the vibration considered is a simple harmonic motion. The above method can be used to analyze vibrations which are truly represented by a simple harmonic motion, or vibrations of small amplitude which can be approximately represented by a simple harmonic motion.

As an example, let us determine the natural frequency of the small oscillations of a uniform semicircular plate of mass m and radius r which is suspended from the midpoint O of its horizontal diameter, as shown in Fig. 11.2(a).

Consider the plate in an arbitrary position defined by the angle φ that the line OC forms

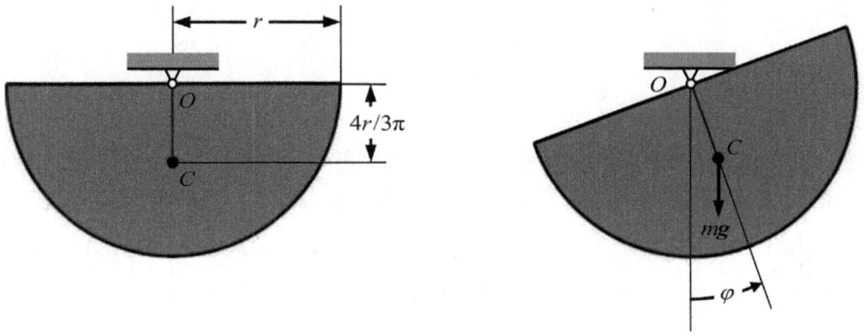

(a) (b)

Fig. 11.2

with the vertical, as shown in Fig. 11.2(b). Using $I_O\alpha = \sum M_O$, we have

$$\left(\frac{1}{2}mr^2\right)\ddot{\varphi} = -(mg)\left(\frac{4r}{3\pi}\sin\varphi\right) \tag{11.10}$$

i.e.,

$$\ddot{\varphi} + \frac{8g}{3\pi r}\sin\varphi = 0 \tag{11.11}$$

For oscillations of small amplitude, we can replace $\sin\varphi$ by φ, expressed in radians, and we write

$$\ddot{\varphi} + \omega_n \varphi = 0 \tag{11.12}$$

where ω_n is the natural frequency, which can be given by

$$\omega_n = \sqrt{\frac{8g}{3\pi r}} \tag{11.13}$$

Example 11.6

A uniform circular disk of mass m and radius r is suspended at its center O from rod AO. Assuming that the torsional spring stiffness of the rod is k_t, determine the period of oscillation of the disk about the axis AO.

Solution

Assuming that the disk is rotated an arbitrary angle φ measured from its position of equilibrium, then the moment of couple acting on the disk exerted by the rod has a magnitude $k_t\varphi$. From $I_O\alpha = \sum M_O$, we have

$$I_O\ddot{\varphi} = -k_t\varphi$$

where $I_O = \frac{1}{2}mr^2$ is the mass moment of inertia of the disk with respect to the axis AO. The above equation can also be rewritten as

$$\ddot{\varphi} + \omega_n^2\varphi = 0$$

where $\omega_n = \sqrt{k_t/I_O}$. Using $T_n = 2\pi/\omega_n$, the period of oscillation of the disk about the axis AO can be given by

$$T_n = 2\pi\sqrt{\frac{I_O}{k_t}} = \pi\sqrt{\frac{2mr^2}{k_t}}$$

Example 11.7

Two small spheres, B and C, each of mass m, are attached to rod AB, which is supported by a pin at A and by a spring CD of stiffness k. Knowing that the mass of the rod is negligible and that the system is in equilibrium when the rod is horizontal, determine the frequency of the small oscillation of the system.

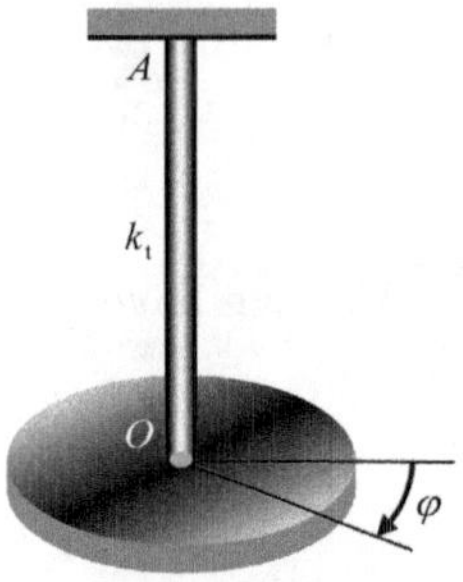

Fig. E11.6

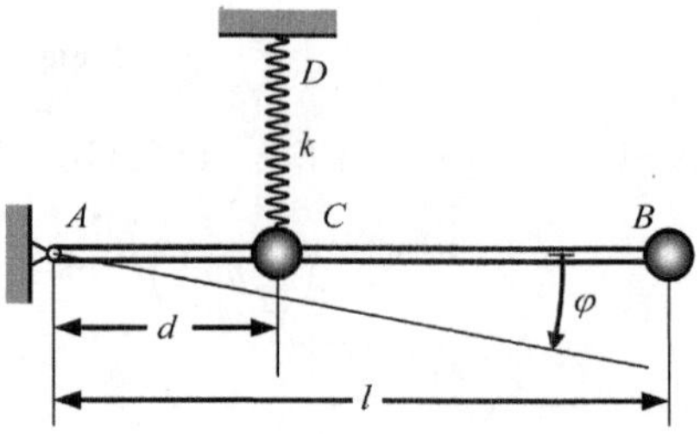

Fig. E11.7

Solution

Assuming that the rod is rotated an arbitrary angle φ measured from its position of equilibrium, then the moment of force acting on the rod exerted by the spring has a magnitude $kd^2\varphi$. From $I_A\alpha = \sum M_A$, we have

$$I_A\ddot{\varphi} = -kd^2\varphi$$

where $I_A = m(d^2 + l^2)$ is the mass moment of inertia of the two spheres with respect to A. The above equation can be rewritten as

$$\ddot{\varphi} + \omega_n^2\varphi = 0$$

where $\omega_n = \sqrt{kd^2/I_A}$. Using $f_n = \omega_n/2\pi$, the frequency of the small oscillation of the system about A can be given by

$$f_n = \frac{1}{2\pi}\sqrt{\frac{kd^2}{I_A}} = \frac{1}{2\pi}\sqrt{\frac{kd^2}{m(d^2+l^2)}}$$

3. Method of Energy for Determination of Natural Frequency

The simple harmonic motion of a body is due only to gravitational or/and elastic restoring forces acting on the body. Since these forces are conservative, it is also possible to use the conservation of energy to obtain the period and frequency of the vibration.

As an example, let us determine the natural frequency of the small oscillations of a uniform semicircular plate of mass m and radius r which is suspended from the midpoint O of its horizontal diameter, as shown in Fig. 11.3(a).

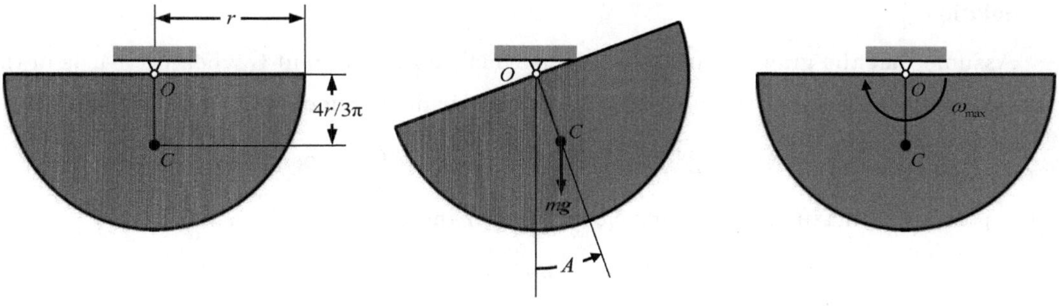

(a) (b) (c)

Fig. 11.3

Assume that the gravitational potential is equal to zero at point C when the plate is in the position of equilibrium. In the position of maximum angular displacement A, as shown in Fig. 11.3(b), we have

$$T_1 = 0,\ V_1 = (mg)\frac{4r}{3\pi}(1-\cos A) \tag{11.14}$$

In the position of maximum angular velocity $\omega_{\max}$, as shown in Fig. 11.3(c), we have

$$T_2 = \frac{1}{2}I_O\omega_{\max}^2,\ V_2 = 0 \tag{11.15}$$

where $I_O = \frac{1}{2}mr^2$. Using the conservation of mechanical energy, $T_1 + V_1 = T_2 + V_2$, we obtain

$$(mg)\frac{4r}{3\pi}(1-\cos A) = \frac{1}{2}\left(\frac{1}{2}mr^2\right)\omega_{\max}^2 \tag{11.16}$$

Using $\cos A \approx 1 - A^2/2$ for oscillations of small amplitude, and $\omega_{\max} = \omega_n A$, we have

$$(mg)\frac{4r}{3\pi}\frac{A^2}{2} = \frac{1}{2}\left(\frac{1}{2}mr^2\right)(\omega_n A)^2 \tag{11.17}$$

i.e., the natural frequency of the small vibration can be written as

$$\omega_n = \sqrt{\frac{8g}{3\pi r}} \tag{11.18}$$

Example 11.8

A uniform disk of radius r and mass m can roll without slipping on a cylindrical surface and is attached to bar ABC of length $2a$ and negligible mass. The bar is attached at point A to a spring of stiffness k and can rotate freely about point B in the vertical plane. Knowing that the spring is undeformed when the bar is vertical and that A is given a small displacement and released, determine the natural frequency of the vibration.

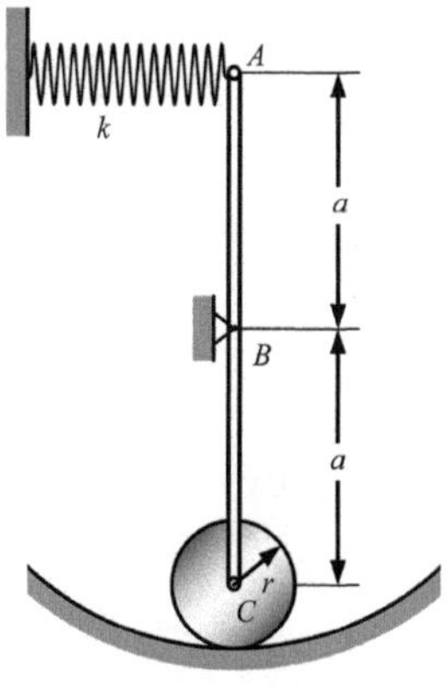

Fig. E11.8

Solution

Assume that the gravitational potential is equal to zero at point C when the bar is in the vertical position. In the position of maximum angular displacement A of the bar, we have

$$T_1 = 0,\ V_1 = \frac{1}{2}k\,(aA)^2 + (mg)a(1-\cos A)$$

In the position of maximum angular velocity ω_{max} of the bar, we have

$$T_2 = \frac{1}{2}\left(\frac{1}{2}mr^2 + mr^2\right)\left(\frac{a\omega_{max}}{r}\right)^2,\ V_2 = 0$$

Using $T_1+V_1=T_2+V_2$, we obtain

$$\frac{1}{2}k\,(aA)^2 + (mg)a(1-\cos A) = \frac{1}{2}\left(\frac{1}{2}mr^2 + mr^2\right)\left(\frac{a\omega_{max}}{r}\right)^2$$

Using $\cos A \approx 1 - A^2/2$ and $\omega_{max} = \omega_n A$, we have

$$\frac{1}{2}k\,(aA)^2 + (mg)a\frac{A^2}{2} = \frac{1}{2}\left(\frac{1}{2}mr^2 + mr^2\right)\left(\frac{a\omega_n A}{r}\right)^2$$

i.e.,

$$\omega_n = \sqrt{\frac{2}{3}\left(\frac{k}{m} + \frac{g}{a}\right)}$$

11.2 Undamped Forced Vibrations

Consider a body of mass m suspended from a spring of stiffness k and subjected to a periodic force of magnitude $F_f = H\sin\omega_f t$, where H is the amplitude of the periodic force acting on the body, and ω_f is the circular frequency of the periodic force, as shown in Fig. 11.4.

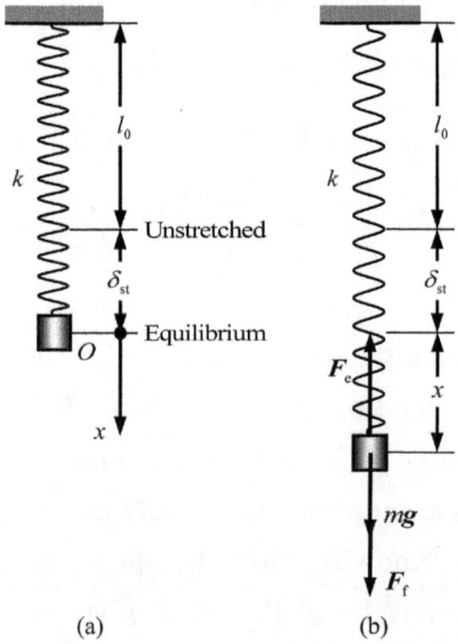

Fig. 11.4

Assuming that the body is located in an arbitrary position x measured from the position of equilibrium O, the forces acting on the body are the gravitational force of magnitude mg, the elastic force of magnitude $F_e = k(\delta_{st} + x)$, and the periodic force of magnitude $F_f = H\sin\omega_f t$. Using $ma_x = \sum F_x$, we have

$$m\ddot{x} = mg - k(\delta_{st} + x) + H\sin\omega_f t \tag{11.19}$$

Using $mg = k\delta_{st}$, we obtain

$$m\ddot{x} + kx = H\sin\omega_f t \tag{11.20}$$

This is a nonhomogeneous second-order differential equation. Its general solution can be obtained by adding a particular solution x_p to the complementary solution x_c (i.e., the general solution of the corresponding homogeneous equation). Thus the general solution of Eq. (11.20) can be expressed as

$$x = x_c + x_p = A\sin(\omega_n t + \alpha) + B\sin\omega_f t \tag{11.21}$$

where

$$B = \frac{H/k}{1 - \lambda^2} \tag{11.22}$$

where $\lambda = \omega_f/\omega_n$ is the ratio of the circular frequency of the periodic force to the natural frequency of the undamped free vibration. Eq. (11.21) describes two superposed vibrations. The complementary solution $x_c = A\sin(\omega_n t + \alpha)$ represents the undamped free vibration. The natural frequency ω_n depends only on the stiffness k of the spring and the mass m of the body, and the constants A and α can be determined from the initial conditions. This undamped free vibration is a transient vibration, since it will soon be damped out by friction forces. The particular solution $x_p = B\sin\omega_f t$ represents a steady-state vibration caused and maintained by the periodic force. The circular frequency of the steady-state vibration is equal to that of the periodic force acting on the body, and its amplitude B depends on the frequency ratio λ. The ratio of the amplitude B of the steady-state vibration to the static deflection H/k caused by force H is called the magnification factor, which can be expressed as

$$\beta = \frac{B}{H/k} = \frac{1}{1 - \lambda^2} \tag{11.23}$$

The magnification factor β has been plotted in Fig. 11.5 against the frequency ratio λ.

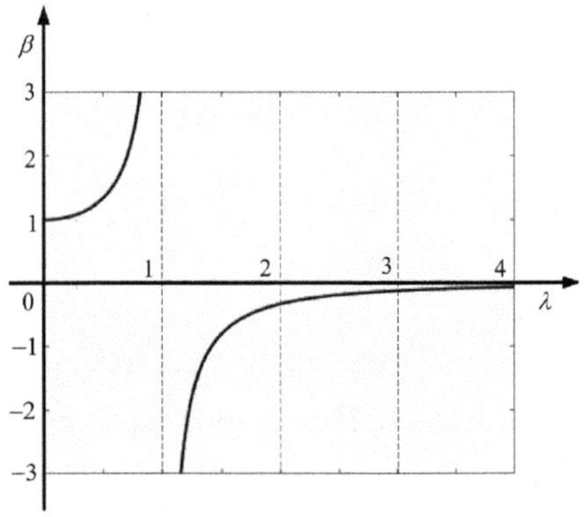

Fig. 11.5

We note from Fig. 11.5 that when $\lambda=1$, the amplitude of the forced vibration become infinite or extremely large. This phenomenon is called resonance. The resonance can be avoided as long as the circular frequency of the steady-state vibration is not chosen too close to the natural frequency of the undamped free vibration. We also note that for $\lambda<1$ B is positive, while for $\lambda>1$ B is negative. In the first case the forced vibration is in phase, while in the second case it is 180° out of phase.

Example 11.9

A small sphere C of mass m is attached to the rod AB of negligible mass which is supported at A by a pin and at D by a spring of stiffness k. The system can move in a vertical plane and is in equilibrium when the rod is horizontal. The rod is acted upon at B by a periodic force of magnitude $F_f = H\sin\omega_f t$, where H is the amplitude of the periodic force. Knowing that $\omega_f = \sqrt{k/m}$, determine the motion of the system corresponding to the steady-state vibration.

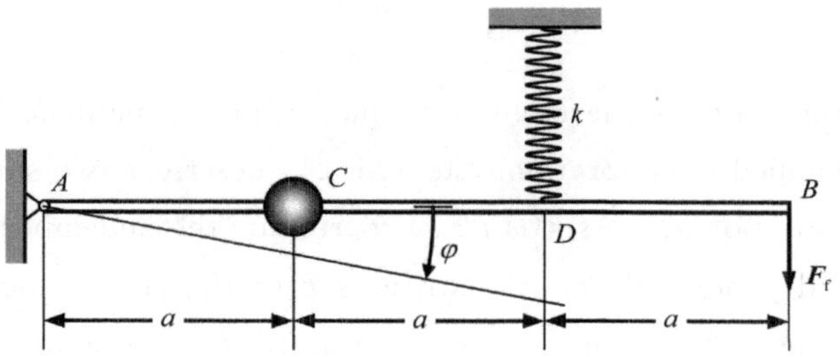

Fig. E11.9

Solution

Assuming that the rod is rotated an arbitrary angle φ measured from the position of equilibrium, we have

$$(ma^2)\ddot{\varphi} + k\,(2a)^2\varphi = H(3a)\sin\omega_f t$$

Using Eq. (11.22) and $\omega_n = \sqrt{k\,(2a)^2/ma^2} = 2\sqrt{k/m}$, the angular amplitude B of the steady-state vibration can be obtained

$$B = \frac{H(3a)/k\,(2a)^2}{1-\lambda^2} = \frac{H(3a)/k\,(2a)^2}{1-(\omega_f/\omega_n)^2} = \frac{H(3a)/k\,(2a)^2}{1-(1/2)^2} = \frac{H}{ka}$$

Thus the equation of steady-state vibration of the system can be obtained

$$\varphi_p = B\,\sin\omega_f t = \frac{H}{ka}\sin\sqrt{\frac{k}{m}}t$$

Example 11.10

A small sphere C of mass m is attached to the rod AB of negligible mass which is supported at A by a pin and connected at B to a moving support D by means of a spring of stiffness k. The system can move in a vertical plane and is in equilibrium when the rod is horizontal. Knowing that support D undergoes a vertical displacement $\delta_f = \Delta\sin\omega_f t$, where

$\omega_f = \sqrt{k/m}$ and Δ is the amplitude of the periodic displacement, determine the motion of the system corresponding to the steady-state vibration.

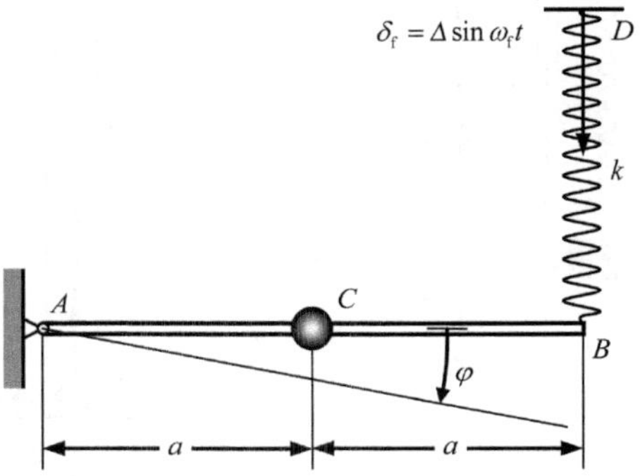

Fig. E11.10

Solution

Assume that the rod is rotated an arbitrary angle φ measured from the position of equilibrium. From $I_A\alpha = \sum M_A$, we obtain

$$(ma^2)\ddot{\varphi} = -k[(2a)\varphi - \Delta\sin\omega_f t](2a)$$

i.e.,

$$ma\ddot{\varphi} + 4ka\varphi = 2k\Delta\sin\omega_f t$$

Using Eq. (11.22) and $\omega_n = \sqrt{4ka/ma} = 2\sqrt{k/m}$, the angular amplitude B of the steady-state vibration can be obtained

$$B = \frac{2k\Delta/4ka}{1-(\omega_f/\omega_n)^2} = \frac{2k\Delta/4ka}{1-(1/2)^2} = \frac{2\Delta}{3a}$$

Therefore, the equation of steady-state vibration of the system can be obtained

$$\varphi_p = B\sin\omega_f t = \frac{2\Delta}{3a}\sin\sqrt{\frac{k}{m}}t$$

11.3 Damped Free Vibrations

The vibrations considered in the previous sections are assumed free of damping. Actually all vibrations are damped by friction forces.

A type of damping of vital importance is called the viscous damping which is caused by fluid friction at low and moderate speeds. Viscous damping is characterized by the fact that the friction force is proportional and opposite to the velocity of the moving body.

Consider a body of mass m suspended from a spring of stiffness k and attached to a dashpot, as shown in Fig. 11.6. The magnitude of the friction force is equal to $F_d = c\dot{x}$, where the constant c is called the coefficient of viscous damping, expressed in N · s/m. The differential equation of motion of the body can be expressed as

$$m\ddot{x} = mg - k(\delta_{st} + x) - c\dot{x} \tag{11.24}$$

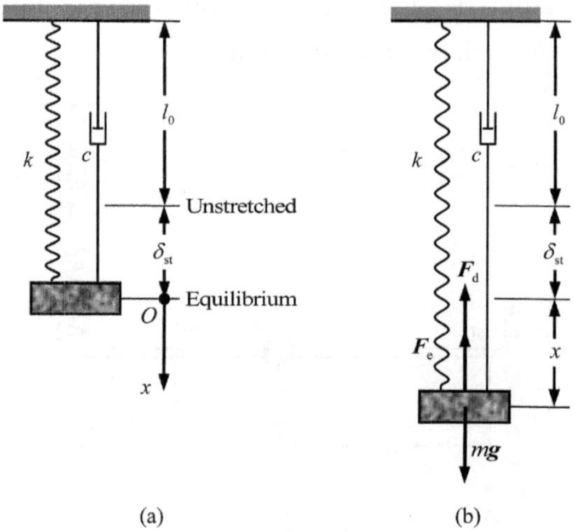

Fig. 11.6

Using $mg=k\delta_{st}$, we obtain

$$m\ddot{x}+c\dot{x}+kx=0 \tag{11.25}$$

Substituting $x=\exp(rt)$ into Eq. (11.25) and dividing Eq. (11.25) by $\exp(rt)$, we obtain the characteristic equation:

$$mr^2+cr+k=0 \tag{11.26}$$

and the roots:

$$r_{1,2}=-\frac{c}{2m}\pm\sqrt{\left(\frac{c}{2m}\right)^2-\frac{k}{m}} \tag{11.27}$$

Defining the critical damping coefficient c_{cr} as the value of c which makes the radical in Eq. (11.27) equal to zero, we have

$$c_{cr}=2\sqrt{mk} \tag{11.28}$$

Defining $\eta=c/c_{cr}$, called the damping factor, and using $\omega_n=\sqrt{k/m}$, Eq. (11.27) can be rewritten as

$$r_{1,2}=(-\eta\pm\sqrt{\eta^2-1})\omega_n \tag{11.29}$$

Depending on the value of the damping factor, three different cases of damping can be outlined as following:

1. Heavy damping ($\eta>1$)

The roots r_1 and r_2 of Eq. (11.26) are real and distinct, and the general solution of Eq. (11.25) is

$$x=\exp(-\eta\omega_n t)[C_1\exp(\sqrt{\eta^2-1}\omega_n t)+C_2\exp(-\sqrt{\eta^2-1}\omega_n t)] \tag{11.30}$$

where C_1 and C_2 are constants. This solution corresponds to a nonvibrating motion. Since r_1 and r_2 are both negative, x approaches zero as t increases, i.e., the body will regain its equilibrium position.

2. Critical damping ($\eta=1$)

Eq. (11.26) has a double root $r=-\omega_n$, and the general solution of Eq. (11.25) is

$$x = \exp(-\omega_n t)(C_1 + C_2 t) \tag{11.31}$$

This solution also corresponds to a nonvibrating motion. In the case of critical damping, the body will regain its equilibrium position in the shortest possible time without oscillation.

3. Light damping ($\eta<1$)

The roots of Eq. (11.26) are complex and conjugate, and the general solution of Eq. (11.25) is

$$x = \exp(-\eta\omega_n t)(C_1\cos\omega_d t + C_2\sin\omega_d t) = A\exp(-\eta\omega_n t)\sin(\omega_d t + \alpha) \tag{11.32}$$

where A and α are constants, which can be determined from the initial conditions, and ω_d is the circular frequency of the damped free vibration, which can be given by

$$\omega_d = \omega_n\sqrt{1-\eta^2} \tag{11.33}$$

The motion defined by Eq. (11.32) is vibratory. The amplitude of this vibration diminishes with each cycle of vibration, since the motion is confined within the bounds of two exponential curves, as shown in Fig. 11.7. The period $T_d = 2\pi/\omega_d$ of the damped free vibration is larger than the period $T_n = 2\pi/\omega_n$ of the corresponding undamped free vibration due to $\omega_d<\omega_n$.

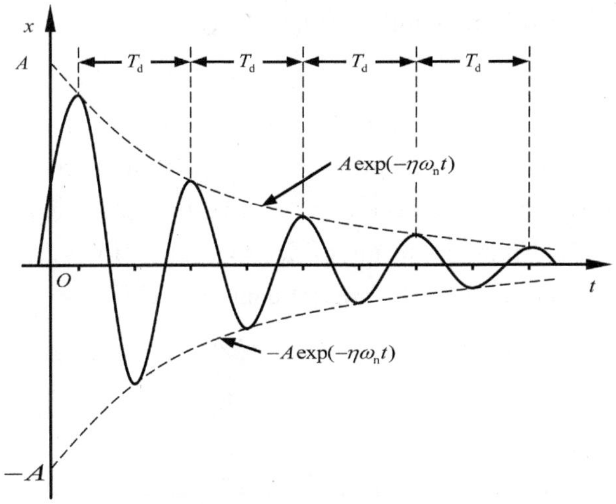

Fig. 11.7

Example 11.11

A 10 N block B is connected by a cord to a 2.5 kg block A which is suspended as shown from two springs, each of stiffness $k=125$ N/m, and a dashpot of damping coefficient $c=30$ N·s/m. Knowing that the system is at rest when the cord connecting A and B is cut, determine the motion of block A.

Solution

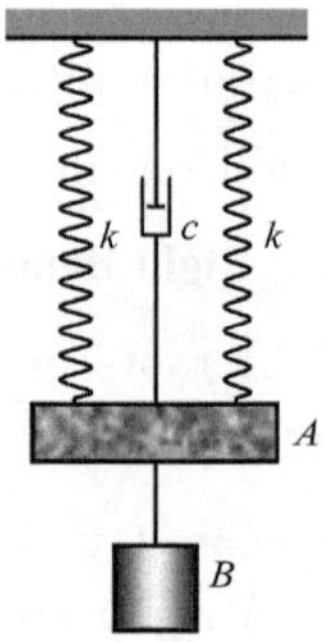

Fig. E11.11

From the initial condition, the initial amplitude and the initial phase can be given by

$$A = \frac{m_B g}{k_{eq}} = 40 \text{ mm}, \alpha = \frac{\pi}{2}$$

Using $k_{eq}=2k$, the critical damping coefficient is equal to

$$c_{cr} = 2\sqrt{m_A k_{eq}} = 50 \text{ N} \cdot \text{s/m}$$

i.e., the damping factor is

$$\eta = \frac{c}{c_{cr}} = 0.6$$

Since $\eta<1$, the motion is a light-damped vibration. Using $\omega_n = \sqrt{k_{eq}/m_A} = 10$ rad/s, the circular frequency of the damped free vibration can be obtained

$$\omega_d = \omega_n \sqrt{1-\eta^2} = 8 \text{ rad/s}$$

Hence, the motion of block A can be expressed as

$$x = A\exp(-\eta\omega_n t)\sin(\omega_d t+\alpha) = 40\exp(-6t)\sin\left(8t+\frac{\pi}{2}\right)$$

where x and t are expressed in mm and s, respectively.

Example 11.12

A small sphere C of mass m is attached to the rod AB of negligible mass which is supported at A by a pin and at D by a spring of stiffness k and is connected at B to a dashpot of damping coefficient c. The system can move in a vertical plane and is in equilibrium when the rod is horizontal. Determine in terms of m, k, and c, for small oscillations, (a) the differential equation of motion, (b) the critical damping coefficient c_{cr}.

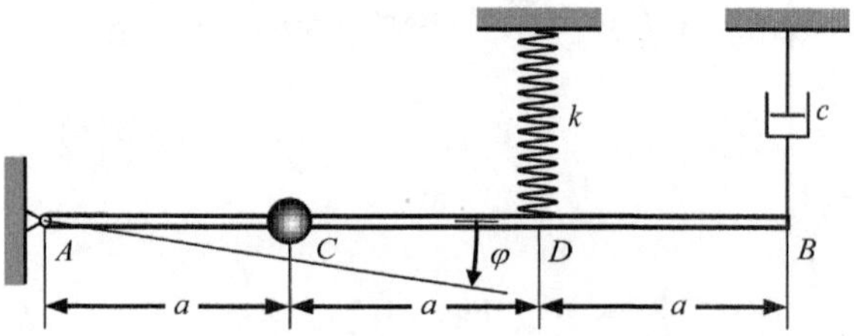

Fig. E11.12

Solution

Assuming that the rod is rotated an arbitrary angle φ measured from the position of equilibrium, we can obtain the differential equation of motion

$$(ma^2)\ddot{\varphi} + c(3a)^2\dot{\varphi} + k(2a)^2\varphi = 0$$

i.e.,

$$\ddot{\varphi} + \frac{9c}{m}\dot{\varphi} + \frac{4k}{m}\varphi = 0$$

The above equation can also be rewritten by

$$\frac{m}{9}\ddot{\varphi}+c\dot{\varphi}+\frac{4k}{9}\varphi=0$$

Comparing this equation with Eq. (11.25) and using Eq. (11.28), we can obtain the critical damping coefficient

$$c_{cr}=2\sqrt{\frac{m}{9}\frac{4k}{9}}=\frac{4}{9}\sqrt{mk}$$

11.4 Damped Forced Vibrations

Consider a body of mass m suspended from a spring of stiffness k, attached to a dashpot of damping coefficient c, and subjected to a periodic force of magnitude $F_f=H\sin\omega_f t$, as shown in Fig. 11.8. The equation of motion of the body can be expressed as

$$m\ddot{x}=mg-k(\delta_{st}+x)-c\dot{x}+H\sin\omega_f t \tag{11.34}$$

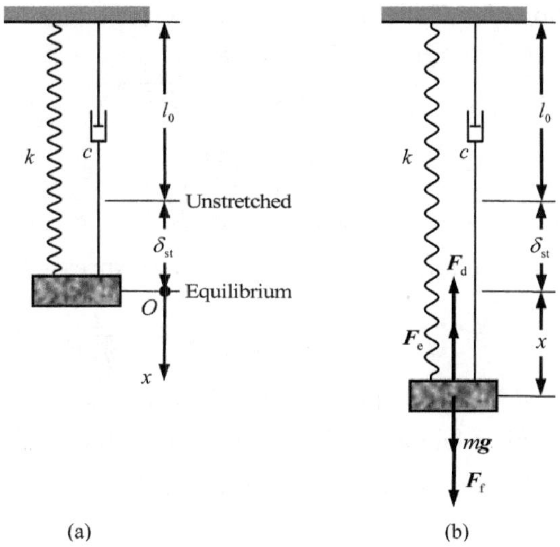

Fig. 11.8

Using $mg=k\delta_{st}$, we have

$$m\ddot{x}+c\dot{x}+kx=H\sin\omega_f t \tag{11.35}$$

The general solution of this equation can be obtained by adding a particular solution x_p to the complementary solution x_c, i.e.,

$$x=x_c+x_p \tag{11.36}$$

where x_c can be written as

$$x_c=\begin{cases}\exp(-\eta\omega_n t)[C_1\exp(\sqrt{\eta^2-1}\omega_n t)+C_2\exp(-\sqrt{\eta^2-1}\omega_n t)] & (\eta>1)\\ \exp(-\omega_n t)(C_1+C_2 t) & (\eta=1)\\ \exp(-\eta\omega_n t)(C_1\cos\omega_d t+C_2\sin\omega_d t) & (\eta<1)\end{cases} \tag{11.37}$$

The complementary solution x_c represents a transient motion which is eventually damped out by friction forces. Thus our interest is centered on the particular solution corresponding to a steady-state vibration. The particular solution x_p can be expressed as

$$x_p = B\sin(\omega_f t - \varepsilon) \tag{11.38}$$

where B and ε are respectively the amplitude of the steady-state vibration and the phase difference between the applied periodic force and the steady-state vibration of the damped system. Substituting x_p for x into Eq. (11.35), we obtain

$$-m\omega_f^2 B\sin(\omega_f t - \varepsilon) + c\omega_f B\cos(\omega_f t - \varepsilon) + kB\sin(\omega_f t - \varepsilon) = H\sin\omega_f t \tag{11.39}$$

Using $\sin\omega_f t = \sin[(\omega_f t - \varepsilon) + \varepsilon] = \sin(\omega_f t - \varepsilon)\cos\varepsilon + \cos(\omega_f t - \varepsilon)\sin\varepsilon$, the above equation can also be rewritten as

$$[-m\omega_f^2 B + kB - F_{\max}\cos\varepsilon]\sin(\omega_f t - \varepsilon) + [c\omega_f B - H\sin\varepsilon]\cos(\omega_f t - \varepsilon) = 0 \tag{11.40}$$

i.e., we have

$$-m\omega_f^2 B + kB - H\cos\varepsilon = 0$$

$$c\omega_f B - H\sin\varepsilon = 0 \tag{11.41}$$

Solving the above equations for B and ε, we obtain, respectively,

$$B = \frac{H/k}{\sqrt{(1-\lambda^2)^2 + (2\eta\lambda)^2}} \tag{11.42}$$

$$\varepsilon = \arctan\frac{2\eta\lambda}{1-\lambda^2} \tag{11.43}$$

where $\eta = c/c_{cr}$ is the damping factor. The amplitude B depends on the frequency ratio λ and the damping factor η. The magnification factor β can be expressed as

$$\beta = \frac{B}{H/m} = \frac{1}{\sqrt{(1-\lambda^2)^2 + (2\eta\lambda)^2}} \tag{11.44}$$

The magnification factor β has been plotted in Fig. 11.9 against the frequency ratio λ for various values of the damping factor η. We observe that the amplitude of a damped forced vibration can be kept small by choosing a large damping factor or by keeping the natural and forced frequencies far apart.

The phase difference ε has also been plotted versus the frequency ratio λ shown in Fig. 11.10 for various values of the damping factor η.

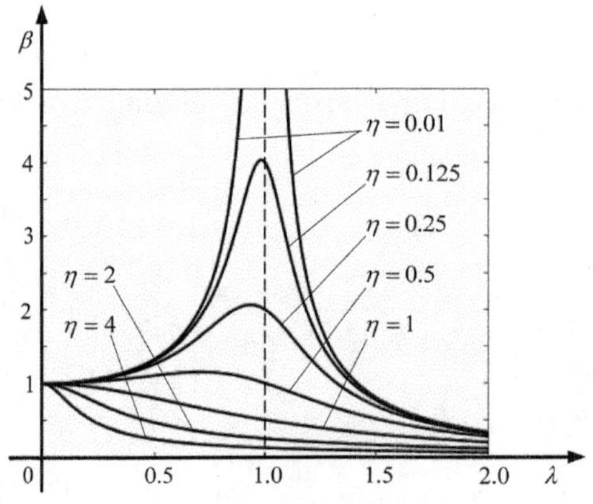

Fig. 11.9

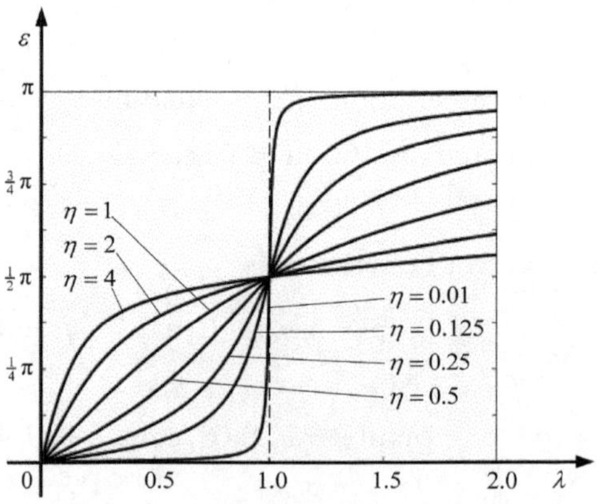

Fig. 11.10

Example 11.13

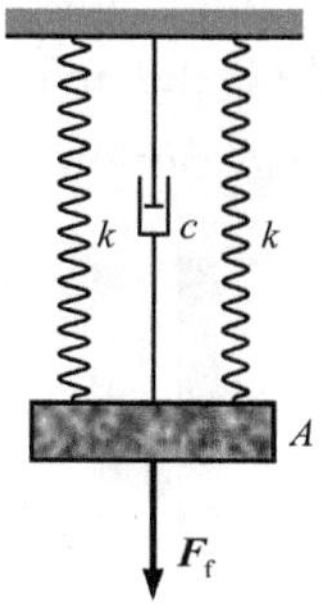

Fig. E11.13

A 5-kg machine element is supported by two springs, each of stiffness 45 N/m. A periodic force of 1.5-N amplitude is applied to the element with a frequency of 2.5 Hz. Knowing that the coefficient of damping is 20 N·s/m, determine the amplitude of the steady-state vibration of the element.

Solution

Using $k_{eq}=2k=90$ N/m and $f_f=2.5$ Hz, we have

$$\omega_n = \sqrt{k_{eq}/m} = 4.243 \text{ rad/s}, \quad \omega_f = 2\pi f_f = 15.71 \text{ rad/s}$$

i.e.,

$$\lambda = \omega_f/\omega_n = 3.703$$

Using $c_{cr}=2\sqrt{mk_{eq}}=42.43$ N·s/m, we have

$$\eta = c/c_{cr} = 0.4714$$

Using these data obtained, we can compute the amplitude of the steady-state vibration as follows:

$$B = \frac{H/k_{eq}}{\sqrt{(1-\lambda^2)^2+(2\eta\lambda)^2}} = 1.264 \text{ mm}$$

Problems

11.1 A 15-kg block is attached to a spring and can move without friction in a slot, as shown in Fig. P11.1. The block is in its equilibrium position when it is struck by a hammer which imparts to the block an initial velocity of 2 m/s. Knowing $k=80$ kN/m, determine (a) the period and frequency of the resulting motion, (b) the amplitude of the motion and the maximum acceleration of the block.

11.2 A 1.4-kg block is supported by a spring of stiffness $k=400$ N/m which can act in tension or compression, as shown in Fig. P11.2. The block is in its equilibrium position when it is struck from below by a hammer which imparts to the block an upward velocity of 2.5 m/s. Determine (a) the time, velocity, and acceleration for the block to move 60 mm upward, (b) the position, velocity, and acceleration of the block 0.90 s after it has been struck by the hammer.

11.3 A 2.5-kg collar rests on but is not attached to the spring, as shown in Fig. P11.3. It is observed that when the collar is pushed down 150 mm or more and released, it loses contact with the spring. Determine (a) the spring stiffness, (b) the position, velocity, and acceleration of the collar 0.2 s after it has been pushed down 150 mm and released.

11.4 A 10-kg collar is released from rest in the position shown in Fig. P11.4 and slides without friction on a vertical rod until it hits a spring of stiffness $k=981$ N/m which it compresses. The velocity of the collar is reduced to zero and the collar reverses the direction of its motion and returns to its initial position. The cycle is then repeated. Determine the period of the motion of the collar. (Note: This is a periodic motion, but not simple harmonic motion.)

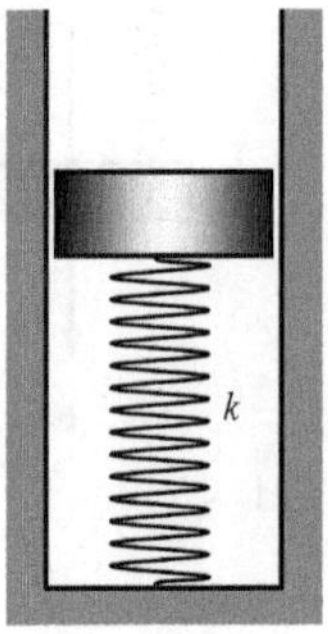

Fig. P11.1

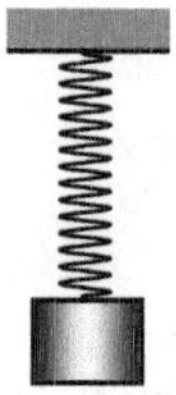

Fig. P11.2

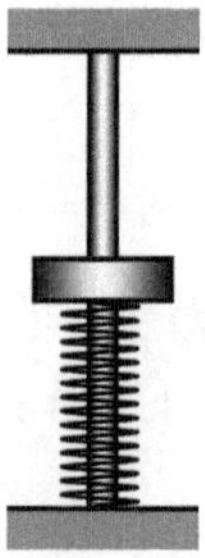

Fig. P11.3

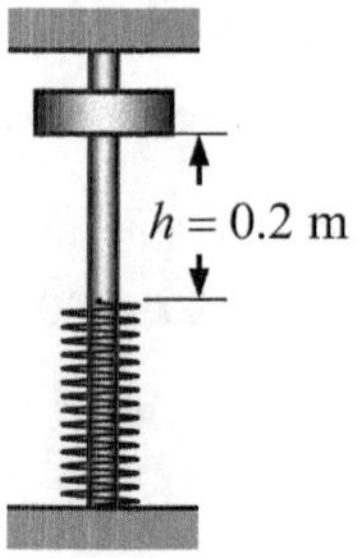

Fig. P11.4

11.5 From mechanics of materials it is known that when a static load F is applied at the end B of a uniform rod fixed at end A, the length of the rod will increase by an amount $\delta = Fl/EA$, where l is the length of the undeformed rod, E is the modulus of elasticity of the material, and A is its cross-sectional area, as shown in Fig. P11.5. Knowing that $l = 250$ mm and $E = 200$ GPa and that the diameter of the rod is $d = 10$ mm, and neglecting the weight of the rod, determine (a) the equivalent spring stiffness of the rod, (b) the frequency of the vertical vibrations of a block of mass $m = 5$ kg attached to end B of the same rod.

11.6 A 5-kg uniform cylinder can roll without sliding on a horizontal surface and is attached by a pin at point C to the 3-kg horizontal bar AB, which is attached to two springs, each of stiffness $k = 4$ kN/m, as shown in Fig. P11.6. Knowing that the bar is moved 20 mm to the right of the equilibrium position and released, determine the period of vibration of the system.

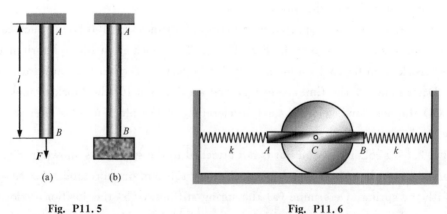

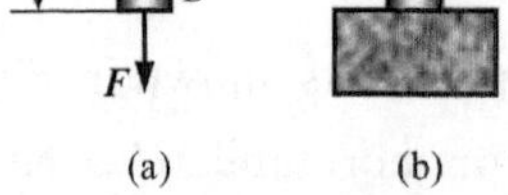

Fig. P11.5

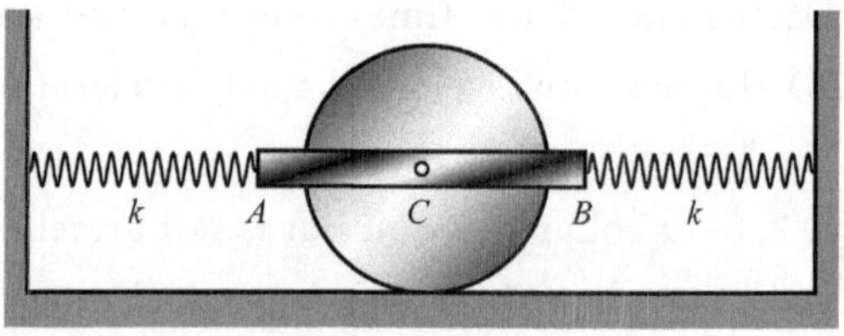

Fig. P11.6

11.7 A uniform rod AB can rotate in a vertical plane about a horizontal axis at O located at a distance d above the mass center C of the rod, as shown in Fig. P11.7. For small oscillations determine the value of d for which the frequency of the motion will be maximum.

11. 8 For the uniform square plate of mass m and side $2a$ shown in Fig. P11. 8, determine the period of small oscillations if the plate is suspended from the midpoint O of one of its sides.

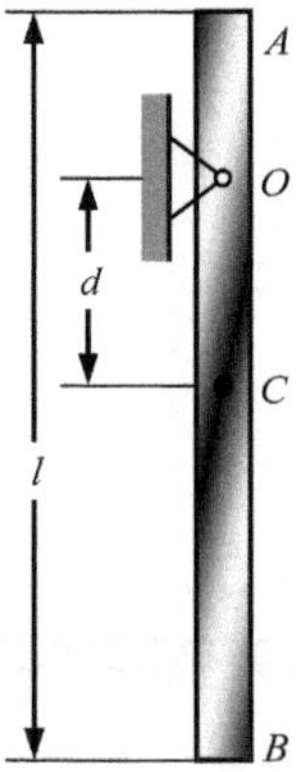

Fig. P11. 7

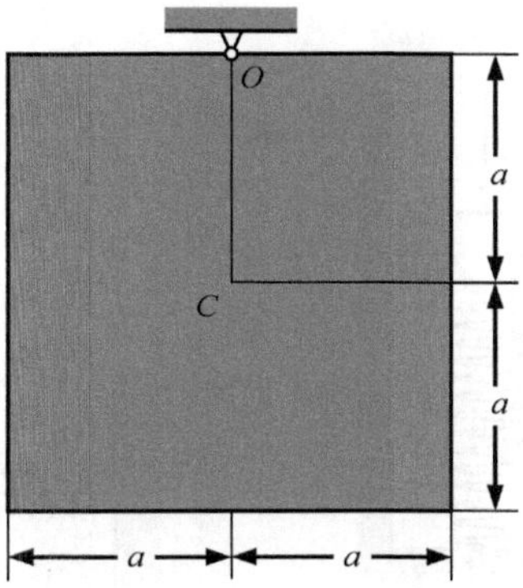

Fig. P11. 8

11. 9 A uniform slender bar AB of mass m and length $2a$ is connected to two collars of negligible mass, as shown in Fig. P11. 9. Collar A is attached to a spring of stiffness k and can slide freely on a horizontal rod, while collar B can slide freely on a vertical rod. Knowing that the system is in equilibrium when bar AB is vertical and that collar A is given a small displacement and released, determine the period of vibration of the system.

11. 10 Three identical uniform slender bars of mass m and length a are connected by pins shown in Fig. P11. 10 and can move in a vertical plane. Knowing that bar BC is given a small displacement and released, determine the period of vibration of the system.

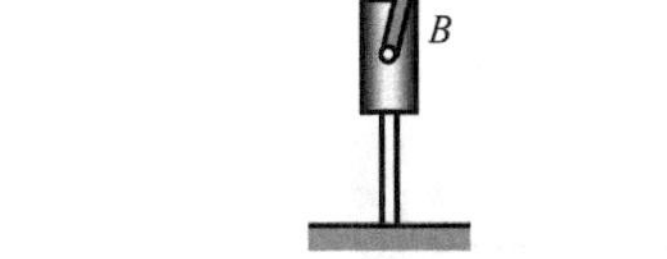

Fig. P11. 9

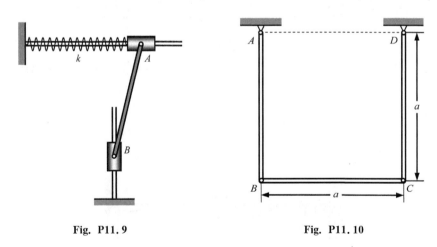

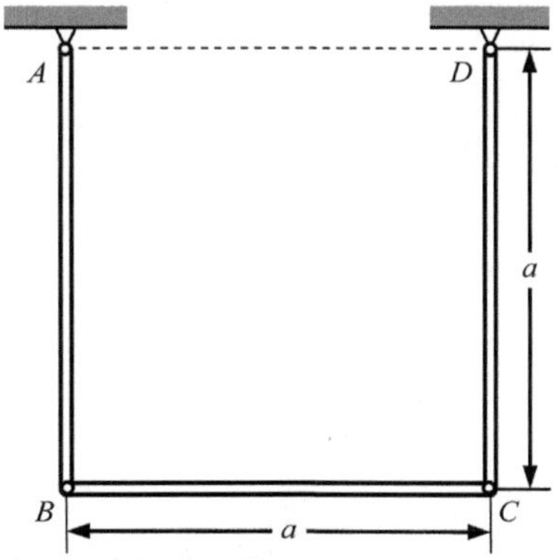

Fig. P11. 10

11. 11 A 20-kg block is attached to a spring of stiffness $k = 8$ kN/m and can move without friction in a vertical slot, as shown in Fig. P11. 11. It is acted upon by a periodic force of magnitude $F_f = H\sin\omega_f t$, where $\omega_f = 10$ rad/s. Knowing that the amplitude of the

motion is 10 mm, determine H.

11.12 A 5-kg collar C can slide on a frictionless horizontal rod AB and is attached to a spring of stiffness 500 N/m, as shown in Fig. P11.12. It is acted upon by a periodic force of magnitude $F_f = H\sin\omega_f t$, where $H = 15$ N. Determine the amplitude of the motion of the collar if (a) $\omega_f = 5$ rad/s, (b) $\omega_f = 10$ rad/s.

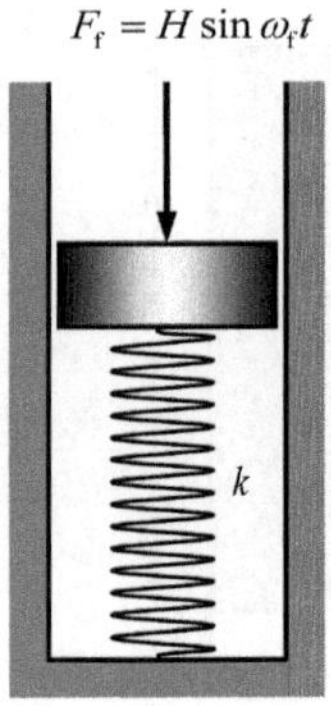

Fig. P11.11

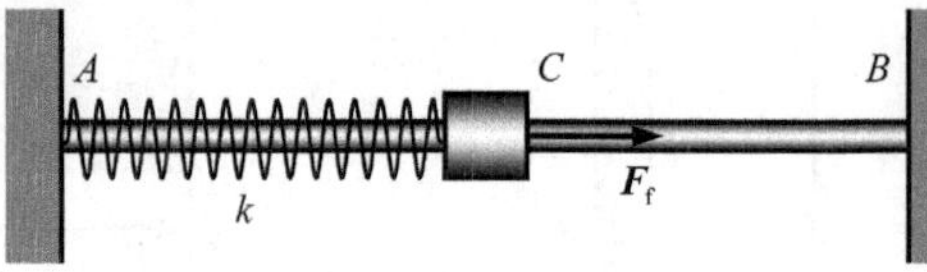

Fig. P11.12

11.13 A 5-kg collar C can slide on a frictionless horizontal rod AB and is attached to a spring of stiffness k, as shown in Fig. P11.13. It is acted upon by a periodic force of magnitude $F_f = H\sin\omega_f t$, where $H = 15$ N and $\omega_f = 5$ rad/s. Determine the value of the spring stiffness k knowing that the motion of the collar has an amplitude of 100 mm and is (a) in phase with the applied force, (b) out of phase with the applied force.

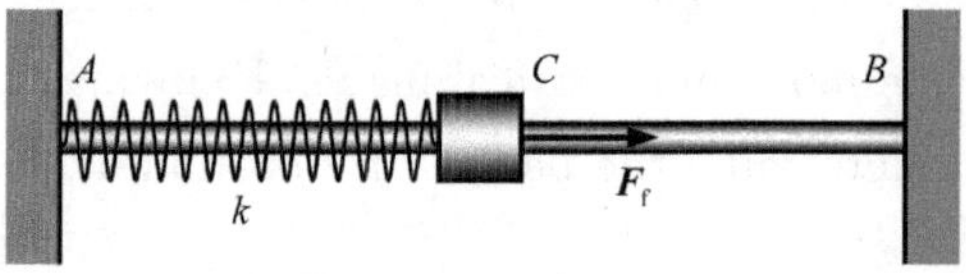

Fig. P11.13

11.14 A collar C of mass m which slides on a frictionless horizontal rod AB is attached to a spring of stiffness k and is acted upon by a periodic force of magnitude $F_f = H\sin\omega_f t$, as shown in Fig. P11.14. Determine the range of values of ω_f for which the amplitude of the vibration exceeds three times the static deflection caused by a constant force of magnitude H.

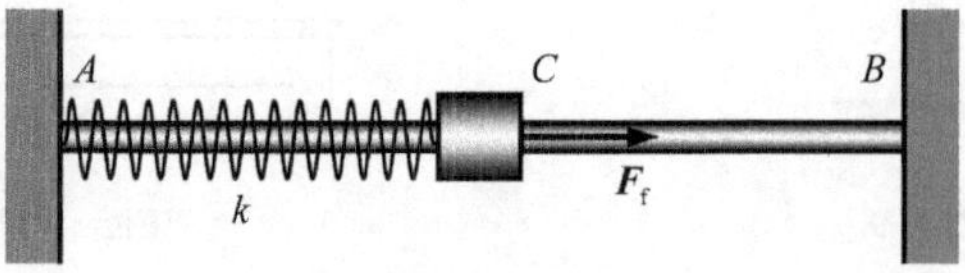

Fig. P11.14

11.15 A small sphere C of mass m is attached to the rod AB of negligible mass which is supported at A by a pin and at B by a spring of stiffness k, as shown in Fig. P11. 15. The system can move in a vertical plane and is in equilibrium when the rod is horizontal. The rod is acted upon at B by a periodic force of magnitude $F_f = H\sin\omega_f t$, where H is the amplitude of the periodic force. Knowing that $\omega_f = 2\omega_n$, where ω_n is the natural frequency of the system, derive the equation of steady-state vibration of the system.

11. 16 A 10-N block B is connected by a cord to a 2. 5-kg block A which is suspended from two springs, each of stiffness $k = 125$ N/m, and a dashpot of damping coefficient $c = 65$ N·s/m, as shown in Fig. P11. 16. Knowing that the system is at rest when the cord connecting A and B is cut, determine the motion of block A.

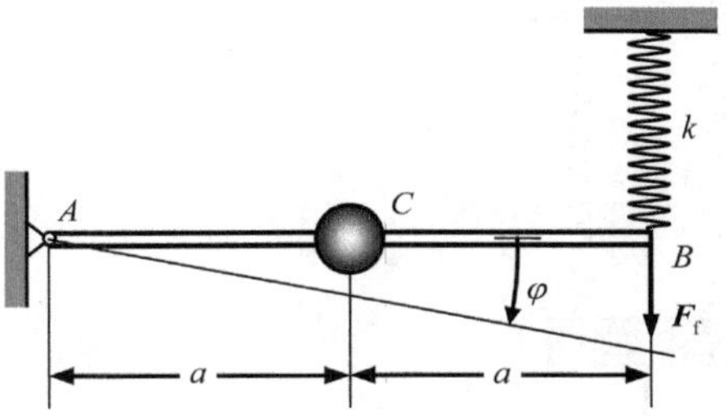

Fig. P11. 15

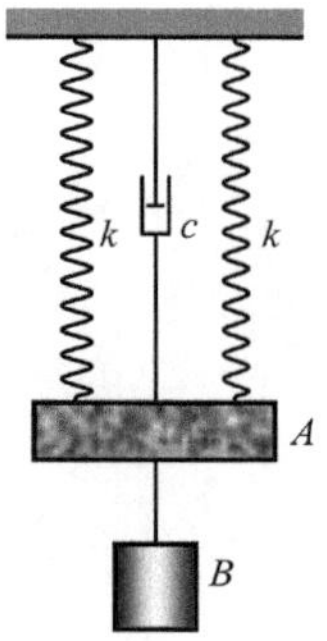

Fig. P11. 16

11. 17 A small sphere C of mass m is attached to the uniform rod AB of mass m which is supported at A by a pin and at D by a spring of stiffness k and is connected at B to a dashpot of damping coefficient c. The system shown in Fig. P11. 17 can move in a vertical plane and is in equilibrium when the rod is horizontal. Determine in terms of m, k, and c, for small oscillations, (a) the differential equation of motion, (b) the critical damping coefficient c_{cr}.

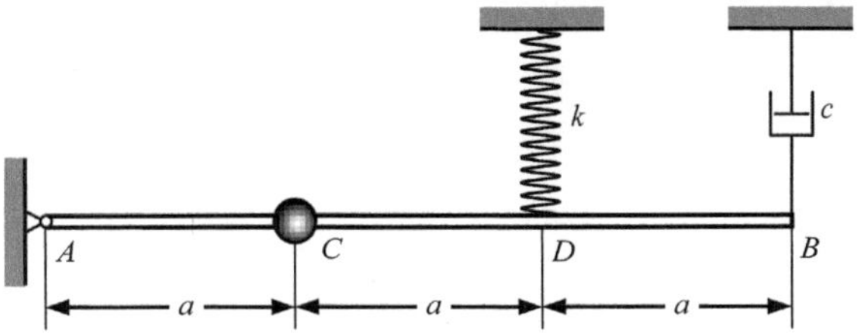

Fig. P11. 17

11. 18 A uniform rod AB of mass m is supported at A by a pin and at C by a spring of stiffness k and is connected at B to a dashpot of damping coefficient c. The system shown in Fig. P11. 18 can move in a vertical plane and is in equilibrium when the rod is horizontal. Determine in terms of m, k, and c, for small oscillations, (a) the differential equation of motion, (b) the critical damping coefficient c_{cr}.

11. 19 A 5-kg machine element is supported by two springs, each of stiffness k, as shown in Fig. P11. 19. A periodic force of 1. 5-N amplitude is applied to the element with a

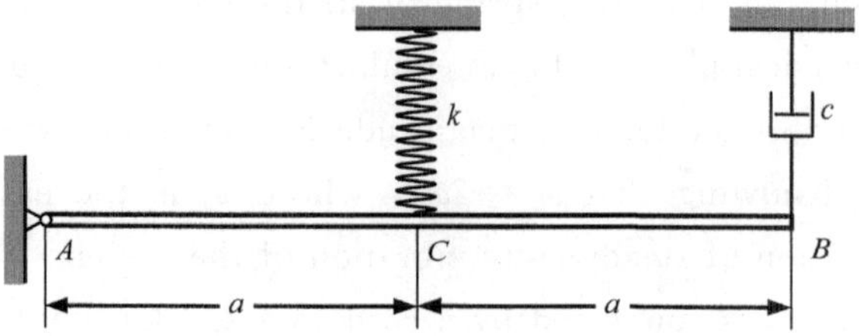

Fig. P11. 18

frequency of 2. 5 Hz. Knowing that the coefficient of damping is 20 N · s/m, determine the required value of the stiffness of each spring if the amplitude of the steady-state vibration is to be 1. 25 mm.

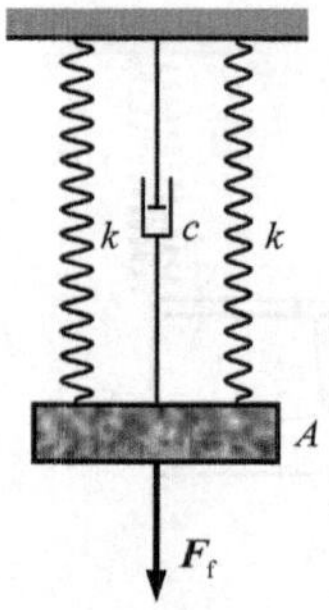

Fig. P11. 19

Chapter 12 Principle of Virtual Work

In the previous chapters, problems involving the equilibrium of bodies are solved by expressing that the external forces acting on the bodies are balanced. The equations of equilibrium are written and solved for the desired unknowns. A different method, which is more effective for solving certain types of equilibrium problems, will now be discussed in this chapter. This method is based on the method of virtual work.

12.1 Constraints and Virtual Work

1. Constraints

A constraint is defined as a geometric or kinematic condition that limits the motion of a body. Considering a single particle and assuming that $\boldsymbol{r}$ is the position vector of the particle and $\dot{\boldsymbol{r}}$ its velocity at time t, then a relation of the form

$$f(\boldsymbol{r},\dot{\boldsymbol{r}},t) \leqslant 0,\ f(\boldsymbol{r},\dot{\boldsymbol{r}},t) = 0,\ \text{or}\ f(\boldsymbol{r},\dot{\boldsymbol{r}},t) \geqslant 0 \tag{12.1}$$

is the mathematical expression for a constraint.

We can classify the constraints according to the following criteria:

(1) Geometric and kinematic constraints: a constraint in which the velocity does not appear is referred to as a geometric or finite constraint (e.g. $f(\boldsymbol{r})=0$), while a velocity-dependent constraint is known as a kinematic or differential constraint (e.g. $f(\boldsymbol{r},\dot{\boldsymbol{r}})=0$). A kinematic or differential constraint which can be put in a finite form is called an integrable constraint (e.g. $\dot{x}-r\dot{\varphi}=0$).

(2) Scleronomous and rheonomous constraints: if the time does not explicitly appear in the constraint equation, this is called a scleronomous or stationary constraint (e.g. $f(\boldsymbol{r})=0$). If the constraint is time-dependent, it is called a rheonomous or nonstationary constraint (e.g. $f(\boldsymbol{r},t)=0$).

(3) Bilateral and unilateral constraints: a constraint expressed by an equality is called a bilateral constraint (e.g. $f(\boldsymbol{r})=0$), while a constraint expressed by an inequality is called unilateral constraint (e.g. $f(\boldsymbol{r})\leqslant 0$).

The geometric constraints, together with the integrable constraints, form the class of holonomic constraints. The nonintegrable constraints, together with the unilateral constraints, are said to be nonholonomic constraints.

A constraint can be characterized simultaneously by all possible criteria. For instance, the constraint expressed by $f(\boldsymbol{r},t)=0$ is bilateral, rheonomous and geometric, while the constraint $f(\boldsymbol{r},\dot{\boldsymbol{r}})=0$ is bilateral, scleronomous and kinematic.

2. Degrees of Freedom

The number of degrees of freedom is defined as the minimum number of variables required to specify completely the configuration of a given system, i.e., the position of all particles of a given system. For example, a particle moving on a line has one degree of freedom, a rigid body in two dimensions has three degrees of freedom (two translational degrees of freedom which give the position of a specified point on the rigid body, and one rotational degree of freedom which gives the orientation of the rigid body), and a rigid body in three dimensions has six degrees of freedom (three of these are translational and correspond to the center of mass, and the other three are rotational and give the orientation of the rigid body).

There is a close relation between the number of constraints and the number of degrees of freedom of a system. A free particle, i.e. a particle subject only to applied forces, has three degrees of freedom. If the coordinates of the particle are connected by a relation $f(\boldsymbol{r},t)=0$, the number of its degrees of freedom reduces to two. In the same way, the existence of the two constraints $f_1(\boldsymbol{r},t)=0$ and $f_2(\boldsymbol{r},t)=0$ implies that the position of the particle is determined by a single parameter, corresponding to a single degree of freedom. In general, each geometric bilateral constraint applied to a system reduces its number of degrees of freedom by one.

3. Virtual Work

A virtual displacement of a body is defined as a fictitious infinitesimal displacement consistent with the constraints imposed on the body. This displacement can be imagined, i.e., it does not actually take place.

Virtual work is defined to be the work done by a force or a couple undergoing a virtual displacement. The virtual work done by a force $\boldsymbol{F}$ undergoing a virtual displacement $\delta\boldsymbol{r}$ can be expressed as

$$\delta W = \boldsymbol{F}\cdot\delta\boldsymbol{r} \tag{12.2}$$

The virtual work done by a couple of magnitude M during a virtual displacement $\delta\theta$ can be expressed as

$$\delta W = M\delta\theta \tag{12.3}$$

4. Ideal Constraints

If the virtual work done by all the constraint forces undergoing an arbitrary virtual displacement is equal to zero, this type of constraints is called the ideal constraints.

12.2 Principle of Virtual Work

Consider a system of n particles. If $\delta\boldsymbol{r}_i$ is the virtual displacement of particle i, consistent with the constraints, then the virtual work of the forces acting on particle i can be expressed as

$$\delta W_i = (\boldsymbol{F}_i+\boldsymbol{R}_i)\cdot\delta\boldsymbol{r}_i \quad (i=1,2,\cdots,n) \tag{12.4}$$

where $\boldsymbol{F}_i$ is the resultant of the applied or active forces acting on particle i and $\boldsymbol{R}_i$ is the resultant of the constraint forces acting on the same particle. If the system of particles is in static equilibrium, i.e., $\boldsymbol{F}_i+\boldsymbol{R}_i=0\ (i=1,2,\cdots,n)$, then Eq. (12.4) can be rewritten as

$$\delta W_i = (\boldsymbol{F}_i+\boldsymbol{R}_i)\cdot\delta\boldsymbol{r}_i = 0 \quad (i=1,2,\cdots,n) \tag{12.5}$$

Summing over i, the total virtual work is equal to

$$\delta W = \sum_{i=1}^{n}\delta W_i = \sum_{i=1}^{n}(\boldsymbol{F}_i+\boldsymbol{R}_i)\cdot\delta\boldsymbol{r}_i = 0 \tag{12.6}$$

If the system is subject to ideal constraints, i.e., $\sum_{i=1}^{n}\boldsymbol{R}_i\cdot\delta\boldsymbol{r}_i=0$, we then have

$$\delta W = \sum_{i=1}^{n}\boldsymbol{F}_i\cdot\delta\boldsymbol{r}_i = 0 \tag{12.7}$$

This relation expresses the principle of virtual work: The necessary and sufficient condition for static equilibrium of a system subject to ideal constraints is that the virtual work of the applied forces, for any virtual displacement consistent with the constraints, is equal to zero.

Example 12.1

An unstretched spring of stiffness $k=800$ N/m is attached to pins at points I and J, as shown in Fig. E12.1. The pin at B is attached to member BCD and can slide freely along the vertical slot in the fixed plate. Determine the force in the spring and the horizontal displacement of point H when a horizontal force of magnitude $F=135$ N, directed to the right, is applied at point G.

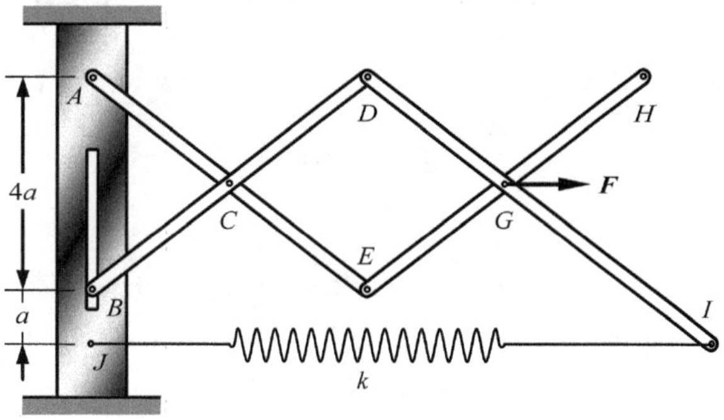

Fig. E12.1

Solution

Using $x_G=3x_C$, $x_H=4x_C$, $x_I=4.5x_C$, we have

$$\delta x_G = 3\delta x_C,\ \delta x_H = 4\delta x_C,\ \delta x_I = 4.5\delta x_C$$

From the principle of virtual work, $\delta W=0$, we have

$$F\delta x_G - F_{\text{spr}}\delta x_I = 0$$

i.e.,

$$F_{\text{spr}} = \frac{\delta x_G}{\delta x_I}F = \frac{3\delta x_C}{4.5\delta x_C}F = 90\ \text{N}$$

Using $F_{\text{spr}}=k\delta x_I$, we obtain

$$x_I = \frac{F_{spr}}{k} = 112.5 \text{ mm}$$

Using $\delta x_H = 4\delta x_C$ and $\delta x_I = 4.5\delta x_C$

$$\delta x_H = \frac{4}{4.5}\delta x_I = 100 \text{ m}$$

Example 12.2

The structure is acted upon by the force $\boldsymbol{F}$ at point G, as shown in Fig. E12.2(a). Neglecting the weight of the structure and knowing $AB = AC = BC = CD = CE = DG = EG = a$, determine the magnitude of horizontal constraint force at point B.

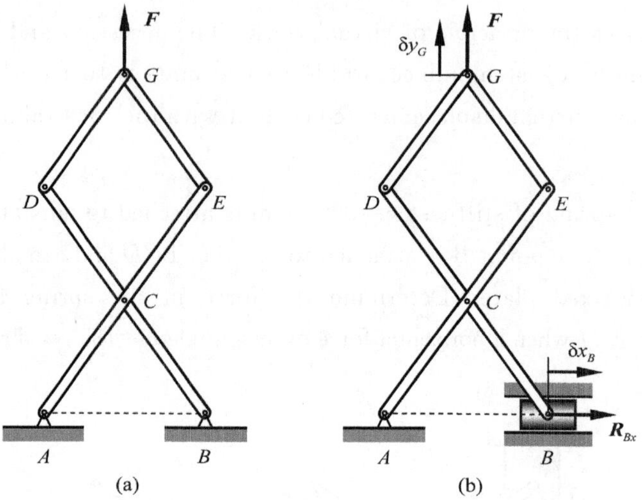

Fig. E12.2

Solution

Replace the horizontal constraint at point B with its horizontal constraint force $\boldsymbol{R}_{Bx}$ and assume that the virtual displacement δx_B at point B is positive to the right and the virtual displacement δy_G at point G is positive upward, as shown in Fig. E12.2(b). Using $x_C^2 + y_C^2 = a^2$, i.e., $\left(\frac{x_B}{2}\right)^2 + \left(\frac{y_G}{3}\right)^2 = a^2$, we have

$$x_B \delta x_B + \frac{4}{9} y_G \delta y_G = 0$$

Using $AB = AC = BC$, i.e. $y_C = \sqrt{3} x_C$, or $y_G = \frac{3\sqrt{3}}{2} x_B$, we have

$$\delta x_B = -\frac{2\sqrt{3}}{3}\delta y_G$$

From the principle of virtual work, $\delta W = 0$, we have

$$R_{Bx}\delta x_B + F\delta y_G = 0$$

i.e.,

$$R_{Bx} = -\frac{\delta y_G}{\delta x_B}F = \frac{\sqrt{3}}{2}F$$

Example 12.3

A homogeneous rod AB, of weight W, is attached to blocks A and B which move freely along the smooth surfaces, as shown in Fig. E12.3 (a). The stiffness of the spring connected to block A is k, and the spring is unstretched when the rod is horizontal. Neglecting the weight of the blocks, derive an equation in W, k, a, and θ which must be satisfied when the rod is in equilibrium.

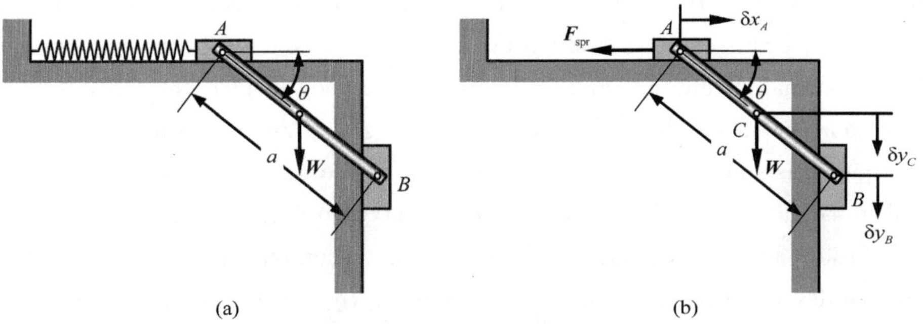

Fig. E12.3

Solution

Denoting by δx_A the virtual displacement at point A, δy_B the virtual displacement at point B, and δy_C the virtual displacement at point C, we have

$$\frac{\delta x_A}{a\sin\theta}=\frac{\delta y_B}{a\cos\theta}=\frac{2\delta y_C}{a\cos\theta}$$

i.e.,

$$\delta x_A = 2\delta y_C \tan\theta$$

From the principle of virtual work, $\delta W=0$, we have

$$-F_{spr}\delta x_A + W\delta y_C = 0, \text{ or } 2F_{spr}\tan\theta = W$$

Using $F_{spr}=ka(1-\cos\theta)$, we have

$$\tan\theta - \sin\theta = \frac{W}{2ka}$$

The above examples have shown that the superiority of the method of virtual work over the conventional method of equilibrium equations is clear due to the fact that all unknown reactions can be eliminated by using the principle of virtual work. It should be noted, however, that the attractiveness of the method of virtual work depends to a large extent upon the existence of simple geometric relations between the various virtual displacements involved in the solution of a given problem. When no such simple relations exist, it is usually advisable to revert to the method of equilibrium equations.

12.3 Generalized Coordinates and Generalized Forces

1. Generalized Coordinates

The independent variables specifying the configuration of a given system are called generalized coordinates. In general, the generalized coordinates can be chosen to be linear and angular coordinates. If a body moves rectilinearly, a linear coordinate can be chosen as the generalized coordinate. If a body rotates around a fixed axis, the natural choice for generalized coordinate is an angular coordinate. For a body in plane motion, the choice for generalized coordinates should be a mixture of linear and angular coordinates.

Sometimes it is suitable to use a larger number of coordinates rather than the number of degrees of freedom. Then these coordinates must be related via some constraints. For example, consider a system consisting of n particles. Assuming that the configuration of the system can be described by generalized coordinates $q_1, q_2, \cdots, q_s$, where s is the number of degrees of freedom of the system, then position vector of particle i can be expressed as

$$\boldsymbol{r}_i = \boldsymbol{r}_i(q_1, q_2, \cdots, q_s, t) \quad (i = 1, 2, \cdots, n) \tag{12.8}$$

Assuming that $\delta \boldsymbol{r}_i$ is the virtual displacement of particle i, consistent with the constraints, then δr_i can be expressed as

$$\delta \boldsymbol{r}_i = \sum_{k=1}^{s} \frac{\partial \boldsymbol{r}_i}{\partial q_k} \delta q_k \quad (i = 1, 2, \cdots, n) \tag{12.9}$$

2. Generalized Forces

Assuming that particle i undergoes a virtual displacement $\delta \boldsymbol{r}_i$, consistent with the constraints, under the action of an applied force $\boldsymbol{F}_i$, then the virtual work δW done by the applied forces acting on the system can be given by

$$\delta W = \sum_{i=1}^{n} \boldsymbol{F}_i \cdot \delta \boldsymbol{r}_i \tag{12.10}$$

Substituting Eq. (12.9) into Eq. (12.10), we have

$$\delta W = \sum_{i=1}^{n} \boldsymbol{F}_i \cdot \left(\sum_{k=1}^{s} \frac{\partial \boldsymbol{r}_i}{\partial q_k} \delta q_k \right) = \sum_{k=1}^{s} \left[\sum_{i=1}^{n} \left(\boldsymbol{F}_i \cdot \frac{\partial \boldsymbol{r}_i}{\partial q_k} \right) \right] \delta q_k = \sum_{k=1}^{s} Q_k \delta q_k \tag{12.11}$$

where

$$Q_k = \sum_{i=1}^{n} \left(\boldsymbol{F}_i \cdot \frac{\partial \boldsymbol{r}_i}{\partial q_k} \right) \quad (k = 1, 2, \cdots, s) \tag{12.12}$$

is called the generalized force associated with the generalized coordinate q_k. Since $Q_k \delta q_k$ has the dimension of work, then Q_k has the dimension of force if q_k is a linear coordinate, and the dimension of moment if q_k is an angular coordinate. There are two methods to calculate the value of Q_k. One of the methods is to use the definition of generalized forces to calculate Q_k. The other method is to determine Q_k from the fact that $Q_k \delta q_k$ is the work done on the system by the applied forces when q_k changes by an arbitrary δq_k (the other generalized coordinates remaining constant), i.e. $Q_k = \delta W / \delta q_k (k = 1, 2, \cdots, s)$.

For a conservative system, the generalized force Q_k can be expressed as

$$Q_k = \sum_{i=1}^{n}\left(\boldsymbol{F}_i \cdot \frac{\partial \boldsymbol{r}_i}{\partial q_k}\right) = -\frac{\partial V}{\partial q_k} \quad (k = 1, 2, \cdots, s) \tag{12.13}$$

where $V=V(q_1, q_2, \cdots, q_s)$ is the potential energy function of the system.

12.4 Conditions of Equilibrium Represented by Generalized Coordinates

According to the principle of virtual work, when the system is in static equilibrium we have

$$\delta W = \sum_{i=1}^{n} \boldsymbol{F}_i \cdot \delta \boldsymbol{r}_i = \sum_{k=1}^{s} Q_k \delta q_k = 0 \tag{12.14}$$

Since $q_1, q_2, \cdots, q_s$ are independent variables and $\delta q_1, \delta q_2, \cdots, \delta q_s$ are arbitrary values, the above equation can be reduced to

$$Q_k = 0 \quad (k = 1, 2, \cdots, s) \tag{12.15}$$

This relation expresses the principle of virtual work in the generalized coordinates: The necessary and sufficient condition for static equilibrium of a system subject to ideal constraints is that all the generalized forces acting on the system are equal to zero.

Problems

12.1 An unstretched spring of stiffness $k=500$ N/m is attached to pins at points I and J, as shown in Fig. P12.1. The pin at B is attached to member BCD and can slide freely along the vertical slot in the fixed plate. Determine the force in the spring and the reaction at point B when a horizontal force of magnitude $F=100$ N, directed to the right, is applied at point H.

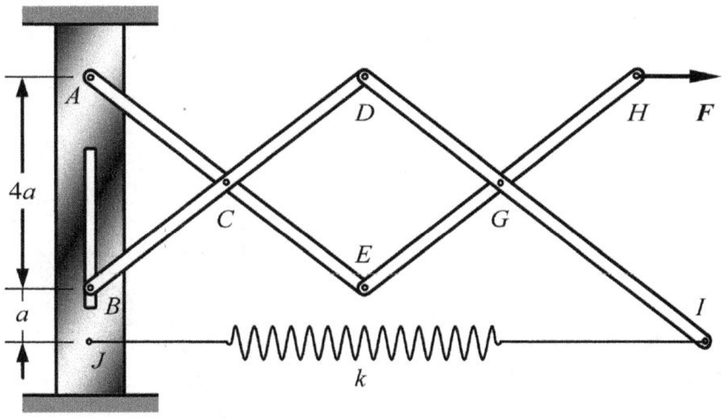

Fig. P12.1

12.2 The mechanism shown in Fig. P12.2 is acted upon by the force $\boldsymbol{F}$. Neglecting the weight of the mechanism and knowing $AC=BC=CD=CE=DG=EG=a$ and $AB=b$, derive an expression for the magnitude of the force R required for equilibrium.

12.3 The pin at B is attached to member ABC and can slide along a slot cut in the fixed plate shown in Fig. P12.3. Neglecting the effect of friction and knowing $AB=BC=CD=a$ and $BD=b$, derive an expression for the magnitude M of the couple required to maintain equilibrium when the force F which acts at A is directed (a) vertically downward, (b) horizontally to the left.

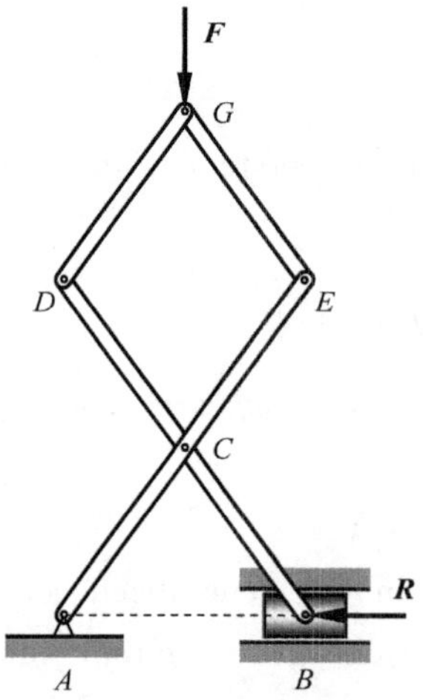

Fig. P12.2

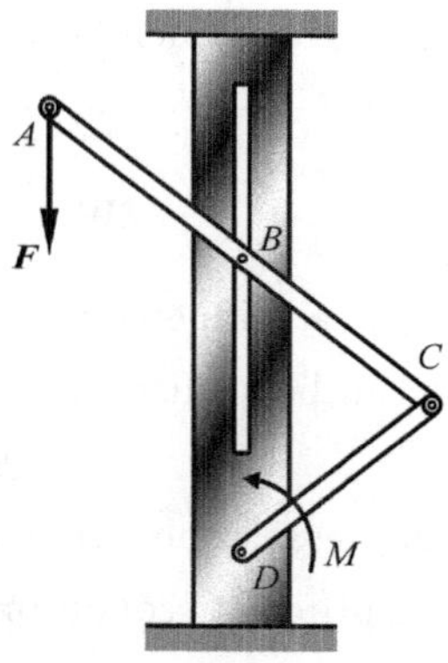

Fig. P12.3

12.4 A slender rod AB of length l is attached to a collar at B and rests on a semicircular cylinder of radius r, as shown in Fig. P12.4. Neglecting the effect of friction and knowing that the mechanism is acted upon by the forces $\boldsymbol{F}_A$ and $\boldsymbol{F}_B$, derive an expression for θ corresponding to the equilibrium position of the mechanism.

12.5 Two rods ABC and CDE are connected by a pin at C and by a spring AE. The stiffness of the spring is k, and the spring is unstretched when $\theta = 30°$, as shown in Fig. P12.5. Knowing $AB=BC=CD=DE=a$, derive an equation in F, θ, l, and k that must be satisfied when the system is in equilibrium under the action of the loading shown.

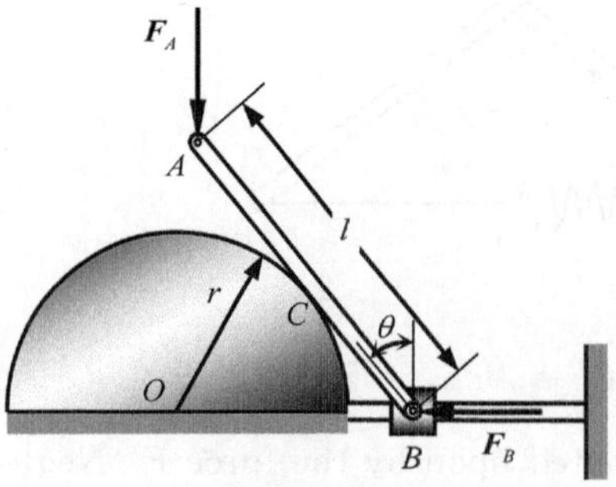

Fig. P12.4

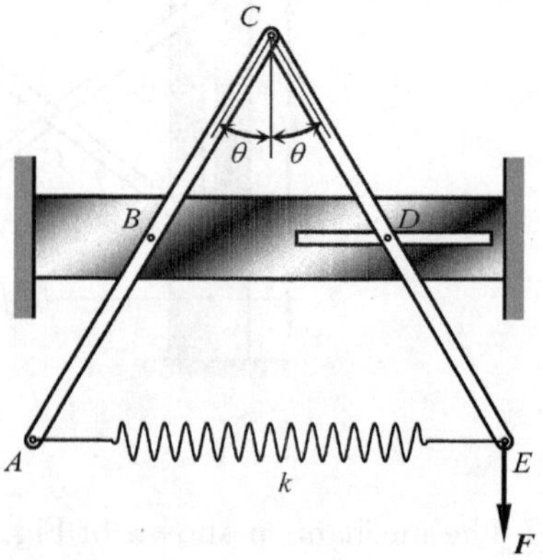

Fig. P12.5

12.6 A horizontal force $\boldsymbol{F}$ of magnitude 150 N is applied to the mechanism at A, as shown in Fig. P12.6. The stiffness of the spring is $k=1.5$ kN/m, and the spring is unstretched when $\theta=0$. Neglecting the mass of the mechanism and knowing $a=250$ mm, $r=150$ mm, determine the value of θ corresponding to equilibrium.

12.7 Determine the vertical displacement of joint I shown in Fig. P12.7 if the length of member AC is increased by 6 mm.

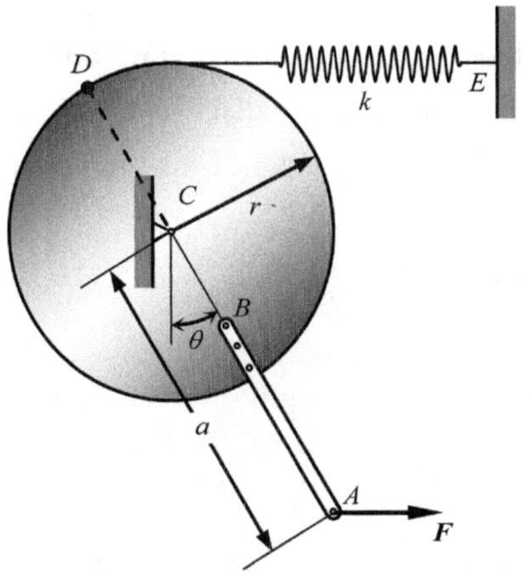

Fig. P12.6

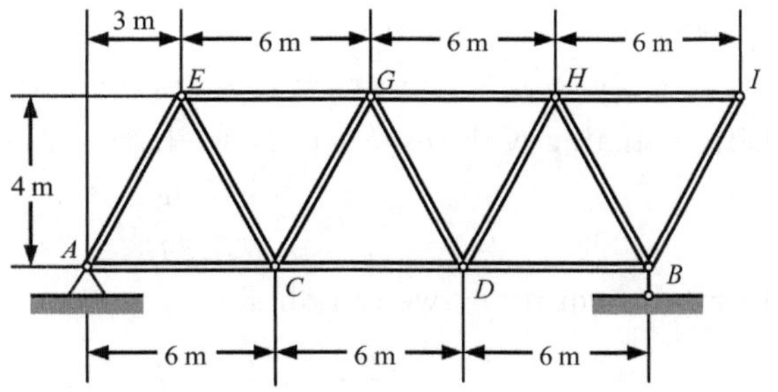

Fig. P12.7

Chapter 13 Lagrange's Equations

13.1 Lagrange's Equations

Consider that a system consisting of n particles is subjected to holonomic constraints. Assuming that the configuration of the system can be described by generalized coordinates $q_1, q_2, \cdots, q_s$, where s is the number of degrees of freedom of the system, then position vector of particle i can be expressed as

$$\boldsymbol{r}_i = \boldsymbol{r}_i(q_1, q_2, \cdots, q_s, t) \quad (i = 1, 2, \cdots, n) \tag{13.1}$$

Differentiating with respect to t, we have

$$\dot{\boldsymbol{r}}_i = \sum_{k=1}^{s} \frac{\partial \boldsymbol{r}_i}{\partial q_k}\dot{q}_k + \frac{\partial \boldsymbol{r}_i}{\partial t} \quad (i = 1, 2, \cdots, n) \tag{13.2}$$

From this equation, we can obtain

$$\frac{\partial \dot{\boldsymbol{r}}_i}{\partial \dot{q}_k} = \frac{\partial \boldsymbol{r}_i}{\partial q_k} \quad (i = 1, 2, \cdots, n;\ k = 1, 2, \cdots, s) \tag{13.3}$$

Multiplying by $\dot{\boldsymbol{r}}_i$ and differentiating with respect to t, we then have

$$\frac{\mathrm{d}}{\mathrm{d}t}\left(\dot{\boldsymbol{r}}_i \cdot \frac{\partial \dot{\boldsymbol{r}}_i}{\partial \dot{q}_k}\right) = \frac{\mathrm{d}}{\mathrm{d}t}\left(\dot{\boldsymbol{r}}_i \cdot \frac{\partial \boldsymbol{r}_i}{\partial q_k}\right) = \ddot{\boldsymbol{r}}_i \cdot \frac{\partial \boldsymbol{r}_i}{\partial q_k} + \dot{\boldsymbol{r}}_i \cdot \frac{\partial \dot{\boldsymbol{r}}_i}{\partial q_k} \quad (i = 1, 2, \cdots, n;\ k = 1, 2, \cdots, s) \tag{13.4}$$

i.e.,

$$\frac{\mathrm{d}}{\mathrm{d}t}\frac{\partial}{\partial \dot{q}_k}\left(\frac{1}{2}m_i \dot{\boldsymbol{r}}_i \cdot \dot{\boldsymbol{r}}_i\right) = (m_i \ddot{\boldsymbol{r}}_i) \cdot \frac{\partial \boldsymbol{r}_i}{\partial q_k} + \frac{\partial}{\partial q_k}\left(\frac{1}{2}m_i \dot{\boldsymbol{r}}_i \cdot \dot{\boldsymbol{r}}_i\right) \quad (i = 1, 2, \cdots, n;\ k = 1, 2, \cdots, s) \tag{13.5}$$

where m_i is the mass of particle i. By summing over i, we obtain

$$\frac{\mathrm{d}}{\mathrm{d}t}\frac{\partial}{\partial \dot{q}_k}\left(\sum_{i=1}^{n} \frac{1}{2}m_i \dot{\boldsymbol{r}}_i \cdot \dot{\boldsymbol{r}}_i\right) = \sum_{i=1}^{n}\left[(m_i \ddot{\boldsymbol{r}}_i) \cdot \frac{\partial \boldsymbol{r}_i}{\partial q_k}\right] + \frac{\partial}{\partial q_k}\left(\sum_{i=1}^{n} \frac{1}{2}m_i \dot{\boldsymbol{r}}_i \cdot \dot{\boldsymbol{r}}_i\right) \quad (k = 1, 2, \cdots, s) \tag{13.6}$$

i.e.,

$$\frac{\mathrm{d}}{\mathrm{d}t}\frac{\partial T}{\partial \dot{q}_k} = \sum_{i=1}^{n}\left(\boldsymbol{F}_i \cdot \frac{\partial \boldsymbol{r}_i}{\partial q_k}\right) + \frac{\partial T}{\partial q_k} \quad (k = 1, 2, \cdots, s) \tag{13.7}$$

where $T = \sum_{i=1}^{n} \frac{1}{2}m_i \dot{\boldsymbol{r}}_i \cdot \dot{\boldsymbol{r}}_i$ is the kinetic energy of the system, $\boldsymbol{F}_i = m_i \ddot{\boldsymbol{r}}_i$ is the force acting on particle i. Using the definition of generalized force, the above equation can be rewritten as

$$\frac{\mathrm{d}}{\mathrm{d}t}\frac{\partial T}{\partial \dot{q}_k} - \frac{\partial T}{\partial q_k} = Q_k \quad (k = 1, 2, \cdots, s) \tag{13.8}$$

where $Q_k = \sum_{i=1}^{n}\left(\boldsymbol{F}_i \cdot \frac{\partial \boldsymbol{r}_i}{\partial q_k}\right)$ is the generalized force associated with the generalized q_k. The above equations are called Lagrange's equations.

For a conservative system, $Q_k = -\dfrac{\partial V}{\partial q_k}$, then Lagrange's equations can be expressed as

$$\frac{\mathrm{d}}{\mathrm{d}t}\frac{\partial T}{\partial \dot{q}_k} - \frac{\partial T}{\partial q_k} = -\frac{\partial V}{\partial q_k} \quad (k = 1,2,\cdots,s) \tag{13.9}$$

Defining

$$L = T - V \tag{13.10}$$

where L is called the Lagrangian function. Using $\dfrac{\partial V}{\partial \dot{q}_k} = 0$, Lagrange's equations for a conservative system can be rewritten as

$$\frac{\mathrm{d}}{\mathrm{d}t}\frac{\partial L}{\partial \dot{q}_k} - \frac{\partial L}{\partial q_k} = 0 \quad (k = 1,2,\cdots,s) \tag{13.11}$$

If part of the generalized forces are not conservative, say Q'_k, and part are derivable from a potential function V, i.e.,

$$Q_k = Q'_k - \frac{\partial V}{\partial q_k} \quad (k = 1,2,\cdots,s) \tag{13.12}$$

we then obtain Lagrange's equations in the most general form:

$$\frac{\mathrm{d}}{\mathrm{d}t}\frac{\partial L}{\partial \dot{q}_k} - \frac{\partial L}{\partial q_k} = Q'_k \quad (k = 1,2,\cdots,s) \tag{13.13}$$

Example 13.1

A mass-spring system consists of two blocks of mass m_1 and m_2 and two springs of stiffness k_1 and k_2. Using the method of Lagrange's equations, determine the differential equations of motion of the system.

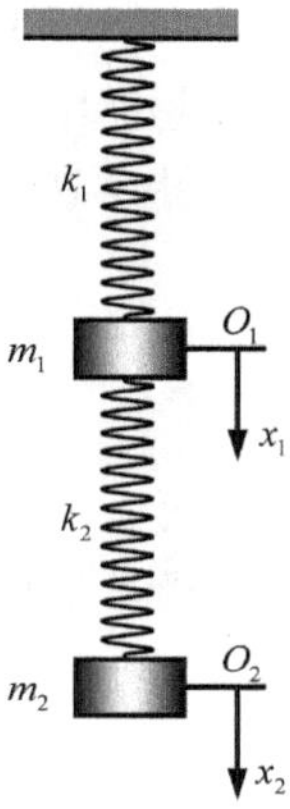

Fig. E13.1

Solution

The system has two degrees of freedom and is subjected to conservative forces. If x_1 and x_2 are chosen as two generalized coordinates measured from the positions of static equilibrium of each block, both are positive downward, then the kinetic and potential energies of the system at an arbitrary position can be given, respectively, by

$$T = \frac{1}{2}m_1\dot{x}_1^2 + \frac{1}{2}m_2\dot{x}_2^2,\ V = \frac{1}{2}k_1(\delta_1 + x_1)^2 + \frac{1}{2}k_2(\delta_2 + x_2 - x_1)^2 - m_1 g x_1 - m_2 g x_2$$

where $\delta_1 = \dfrac{(m_1 + m_2)g}{k_1}$ and $\delta_2 = \dfrac{m_2 g}{k_2}$. Hence, we can obtain the Lagrangian function

$$L = T - V = \frac{1}{2}m_1\dot{x}_1{}^2 + \frac{1}{2}m_2\dot{x}_2{}^2 - \frac{1}{2}k_1(\delta_1 + x_1)^2$$
$$- \frac{1}{2}k_2(\delta_2 + x_2 - x_1)^2 + m_1 g x_1 + m_2 g x_2$$

For the block m_1, we have

$$\frac{\partial L}{\partial \dot{x}_1}=m_1\dot{x}_1,\ \frac{\partial L}{\partial x_1}=-k_1(\delta_1+x_1)+k_2(\delta_2+x_2-x_1)+m_1 g=-(k_1+k_2)x_1+k_2x_2$$

Substituting into Lagrange's equation, we obtain

$$m_1\ddot{x}_1+(k_1+k_2)x_1-k_2x_2=0$$

Similarly, we have

$$\frac{\partial L}{\partial \dot{x}_2}=m_2\dot{x}_2,\ \frac{\partial L}{\partial x_2}=-k_2(\delta_2+x_2-x_1)+m_2 g=k_2x_1-k_2x_2$$

and the differential equation of motion for the block m_2

$$m_2\ddot{x}_2-k_2x_1+k_2x_2=0$$

Example 13.2

A two degree of freedom system consists of two blocks and three pulleys. The center pulley is free to move vertically and it has a mass m. The string connecting the three masses is weightless. Masses m_1 and m_2 hang on the left and right respectively from the fixed pulleys. All three pulleys are frictionless, so that the string slides freely over them. Using the method of Lagrange's equations, determine the differential equations of motion of the system.

Solution

This is a conservative system having two degrees of freedom. If x_1 and x_2 are chosen as two generalized coordinates measured from the reference plane, both are positive upward, then the kinetic and potential energies of the system at an arbitrary position can be given, respectively, by

$$T=\frac{1}{2}m_1\dot{x}_1{}^2+\frac{1}{2}m_2\dot{x}_2{}^2+\frac{1}{2}m\left[\frac{1}{2}(\dot{x}_1+\dot{x}_2)\right]^2,$$

$$V=m_1gx_1+m_2gx_2+mg\left[h-\frac{1}{2}(x_1+x_2)\right]$$

where h is a constant. The Lagrangian function is equal to

$$L=T-V=\frac{1}{2}m_1\dot{x}_1^2+\frac{1}{2}m_2\dot{x}_2^2+\frac{1}{2}m\left[\frac{1}{2}(\dot{x}_1+\dot{x}_2)\right]^2$$

$$-m_1gx_1-m_2gx_2-mg\left[h-\frac{1}{2}(x_1+x_2)\right]$$

For the block m_1, we have

$$\frac{\partial L}{\partial \dot{x}_1}=m_1\dot{x}_1+\frac{1}{4}m(\dot{x}_1+\dot{x}_2),\ \frac{\partial L}{\partial x_1}=-m_1g+\frac{1}{2}mg$$

Substituting into Lagrange's equation, we obtain

$$\left(m_1+\frac{1}{4}m\right)\ddot{x}_1+\frac{1}{4}m\ddot{x}_2+\left(m_1-\frac{1}{2}m\right)g=0$$

Similarly, we have the differential equation of motion for the block m_2

$$\frac{1}{4}m\ddot{x}_1+\left(m_2+\frac{1}{4}m\right)\ddot{x}_2+\left(m_2-\frac{1}{2}m\right)g=0$$

Example 13.3

A block of mass m_1 slides on a smooth inclined plane, which itself has a mass m_2 and slides on a smooth horizontal surface. Determine the horizontal acceleration of the block and the acceleration of the inclined plane.

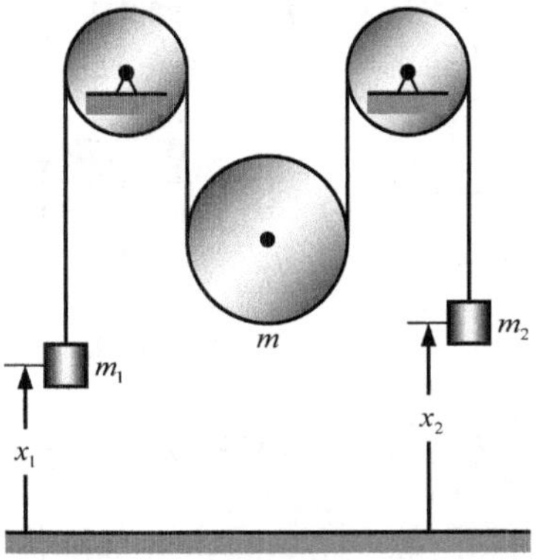

Fig. E13.2

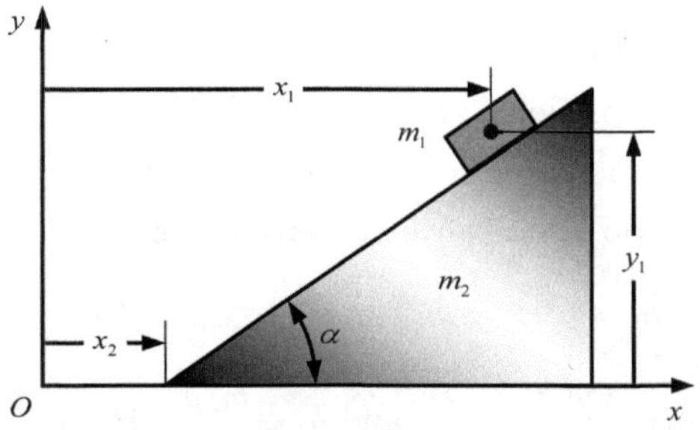

Fig. E13.3

Solution

Choosing x_1 and x_2 as the generalized coordinates, then $y_1 = \tan\alpha(x_1 - x_2)$ and $\dot{y}_1 = \tan\alpha(\dot{x}_1 - \dot{x}_2)$. Hence, the kinetic and potential energies of the system can be expressed, respectively, as

$$T = \frac{1}{2}m_1[\dot{x}_1^2 + \tan^2\alpha\,(\dot{x}_1 - \dot{x}_2)^2] + \frac{1}{2}m_2\dot{x}_2^2,\quad V = m_1 g\tan\alpha(x_1 - x_2) + m_2 gh$$

where h is a constant. The corresponding Lagrangian function is equal to

$$L = T - V = T = \frac{1}{2}m_1[\dot{x}_1^2 + \tan^2\alpha\,(\dot{x}_1 - \dot{x}_2)^2] + \frac{1}{2}m_2\dot{x}_2^2 - m_1 g\tan\alpha(x_1 - x_2) - m_2 gh$$

For the block of mass m_1, we have

$$\frac{\partial L}{\partial \dot{x}_1} = m_1[\dot{x}_1 + \tan^2\alpha(\dot{x}_1 - \dot{x}_2)],\quad \frac{\partial L}{\partial x_1} = -m_1 g\tan\alpha$$

Substituting into Lagrange's equation, we obtain

$$m_1[\ddot{x}_1 + \tan^2\alpha(\ddot{x}_1 - \ddot{x}_2)] + m_1 g\tan\alpha = 0$$

Similarly, for the inclined plane of mass m_2, we have

$$\frac{\partial L}{\partial \dot{x}_2} = -m_1\tan^2\alpha(\dot{x}_1 - \dot{x}_2) + m_2\dot{x}_2,\quad \frac{\partial L}{\partial x_2} = m_1 g\tan\alpha$$

Substituting into Lagrange's equation, we obtain

$$-m_1\tan^2\alpha(\ddot{x}_1 - \ddot{x}_2) + m_2\ddot{x}_2 - m_1 g\tan\alpha = 0$$

Solving the equations obtained, we have

$$\ddot{x}_1 = -\frac{m_2 g\sin\alpha\cos\alpha}{m_1\sin^2\alpha + m_2},\quad \ddot{x}_2 = \frac{m_1 g\sin\alpha\cos\alpha}{m_1\sin^2\alpha + m_2}$$

Example 13.4

A two degrees of freedom system of damped forced vibration, consisting of two blocks of mass m_1 and m_2, two springs of stiffness k_1 and k_2, and two dashpots of damping coefficient c_1 and c_2, is subjected to two periodic forces of magnitude $H_1\sin\omega_1 t$ and $H_2\sin\omega_2 t$, where H_1 and H_2 are the amplitudes of the periodic forces, and ω_1 and ω_2 are the circular frequencies of the periodic forces, respectively. Determine the differential equations of motion of the system.

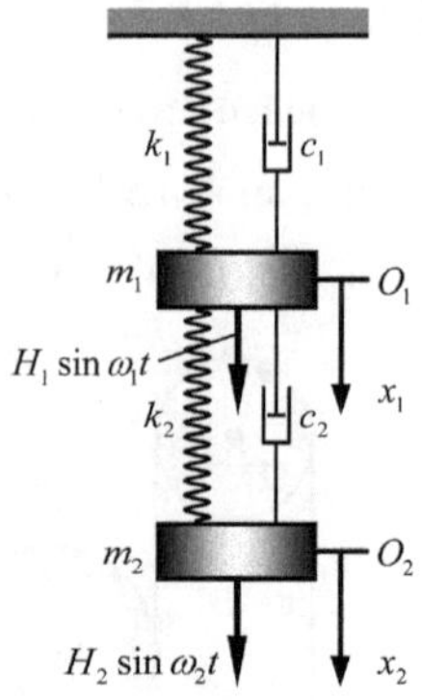

Fig. E13. 4

Solution

Choosing x_1 and x_2 as the generalized coordinates measured from the positions of static equilibrium of the block, then the kinetic and potential energies of the system located at an arbitrary position can be given, respectively, by

$$T=\frac{1}{2}m_1\dot{x}_1^{\ 2}+\frac{1}{2}m_2\dot{x}_2^{\ 2},\ V=\frac{1}{2}k_1(\delta_1+x_1)^2+\frac{1}{2}k_2(\delta_2+x_2-x_1)^2-m_1gx_1-m_2gx_2$$

where $\delta_1=\dfrac{(m_1+m_2)g}{k_1}$ and $\delta_2=\dfrac{m_2 g}{k_2}$. Hence, we can obtain the Lagrangian function

$$L=T-V=\frac{1}{2}m_1\dot{x}_1^{\ 2}+\frac{1}{2}m_2\dot{x}_2^{\ 2}-\frac{1}{2}k_1(\delta_1+x_1)^2$$
$$-\frac{1}{2}k_2(\delta_2+x_2-x_1)^2+m_1gx_1+m_2gx_2$$

For the block m_1, we have

$$\frac{\partial L}{\partial\dot{x}_1}=m_1\dot{x}_1,\ \frac{\partial L}{\partial x_1}=-k_1(\delta_1+x_1)+k_2(\delta_2+x_2-x_1)+m_1g=-(k_1+k_2)x_1+k_2x_2$$

Using Lagrange's equation, $\dfrac{\mathrm{d}}{\mathrm{d}t}\dfrac{\partial L}{\partial\dot{x}_1}-\dfrac{\partial L}{\partial x_1}=Q_1'$, we have

$$m_1\ddot{x}_1+(k_1+k_2)x_1-k_2x_2=Q_1'$$

where $Q_1'=H_1\sin\omega_1 t-c_1\dot{x}_1+c_2(\dot{x}_2-\dot{x}_1)$. Hence, the differential equation of motion for the block m_1 can be rewritten as

$$m_1\ddot{x}_1+(c_1+c_2)\dot{x}_1-c_2\dot{x}_2+(k_1+k_2)x_1-k_2x_2=H_1\sin\omega_1 t$$

Similarly, we have

$$\frac{\partial L}{\partial\dot{x}_2}=m_2\dot{x}_2,\ \frac{\partial L}{\partial x_2}=-k_2(\delta_2+x_2-x_1)+m_2g=k_2x_1-k_2x_2$$

and the differential equation for the block m_2

$$m_2\ddot{x}_2-k_2x_1+k_2x_2=Q_2'$$

where $Q_2'=H_2\sin\omega_2 t-c_2(\dot{x}_2-\dot{x}_1)$. Substituting Q_2' into the above equation, we have the differential equation of motion for the block m_2

$$m_2\ddot{x}_2-c_2\dot{x}_1+c_2\dot{x}_2-k_2x_1+k_2x_2=H_2\sin\omega_2 t$$

13. 2 First Integrals of Lagrange's Equations

1. Energy Integral

Consider that a conservative system consisting of n particles is subjected to scleronomous constraints. Assuming that the configuration of the system can be described by generalized coordinates $q_1, q_2, \cdots, q_s$, where s is the number of degrees of freedom of the system, then position vector of particle i can be expressed as

$$\boldsymbol{r}_i = \boldsymbol{r}_i(q_1, q_2, \cdots, q_s) \quad (i = 1, 2, \cdots, n) \tag{13.14}$$

Differentiating with respect to t, we have

$$\dot{\boldsymbol{r}}_i = \sum_{k=1}^{s} \frac{\partial \boldsymbol{r}_i}{\partial q_k} \dot{q}_k \quad (i = 1, 2, \cdots, n) \tag{13.15}$$

Using the above expression, the kinetic energy of the system can be expressed as

$$T = \sum_{i=1}^{n} \frac{1}{2} m_i \dot{\boldsymbol{r}}_i \cdot \dot{\boldsymbol{r}}_i = \sum_{i=1}^{n} \frac{1}{2} m_i \left(\sum_{k=1}^{s} \frac{\partial \boldsymbol{r}_i}{\partial q_k} \dot{q}_k \right) \cdot \left(\sum_{l=1}^{s} \frac{\partial \boldsymbol{r}_i}{\partial q_l} \dot{q}_l \right) = \sum_{k,l=1}^{s} \frac{1}{2} m_{kl} \dot{q}_k \dot{q}_l \tag{13.16}$$

where

$$m_{kl} = \sum_{i=1}^{n} m_i \frac{\partial \boldsymbol{r}_i}{\partial q_k} \cdot \frac{\partial \boldsymbol{r}_i}{\partial q_l} \quad (k, l = 1, 2, \cdots, s) \tag{13.17}$$

is called the generalized mass. Differentiating Eq. (13.16) with respect to the generalized velocity $\dot{q}_k$, we have

$$\frac{\partial T}{\partial \dot{q}_k} = \frac{\partial}{\partial \dot{q}_k} \sum_{k,l=1}^{s} \frac{1}{2} m_{kl} \dot{q}_k \dot{q}_l = 2 \sum_{l=1}^{s} \frac{1}{2} m_{kl} \dot{q}_l \quad (k = 1, 2, \cdots, s) \tag{13.18}$$

Multiplying by $\dot{q}_k$ and summing over k, we then have

$$\sum_{k=1}^{s} \frac{\partial T}{\partial \dot{q}_k} \dot{q}_k = 2 \sum_{k,l=1}^{s} \frac{1}{2} m_{kl} \dot{q}_k \dot{q}_l = 2T \tag{13.19}$$

Using $\partial T / \partial \dot{q}_k = \partial L / \partial \dot{q}_k$, the above expression can be rewritten as

$$2T = \sum_{k=1}^{s} \frac{\partial L}{\partial \dot{q}_k} \dot{q}_k \tag{13.20}$$

Differentiating with respect to t, we have

$$2\dot{T} = \frac{\mathrm{d}}{\mathrm{d}t} \sum_{k=1}^{s} \frac{\partial L}{\partial \dot{q}_k} \dot{q}_k = \sum_{k=1}^{s} \left[\frac{\mathrm{d}}{\mathrm{d}t} \left(\frac{\partial L}{\partial \dot{q}_k} \right) \dot{q}_k + \frac{\partial L}{\partial \dot{q}_k} \ddot{q}_k \right] = \sum_{k=1}^{s} \left(\frac{\partial L}{\partial q_k} \dot{q}_k + \frac{\partial L}{\partial \dot{q}_k} \ddot{q}_k \right) = \dot{L} \tag{13.21}$$

i.e.,

$$2\dot{T} - \dot{L} = 0 \tag{13.22}$$

Integrating the above equation, we obtain

$$2T - L = T + V = \text{constant} \tag{13.23}$$

The above expression shows that when a conservative system is subjected to scleronomous constraints, the sum of the kinetic energy and of the potential energy of the system remains constant.

2. Cyclic Integral

If a conservative system consisting of n particles can be described by the generalized coordinates $q_1, q_2, \cdots, q_s$, where s is the number of degrees of freedom of the system, then the generalized momenta are defined as

$$p_k = \frac{\partial L}{\partial \dot{q}_k} \quad (k = 1, 2, \cdots, s) \tag{13.24}$$

Using Lagrange's equation for a conservative system, we have

$$\dot{p}_k = \frac{\partial L}{\partial q_k} \quad (k = 1, 2, \cdots, s) \tag{13.25}$$

Assuming that one of the generalized coordinates, say q_λ, is not explicitly contained in the Lagrangian function L, then we have

$$\dot{p}_\lambda = \frac{\partial L}{\partial q_\lambda} = 0 \tag{13.26}$$

where q_λ is called the ignorable or cyclic coordinate. Integrating the above equation, we obtain

$$p_\lambda = \text{constant} \tag{13.27}$$

The above expression shows that the generalized momentum associated with an ignorable or cyclic coordinate is a constant.

Problems

13.1 The mass-spring system shown in Fig. P13.1 consists of two blocks of mass m_1 and m_2 and two springs of stiffness k_1 and k_2. Using the method of Lagrange's equations, determine the differential equations of motion of the system.

13.2 The system shown in Fig. P13.2 consists of two blocks of mass m_1 and m_2, respectively, connected by a light inextensible cord of length l which passes over a uniform pulley of mass m and radius r. There is no relative motion between the cord and the pulley. Using the method of Lagrange's equations, determine the differential equations of motion of the system.

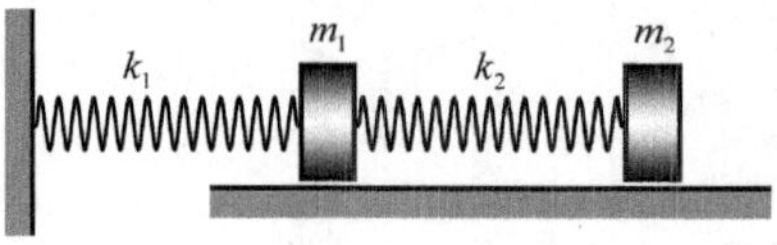

Fig. P13.1

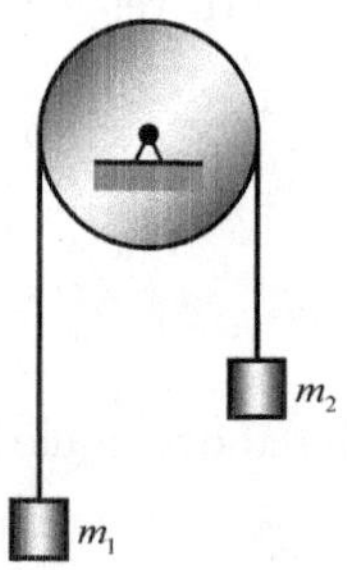

Fig. P13.2

13.3 A small sphere of mass m_1 attached to a string of length l is suspended from the boundary of a uniform disk of mass m_2 and radius r, as shown in Fig. P13.3. Assuming that the disk can rotate freely about O, determine the differential equations of motion of the system.

13.4 A block of mass m_1 slides on a smooth inclined plane, which itself has a mass m_2 and slides on a smooth horizontal surface, as shown in Fig. P13.4. Determine the relative acceleration of the block with respect to the inclined plane and the acceleration of the inclined plane.

13.5 A body of mass m suspended from a spring of stiffness k and a dashpot of damping coefficient c is subjected to a periodic force of magnitude $H\sin\omega_f t$, where H and ω_f are respectively the amplitude and the circular frequency of the periodic force, as shown in Fig. P13.5. Determine the differential equation of motion of the body.

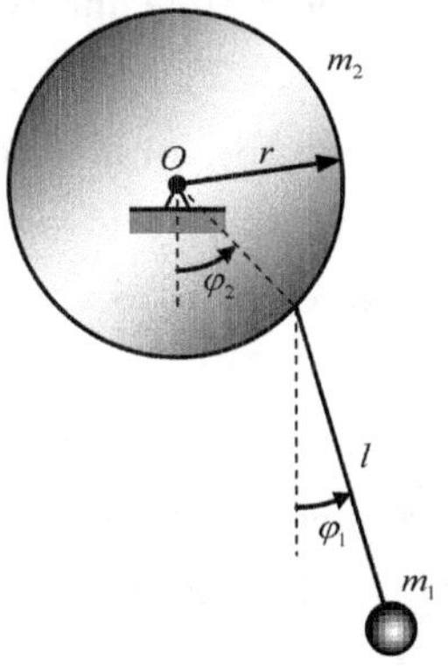

Fig. P13.3

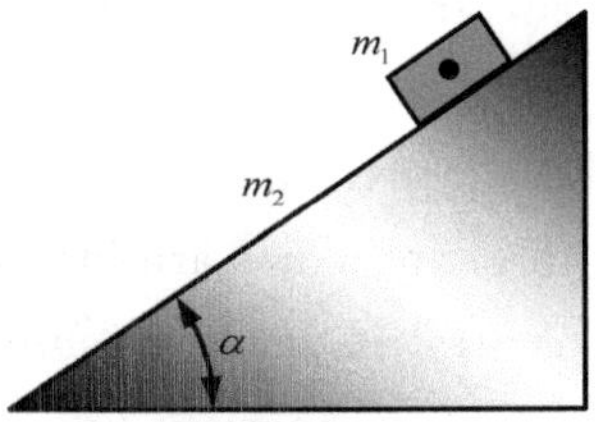

Fig. P13.4

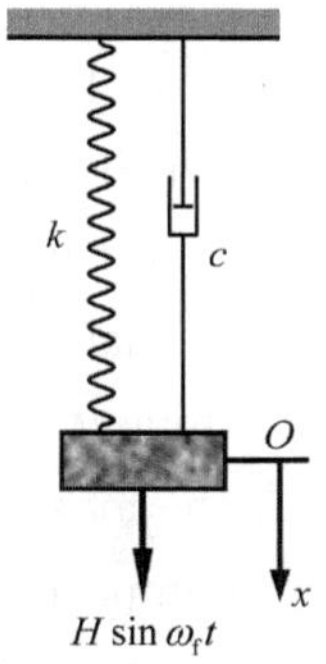

Fig. P13.5

13.6 A two degrees of freedom system of damped free vibration, as shown in Fig. P13.6, consists of two block of mass m_1 and m_2, two springs of stiffness k_1 and k_2, and two dashpots of damping coefficient c_1 and c_2. Determine the differential equations of motion of the system.

13.7 A two degrees of freedom system of undamped forced vibration, as shown in Fig. P13.7, consisting of two block, each of mass m, and four springs, each of stiffness k, is subjected to two periodic forces, each of magnitude $H\sin\omega_f t$, where H and ω_f are the amplitude and the circular frequency of the periodic force, respectively. Determine the differential equations of motion of the system.

13.8 A collar of mass m_1 can slide on a horizontal beam and is attached to a spring of stiffness k, as shown in Fig. P13.8. From the collar hangs a homogeneous rod of length l and mass m_2 which can swing in a vertical plane through the beam. Neglecting friction, find Lagrange's equations of motion for the system.

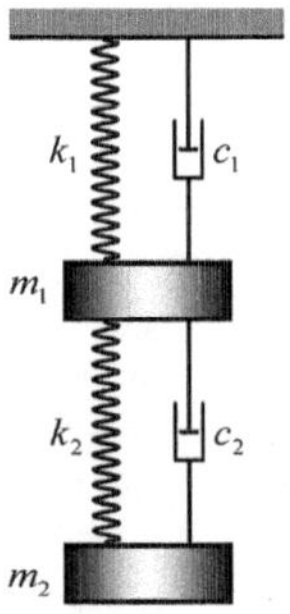

Fig. P13.6

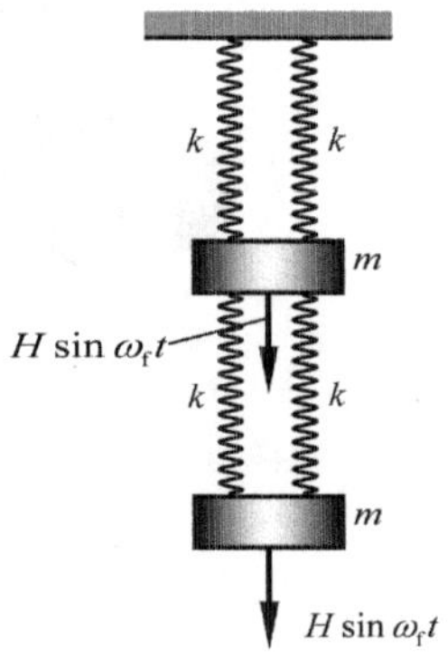

Fig. P13.7

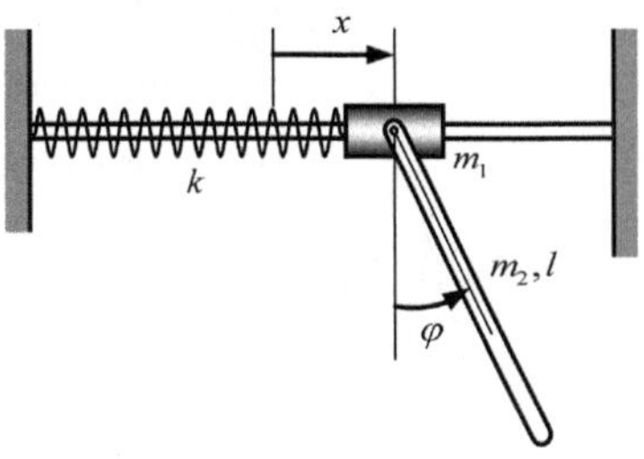

Fig. P13.8

Chapter 14 Impact

An interaction between two bodies, occurring during a very short time interval and causing very large acting forces between the bodies, is called an impact. The striking of a hammer on a nail is a typical example of impact. When a nail is struck, the contact between hammer and nail takes place during a very short time interval. But the force exerted by the hammer on the nail is very large, and the resulting impulse is large enough to change the motion of the nail.

The common normal to the surfaces in contact during the impact is called the line of impact. If the mass centers of the two bodies are located on the common normal during the impact, the impact is called a central impact, as shown in Fig. 14.1(a). Otherwise the impact is called an eccentric impact, as shown in Fig. 14.1(b).

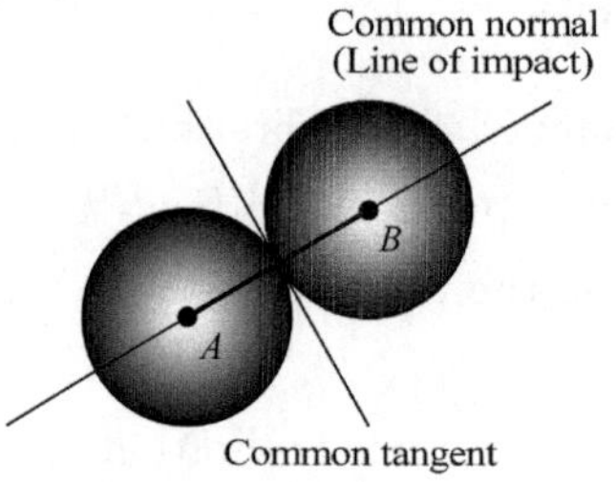

(a) Central impact

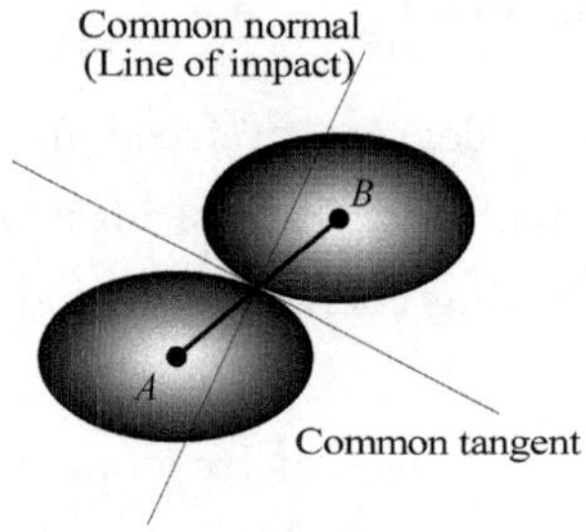

(b) Eccentric impact

Fig. 14.1

If the velocities of the two bodies are along the common normal, the impact is called a direct impact, as shown in Fig. 14.2. If either or both bodies move along a line other than the common normal, the impact is called an oblique impact, as shown in Fig. 14.3.

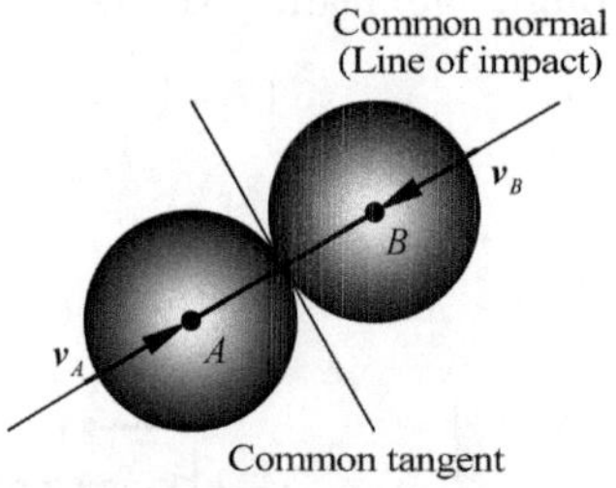

(a) Direct central impact

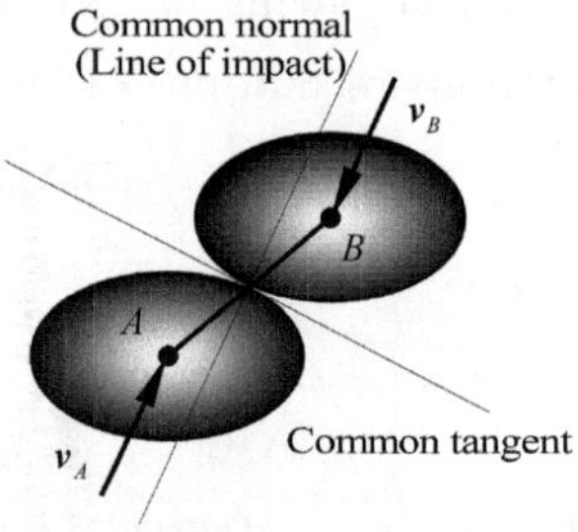

(b) Direct eccentric impact

Fig. 14.2

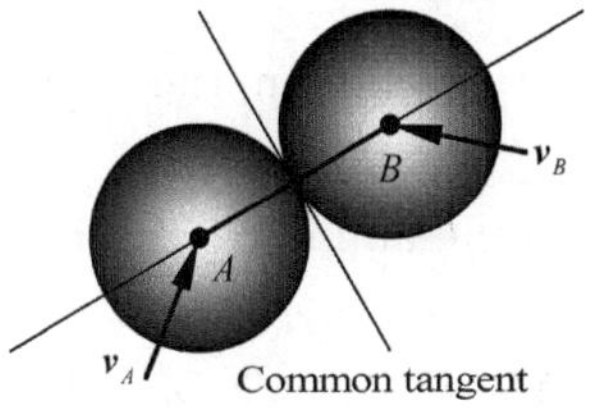

(a) Oblique central impact

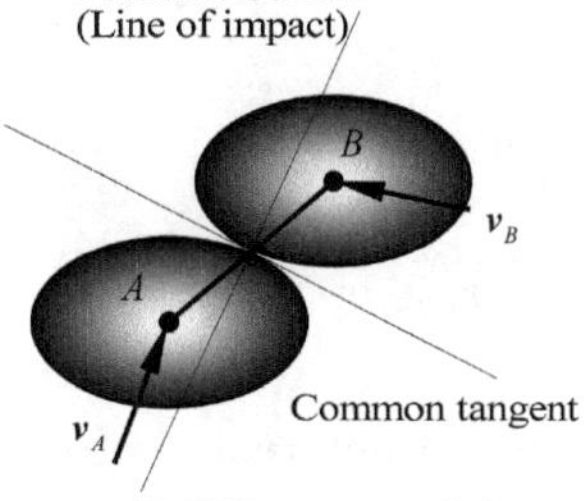

(b) Oblique eccentric impact

Fig. 14.3

14.1 Fundamental Principles Used for Impact

1. Impact for Particle

Considering that a particle of mass m has a velocity $\boldsymbol{v}$ before impact and $\boldsymbol{v}'$ after impact, the principle of impulse and momentum during impact can be expressed as

$$m\boldsymbol{v}' - m\boldsymbol{v} = \boldsymbol{I} \tag{14.1}$$

where $\boldsymbol{I}$ is the impulse of the impulsive forces acting on the particle during impact. An impulsive force is defined as a force acting on a particle, which is large enough to cause an abrupt change in momentum of the particle during a very short interval of time. A nonimpulsive force can be neglected during impact due to the impulse produced by this force is usually very small. The gravitational and elastic forces can be considered as nonimpulsive forces during impact.

2. Impact for Rigid Body in Plane Motion

For a rigid body in plane motion, the principle of impulse and momentum during impact can be written as

$$m\,\boldsymbol{v}'_C - m\,\boldsymbol{v}_C = \boldsymbol{I},\ I_C\,\omega' - I_C\,\omega = M_C(\boldsymbol{I}) \tag{14.2}$$

where m is the mass of the rigid body, $\boldsymbol{v}_C$ and $\boldsymbol{v}'_C$ are the velocities of the mass center of the rigid body before and after impact, $\boldsymbol{I}$ is the impulse of the impulsive forces acting on the rigid body during the impact, I_C is the mass moment of inertia about the mass center of the rigid body, ω and ω' are the angular velocities of the rigid body before and after impact, and $M_C(\boldsymbol{I})$ is the angular impulse about the mass center C of the impulsive forces acting on the rigid body during the impact. The change in position of the mass center of the rigid body during the impact has been neglected since it is very small during a very short interval of time.

3. Impact for System of Particles

For a system of n particles, the principle of impulse and momentum during impact can be given by

$$\boldsymbol{L}' - \boldsymbol{L} = \sum_{i=1}^{n} \boldsymbol{I}_i^{(e)},\ \boldsymbol{H}'_O - \boldsymbol{H}_O = \sum_{i=1}^{n} \boldsymbol{M}_O(\boldsymbol{I}_i^{(e)}) \tag{14.3}$$

where $\boldsymbol{L} = \sum_{i=1}^{n} m_i \boldsymbol{v}_i$ and $\boldsymbol{L}' = \sum_{i=1}^{n} m_i \boldsymbol{v}'_i$ are the momenta of the system before and after impact, $\sum_{i=1}^{n} \boldsymbol{I}_i^{(e)}$ is the impulse of the external impulsive forces acting on the system during the impact, $\boldsymbol{H}_O = \sum_{i=1}^{n} \boldsymbol{r}_i \times m_i \boldsymbol{v}_i$ and $\boldsymbol{H}'_O = \sum_{i=1}^{n} \boldsymbol{r}_i \times m_i \boldsymbol{v}'_i$ are the angular momenta about a fixed point O of the system before and after impact, and $\sum_{i=1}^{n} \boldsymbol{M}_O(\boldsymbol{I}_i^{(e)}) = \sum_{i=1}^{n} \boldsymbol{r}_i \times \boldsymbol{I}_i^{(e)}$ is the angular impulse about the same point O of the external impulsive forces acting on the system during the impact. The displacement of each particle in the system during the impact has been neglected, since the interval of time is so short that the displacement of each particle is very small.

14.2 Coefficient of Restitution

1. Central Impact

Consider a direct central impact in which two particles of mass m_A and m_B have velocities $\boldsymbol{v}_A$ and $\boldsymbol{v}_B$ before impact as shown in Fig. 14.4(a). If $v_A > v_B$, particle A will eventually collide with particle B. Assuming that the velocities of the two particles after impact are respectively $\boldsymbol{v}'_A$ and $\boldsymbol{v}'_B$ as shown in Fig. 14.4(b), then the coefficient of restitution can be expressed as

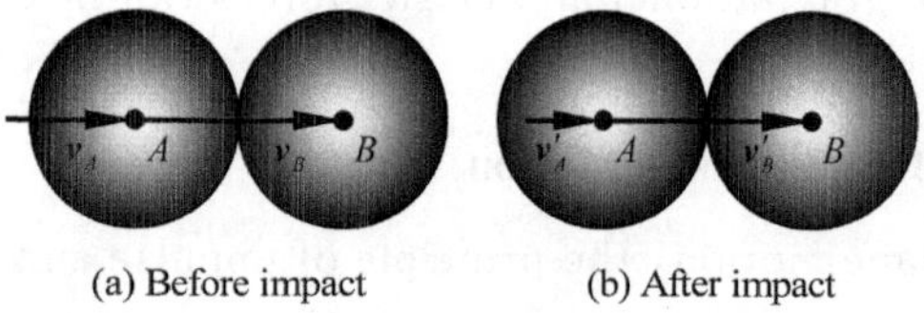

Fig. 14.4

$$e = \frac{v'_B - v'_A}{v_A - v_B} \tag{14.4}$$

where $v_A - v_B$ represents the relative velocity of approach of the two particles before impact, and $v'_B - v'_A$ represents the relative velocity of separation of the two particles after impact. According to the value of the coefficient of restitution, an impact can be classified as follows:

(1) Perfectly Elastic Impact ($e=1$)

In this case, we have $v'_B - v'_A = v_A - v_B$, i.e. the relative velocities of two particles before and after impact are equal. In the case of a perfectly elastic impact, the total energy of two particles, as well as their total momentum, is conserved.

(2) Perfectly Plastic Impact ($e=0$)

In this case, we obtain $v'_B = v'_A$, i. e. two particles stay together after impact. In the case of a perfectly plastic impact, although the total momentum is conserved, the total energy is no longer conserved.

(3) Elastic Plastic Impact ($0<e<1$)

In the case of an elastic plastic impact, it should be noted that the total energy of the particles is not conserved, but their total momentum is still conserved.

For an oblique central impact shown in Fig. 14.5, assuming that two particles have velocities $\boldsymbol{v}_A$ and $\boldsymbol{v}_B$ before impact and $\boldsymbol{v}'_A$ and $\boldsymbol{v}'_B$ after impact, then the coefficient of restitution can be given by

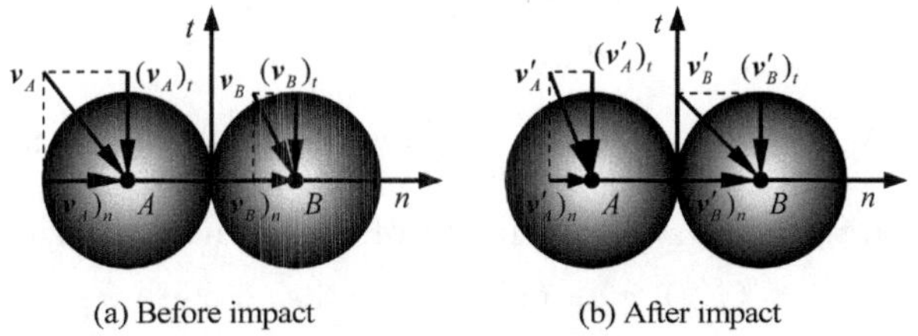

Fig. 14.5

$$e=\frac{(v'_B)_n-(v'_A)_n}{(v_A)_n-(v_B)_n} \tag{14.5}$$

where $(v_A)_n-(v_B)_n$ and $(v'_B)_n-(v'_A)_n$ represent the n components of the relative velocities of the two particles before and after impact, respectively.

Example 14.1

Two uniform spheres, of mass $m_A=0.5$ kg and $m_B=0.3$ kg, have the same size and are sliding on a frictionless horizontal plane with velocities of magnitude $v_A=10$ m/s and $v_B=5$ m/s. Knowing that the coefficient of restitution is $e=0.6$, determine the velocity of each sphere after impact.

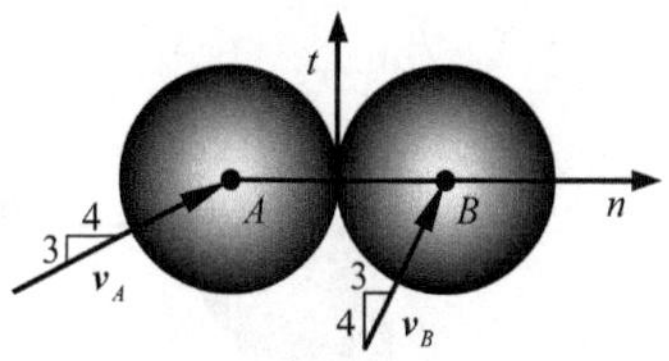

Fig. E14.1

Solution

Resolving the velocities of each sphere before impact into components along the t and n axes, we have

$$(v_A)_t=6\ \text{m/s},\ (v_A)_n=8\ \text{m/s},\ (v_B)_t=4\ \text{m/s},\ (v_B)_n=3\ \text{m/s}$$

Applying the principle of impulse and momentum to each sphere and noting that the impulse of the impulsive force acting on each sphere during the impact is along the n axis, we obtain

$$m_A[(v'_A)_t\boldsymbol{e}_t+(v'_A)_n\boldsymbol{e}_n]-m_A[(v_A)_t\boldsymbol{e}_t+(v_A)_n\boldsymbol{e}_n]=-I\boldsymbol{e}_n$$

$$m_B[(v'_B)_t\boldsymbol{e}_t+(v'_B)_n\boldsymbol{e}_n]-m_B[(v_B)_t\boldsymbol{e}_t+(v_B)_n\boldsymbol{e}_n]=I\boldsymbol{e}_n$$

where $\boldsymbol{e}_t$ and $\boldsymbol{e}_n$ are the unit vectors respectively along the t and n axes, I is the value of the impulse of the impulsive force exerted by one sphere on another.

Since the t component of the impulse of the impulsive force acting on each sphere is equal to zero, the t component of the velocity of each sphere remains unchanged during the impact, i.e. we have

$$(v'_A)_t = (v_A)_t = 6 \text{ m/s}, \ (v'_B)_t = (v_B)_t = 4 \text{ m/s}$$

Considering the two spheres as a single system, the n component of the impulse of the impulsive forces acting on the system is still equal to zero, though the n component of the impulse of the impulsive force acting on each sphere is not equal to zero. Hence we obtain

$$m_A[(v'_A)_n - (v_A)_n] + m_B[(v'_B)_n - (v_B)_n] = 0$$

Using $(v'_B)_n - (v'_A)_n = e[(v_A)_n - (v_B)_n]$, we obtain

$$(v'_A)_n = (v_A)_n - \frac{(1+e)m_B}{m_A + m_B}[(v_A)_n - (v_B)_n] = 5 \text{ m/s}$$

$$(v'_B)_n = (v_B)_n + \frac{(1+e)m_A}{m_A + m_B}[(v_A)_n - (v_B)_n] = 8 \text{ m/s}$$

Using the above computing results, the velocity of each sphere after impact can be given by

$$\boldsymbol{v}'_A = (v'_A)_t \boldsymbol{e}_t + (v'_A)_n \boldsymbol{e}_n = 6\, \boldsymbol{e}_t + 5\, \boldsymbol{e}_n \text{ m/s}$$

$$\boldsymbol{v}'_B = (v'_B)_t \boldsymbol{e}_t + (v'_B)_n \boldsymbol{e}_n = 4\, \boldsymbol{e}_t + 8\, \boldsymbol{e}_n \text{ m/s}$$

2. Eccentric Impact

Consider that an eccentric impact takes place between two bodies shown in Fig. 14. 6. Assuming that the velocities before impact of the points of contact A and B are $\boldsymbol{v}_A$ and $\boldsymbol{v}_B$, and that the velocities of the same points are v'_A and $\boldsymbol{v}'_B$ after impact, the coefficient of restitution can be defined as

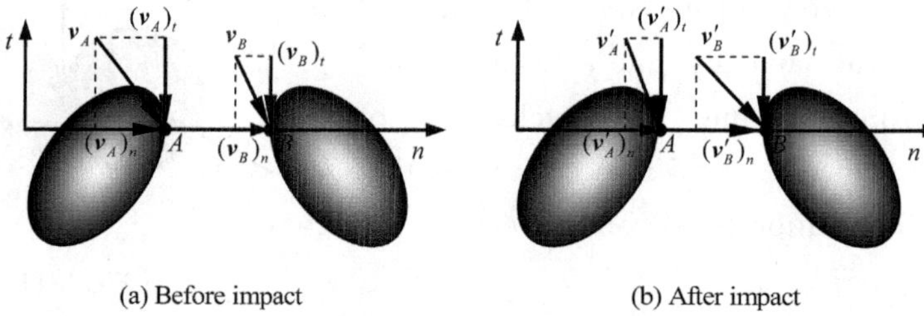

Fig. 14. 6

$$e = \frac{(v'_B)_n - (v'_A)_n}{(v_A)_n - (v_B)_n} \tag{14. 6}$$

where $(v_A)_n - (v_B)_n$ and $(v'_B)_n - (v'_A)_n$ are respectively the components of the relative velocities along the common normal of the surfaces in contact of the bodies before and after impact.

Example 14. 2

A uniform slender rod AB of mass m and length l forms an angle θ with the horizontal direction as it strikes the frictionless surface shown with a vertical velocity $\boldsymbol{v}$ and no angular velocity. Assuming that the impact is perfectly elastic, derive expressions for both the angular velocity of the rod immediately after the impact and the impulse of the impulsive force exerted by the surface on the rod during the impact.

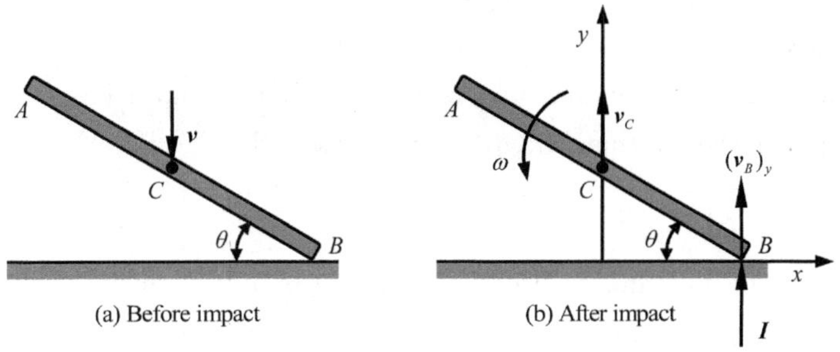

Fig. E14.2

Solution

Establishing the coordinate system shown in Fig. E14.2, applying the principle of impulse and momentum to the rod, and noting that the velocity of the mass center of the rod after impact is vertical, and that the impulse of the impulsive force acting on the rod during the impact is also vertical, thus we have

$$mv_C - m(-v) = I, \quad I_C\omega - 0 = I\left(\frac{1}{2}l\cos\theta\right)$$

where I_C is the mass moment of inertia about the mass center of the rod, and I is the value of the impulse of the impulsive force exerted by the surface on the rod. Using $I_C = \frac{1}{12}ml^2$, $(v_B)_y = ev = v$, and $v_C = (v_B)_y - \omega\left(\frac{1}{2}l\right)\cos\theta = v - \frac{1}{2}\omega l\cos\theta$, we solve the above equations for ω and I, and obtain

$$\omega = \frac{12v\cos\theta}{(1+3\cos^2\theta)l}, \quad I = \frac{2mv}{1+3\cos^2\theta}$$

Example 14.3

A bullet of mass m_1 is fired with a horizontal velocity of magnitude $\boldsymbol{v}$ into a uniform slender bar of mass m_2 and length l. Knowing that the bar is initially at rest, determine (a) the angular velocity of the bar immediately after the bullet becomes embedded, (b) the impulse of the impulsive reaction at A, and (c) the distance a if the impulsive reaction at A is equal to zero.

Solution

Applying the principle of impulse and momentum to the system consisting of the bullet and the bar, and noting that the impulse of the impulsive reaction acting on the bar during the impact is horizontal, we have

$$m_1 a\omega + \frac{1}{2}m_2 l\omega - m_1 v = I_A, \quad m_1 a^2\omega + \frac{1}{3}m_2 l^2\omega - m_1 av = 0$$

where ω is the angular velocity of the bar immediately after the bullet becomes embedded, and I_A is the impulse of the impulsive reaction exerted by the support on the bar at A.

Solving the above equations for ω and I_A, we obtain

$$\omega = \frac{3m_1 av}{3m_1 a^2 + m_2 l^2}, \quad I_A = \frac{(3a - 2l) l m_1 m_2 v}{6m_1 a^2 + 2m_2 l^2}$$

It can be seen from the second expression above that if the impulsive reaction at A is equal to zero, then the distance a is equal to

$$a = \frac{2}{3}l$$

The point located on the bar and having a distance $a = 2l/3$ from the pivot point is called the center of percussion. The center of percussion is defined as the point located on an object where a perpendicular impact will not cause impulsive forces at a given pivot point.

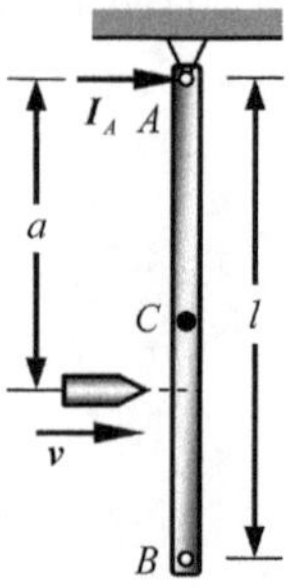

Fig. E14.3

Problems

14.1 A 1-kg sphere A is moving with a velocity $\boldsymbol{v}_A$ when it is struck by a 2-kg sphere B which has a velocity $\boldsymbol{v}_B$ of magnitude 5 m/s, as shown in Fig. P14.1. Knowing that the velocity of sphere B is zero after impact and that the coefficient of restitution is 0.5, determine the velocity of sphere A (a) before impact, (b) after impact.

14.2 Four spheres of equal mass as shown in Fig. P14.2, are suspended from the ceiling by cords of equal length which are spaced at a distance slightly greater than the diameter of the spheres. After sphere A is pulled back and released, three collisions will take place between adjacent spheres A, B, C, and D. Denoting by e the coefficient of restitution between the spheres and by $\boldsymbol{v}$ the velocity of A just before it hits B, determine (a) the velocities of A and B immediately after the first collision, (b) the velocities of B and C immediately after the second collision, and (c) the velocities of C and D immediately after the third collision.

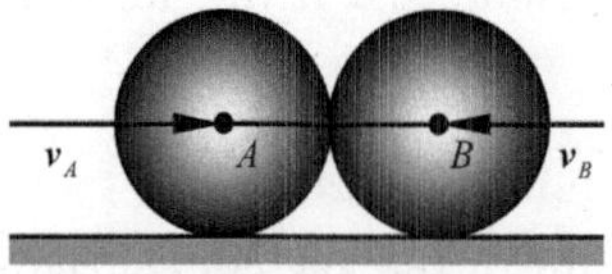

Fig. P14.1

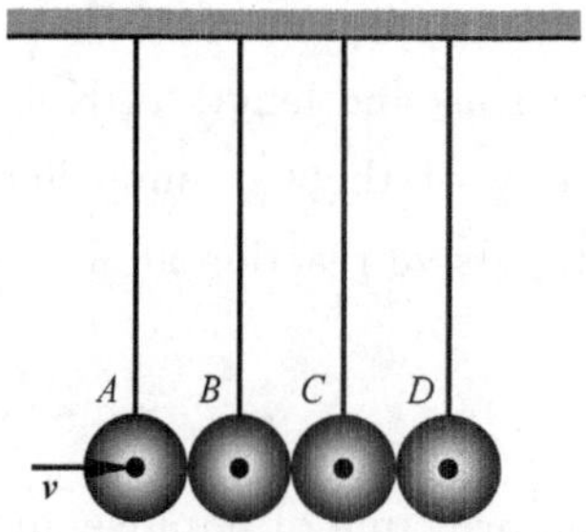

Fig. P14.2

14.3 A sphere A of mass m_A moving with a velocity $\boldsymbol{v}$ parallel to the ground strikes the inclined face of a block B of mass m_B, which can slide freely on the ground and is initially at rest, as shown in Fig. P14.3. Knowing that the coefficient of restitution between the sphere and the block is e, determine the velocity of the block immediately after impact.

14.4 A 2-kg block A is moving with a velocity of magnitude $v=1$ m/s as it hits the 1-kg block B, which is at rest and hanging from a cord attached at O, as shown in Fig. P14.4. Knowing that $e=0.8$ between the blocks and $\mu_k=0.5$ between the block and the horizontal surface, determine after impact, (a) the maximum height reached by the block B, (b) the distance traveled by the block A.

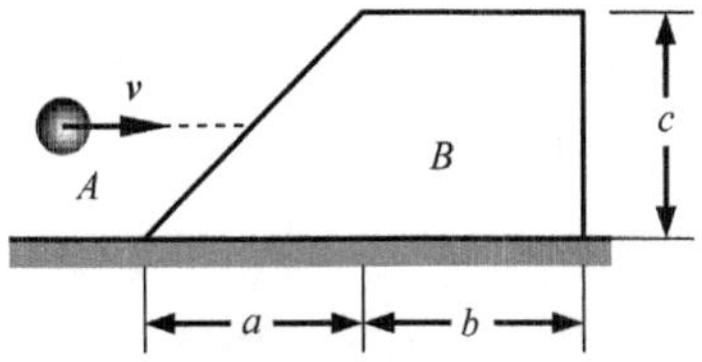

Fig. P14.3

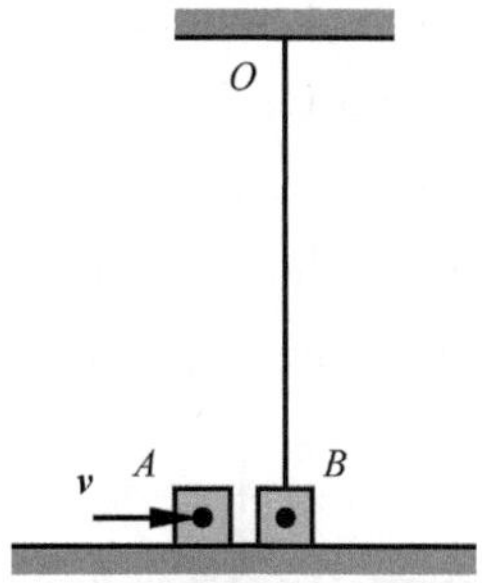

Fig. P14.4

14.5 Three spheres, each of mass m, can slide on a frictionless, horizontal surface. Spheres A and B are attached to an inextensible cord and are at rest in the position shown in Fig. P14.5 when sphere B is struck by sphere C which is moving with a velocity $\boldsymbol{v}$. Knowing that the cord is taut when sphere B is struck by sphere C and assuming that the coefficient of restitution $e=1$ between B and C, determine the velocity of each sphere immediately after impact.

14.6 Three spheres, each of mass m, can slide on a frictionless, horizontal surface. Spheres A and B are attached to an inextensible cord and are at rest in the position shown in Fig. P14.6 when sphere B is struck by sphere C which is moving with a velocity $\boldsymbol{v}$. Knowing that the cord is slack when sphere B is struck by sphere C and assuming that the coefficient of restitution is equal to e between B and C, determine the velocity of each sphere immediately after the cord becomes taut.

14.7 A bullet of mass $m_1=10$ g is fired with a horizontal velocity of magnitude $v=500$ m/s into the lower end of a uniform slender bar of mass $m_2=5$ kg and length $l=1$ m, as shown in Fig. P14.7. Knowing that the bar is initially at rest, determine (a) the angular velocity of the bar immediately after the bullet becomes embedded, (b) the impulse exerted on the bar at point A.

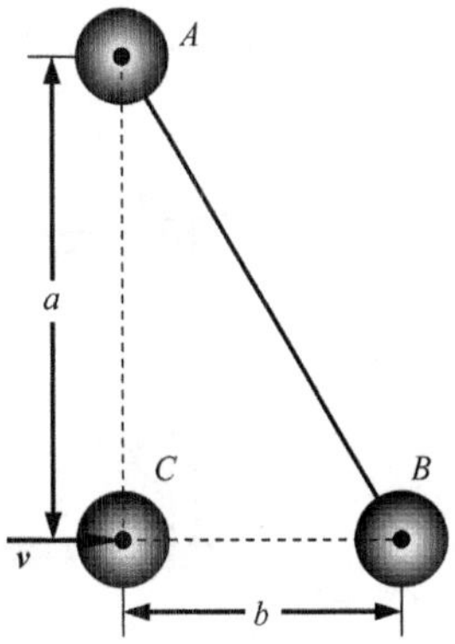

Fig. P14.5

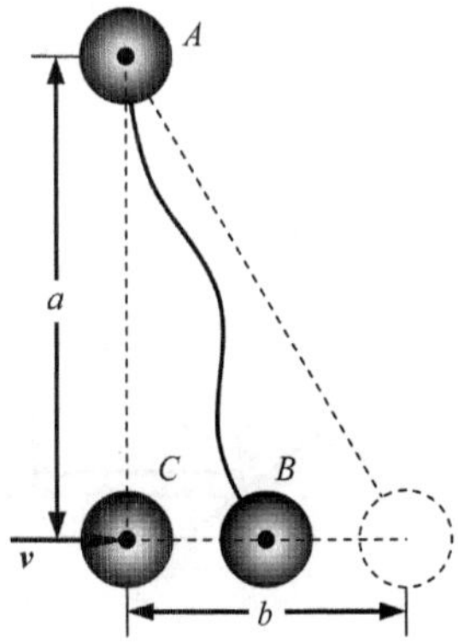

Fig. P14.6

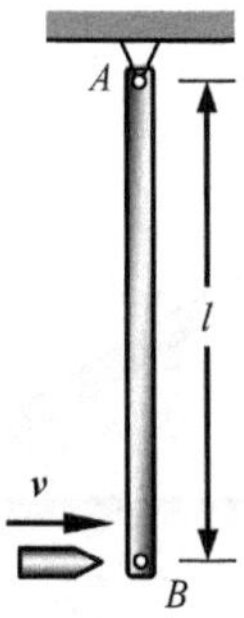

Fig. P14.7

14. 8 As shown in Fig. P14. 8 , a uniform slender rod AB of mass m and length l has a vertical velocity of magnitude v and no angular velocity when it strikes a rigid frictionless support at point D. Knowing that $a=l/5$ and assuming that $e=0$, determine (a) the angular velocity of the rod and the velocity of its mass center immediately after the impact, (b) the impulse exerted on the rod at point D.

14. 9 As shown in Fig. P14. 9, a uniform slender rod $AB$of mass m and length l has an angular velocity ω counterclockwise and a zero velocity of the mass center when it strikes a rigid frictionless support at point D. Knowing that $a = l/5$ and assuming that $e = 1$, determine (a) the angular velocity of the rod and the velocity of its mass center immediately after the impact, (b) the impulse exerted on the rod at point D.

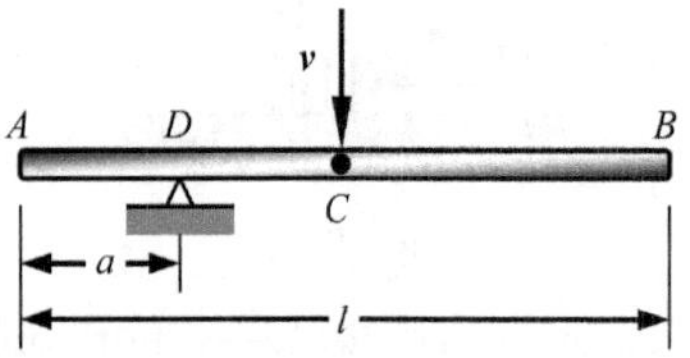

Fig. P14. 8

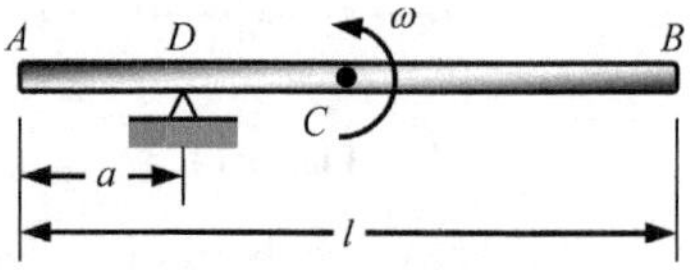

Fig. P14. 9

14. 10 A uniform slender rod AB, as shown in Fig. P14. 10, of mass m and length l forms an angle θ with the horizontal direction as it strikes the frictionless surface with a vertical velocity $\boldsymbol{v}$ and no angular velocity. Assuming that the coefficient of restitution between the rod and the surface is e, derive an expression for the angular velocity of the rod immediately after the impact.

14. 11 A uniform slender rod AB, as shown in Fig. P14. 11, of mass m and length l is falling freely with a velocity $\boldsymbol{v}$ when end B strikes a smooth inclined surface with an angle of inclination θ as shown. Assuming that the impact is perfectly elastic, determine the angular velocity of the rod and the velocity of its mass center immediately after the impact.

14. 12 A uniform slender rod AB of mass m_1 and length l is released from rest in the horizontal position shown in Fig. P14. 12. It swings down to a vertical position and strikes a block C of mass m_2 which is resting on a frictionless surface. Assuming that the coefficient of restitution between the rod and the block is e, determine the velocity of the block immediately after the impact.

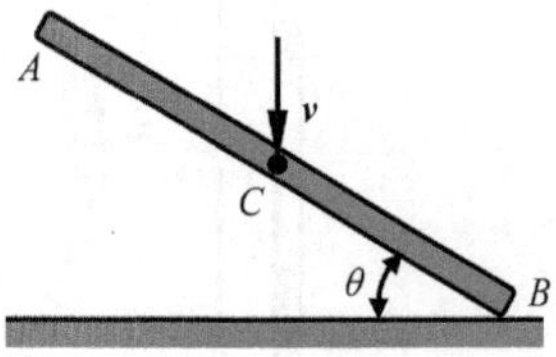

Fig. P14. 10

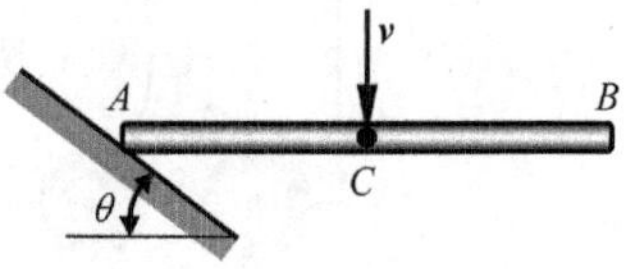

Fig. P14. 11

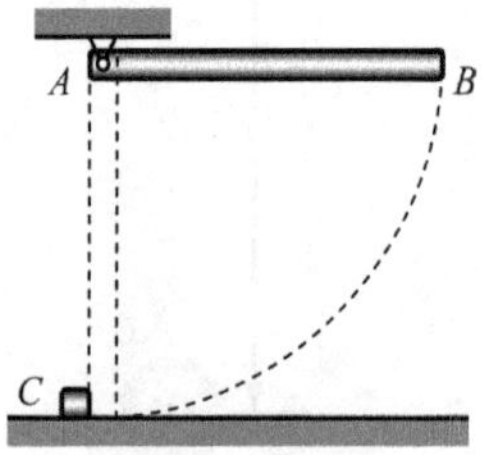

Fig. P14. 12

Appendix Ⅰ Centers of Gravity and Centroids

Ⅰ.1 Center of Gravity and Centroid of Plate

1. Center of Gravity of Plate

Considering a plate shown in Fig. Ⅰ.1, the center of gravity of this plate can be defined as

$$\bar{x} = \frac{\int x\mathrm{d}W}{W},\quad \bar{y} = \frac{\int y\mathrm{d}W}{W} \tag{Ⅰ.1}$$

where $W = \int \mathrm{d}W$ is the weight of the plate.

2. Centroid of Plate

If the plate has both homogeneous density ρ and uniform thickness t, the center of gravity G of the plate will coincide with its centroid C, as shown in Fig. Ⅰ.2. Using $W=\rho gAt$, the centroid of the plate can be expressed as

$$\bar{x} = \frac{\int x\mathrm{d}A}{A},\quad \bar{y} = \frac{\int y\mathrm{d}A}{A} \tag{Ⅰ.2}$$

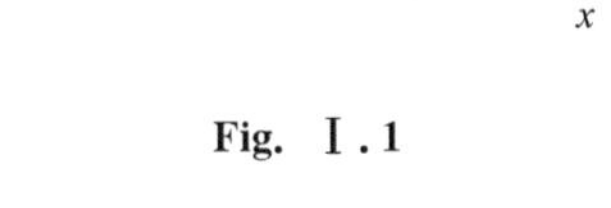

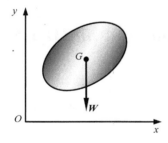

Fig. Ⅰ.1

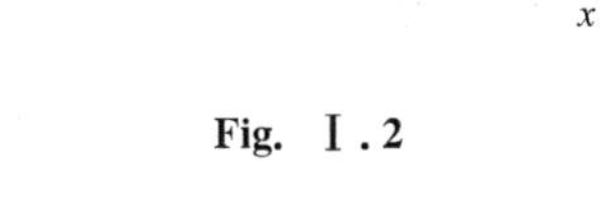

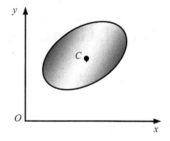

Fig. Ⅰ.2

where $A = \int \mathrm{d}A$ is the area of the plate.

Ⅰ.2 Center of Gravity and Centroid of Composite Plate

1. Center of Gravity of Composite Plate

Considering a composite plate shown in Fig. Ⅰ.3, the center of gravity of this composite plate can be determined by

$$\bar{X}=\frac{\sum \bar{x}_i W_i}{W},\ \bar{Y}=\frac{\sum \bar{y}_i W_i}{W} \qquad (\text{I}.3)$$

where $W=\sum W_i$ is the weight of the composite plate.

2. Centroid of Composite Plate

If the composite plate has both homogeneous density ρ and uniform thickness t, the center of gravity G of the composite plate will coincide with its centroid C, as shown in Fig. Ⅰ.4. Using $W_i=\rho g A_i t$, the centroid of the composite plate can be given by

$$\bar{X}=\frac{\sum \bar{x}_i A_i}{A},\ \bar{Y}=\frac{\sum \bar{y}_i A_i}{A} \qquad (\text{I}.4)$$

where $A=\sum A_i$ is the area of the composite plate.

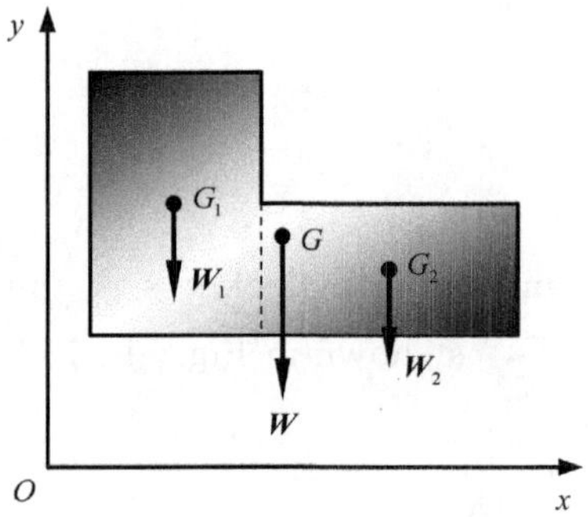

Fig. Ⅰ.3

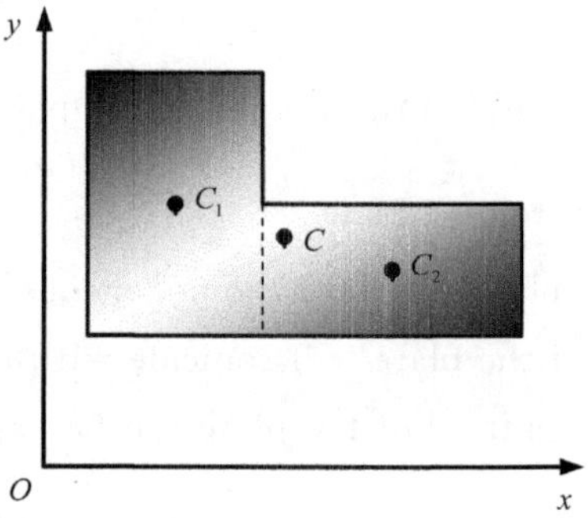

Fig. Ⅰ.4

Ⅰ.3 Center of Gravity and Centroid of 3D Body

1. Center of Gravity of 3D Body

Considering a three-dimensional body shown in Fig. Ⅰ.5, the center of gravity of this 3D body can be defined as

$$\bar{x}=\frac{\int x\mathrm{d}W}{W},\ \bar{y}=\frac{\int y\mathrm{d}W}{W},\ \bar{z}=\frac{\int z\mathrm{d}W}{W} \qquad (\text{I}.5)$$

where $W=\int \mathrm{d}W$ is the weight of the 3D body.

2. Centroid of 3D Body

If the 3D body is made of a homogeneous material of density ρ, the center of gravity G of the 3D body will coincide with its centroid C, as shown in Fig. Ⅰ.6. Using $W=\rho g V$, the centroid of the 3D body can be expressed as

$$\bar{x}=\frac{\int x\mathrm{d}V}{V},\ \bar{y}=\frac{\int y\mathrm{d}V}{V},\ \bar{z}=\frac{\int z\mathrm{d}V}{V} \qquad (\text{I}.6)$$

where $V = \int dV$ is the volume of the 3D body.

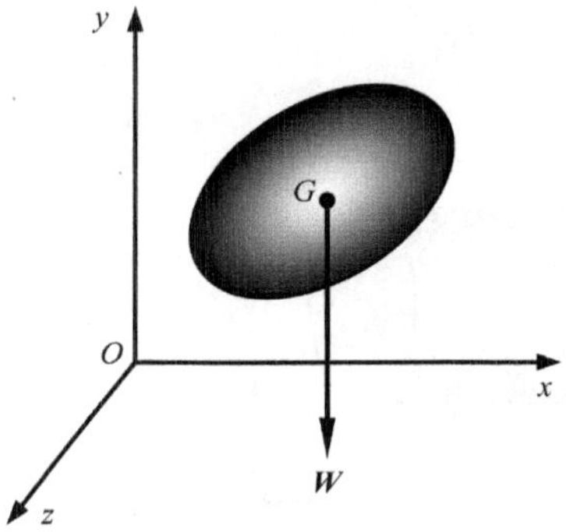

Fig. Ⅰ.5

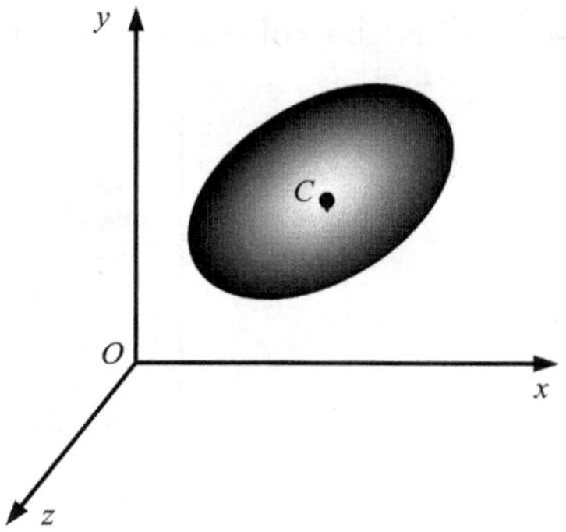

Fig. Ⅰ.6

Ⅰ.4 Center of Gravityand Centroid of 3D Composite Body

1. Center of Gravity of 3D Composite Body

Considering a three-dimensional composite body shown in Fig. Ⅰ.7, the center of gravity of this 3D composite body can be defined as

$$\bar{X} = \frac{\sum \bar{x}_i W_i}{W},\ \bar{Y} = \frac{\sum \bar{y}_i W_i}{W},\ \bar{Z} = \frac{\sum \bar{z}_i W_i}{W} \qquad (\text{Ⅰ}.7)$$

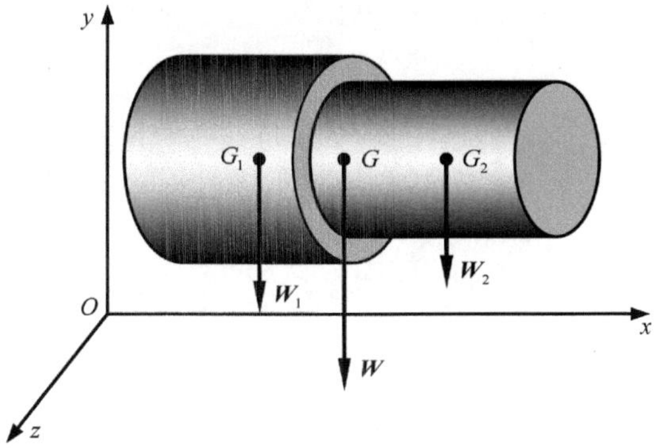

Fig. Ⅰ.7

where $W = \sum W_i$ is the weight of the 3D composite body.

2. Centroid of 3D Composite Body

If the 3D composite body is made of a homogeneous material of density ρ, the center of gravity G of the 3D composite body will coincide with its centroid C, as shown in Fig. Ⅰ.8. Using $W_i = \rho g V_i$, the centroid of the 3D composite body can be expressed as

$$\bar{X} = \frac{\sum \bar{x}_i V_i}{V}, \bar{Y} = \frac{\sum \bar{y}_i V_i}{V}, \bar{Z} = \frac{\sum \bar{z}_i V_i}{V} \tag{Ⅰ.8}$$

where $V = \sum V_i$ is the volume of the 3D composite body.

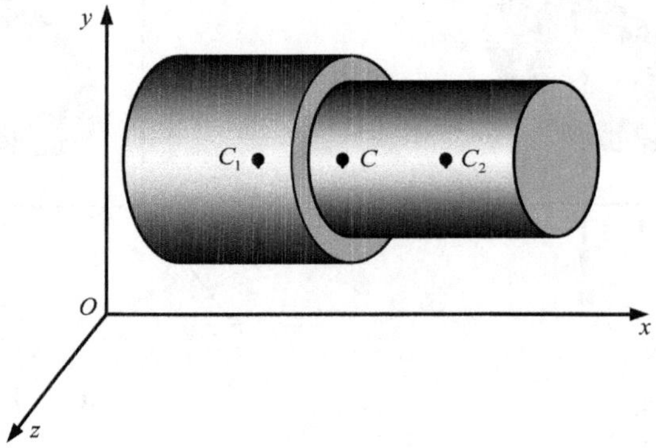

Fig. Ⅰ.8

Appendix Ⅱ Mass Moments of Inertia

Ⅱ.1 Moment of Inertia and Radius of Gyration

Considering a three-dimensional body of mass m, as shown in Fig. Ⅱ.1, the mass moments of inertia of the body with respect to the x, y and z axes can be defined as

$$I_x = \int (y^2 + z^2)\mathrm{d}m,\ I_y = \int (z^2 + x^2)\mathrm{d}m,\ I_z = \int (x^2 + y^2)\mathrm{d}m \tag{Ⅱ.1}$$

In SI units, the mass moments of inertia, I_x, I_y, and I_z, are expressed in kg · m^2.

The mass radius of gyration of the body with respect to the x, y and z axes can be defined, respectively, as

$$i_x = \sqrt{\frac{I_x}{m}},\ i_y = \sqrt{\frac{I_y}{m}},\ i_z = \sqrt{\frac{I_z}{m}} \tag{Ⅱ.2}$$

In SI units, the mass radius of gyration, i_x, i_y, and i_z, are expressed in m.

Ⅱ.2 Parallel-Axis Theorem

Considering a three-dimensional body of mass m, Fig. as shown in Ⅱ.2, the mass moments of inertia of the body with respect to the x and x_C can be defined, respectively, as

$$I_x = \int (y^2 + z^2)\mathrm{d}m,\ I_{x_C} = \int (y_C^2 + z_C^2)\mathrm{d}m \tag{Ⅱ.3}$$

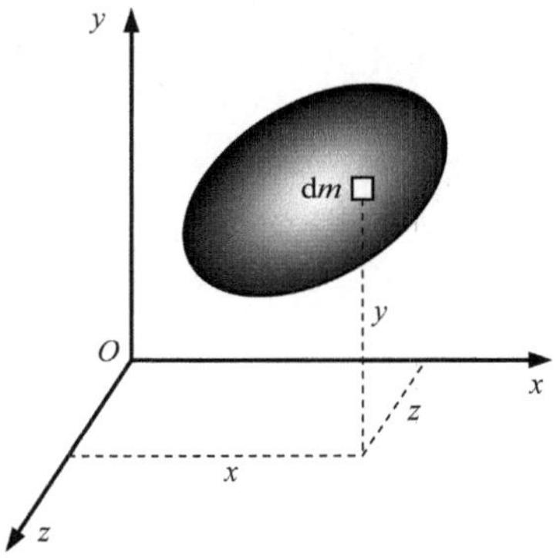

Fig. Ⅱ.1

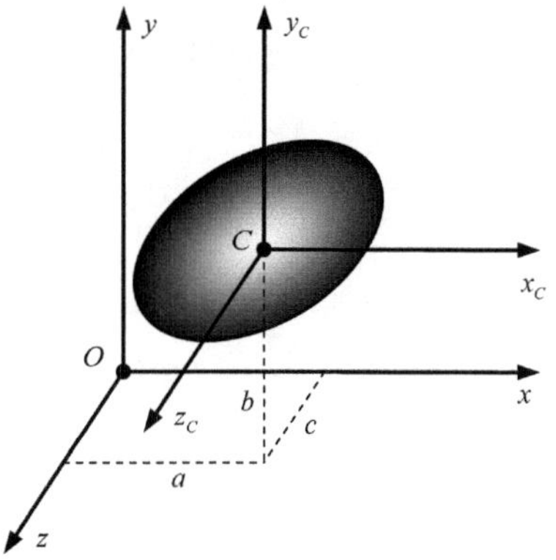

Fig. Ⅱ.2

Using $y = y_C + b$ and $z = z_C + c$, we have

$$I_x = \int (y^2 + z^2)\mathrm{d}m = \int (y_C^2 + z_C^2)\mathrm{d}m + 2\int (by_C + cz_C)\mathrm{d}m + (b^2 + c^2)\int \mathrm{d}m \tag{Ⅱ.4}$$

Using $\int (by_C + cz_C)\mathrm{d}m = 0$, the equation above can be simplified as

$$I_x = I_{x_C} + m(b^2 + c^2) \tag{Ⅱ.5}$$

Similarly, we can obtain

$$I_y = I_{y_C} + m(c^2 + a^2),\ I_z = I_{z_C} + m(a^2 + b^2) \tag{Ⅱ.6}$$

The relations expressed by Eqs. (Ⅱ. 5) and (Ⅱ. 6) are called the parallel-axis theorem. These relations are often used to determine the mass moment of inertia of a body with respect to an arbitrary axis if the mass moment of inertia with respect to the centroidal axis is known.

References

[1] Pytel A, et al. Engineering Mechanics: Statics. 2nd ed. Beijing: Tsinghua University Press, 2001.

[2] Pytel A , et al. Engineering Mechanics: Dynamics. 2nd ed. Beijing: Tsinghua University Press, 2001.

[3] Beer F P, et al. Vector Mechanics for Engineers: Statics. 3rd SI Metric ed. Beijing: Tsinghua University Press, 2003.

[4] Beer F P, et al. Vector Mechanics for Engineers: Dynamics. 3rd SI Metric ed. Beijing: Tsinghua University Press, 2003.

[8] Hibbeler R C. Engineering Mechanics: Statics. 10th ed. Beijing: Higher Education Press, 2004.

[9] Hibbeler R C. Engineering Mechanics: Dynamics. 10th ed. Beijing: Higher Education Press, 2004.

[10] Wang K. Engineering Mechanics. Beijing: Science Press, 2012.

第1章 基本概念与普遍原理

理论力学研究物体的平衡或运动，由静力学、运动学和动力学组成。静力学研究物体的静止和平衡；运动学在不涉及作用力的情况下研究物体的运动；动力学研究物体的运动与作用力之间的关系。

在理论力学中，物体被假设为刚体。虽然实际物体不可能具有完全刚性，即在力作用下总要发生或大或小的变形，但是这些变形通常很小，因而不影响物体的平衡或运动。

1.1 基本概念

1. 长 度

长度用于确定空间点的位置。一个点的位置可以通过距一个确定的参考点沿三个给定方向的三个长度进行定义。

2. 时 间

时间用于表示事件从过去到现在至未来以不可逆转的顺序发生的非空间性的延续。为了定义事件，仅指出事件的空间位置还不够，还需要给出事件的时间。

3. 质 量

质量用于表征物体含有的物质的量。物体的质量并不依赖于重力，因而质量不同于重量，但正比于重量。

4. 力

力用于表示一个物体对另一个物体的作用。力具有使物体在力的作用方向产生加速度的趋势。力的效果完全由力的大小、方向和作用点确定。

5. 质 点

如果物体的尺寸和形状并不影响所考虑问题的解，则该物体可以理想化为质点。质点仅有质量，其尺寸和形状可以忽略。

6. 刚 体

刚体可以认为是大量质点的集合，所有质点在力的作用前后都具有相对固定的位置，即刚体定义为在力作用下不发生变形的物体。

7. 标 量

标量仅具有大小，如长度、时间、质量、功和能等。标量通过代数法进行相加。

8. 矢 量

矢量既具有大小，又具有方向，如力、位移、冲量和动量等。矢量通过平行四边形定律进行相加。

9. 自由矢量

自由矢量可以移动到空间任何位置，只要保持矢量的大小和方向不变。

10. 滑移矢量

滑移矢量可以移动到矢量作用线上的任何点。

11. 固定矢量

固定矢量必须保持在相同的作用点。

1.2　普遍原理

1. 平行四边形定律

该定律认为，作用于质点上的两个力可以由一个合力来代替，该合力通过以给定力为边所画的平行四边形的对角线来确定。

例如，图 1.1(a)所示的作用于质点 O 的力 $\boldsymbol{F}_1$ 和 $\boldsymbol{F}_2$ 可以通过图 1.1(b)所示的力 $\boldsymbol{R}$ 来代替。力 $\boldsymbol{R}$ 对质点具有相同的效果，因而称为力 $\boldsymbol{F}_1$ 和 $\boldsymbol{F}_2$ 的合力。合力 $\boldsymbol{R}$ 通过以力 $\boldsymbol{F}_1$ 和 $\boldsymbol{F}_2$ 为邻边所画的平行四边形来确定。通过 O 点的对角线就表示合力 $\boldsymbol{R}$，即 $\boldsymbol{R}=\boldsymbol{F}_1+\boldsymbol{F}_2$。这种求合力的方法称为平行四边形定律。

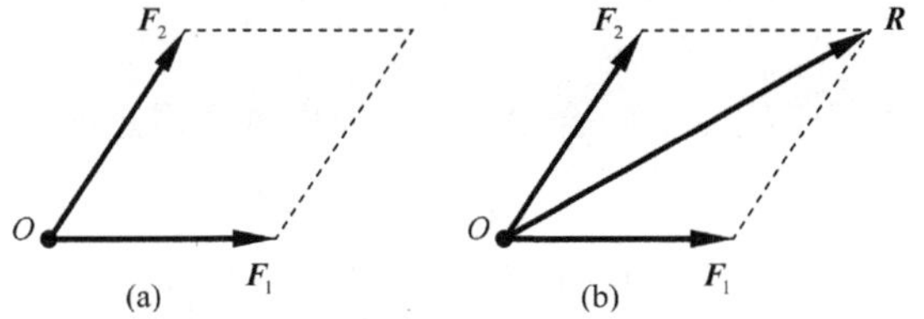

图 1.1

根据平行四边形定律，通过图 1.2(b)所示三角形可以得到求合力的替代方法。把力 $\boldsymbol{F}_1$ 和 $\boldsymbol{F}_2$ 进行首尾相连，并把力 $\boldsymbol{F}_1$ 的尾连接力 $\boldsymbol{F}_2$ 的头，则可得到合力 $\boldsymbol{R}$，即 $\boldsymbol{R}=\boldsymbol{F}_1+\boldsymbol{F}_2$。这种方法称为三角形法则。

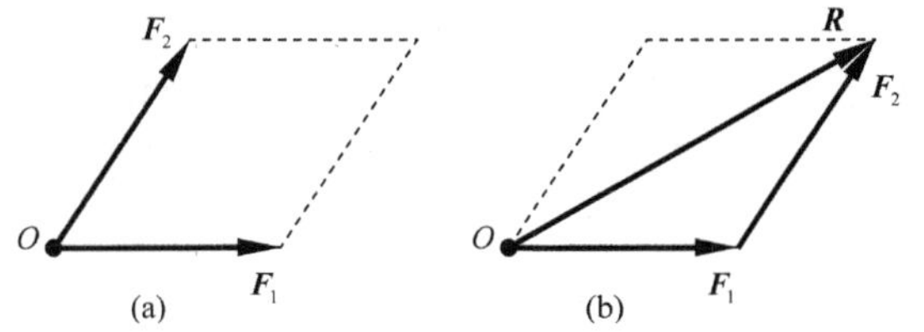

图 1.2

2. 可传性原理

该原理认为，作用于刚体上给定点的一个力如果由具有相同大小、相同方向和相同作用线但作用于不同点的另一个力来代替，则刚体的平衡或运动状态将保持不变。

例如，图 1.3(a)所示的作用于给定点 O 的力 F 可以通过图 1.3(b)所示的具有相同大小、相同方向和相同作用线但作用于不同点 O' 的力 $\boldsymbol{F}'$ 来代替。力 $\boldsymbol{F}$ 和 $\boldsymbol{F}'$ 对刚体具有相同效果，因而称为等效。该原理表明，只要作用于刚体上的力沿其作用线移动，则力对刚体的效果保持不变。因此，作用于刚体上的力是滑移矢量。

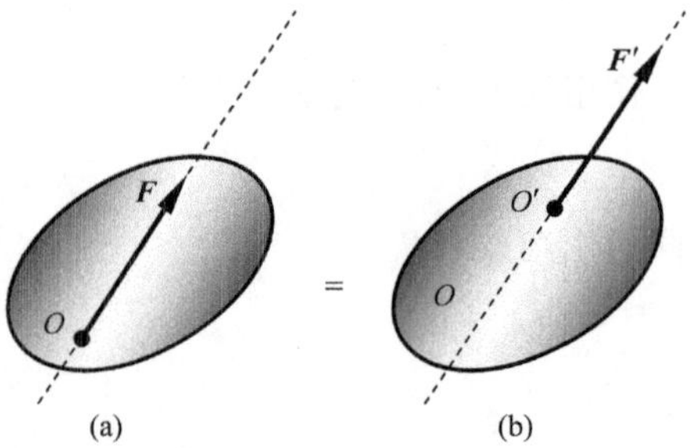

图 1.3

3. 牛顿第一定律

该定律认为，如果作用于质点上的合力等于零，则质点将保持静止(如果质点原来静止)或匀速直线运动(如果质点原来运动)。

4. 牛顿第二定律

该定律认为，如果作用于质点上的合力不等于零，则质点将具有与合力大小成正比并沿合力方向的加速度。数学上，该定律可表示为

$$\boldsymbol{F} = m\boldsymbol{a} \tag{1.1}$$

式中，$\boldsymbol{F}$，m 和 $\boldsymbol{a}$ 分别为作用于质点上的合力，质点的质量和质点的加速度。

5. 牛顿第三定律

该定律认为，相互接触物体之间的作用和反作用力大小相等，作用线共线，但方向相反。

6. 牛顿引力定律

该定律认为，两个质点通过等值反向的两个力相互吸引。这两个力的大小可表示为

$$F = G\frac{m_1 m_2}{r^2} \tag{1.2}$$

式中，F 为质点之间的引力，G 为引力常数，m_1 和 m_2 分别为两个质点的质量，r 为质点之间的距离。

当一个质点位于地球表面或表面附近时，地球作用于质点上的力定义为质点的重量。取 m_1 等于地球质量 M，m_2 等于质点质量 m，r 等于地球半径 R，并设

$$g = G\frac{M}{R^2} \tag{1.3}$$

式中，g 为重力加速度，则质点的重量大小可表示为

$$W = mg \tag{1.4}$$

在国际单位制中，只要质点位于地球表面或表面附近，则 g 的数值近似等于 9.81 $\mathrm{m/s^2}$。

第 2 章　质点静力学

如果物体的尺寸和形状能够忽略不计，则可以把物体理想化为质点。作用于质点上的所有的力都将汇交于同一点，因而形成汇交力系。

2.1　平面汇交力的合成

平面汇交力系是由位于同一平面内的汇交力构成。

1. 力合成的图解法

利用图解法可以得到作用于质点上的平面汇交力系的合力。如果有三个或三个以上的平面汇交力作用于质点，则通过重复应用三角形法则即可获得合力。

图 2.1(a)所示的平面汇交力 $\boldsymbol{F}_1$、$\boldsymbol{F}_2$ 和 $\boldsymbol{F}_3$ 作用于质点 O，把作用于质点上的所有的力进行首尾相连，由第一个力的尾指向最后一个力的头的矢量即表示平面汇交力的合力 $\boldsymbol{R}$，如图 2.1(b)所示。这种方法称为多边形法则。

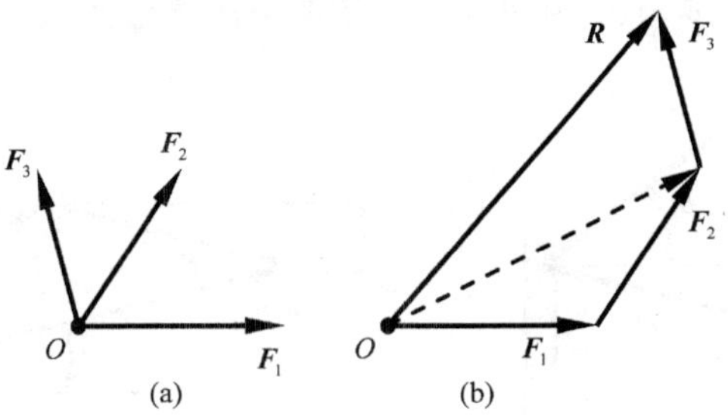

图 2.1

由此得出结论，作用于质点上的平面汇交力系可由通过汇交点的合力代替，合力等于平面汇交力的矢量和，即

$$\boldsymbol{R} = \boldsymbol{F}_1 + \boldsymbol{F}_2 + \boldsymbol{F}_3 + \cdots = \sum \boldsymbol{F} \tag{2.1}$$

例 2.1

如图 E2.1(a)所示，两根杆 AC 和 AD 连接于柱 AB 的 A 点。已知左杆作用力 $F_1=150$ N，两杆的倾角 $\theta_1=30°$和 $\theta_2=15°$，用图解法求：(a)如果两杆作用于柱的力的合力在竖直方向，右杆作用力 F_2；(b)相应合力的大小。

解

根据平行四边形定律，作用于 A 点的力 $\boldsymbol{F}_1$ 和 $\boldsymbol{F}_2$ 可以由合力 $\boldsymbol{R}$ 代替，如图 E2.1(b)所示。根据图 E2.1(b)所示阴影三角形，并利用正弦定理，有

$$\frac{F_1}{\sin(90° - \theta_2)} = \frac{F_2}{\sin(90° - \theta_1)} = \frac{R}{\sin(\theta_1 + \theta_2)}$$

代入 $F_1=150$ N、$\theta_1=30°$和 $\theta_2=15°$，得

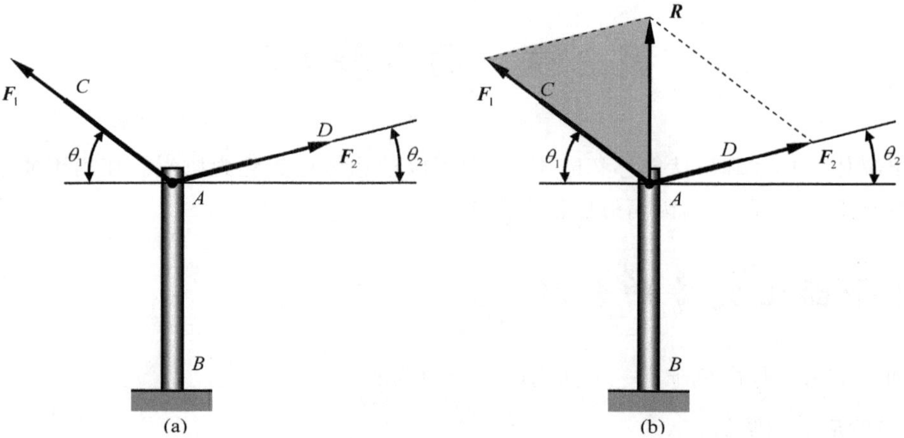

图 E2.1

$$F_2=\frac{\sin(90^\circ-\theta_1)}{\sin(90^\circ-\theta_2)}F_1=134.49\ \mathrm{N},\ R=\frac{\sin(\theta_1+\theta_2)}{\sin(90^\circ-\theta_2)}F_1=109.81\ \mathrm{N}$$

例 2.2

如图 E2.2(a)所示,两根杆 AC 和 AD 连接于柱 AB 的 A 点。已知两杆作用力 $F_1=120$ N 和 $F_2=100$ N,两杆的倾角 $\theta_1=35^\circ$和 $\theta_2=20^\circ$,用图解法求合力。

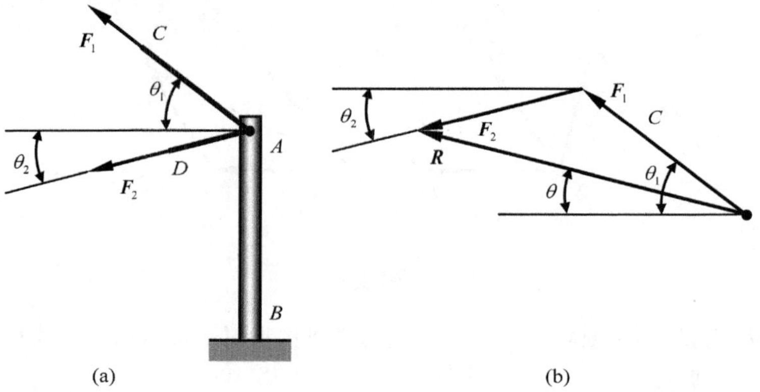

图 E2.2

解

根据三角形法则所画的力三角形如图 E2.2(b)所示。利用余弦和正弦定理,有

$$R^2=F_1{}^2+F_2{}^2-2F_1F_2\cos[180^\circ-(\theta_1+\theta_2)]$$

$$\frac{F_2}{\sin(\theta_1-\theta)}=\frac{R}{\sin[180^\circ-(\theta_1+\theta_2)]}$$

代入 $F_1=120$ N、$F_2=100$ N、$\theta_1=35^\circ$和 $\theta_2=20^\circ$,得

$$R=\sqrt{F_1{}^2+F_2{}^2+2F_1F_2\cos(\theta_1+\theta_2)}=195.36\ \mathrm{N}$$

$$\theta=\theta_1-\arcsin\left[\frac{F_2}{R}\sin(\theta_1+\theta_2)\right]=17.93^\circ$$

2. 力的分量

作用于质点上的两个或两个以上的力能够由对质点有相同效果的单一力代替。反之，作用于质点上的单一力也能够由合成后对质点具有相同效果的两个或两个以上的力代替。

例如，图 2.2(a)所示的作用于质点 O 上的力 $\boldsymbol{F}$ 能够由 $\boldsymbol{F}_1$ 和 $\boldsymbol{F}_2$ 代替。$\boldsymbol{F}_1$ 和 $\boldsymbol{F}_2$ 称为力 $\boldsymbol{F}$ 的矢量分量，用 $\boldsymbol{F}_1$ 和 $\boldsymbol{F}_2$ 代替 $\boldsymbol{F}$ 的方法称为力的分解。显然，对于力 $\boldsymbol{F}$ 存在无穷多组的矢量分量，如图 2.2(b)所示。

3. 力的直角分量

把一个力分解为相互垂直的分量通常很方便。例如，如图 2.3 所示作用于质点 O 的力 $\boldsymbol{F}$ 可以分解为分别沿 x 和 y 轴的两个矢量分量 $\boldsymbol{F}_x$ 和 $\boldsymbol{F}_y$，其中 $\boldsymbol{F}_x$ 和 $\boldsymbol{F}_y$ 称为力 $\boldsymbol{F}$ 的直角分量。因此，有

$$\boldsymbol{F} = \boldsymbol{F}_x + \boldsymbol{F}_y \tag{2.2}$$

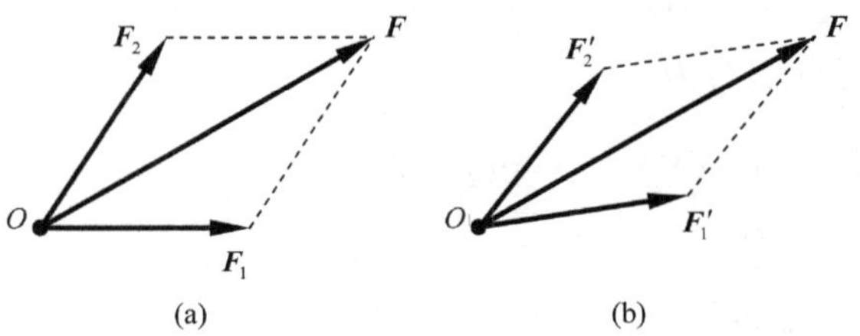

图 2.2

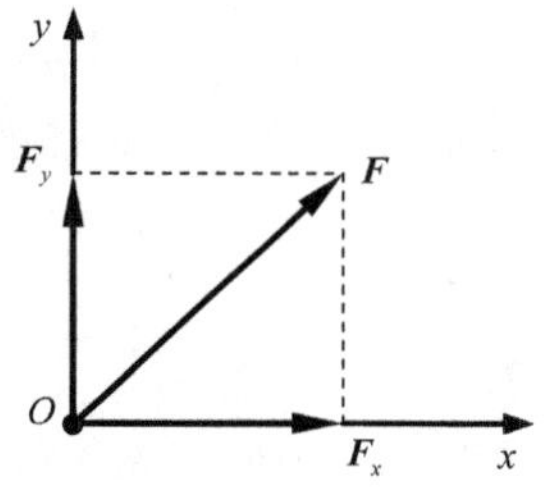

图 2.3

如图 2.4 所示，引进分别指向正 x 和 y 轴的两个单位矢量 $\boldsymbol{i}$ 和 $\boldsymbol{j}$，则 $\boldsymbol{F}$ 也可表示为

$$\boldsymbol{F} = F_x\boldsymbol{i} + F_y\boldsymbol{j} \tag{2.3}$$

式中，F_x 和 F_y 称为力 $\boldsymbol{F}$ 分别沿 x 和 y 轴的标量分量。F_x 和 F_y 依据 $\boldsymbol{F}_x$ 和 $\boldsymbol{F}_y$ 的指向可能为正，也可能为负。当 $\boldsymbol{F}_x$ 与正 x 轴同向，则 F_x 为正，反之为负。关于 F_y 的符号，可得类似结论。

如图 2.5 所示，用 F 表示力 $\boldsymbol{F}$ 的大小，θ 表示力 $\boldsymbol{F}$ 与正 x 轴的夹角，则力 $\boldsymbol{F}$ 的标量分量 F_x 和 F_y 可表示为

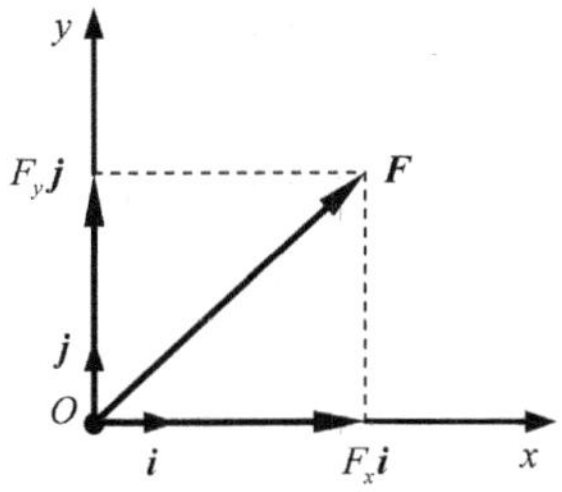

图 2.4

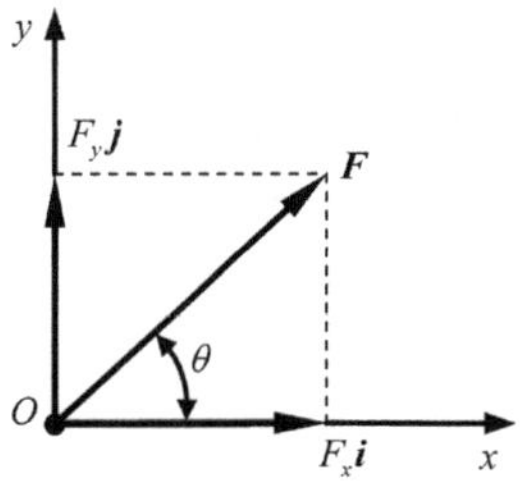

图 2.5

$$F_x = F\cos\theta,\ F_y = F\sin\theta \tag{2.4}$$

4. 力合成的解析法

采用图解法求平面汇交力的合力常常需要进行大量的几何或三角计算，尤其求三个或三

个以上平面汇交力的合力更是如此。用解析法代替,这类问题就很容易求解。

如图 2.6 所示,考虑作用于质点 O 的平面汇交力 $\boldsymbol{F}_1$、$\boldsymbol{F}_2$ 和 $\boldsymbol{F}_3$,利用图解法这些力的合力 $\boldsymbol{R}$ 可表示为

$$\boldsymbol{R} = \boldsymbol{F}_1 + \boldsymbol{F}_2 + \boldsymbol{F}_3 \tag{2.5}$$

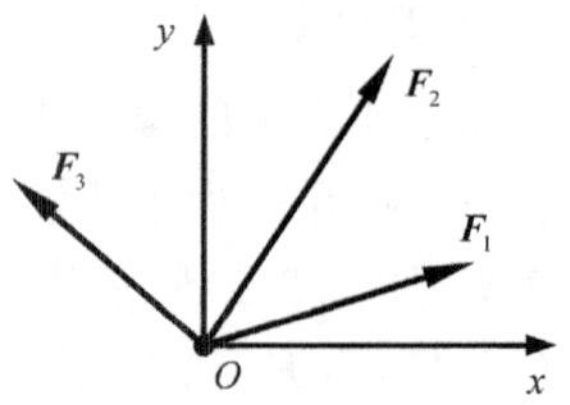

图 2.6

把合力包括在内的每个力分解为直角分量,有

$$R_x\boldsymbol{i} + R_y\boldsymbol{j} = (F_{1x} + F_{2x} + F_{3x})\boldsymbol{i} + (F_{1y} + F_{2y} + F_{3y})\boldsymbol{j} \tag{2.6}$$

由式(2.6)得

$$R_x = F_{1x} + F_{2x} + F_{3x},\ R_y = F_{1y} + F_{2y} + F_{3y} \tag{2.7}$$

由此得出结论,作用于质点上的平面汇交力的合力沿 x 和 y 轴的标量分量分别等于给定力在相同轴上的标量分量的代数和,即

$$R_x = \sum F_x,\ R_y = \sum F_y \tag{2.8}$$

合力大小 R 和合力与正 x 轴的夹角 θ 可分别写为

$$R = \sqrt{R_x^2 + R_y^2},\ \theta = \arctan\frac{R_y}{R_x} \tag{2.9}$$

例 2.3

如图 E2.3(a)所示,两根杆 AC 和 AD 连接于柱 AB 的 A 点。已知左杆作用力 $F_1 = 150$ N,两杆的倾角 $\theta_1 = 30°$和 $\theta_2 = 15°$,用解析法求:(a)如果两杆作用于柱的力的合力在竖直方向,右杆作用力 F_2;(b)相应合力的大小。

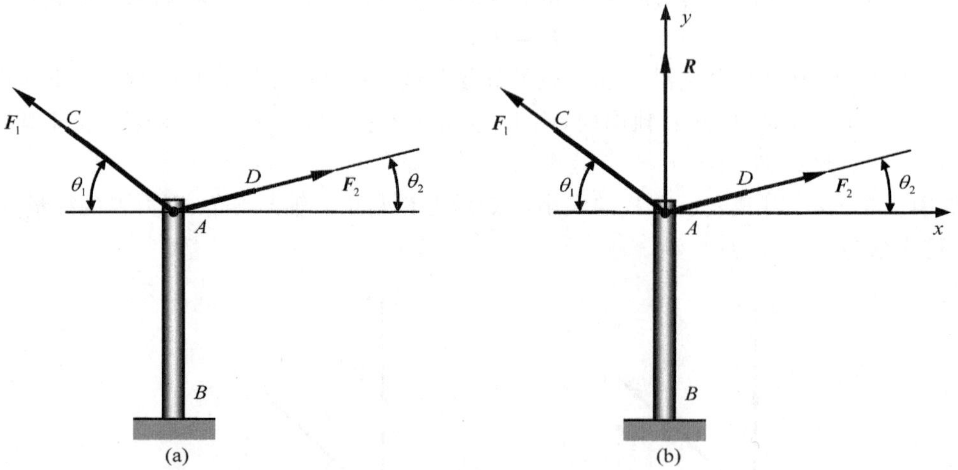

图 E2.3

解

建立图 E2.3(b)所示参考系,则合力的标量分量可表示为

$$R_x = F_2\cos\theta_2 - F_1\cos\theta_1,\ R_y = F_2\sin\theta_2 + F_1\sin\theta_1$$

因合力在竖直方向,即 $R_x = 0$,则有

$$F_2 = \frac{\cos\theta_1}{\cos\theta_2}F_1 = 134.49\ \text{N},\ R = R_y = F_2\sin\theta_2 + F_1\sin\theta_1 = 109.81\ \text{N}$$

例 2.4

受三个力作用的物块放在倾角 $\alpha=25°$ 的斜面上，如图 E2.4(a)所示。假设 $\theta=40°$、$F_1=150$ N、$F_2=250$ N 和 $F_3=200$ N，用解析法求作用于物块上的力的合力。

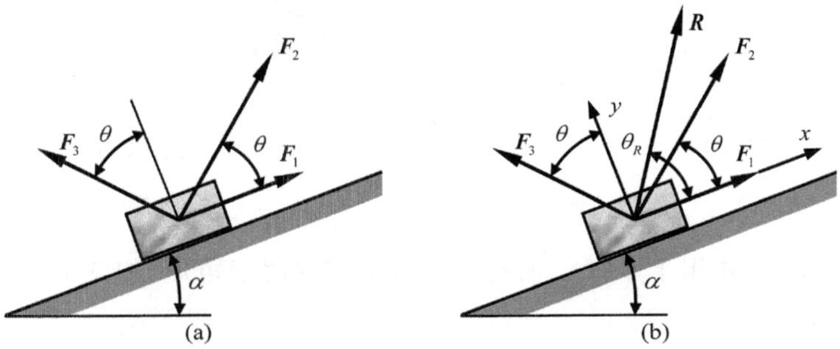

图 E2.4

解

建立图 E2.4(b)所示参考系，则合力的标量分量可表示为

$$R_x = F_1 + F_2\cos\theta - F_3\sin\theta = 212.95\ \text{N},\ R_y = F_2\sin\theta + F_3\cos\theta = 313.91\ \text{N}$$

利用上述标量分量，可得合力的大小和方向

$$R = \sqrt{R_x^2 + R_y^2} = 379.32\ \text{N},\ \theta_R = \arctan\frac{R_y}{R_x} = 55.85°$$

由此得，合力大小为 379.32 N，倾角为 $\alpha+\theta_R=80.85°$。

2.2　平面汇交力的平衡

1. 受力图

在求解有关质点的平衡问题时，必须考虑作用于质点上的所有的力。为了做到这一点，可选择所考虑的质点，绘制单独简图表示该质点和作用于该质点上的所有的力。这种简图称为受力图。

2. 力平衡的图解法

如果作用于质点上的力的合力等于零，则质点保持平衡。因此，受平面汇交力系作用的质点保持平衡的充要条件可表示为

$$\sum \boldsymbol{F} = 0 \tag{2.10}$$

从式(2.10)可看出，如果作用于质点上的力形成封闭多边形，则质点保持平衡。

如图 2.7(a)所示，质点 O 受力 $\boldsymbol{F}_1$、$\boldsymbol{F}_2$ 和 $\boldsymbol{F}_3$ 作用，利用多边形法则可得给定力的合力。从点 O 处由 F_1 开始，把力首尾相连，即可发现 F_3 的头与起点 O 重合，如图 2.7(b)所示。因此，给定力的合力为零，质点保持平衡。

例 2.5

三根绳索在 A 处连接，按图 E2.5(a)加载。用图解法求绳索 AB 和 AC 中的拉力。

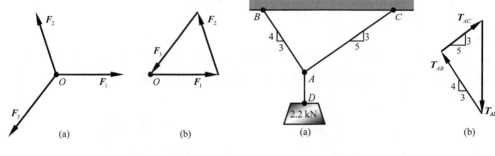

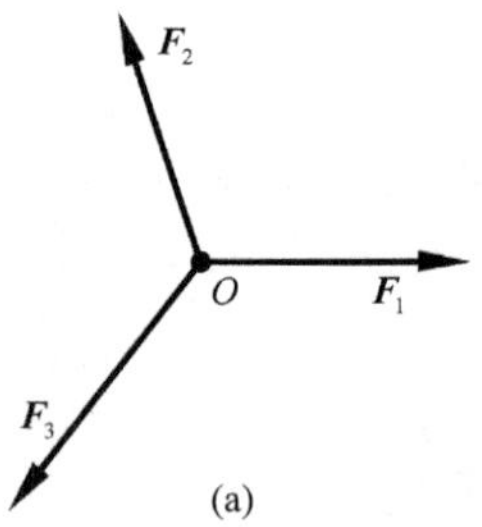

图 2.7

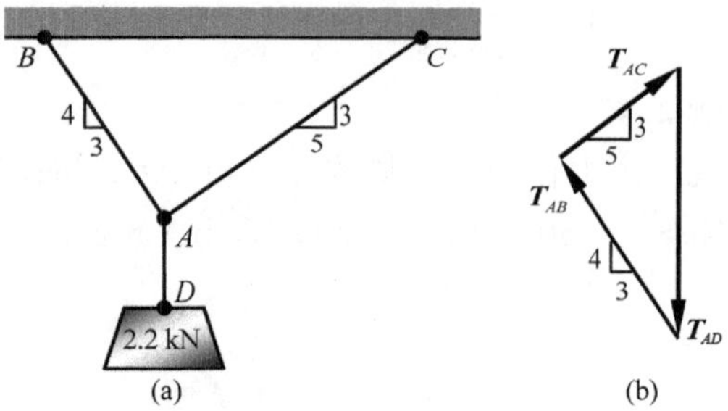

图 E2.5

解

因节点 A 平衡,故作用于点 A 的所有的力将形成封闭三角形,如图 E2.5(b)所示。根据正弦定理,有

$$\frac{T_{AB}}{\sin[90°-\arctan(3/5)]}=\frac{T_{AC}}{\sin[90°-\arctan(4/3)]}=\frac{T_{AD}}{\sin[\arctan(4/3)+\arctan(3/5)]}$$

代入 $T_{AD}=2.2\ \text{kN}$,得

$$T_{AB}=1.90\ \text{kN},\ T_{AC}=1.33\ \text{kN}$$

3. 力平衡的解析法

受平面汇交力作用的质点保持平衡的充要条件表示为

$$\sum \boldsymbol{F}=0 \tag{2.11}$$

把每个力都分解为直角分量,有

$$(\sum F_x)\boldsymbol{i}+(\sum F_y)\boldsymbol{j}=0 \tag{2.12}$$

由此得出结论,受平面汇交力作用的质点保持平衡的充要条件可表示为

$$\sum F_x=0,\ \sum F_y=0 \tag{2.13}$$

式(2.13)称为平面汇交力系的平衡方程。

例 2.6

三根绳索在 A 处连接,按图 E2.6(a)加载。用解析法求绳索 AB 和 AC 中的拉力。

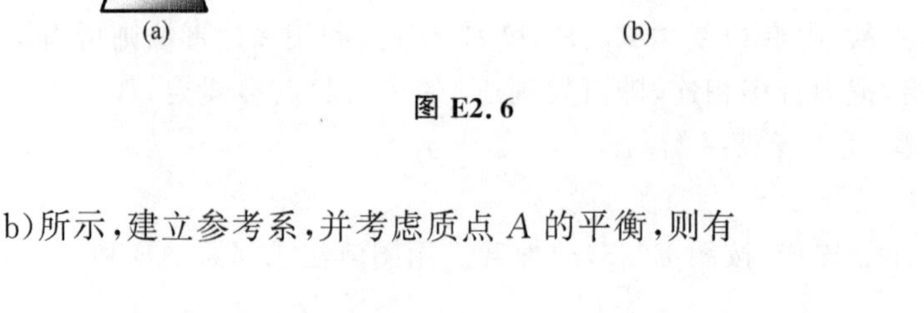

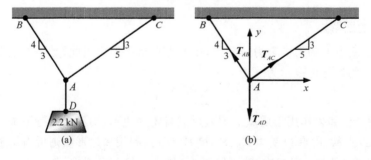

图 E2.6

解

如图 E2.6(b)所示,建立参考系,并考虑质点 A 的平衡,则有

$$\sum F_x = 0,\ T_{AC} \times \frac{5}{\sqrt{5^2+3^2}} - T_{AB} \times \frac{3}{\sqrt{3^2+4^2}} = 0$$

$$\sum F_y = 0,\ T_{AC} \times \frac{3}{\sqrt{5^2+3^2}} + T_{AB} \times \frac{4}{\sqrt{3^2+4^2}} - T_{AD} = 0$$

解上述方程，得

$$T_{AB} = 1.90\ \text{kN},\ T_{AC} = 1.33\ \text{kN}$$

2.3 空间汇交力的合成

1. 空间力的分量

如图 2.8 所示，空间力 $\boldsymbol{F}$ 可以分解为分别沿 x、y 和 z 轴的三个矢量分量 $\boldsymbol{F}_x$、$\boldsymbol{F}_y$ 和 $\boldsymbol{F}_z$，其中 $\boldsymbol{F}_x$、$\boldsymbol{F}_y$ 和 $\boldsymbol{F}_z$ 称为力 $\boldsymbol{F}$ 的直角分量。因此，有

$$\boldsymbol{F} = \boldsymbol{F}_x + \boldsymbol{F}_y + \boldsymbol{F}_z \tag{2.14}$$

如图 2.9 所示，引进分别指向正 x、y 和 z 轴的三个单位矢量 $\boldsymbol{i}$、$\boldsymbol{j}$ 和 $\boldsymbol{k}$，则 $\boldsymbol{F}$ 也可表示为

$$\boldsymbol{F} = F_x\boldsymbol{i} + F_y\boldsymbol{j} + F_z\boldsymbol{k} \tag{2.15}$$

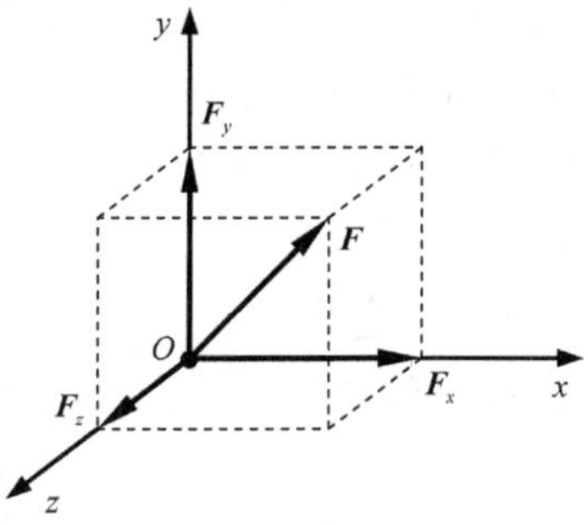

图 2.8

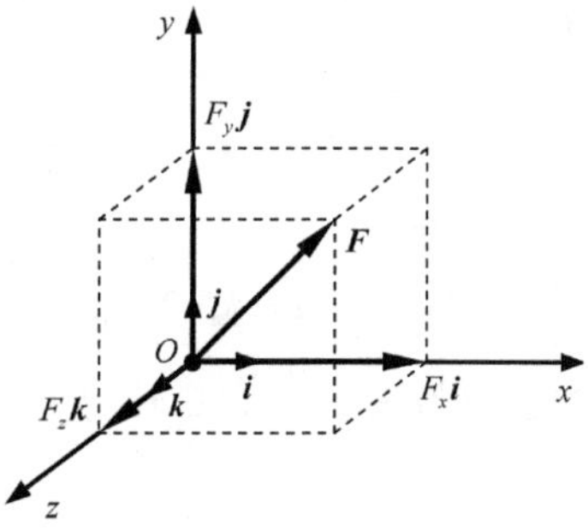

图 2.9

式中，F_x、F_y 和 F_z 称为力 $\boldsymbol{F}$ 分别沿 x、y 和 z 轴的标量分量。F_x、F_y 和 F_z 依据 $\boldsymbol{F}_x$、$\boldsymbol{F}_y$ 和 $\boldsymbol{F}_z$ 的指向可能为正，也可能为负。当 $\boldsymbol{F}_x$ 与正 x 轴同向，则 F_x 为正，反之为负。关于 F_y 和 F_z 的符号，可得类似结论。

如图 2.10 所示，用 F 表示力 $\boldsymbol{F}$ 的大小，θ_x、θ_y 和 θ_z 表示力 $\boldsymbol{F}$ 与正 x、y 和 z 轴的夹角，则力 $\boldsymbol{F}$ 的标量分量 F_x、F_y 和 F_z 可表示为

$$F_x = F\cos\theta_x,\ F_y = F\cos\theta_y,\ F_z = F\cos\theta_z \tag{2.16}$$

式中，$\cos\theta_x$、$\cos\theta_y$ 和 $\cos\theta_z$ 为力 $\boldsymbol{F}$ 的方向余弦。这些方向余弦满足如下关系：

$$\cos^2\theta_x + \cos^2\theta_y + \cos^2\theta_z = 1 \tag{2.17}$$

如图 2.11 所示，如果已知力 $\boldsymbol{F}$ 与正 y 轴之间的夹角为 γ，正 z 轴与由 $\boldsymbol{F}$ 和 y 轴构成的平面之间的夹角为 φ，则力 $\boldsymbol{F}$ 的对应标量分量 F_x、F_y 和 F_z 可表示为

$$\left.\begin{aligned} F_x &= F_{xz}\sin\varphi = F\sin\gamma\sin\varphi \\ F_y &= F\cos\gamma \\ F_z &= F_{xz}\cos\varphi = F\sin\gamma\cos\varphi \end{aligned}\right\} \tag{2.18}$$

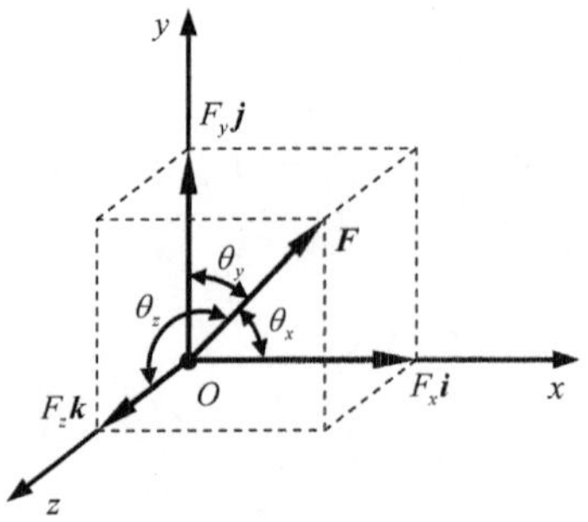

图 2.10

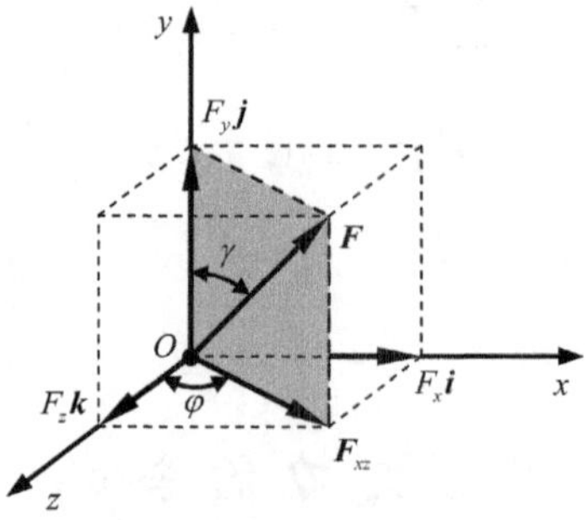

图 2.11

2. 空间力的合成

如图 2.12 所示,考虑作用于质点 O 上的三个力 $\boldsymbol{F}_1$、$\boldsymbol{F}_2$ 和 $\boldsymbol{F}_3$,采用图解法,这些力的合力 $\boldsymbol{R}$ 可表示为

$$\boldsymbol{R} = \boldsymbol{F}_1 + \boldsymbol{F}_2 + \boldsymbol{F}_3 \tag{2.19}$$

把合力包括在内的每个力分解为直角分量,有

$$R_x\boldsymbol{i} + R_y\boldsymbol{j} + R_z\boldsymbol{k} = (F_{1x} + F_{2x} + F_{3x})\boldsymbol{i} + (F_{1y} + F_{2y} + F_{3y})\boldsymbol{j} + (F_{1z} + F_{2z} + F_{3z})\boldsymbol{k} \tag{2.20}$$

由式(2.20)得

$$R_x = F_{1x} + F_{2x} + F_{3x},\ R_y = F_{1y} + F_{2y} + F_{3y},\ R_z = F_{1z} + F_{2z} + F_{3z} \tag{2.21}$$

由此得出结论,作用于质点上的力的合力沿任意轴的标量分量等于给定力在相同轴上的标量分量的代数和,即

$$R_x = \sum F_x,\ R_y = \sum F_y,\ R_z = \sum F_z \tag{2.22}$$

合力大小 R 和合力与正 x、y 和 z 轴的夹角 θ_x、θ_y 和 θ_z 可分别写为

$$R = \sqrt{R_x^2 + R_y^2 + R_z^2},\ \theta_x = \arccos\frac{R_x}{R},\ \theta_y = \arccos\frac{R_y}{R},\ \theta_z = \arccos\frac{R_z}{R} \tag{2.23}$$

例 2.7

如图 E2.7 所示,已知 $F_1 = 300$ N、$F_2 = 200$ N 和 $F_3 = 100$ N,求合力的大小和方向。

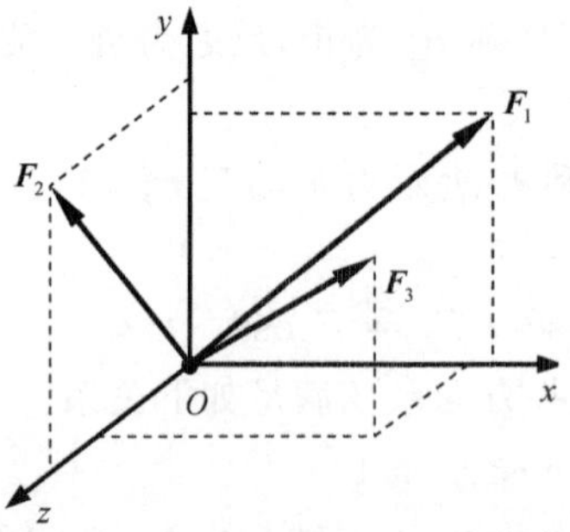

图 2.12

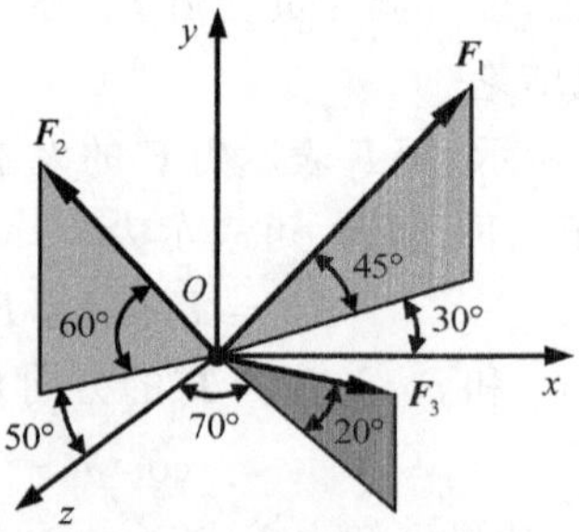

图 E2.7

解

合力的三个标量分量分别为

$$R_x=\sum F_x=F_1\cos45°\cos30°-F_2\cos60°\sin50°+F_3\cos20°\sin70°=195.41\ \text{N}$$

$$R_y=\sum F_y=F_1\sin45°+F_2\sin60°+F_3\sin20°=419.54\ \text{N}$$

$$R_z=\sum F_z=-F_1\cos45°\sin30°+F_2\cos60°\cos50°+F_3\cos20°\cos70°=-9.65\ \text{N}$$

由此，合力的大小和方向为

$$R=\sqrt{R_x^2+R_y^2+R_z^2}=462.92\ \text{N}$$

$$\theta_x=\arccos\frac{R_x}{R}=65.0°,\ \theta_y=\arccos\frac{R_y}{R}=25.0°,\ \theta_z=\arccos\frac{R_z}{R}=91.2°$$

2.4　空间汇交力的平衡

受空间汇交力作用的质点保持平衡的充要条件为

$$\sum \boldsymbol{F}=0 \tag{2.24}$$

把每个力都分解为直角分量，有

$$(\sum F_x)\boldsymbol{i}+(\sum F_y)\boldsymbol{j}+(\sum F_z)\boldsymbol{k}=0 \tag{2.25}$$

由此得出结论，受空间汇交力作用的质点保持平衡的充要条件可表示为

$$\sum F_x=0,\ \sum F_y=0,\ \sum F_z=0 \tag{2.26}$$

式(2.26)称为空间汇交力系的平衡方程。

例 2.8

如图 E2.8(a)所示，重量 $W=300$ N 的水平均质圆形薄板通过三根金属丝悬挂于 A 处，金属丝与竖直方向形成的夹角均为 30°，求每根金属丝的拉力。

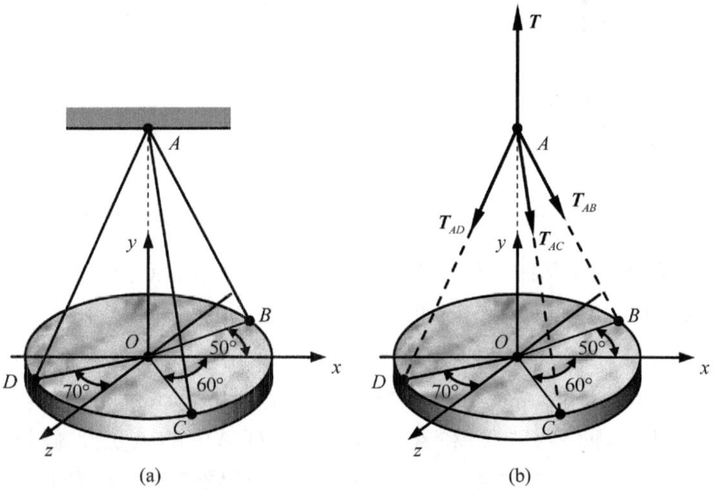

图 E2.8

解

考虑整个系统平衡,有

$$T = W = 300\ \text{N}$$

如图 E2.8(b)所示,取质点 A 为自由体,画受力图,有

$$\sum F_x = 0,\ T_{AB}\sin30°\cos50° + T_{AC}\sin30°\cos60° - T_{AD}\sin30°\sin70° = 0$$

$$\sum F_y = 0,\ T - T_{AB}\cos30° - T_{AC}\cos30° - T_{AD}\cos30° = 0$$

$$\sum F_z = 0,\ -T_{AB}\sin30°\sin50° + T_{AC}\sin30°\sin60° + T_{AD}\sin30°\cos70° = 0$$

解方程,得

$$T_{AB} = 140.71\ \text{N},\ T_{AC} = 71.44\ \text{N},\ T_{AD} = 134.26\ \text{N}$$

习　题

2.1 如图 P2.1 所示,两根杆 AC 和 AD 连接于柱 AB 的 A 点。已知右杆作用力 $F_2 = 100$ N,两杆的倾角 $\theta_1 = 20°$ 和 $\theta_2 = 10°$,求:(a)如果两杆作用于柱的力的合力在竖直方向,左杆作用力 F_1;(b)相应合力的大小。

2.2 如图 P2.2 所示,两根杆 AC 和 AD 连接于柱 AB 的 A 点。已知上杆作用力 $F_1 = 150$ N,两杆的倾角 $\theta_1 = 30°$ 和 $\theta_2 = 15°$,求:(a)如果两杆作用于柱的力的合力在水平方向,下杆作用力 F_2;(b)相应合力的大小。

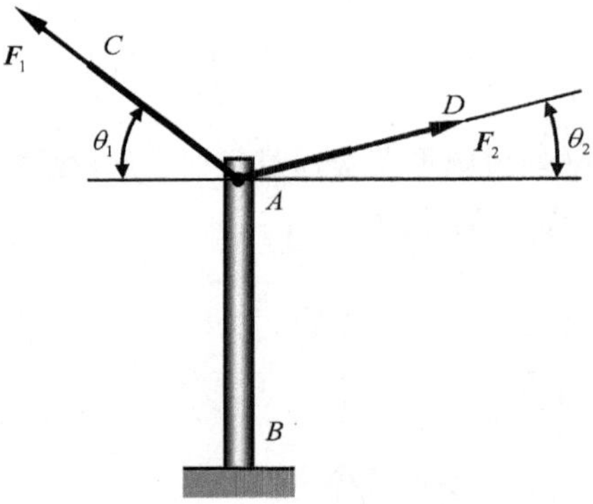

图 P2.1

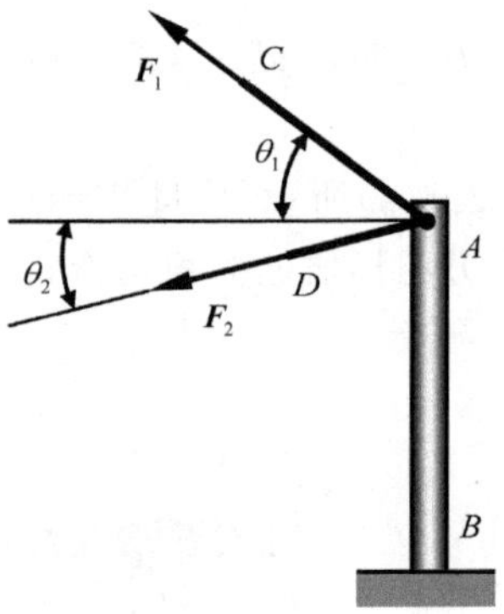

图 P2.2

2.3 如图 P2.3 所示,两根杆 AC 和 AD 连接于柱 AB 的 A 点。已知两杆作用力 $F_1 = 110$ N和 $F_2 = 90$ N,两杆的倾角 $\theta_1 = 40°$ 和 $\theta_2 = 25°$,求合力。

2.4 已知 BC 绳的拉力为 650 N,如图 P2.4 所示,求作用于梁 AB 上点 B 处的三个力的合力。

2.5 三根绳索在 A 处连接,按图 P2.5 加载。求(a)绳索 AB 和(b)绳索 AC 中的拉力。

2.6 如图 P2.6 所示,两根杆 AC 和 AD 连接于杠杆 AB 的 A 点。假设杠杆保持平衡状态,并已知 $F_1 = 200$ N、$F_2 = 175$ N 和 $\theta_1 = 30°$,求:(a)右杆倾角 θ_2;(b)两杆作用于杠杆的力的合力。

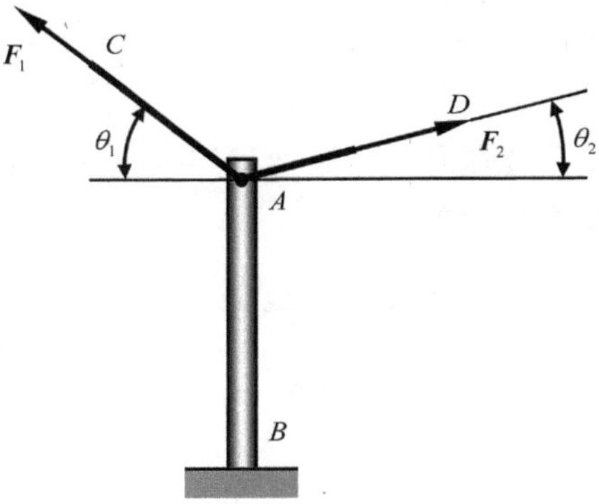

图 P2.3

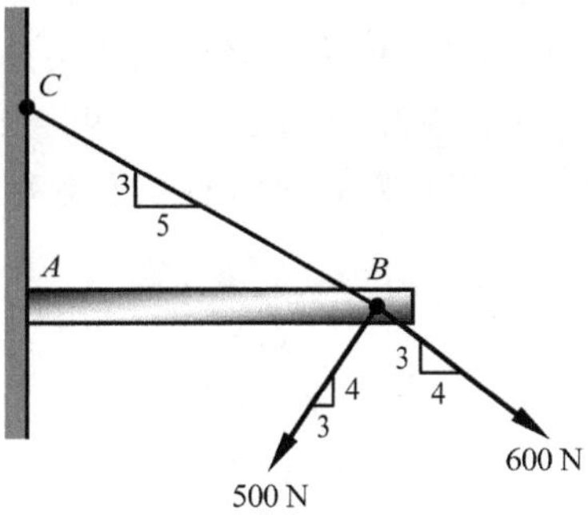

图 P2.4

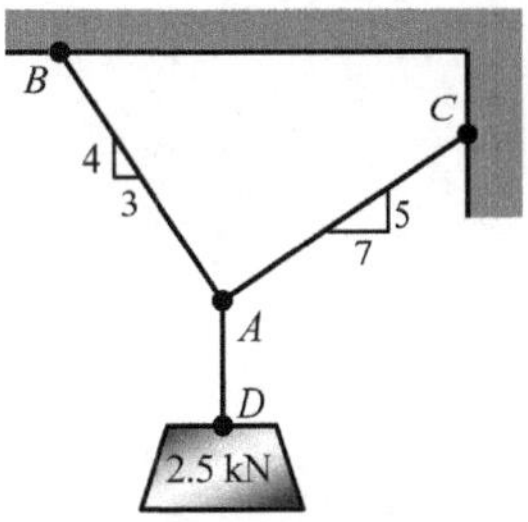

图 P2.5

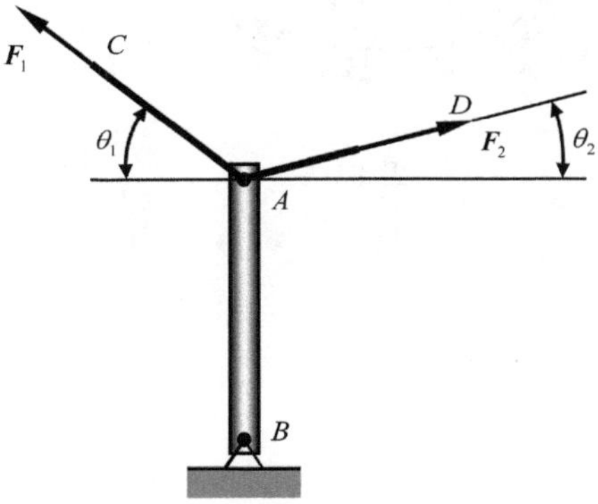

图 P2.6

2.7 环 A 可沿无摩擦竖直杆滑动，并穿过无摩擦定滑轮 B 与弹簧 C 相连，如图 P2.7 所示。假设当 $h=0.3$ m 时弹簧未伸长，当 $h=0.4$ m 时环平衡，并已知 $a=0.4$ m 和弹簧刚度 $k=500$ N/m，求环的重量和杆给环的作用力。

2.8 重量为 W 的环 B 可沿竖直杆自由运动，如图 P2.8 所示，。假设弹簧刚度为 k，当 $\theta=0$ 时弹簧未伸长，并已知 $W=13.5$ N、$l=150$ mm 和 $k=120$ N/m，求与平衡对应的 θ 值。

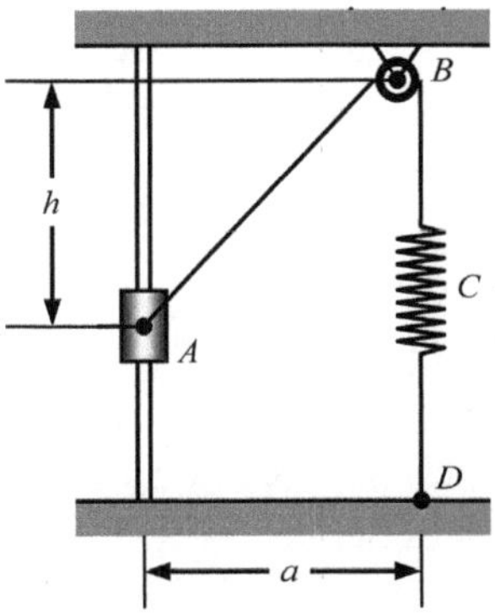

图 P2.7

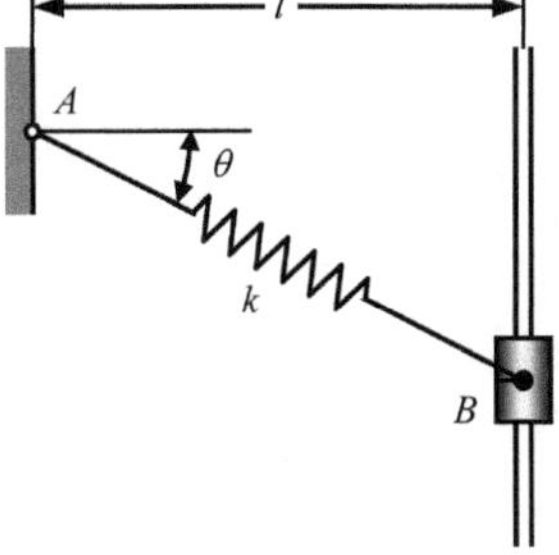

图 P2.8

2.9 如图 P2.9 所示,水平圆形薄板通过三根金属丝悬挂于 A 处,金属丝与竖直方向形成的夹角均为 30°。已知金属丝 AB 施加于薄板的力的 x 分量为 50 N,求(a)金属丝 AB 的拉力和(b)作用于 B 处的力与坐标轴的夹角 θ_x、θ_y 和 θ_z。

2.10 如图 P2.10 所示,重量 $W=200$ N 的水平均质圆形薄板通过三根金属丝悬挂于 A 处,金属丝与竖直方向形成的夹角均为 35°,求每根金属丝的拉力。

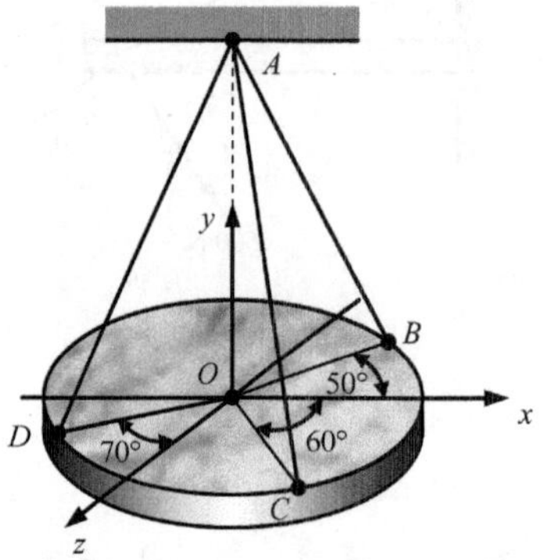

图 P2.9

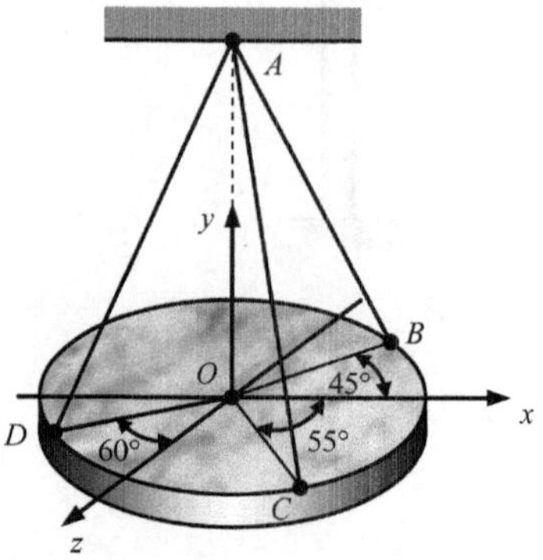

图 P2.10

第3章 力系简化

3.1 力对点之矩

如图 3.1 所示，考虑作用于刚体上 A 点的力 $\boldsymbol{F}$，作用点 A 的位置可通过连接参考点 O 与作用点 A 的矢量 $\boldsymbol{r}$ 进行表示。$\boldsymbol{r}$ 称为作用点 A 的位置矢量。力 $\boldsymbol{F}$ 对点 O 之矩 $\boldsymbol{M}_O(\boldsymbol{F})$ 定义为位置矢量 $\boldsymbol{r}$ 与力 $\boldsymbol{F}$ 的矢量积，即

$$\boldsymbol{M}_O(\boldsymbol{F}) = \boldsymbol{r} \times \boldsymbol{F} \tag{3.1}$$

式中，$\boldsymbol{M}_O(\boldsymbol{F})$ 满足：

(1) 大小等于

$$M_O(\boldsymbol{F}) = rF\sin\theta = Fd \tag{3.2}$$

式中，θ 为 $\boldsymbol{r}$ 和 $\boldsymbol{F}$ 之间的夹角（$\theta \leqslant 180°$），d 为从点 O 到力 $\boldsymbol{F}$ 作用线的垂直距离。

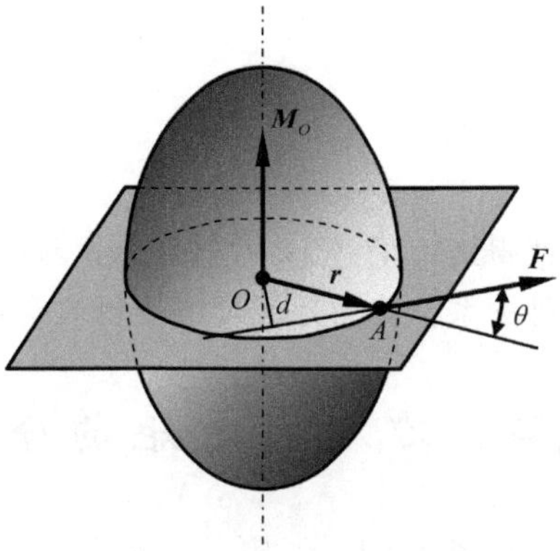

图 3.1

(2) 作用线垂直于 O 和 F 确定的平面。

(3) 方向通过右手法则确定。右手法则表述如下：如果你的右手四指从 $\boldsymbol{r}$ 向 $\boldsymbol{F}$ 卷曲，那么你的右手拇指则指向 $\boldsymbol{M}_O(\boldsymbol{F})$ 的方向。

3.2 力对轴之矩

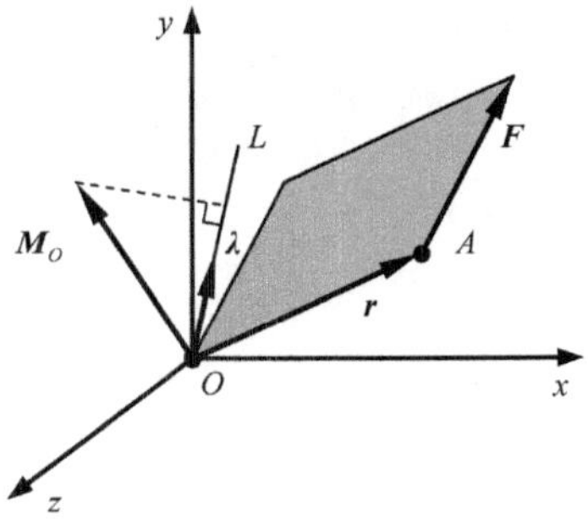

图 3.2

如图 3.2 所示，考虑作用于刚体上 A 点的力 $\boldsymbol{F}$ 和力 $\boldsymbol{F}$ 对点 O 之矩 $\boldsymbol{M}_O(\boldsymbol{F})$。假设 OL 为通过点 O 之轴，则力 $\boldsymbol{F}$ 对轴 OL 之矩 M_{OL} 定义为矩 $\boldsymbol{M}_O(\boldsymbol{F})$ 在轴 OL 上的投影，即

$$M_{OL} = \lambda \cdot \boldsymbol{M}_O(\boldsymbol{F}) = \lambda \cdot (\boldsymbol{r} \times \boldsymbol{F}) \tag{3.3}$$

式中，λ 为沿 OL 的单位矢量。

3.3 力矩定理

如果汇交力 $\boldsymbol{F}_1$，$\boldsymbol{F}_2$，$\boldsymbol{F}_3$，… 作用于同一点 A，并且点 A 相对固定参考点 O 的位置矢量由 $\boldsymbol{r}$ 表示，如图 3.3 所示，则有

$$\boldsymbol{r} \times \boldsymbol{R} = \boldsymbol{r} \times (\boldsymbol{F}_1 + \boldsymbol{F}_2 + \boldsymbol{F}_3 + \cdots) = \boldsymbol{r} \times \boldsymbol{F}_1 + \boldsymbol{r} \times \boldsymbol{F}_2 + \boldsymbol{r} \times \boldsymbol{F}_3 + \cdots \tag{3.4}$$

或

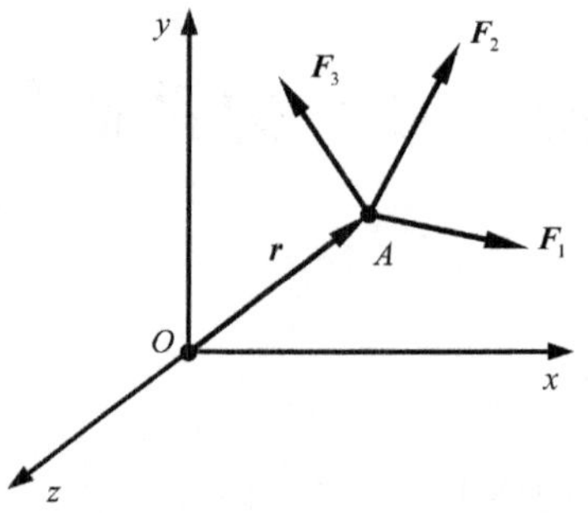

图 3.3

$$\boldsymbol{M}_O(\boldsymbol{R}) = \sum \boldsymbol{M}_O(\boldsymbol{F}) \tag{3.5}$$

因此得到结论：汇交力的合力相对给定点的力矩等于各个汇交力相对同一点的力矩的矢量和。这种关系称为力矩原理或伐里农定理。

3.4 力对点之矩的分量

如图 3.4 所示，假设力 $\boldsymbol{F}$ 作用于点 A，$\boldsymbol{r}$ 表示点 A 相对定系 $Oxyz$ 坐标原点 O 的位置矢量。

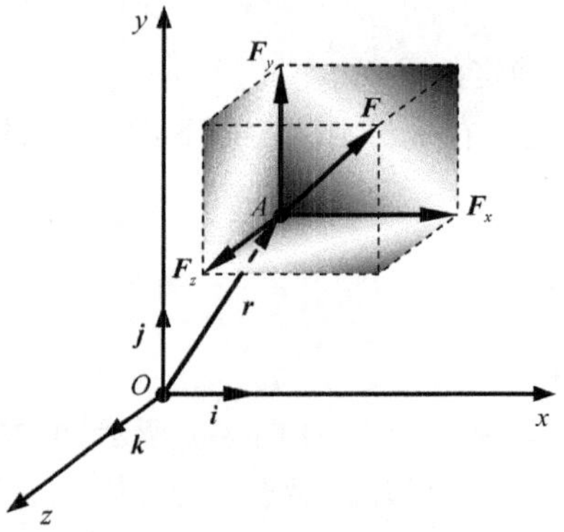

图 3.4

引用式(3.1)，$\boldsymbol{F}$ 对 O 之矩 $\boldsymbol{M}_O(\boldsymbol{F})$ 可表示为

$$\boldsymbol{M}_O(\boldsymbol{F}) = \begin{vmatrix} \boldsymbol{i} & \boldsymbol{j} & \boldsymbol{k} \\ x & y & z \\ F_x & F_y & F_z \end{vmatrix} = (yF_z - zF_y)\boldsymbol{i} + (zF_x - xF_z)\boldsymbol{j} + (xF_y - yF_x)\boldsymbol{k} \tag{3.6}$$

式中，x, y, z 和 F_x, F_y, F_z 分别为位置矢量 $\boldsymbol{r}$ 和力 $\boldsymbol{F}$ 的标量分量，$\boldsymbol{i}$, $\boldsymbol{j}$, $\boldsymbol{k}$ 为单位矢量。根据式(3.6)，$\boldsymbol{M}_O$ 的标量分量可表示为

$$[\boldsymbol{M}_O(\boldsymbol{F})]_x = yF_z - zF_y,\ [\boldsymbol{M}_O(\boldsymbol{F})]_y = zF_x - xF_z,\ [\boldsymbol{M}_O(\boldsymbol{F})]_z = xF_y - yF_x \tag{3.7}$$

引用式(3.3)并利用式(3.6)，则 $\boldsymbol{F}$ 相对 x、y 和 z 之矩分别表示为

$$\left.\begin{aligned} M_x(\boldsymbol{F}) &= \boldsymbol{i} \cdot \boldsymbol{M}_O = yF_z - zF_y, \\ M_y(\boldsymbol{F}) &= \boldsymbol{j} \cdot \boldsymbol{M}_O = zF_x - xF_z, \\ M_z(\boldsymbol{F}) &= \boldsymbol{k} \cdot \boldsymbol{M}_O = xF_y - yF_x \end{aligned}\right\} \tag{3.8}$$

根据式(3.7)和(3.8)，显然有

$$M_x(\boldsymbol{F}) = [\boldsymbol{M}_O(\boldsymbol{F})]_x,\ M_y(\boldsymbol{F}) = [\boldsymbol{M}_O(\boldsymbol{F})]_y,\ M_z(\boldsymbol{F}) = [\boldsymbol{M}_O(\boldsymbol{F})]_z \qquad (3.9)$$

例 3.1

如图 E3.1(a)所示，平板由两根链条悬挂。已知 BH 中的拉力为 200 N，求：(a)链条 BH 施加到平板上的力对点 A 之矩；(b)如果对点 A 产生相同力矩，则作用于点 E 的最小力。

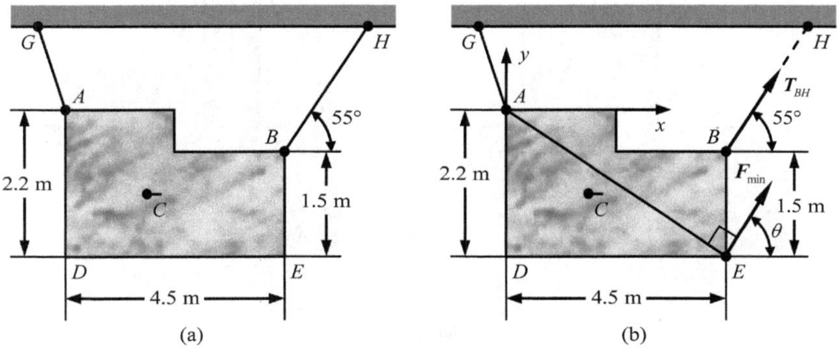

图 E3.1

解

(a) 建立图 E3.1(b)所示坐标系 Axy，则有

$$x_{B/A} = 4.5\ \text{m},\ y_{B/A} = -0.7\ \text{m};$$

$$(T_{BH})_x = T_{BH}\cos 55^\circ = 114.72\ \text{N},\ (T_{BH})_y = T_{BH}\sin 55^\circ = 163.83\ \text{N}$$

因而得

$$M_A(\boldsymbol{T}_{BH}) = x_{B/A}(T_{BH})_y - y_{B/A}(T_{BH})_x = 817.54\ \text{N}\cdot\text{m}$$

(b) 如果对点 A 产生相同力矩，则作用于点 E 的最小力应该垂直于 A 和 E 两点连线。利用 $M_A(\boldsymbol{F}_{\min})=r_{E/A}F_{\min}=M_A(\boldsymbol{T}_{BH})$，得

$$F_{\min} = \frac{M_A(\boldsymbol{T}_{BH})}{r_{E/A}} = \frac{817.54}{\sqrt{2.2^2+4.5^2}} = 163.21\ \text{N},\ \theta = \arctan\frac{4.5}{2.2} = 63.95^\circ$$

例 3.2

如图 E3.2(a)，缆绳 AB 和 AC 连在混凝土立柱上。已知缆绳 AB 和 AC 中的拉力分别为 800 N 和 500 N，求缆绳作用于立柱上点 A 的合力对点 O 的力矩。

解

假设沿 x、y 和 z 轴的单位矢量分别为 $\boldsymbol{i}$、$\boldsymbol{j}$ 和 $\boldsymbol{k}$，则有

$$\boldsymbol{r}_{A/O} = 9\boldsymbol{j}\ \text{m},\ \boldsymbol{r}_{B/A} = 4\boldsymbol{i} - 9\boldsymbol{j} + \boldsymbol{k}\ \text{m},\ \boldsymbol{r}_{C/A} = -2\boldsymbol{i} - 9\boldsymbol{j} + 3\boldsymbol{k}\ \text{m}$$

$$\boldsymbol{T}_{AB} = T_{AB}\frac{\boldsymbol{r}_{B/A}}{r_{B/A}} = 800\frac{4\boldsymbol{i}-9\boldsymbol{j}+\boldsymbol{k}}{\sqrt{4^2+(-9)^2+1^2}} = 80.81(4\boldsymbol{i}-9\boldsymbol{j}+\boldsymbol{k})\ \text{N}$$

$$\boldsymbol{T}_{AC} = T_{AC}\frac{\boldsymbol{r}_{C/A}}{r_{C/A}} = 500\frac{-2\boldsymbol{i}-9\boldsymbol{j}+3\boldsymbol{k}}{\sqrt{(-2)^2+(-9)^2+3^2}} = 51.57(-2\boldsymbol{i}-9\boldsymbol{j}+3\boldsymbol{k})\ \text{N}$$

利用 $\boldsymbol{R}=\boldsymbol{T}_{AB}+\boldsymbol{T}_{AC}$，得

$$\boldsymbol{R} = 220.1\boldsymbol{i} - 1\,191.4\boldsymbol{j} + 235.5\boldsymbol{k}\ \text{N}$$

因此，缆绳作用于立柱上点 A 的合力对点 O 的力矩等于

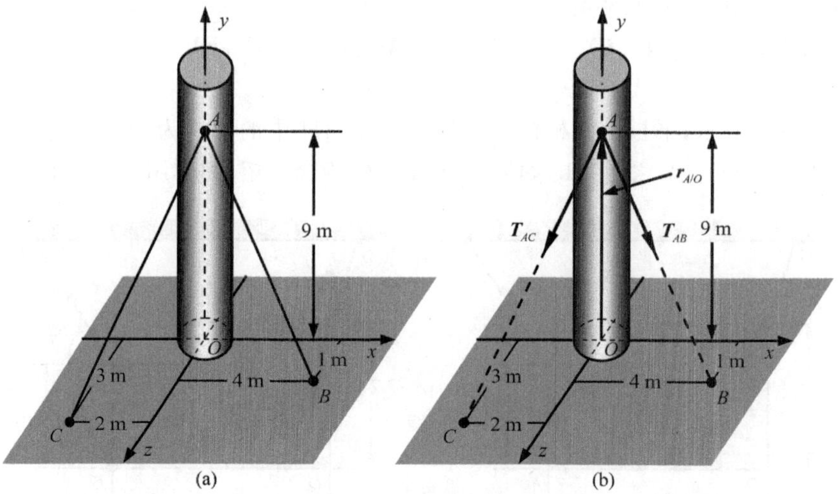

图 E3.2

$$\boldsymbol{M}_O(\boldsymbol{R})=\boldsymbol{r}_{A/O}\times\boldsymbol{R}=\begin{vmatrix}\boldsymbol{i} & \boldsymbol{j} & \boldsymbol{k}\\ 0 & 9 & 0\\ 220.1 & -1\,191.4 & 235.5\end{vmatrix}\text{N}\cdot\text{m}=2\,119.5\boldsymbol{i}-1\,980.9\boldsymbol{k}\ \text{N}\cdot\text{m}$$

3.5 力偶矩

大小相同、作用线平行和方向相反的两个力 $\boldsymbol{F}$ 和 $\boldsymbol{F}'$将形成力偶,如图 3.5 所示。显然,两个力在任何方向的分量之和等于零。然而,两个力相对给定点的力矩之和并不等于零。因此,两个力不会使物体发生平移,但会使物体具有转动趋势。

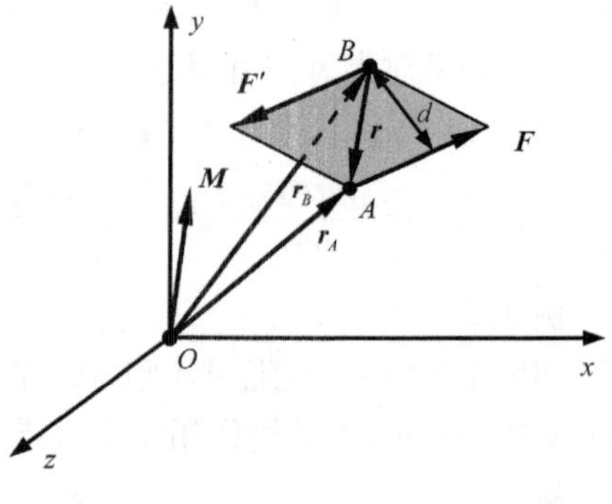

图 3.5

假设$\boldsymbol{r}_A$ 和$\boldsymbol{r}_B$ 分别为 $\boldsymbol{F}$ 和 $\boldsymbol{F}'$作用点的位置矢量,则 $\boldsymbol{F}$ 和 $\boldsymbol{F}'$对 O 的力矩之和为

$$\begin{aligned}\boldsymbol{M}_O(\boldsymbol{F},\boldsymbol{F}')&=\boldsymbol{r}_A\times\boldsymbol{F}+\boldsymbol{r}_B\times\boldsymbol{F}'\\&=\boldsymbol{r}_A\times\boldsymbol{F}+\boldsymbol{r}_B\times(-\boldsymbol{F})\\&=(\boldsymbol{r}_A-\boldsymbol{r}_B)\times\boldsymbol{F}=\boldsymbol{r}\times\boldsymbol{F}\end{aligned}\tag{3.10}$$

式中,$\boldsymbol{r}=\boldsymbol{r}_A-\boldsymbol{r}_B$ 为 $\boldsymbol{F}$ 作用点 A 相对 $\boldsymbol{F}'$作用点 B 的位置矢量。显然,$\boldsymbol{M}_O(\boldsymbol{F},\boldsymbol{F}')=\boldsymbol{r}\times\boldsymbol{F}$ 与参考点 O 无关,因此上式可重写为

$$\boldsymbol{M}=\boldsymbol{r}\times\boldsymbol{F}\tag{3.11}$$

式中,$\boldsymbol{M}$ 为力偶矩。力偶矩是矢量,垂直于由两个力构成的平面,大小等于

$$M=rF\sin\theta=Fd\tag{3.12}$$

式中,d 为 $\boldsymbol{F}$ 和 $\boldsymbol{F}'$作用点之间的垂直距离。$\boldsymbol{M}$ 的方向由右手法则确定。

因 $\boldsymbol{M}$ 与参考点的选择无关,故力偶矩 $\boldsymbol{M}$ 是自由矢量,可作用于刚体上任何位置。因此,如果两个力偶分别位于两个平行平面内(或处于同一平面内),并具有相同大小和方向,那么这两个力偶具有相等的力偶矩。

3.6　力偶的合成

假设 P_1 和 P_2 是两个平面，AA' 是 P_1 和 P_2 的交线，如图 3.6 所示。为不失一般性，垂直于 P_1 的力偶 M_1 可认为由位于 P_1 平面内垂直于 AA' 交线的分别作用于 A 和 A' 点的力 $\boldsymbol{F}_1$ 和 $\boldsymbol{F}'_1$ 构成。同理，垂直于 P_2 的力偶 $\boldsymbol{M}_2$ 为由位于 P_2 平面内的力 $\boldsymbol{F}_2$ 和 $\boldsymbol{F}'_2$ 构成。显然，A 点处 $\boldsymbol{F}_1$ 和 $\boldsymbol{F}_2$ 的合力 $\boldsymbol{R}$ 与 A' 点处 $\boldsymbol{F}'_1$ 和 $\boldsymbol{F}'_2$ 的合力 $\boldsymbol{R}'$ 将形成力偶。设 $\boldsymbol{r}$ 为由 A' 指向 A 的位置矢量，利用力矩原理，合成力偶矩 $\boldsymbol{M}$ 可表示如下：

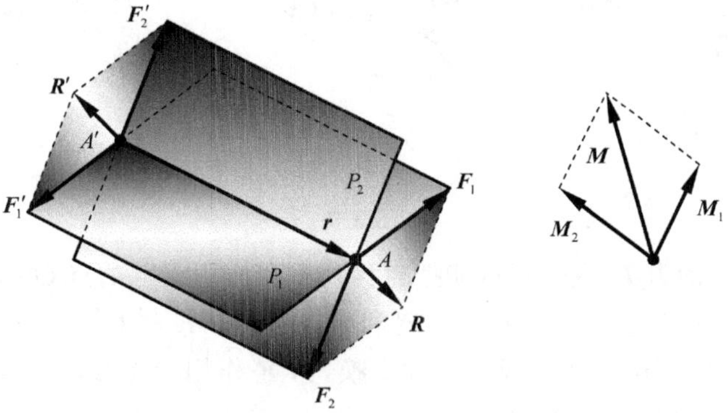

图 3.6

$$\boldsymbol{M} = \boldsymbol{r} \times \boldsymbol{R} = \boldsymbol{r} \times (\boldsymbol{F}_1 + \boldsymbol{F}_2) = \boldsymbol{r} \times \boldsymbol{F}_1 + \boldsymbol{r} \times \boldsymbol{F}_2 = \boldsymbol{M}_1 + \boldsymbol{M}_2 \tag{3.13}$$

式中，$\boldsymbol{M}_1 = \boldsymbol{r} \times \boldsymbol{F}_1$ 和 $\boldsymbol{M}_2 = \boldsymbol{r} \times \boldsymbol{F}_2$ 分别是位于平面 P_1 和 P_2 内的力偶。由此可得出结论，合成力偶的力偶矩 $\boldsymbol{M}$ 等于两个力偶力偶矩 $\boldsymbol{M}_1$ 和 $\boldsymbol{M}_2$ 的矢量和。

例 3.3

如图 E3.3 所示，三个力偶作用于物块。已知 $M_1 = 10\ \text{N} \cdot \text{m}$、$M_2 = 15\ \text{N} \cdot \text{m}$ 和 $M_3 = 8\ \text{N} \cdot \text{m}$，试用一个等效力偶代替这三个力偶，并求等效力偶的大小和方向。

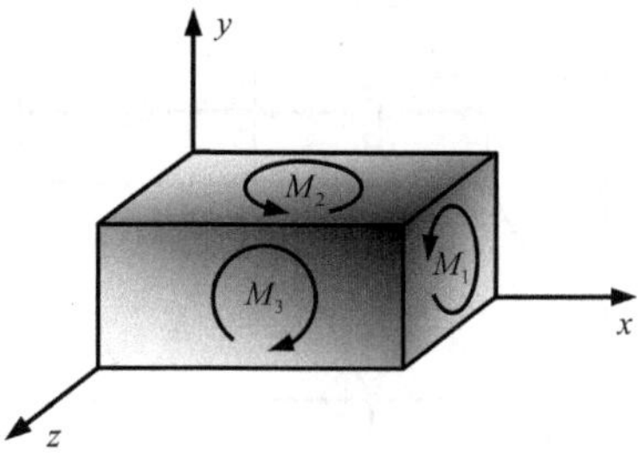

图 E3.3

解

设沿 x、y 和 z 轴的单位矢量分别为 $\boldsymbol{i}$、$\boldsymbol{j}$ 和 $\boldsymbol{k}$，则三个力偶矩可写为

$$\boldsymbol{M}_1 = 10\boldsymbol{i}\ \text{N} \cdot \text{m},\ \boldsymbol{M}_2 = 15\boldsymbol{j}\ \text{N} \cdot \text{m},\ \boldsymbol{M}_3 = -8\boldsymbol{k}\ \text{N} \cdot \text{m}$$

因此，合成力偶可表示为

$$\boldsymbol{M} = \boldsymbol{M}_1 + \boldsymbol{M}_2 + \boldsymbol{M}_3 = 10\boldsymbol{i} + 15\boldsymbol{j} - 8\boldsymbol{k}\ \text{N} \cdot \text{m}$$

合成力偶的大小和方向分别为

$$M = \sqrt{10^2 + 15^2 + (-8)^2} = 19.72\ \text{N} \cdot \text{m}$$
$$\theta_x = \arccos(10/19.72) = 59.53^\circ$$
$$\theta_y = \arccos(15/19.72) = 40.48^\circ$$
$$\theta_z = \arccos(-8/19.72) = 113.93^\circ$$

3.7 作用于刚体上力的等效

作用于刚体 A 点的力为 $\boldsymbol{F}$，A 点的位置矢量为 $\boldsymbol{r}$，如图 3.7(a)所示。利用可传性原理，$\boldsymbol{F}$ 可以沿着其作用线移动而不改变对刚体的作用。

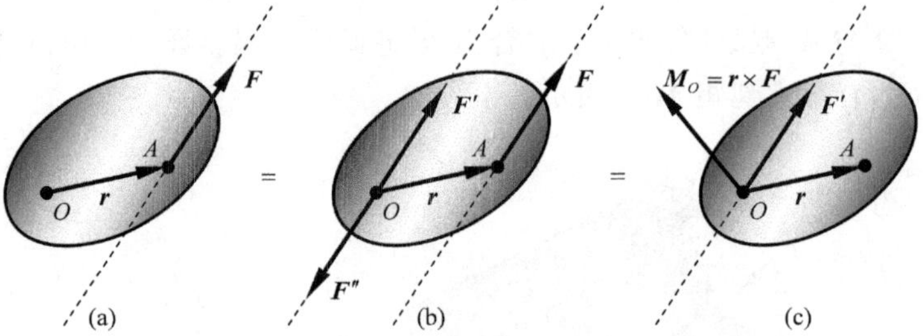

图 3.7

对作用于刚体上的力 $\boldsymbol{F}$，我们不可能把力移到不在作用线上的点 O 而不改变力对刚体的作用。然而，我们可以在点 O 附加两个力 $\boldsymbol{F}'$ 和 $\boldsymbol{F}''$，使 $\boldsymbol{F}'=F$ 和 $\boldsymbol{F}''=-\boldsymbol{F}$，而不改变原来的力对刚体的作用，如图 3.7(b)所示。通过这种变换，力 $\boldsymbol{F}'$ 作用于点 O，其余两个力 $\boldsymbol{F}$ 和 $\boldsymbol{F}''$ 将形成矩为 $\boldsymbol{M}_O=\boldsymbol{r}\times\boldsymbol{F}$ 的力偶，如图 3.7(c)所示。

因此，我们得出结论：只要附加一个力偶，则作用于刚体上的力 $\boldsymbol{F}$ 可以移到任意点 O，附加力偶的力偶矩等于力 $\boldsymbol{F}$ 对点 O 的力矩 $\boldsymbol{M}_O$。因 $\boldsymbol{M}_O$ 是自由矢量，故 $\boldsymbol{M}_O$ 可以附加在任意位置，但为方便起见，$\boldsymbol{M}_O$ 通常附加在点 O。作用于点 O 的力和力偶称为力-力偶系。

例 3.4

如图 E3.4(a)所示，垂直力 $\boldsymbol{F}$ 作用于平面桁架的点 C。已知 $F=80$ N，试用作用于点 G 的力-力偶系代替力 $\boldsymbol{F}$。

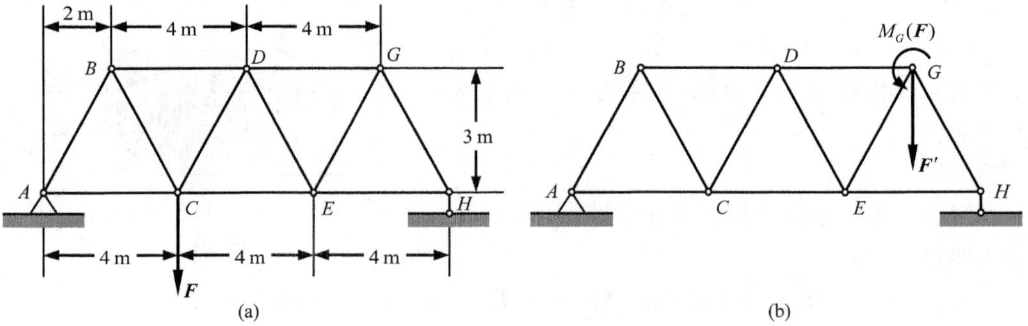

图 E3.4

解

如图 E3.4(b)所示，等效力-力偶系为

$$F' = F = 80\ \text{N}$$

$$M_G(\boldsymbol{F}) = 80\times 6\ \text{N}\cdot\text{m} = 480\ \text{N}\cdot\text{m}$$

3.8　力系简化

1. 力系简化为力-力偶系

力系 $\boldsymbol{F}_1$，$\boldsymbol{F}_3$，$\boldsymbol{F}_3$，…分别作用于刚体上的点 A_1，A_3，A_3，…，点 A_1，A_3，A_3，…的位置矢量分别为$\boldsymbol{r}_1$，$\boldsymbol{r}_3$，$\boldsymbol{r}_3$，…，如图 3.8(a)所示。

只要把力偶$(M)_O(\boldsymbol{F}_1)=\boldsymbol{r}_1\times\boldsymbol{F}_1$ 附加到力系。$\boldsymbol{F}_1$ 可以从 A_1 移到 O 对 $\boldsymbol{F}_3$，$\boldsymbol{F}_3$，…重复该过程，则得到一个新的力系，该力系由作用于 O 的力$\boldsymbol{F}'_1$，$\boldsymbol{F}'_3$，$\boldsymbol{F}'_3$，…和力偶$\boldsymbol{M}_O(\boldsymbol{F}_1)=\boldsymbol{r}_1\times\boldsymbol{F}_1$，$M_O(\boldsymbol{F}_2)=\boldsymbol{r}_2\times\boldsymbol{F}_2$，$M_O(\boldsymbol{F}_3)=\boldsymbol{r}_3\times\boldsymbol{F}_3$，…构成，如图 3.8(b)所示。

因力$\boldsymbol{F}'_1$，$\boldsymbol{F}'_3$，$\boldsymbol{F}'_3$，…形成汇交力系，故可进行矢量相加，并可由合力 $\boldsymbol{R}'$代替。同理，力偶$\boldsymbol{M}_O(\boldsymbol{F}_1)=\boldsymbol{r}_1\times\boldsymbol{F}_1$，$M_O(\boldsymbol{F}_2)=\boldsymbol{r}_2\times\boldsymbol{F}_2$，$\boldsymbol{M}_O(\boldsymbol{F}_3)=\boldsymbol{r}_3\times\boldsymbol{F}_3$，…也可进行矢量相加，并可由合力偶$\boldsymbol{M}_O$ 代替，如图 3.8(c)所示。其中 $\boldsymbol{R}'$和$\boldsymbol{M}_O$ 可分别表示为

$$\boldsymbol{R}'=\sum\boldsymbol{F}'=\sum\boldsymbol{F},\ \boldsymbol{M}_O=\sum\boldsymbol{M}_O(\boldsymbol{F}) \tag{3.14}$$

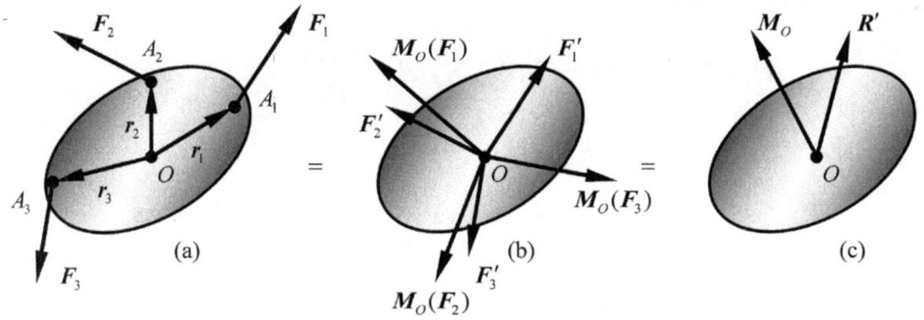

图 3.8

上式表明，力 $\boldsymbol{R}'$可通过相加所有力而得到，力偶$\boldsymbol{M}_O$ 可通过相加所有力对 O 之矩而得到。

2. 力-力偶系的进一步简化

作用于刚体上的任何力系都可简化为作用于 O 的力 $\boldsymbol{R}'$和力偶$\boldsymbol{M}_O$。

如果$\boldsymbol{M}_O=0$ 和 $\boldsymbol{R}'=0$，则给定力系平衡。

如果 $\boldsymbol{R}'\neq0$，但$\boldsymbol{M}_O=0$，则给定力系简化为合力 $\boldsymbol{R}=\boldsymbol{R}'$。

如果$\boldsymbol{M}_O\neq0$，但 $\boldsymbol{R}'=0$，则给定力系简化为合力偶 $\boldsymbol{M}=\boldsymbol{M}_O$。

如果 $\boldsymbol{R}'\neq0$ 和$\boldsymbol{M}_O\neq0$，且 $\boldsymbol{R}'$和$\boldsymbol{M}_O$ 互相垂直，则给定力系可进一步简化为合力 $\boldsymbol{R}=\boldsymbol{R}'$。对汇交力系、平面力系或平行力系，$\boldsymbol{R}'$和$\boldsymbol{M}_O$ 总是相互垂直。

如果 $\boldsymbol{R}'\neq0$ 和$\boldsymbol{M}_O\neq0$，但 $\boldsymbol{R}'$和$\boldsymbol{M}_O$ 不互相垂直，则给定力系可简化为力螺旋。

例 3.5

桁架承受图 E3.5(a)所示载荷作用。已知 $\boldsymbol{F}_1=160$ N，$\boldsymbol{F}_2=150$ N 和 $\boldsymbol{F}_3=80$ N，求等效力以及等效力与直线 AH 的交点。

解

(1) 原始力系可以简化为作用于 A 处的等效力-力偶系，如图 E3.5(b)。设沿 x 和 y 轴的单位矢量分别为 $\boldsymbol{i}$ 和 $\boldsymbol{j}$，利用 $\boldsymbol{R}'=\sum\boldsymbol{F}$ 和 $M_A=\sum M_A(\boldsymbol{F})$，得

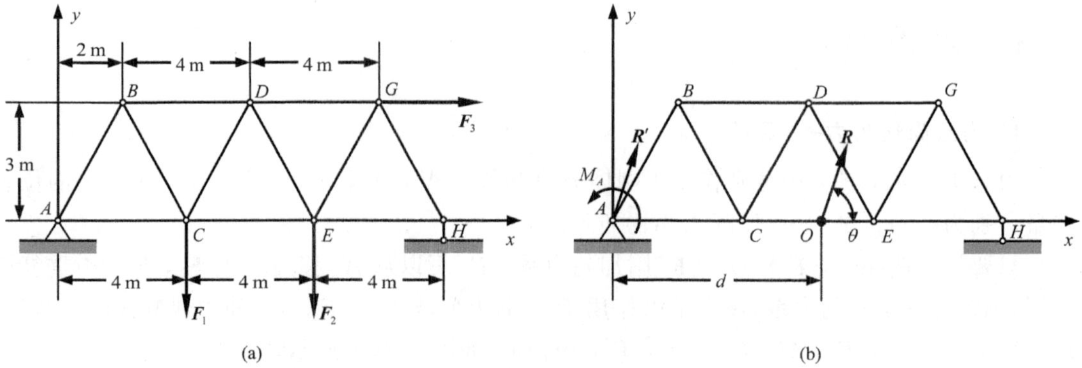

图 E3.5

$$\boldsymbol{R}' = \boldsymbol{F}_1 + \boldsymbol{F}_2 + \boldsymbol{F}_3 = 80\boldsymbol{i} - 310\boldsymbol{j} \text{ N}$$

$$M_A = M_A(\boldsymbol{F}_1) + M_A(\boldsymbol{F}_2) + M_A(\boldsymbol{F}_3) = -2\,080 \text{ N} \cdot \text{m}$$

(2) 上述作用于 A 处的等效力 - 力偶系可简化为作用于 O 处的等效力,如图 E3.5(b)所示。利用 $\boldsymbol{R}=\boldsymbol{R}'$ 和 $M_A=R'_y d$,得

$$\boldsymbol{R} = \boldsymbol{R}' = 80\boldsymbol{i} - 310\boldsymbol{j} \text{ N},\ d = \frac{M_A}{R'_y} = 6.71 \text{ m}$$

或

$$R = \sqrt{R_x^2 + R_y^2} = 320.16 \text{ N},\ \theta = \arctan\frac{R_y}{R_x} = -75.53°,\ d = \frac{M_A}{R'_y} = 6.71 \text{ m}$$

习　题

3.1 如图 P3.1 所示,平板由两根链条 AG 和 BH 悬挂。已知 AG 中的拉力为 300 N,求:(a)链条 AG 施加到平板上的力对点 B 之矩;(b)如果对点 B 产生相同力矩,则作用于点 D 的最小力。

3.2 如图 P3.2 所示,缆绳 AB 连在混凝土立柱上。已知缆绳 AB 中的拉力为 500 N,求缆绳作用于立柱上点 A 的力对点 O 的力矩。

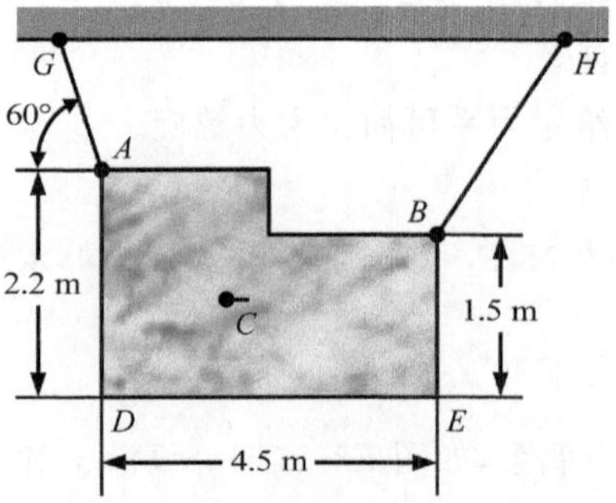

图 P3.1

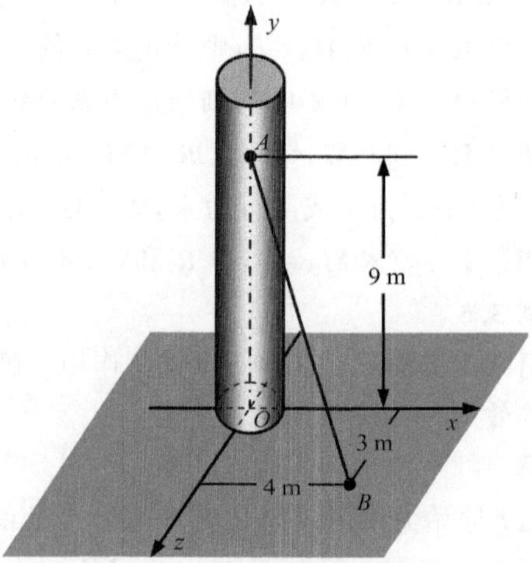

图 P3.2

3.3 缆绳 AB、AC 和 AD 连在混凝土立柱上，如图 P3.3 所示。已知缆绳 AB、AC 和 AD 中的拉力分别为 800 N、700 N 和 500 N，求缆绳作用于立柱上点 A 的合力对点 O 的力矩。

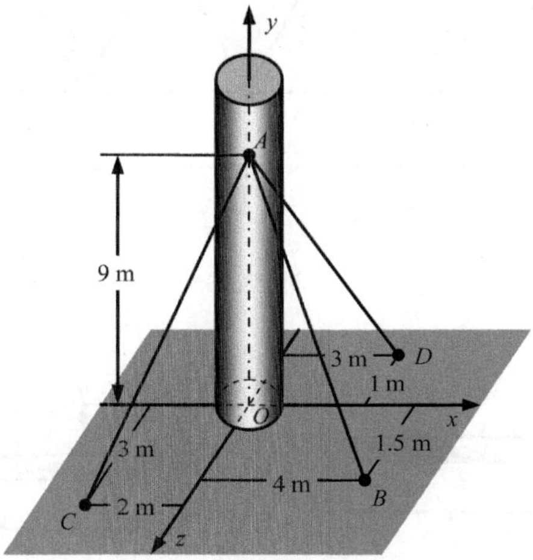

图 P3.3

3.4 三个力偶作用于物块，如图 P3.4 所示。已知 $M_1=10$ N·m，$M_2=15$ N·m 和 $M_3=8$ N·m，试用一个等效力偶代替这三个力偶，并求等效力偶的大小和方向。

3.5 垂直力 $\boldsymbol{F}$ 作用于平面桁架的点 C，如图 P3.5 所示。已知 $F=100$ N，试用作用于点 B 的力-力偶系代替力 $\boldsymbol{F}$。

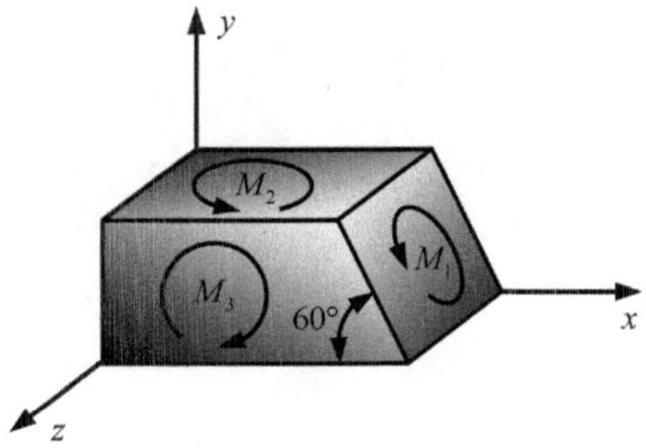

图 P3.4

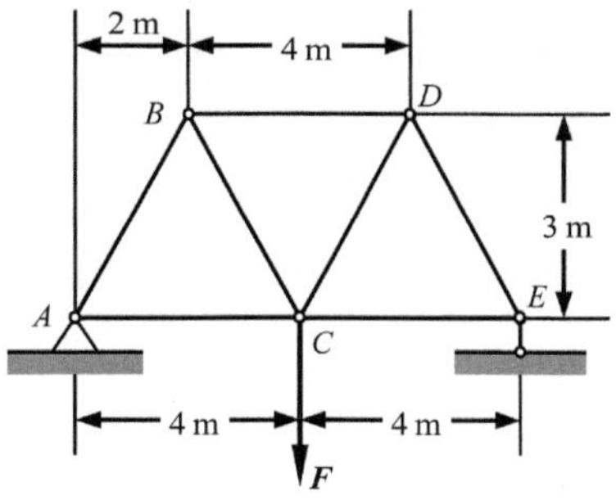

图 P3.5

3.6 桁架承受图 P3.6 所示载荷作用。已知 $F_1=F_2=100$ N 和 $F_3=90$ N，求等效力以及等效力与直线 AH 的交点。

3.7 悬臂梁 AB 承受图 P3.7 所示分布载荷。已知 $a=2$ m 和 $q=100$ N/m，试用 B 处等效力-力偶系代替分布载荷。

3.8 悬臂梁 AB 承受图 P3.8 所示分布载荷。已知 $l=8$ m 和 $q=100$ N/m，试求等效力以及等效力与梁的交点。

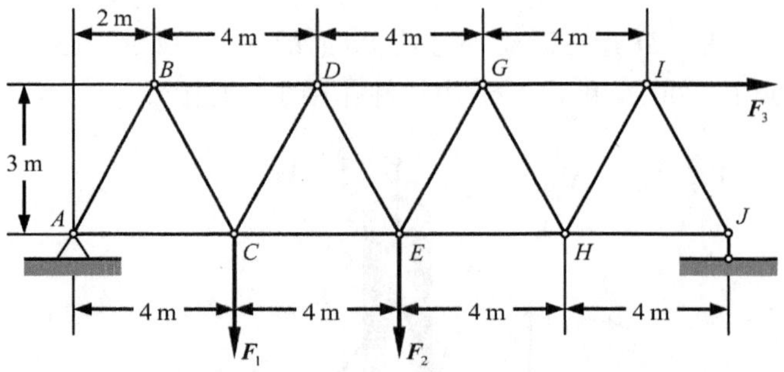

图 P3.6

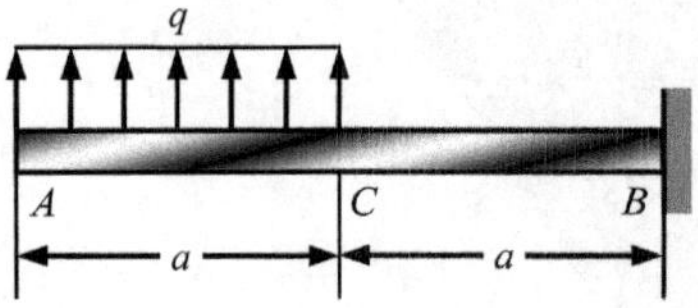

图 P3.7

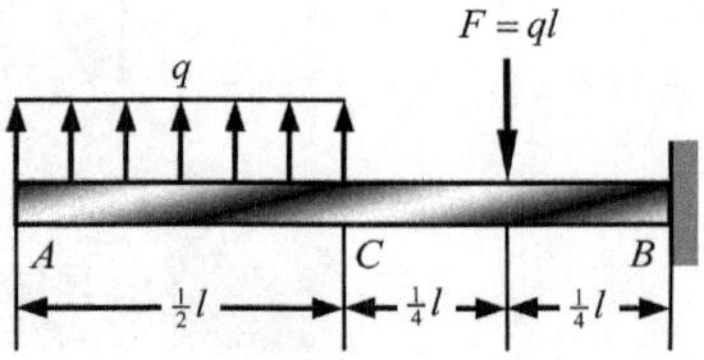

图 P3.8

第4章　刚体静力学

作用于刚体上的外力可以简化为作用于任意点 O 的力-力偶系。当力-力偶系中力和力偶都等于零，则外力将形成等于零的力系，因而刚体处于平衡状态。

刚体平衡的充分必要条件为

$$\sum \boldsymbol{F} = 0,\ \sum \boldsymbol{M}_O(\boldsymbol{F}) = 0 \tag{4.1}$$

把矢量方程分解，则刚体平衡的充分必要条件可通过六个标量平衡方程进行表示：

$$\sum F_x = 0,\ \sum F_y = 0,\ \sum F_z = 0,\ \sum M_x(\boldsymbol{F}) = 0,\ \sum M_y(\boldsymbol{F}) = 0,\ \sum M_z(\boldsymbol{F}) = 0 \tag{4.2}$$

式(4.2)可用于求解作用于刚体上的未知力或支撑产生的未知反力。

求解刚体平衡问题，需要考虑作用于刚体上的所有力，并排除不直接作用于刚体的力。应该选择刚体作为自由体，并画出单独的自由体受力图以显示刚体及作用于其上的所有力。

4.1　二维刚体平衡

1. 平衡条件

根据式(4.2)，二维刚体的平衡方程可表示为

$$\sum F_x = 0,\ \sum F_y = 0,\ \sum M_A(\boldsymbol{F}) = 0 \tag{4.3}$$

式中，A 为二维刚体所在平面内的任意一点。上述三个方程可以求解不超过三个的未知量。

虽然三个平衡方程不可能增加，但是其中的任何一个方程可以用其他方程代替。因此，平衡方程的第二种形式可表示为

$$\sum F_x = 0,\ \sum M_A(\boldsymbol{F}) = 0,\ \sum M_B(\boldsymbol{F}) = 0 \tag{4.4}$$

式中，A 和 B 连线不能垂直于 x 轴。

平衡方程的第三种形式可表示为

$$\sum M_A(\boldsymbol{F}) = 0,\ \sum M_B(\boldsymbol{F}) = 0,\ \sum M_C(\boldsymbol{F}) = 0 \tag{4.5}$$

式中，A、B 和 C 不能共线。

2. 二维刚体的约束反力

限制刚体运动的支撑或连接称为约束。对应于三种类型约束，限制二维刚体的约束产生的反力分为三种：

(1) 等效于作用线已知的力。引起这类反力的约束包括：滚轮、摇杆、缆绳、连杆、光滑表面、杆上光滑轴环和槽中光滑销钉等。

(2) 等效于方向和大小都未知的力。引起这类反力的约束包括：孔中光滑销钉、光滑铰链和粗糙表面等。

(3) 等效于力和力偶。引起这类反力的约束包括：固定支撑和固定连接等。

例 4.1

T 型构件 $ABCD$ 由 C 处销钉支撑,并由穿过定滑轮 E 的缆绳 AED 连接。已知 $F=150$ N,不计摩擦,试求缆绳拉力和 C 处反力。

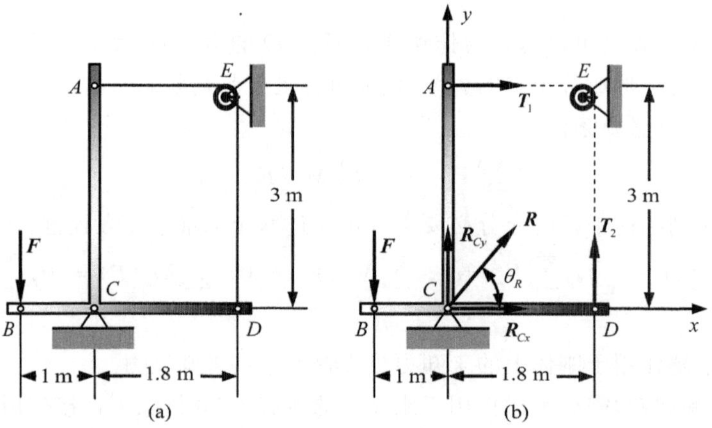

图 E4.1

解

如图 E4.1(b),取 T 型构件为自由体,画受力图。根据 T 型构件平衡,得

$$\sum F_x = 0,\ R_{Cx} + T_1 = 0$$

$$\sum F_y = 0,\ R_{Cy} + T_2 - F = 0$$

$$\sum M_C = 0,\ F\times 1 - T_1 \times 3 + T_2 \times 1.8 = 0$$

利用 $F=150$ N 和 $T_1=T_2$,解上述方程,得

$$R_{Cx} = -125\ \text{N},\ R_{Cy} = 25\ \text{N},\ T_1 = T_2 = 125\ \text{N}$$

因此,缆绳拉力为 125 N,C 处反力为

$$R = \sqrt{R_{Cx}^2 + R_{Cy}^2} = 127.48\ \text{N},\ \theta_R = 180^\circ + \arctan\frac{R_{Cy}}{R_{Cx}} = 168.69^\circ$$

例 4.2

如图 E4.2(a)所示,重量为 W 的均质杆 AB 连接到可沿光滑表面自由运动的物块 A 和 B 上。与物块 A 相连弹簧的刚度为 k,当杆水平时弹簧未伸长。(a)忽略物块重量,试推导当杆平衡时 W、k、a 和 θ 需要满足的方程。(b)当 $W=45$ N、$a=1$ m 和 $k=50$ N/m 时,求 θ 值。

解

(a) 如图 E4.2(b),取杆为自由体,画受力图。取 C 为矩心,根据杆相对 C 的力矩平衡,得

$$\sum M_C = 0,\ Fa\sin\theta - \frac{1}{2}Wa\cos\theta = 0$$

利用 $F=ka(1-\cos\theta)$,解上述方程,得

$$\tan\theta - \sin\theta = \frac{W}{2ka}$$

(b) 利用 $W=45$ N、$a=1$ m 和 $k=50$ N/m,得

$$\tan\theta - \sin\theta = 0.45$$

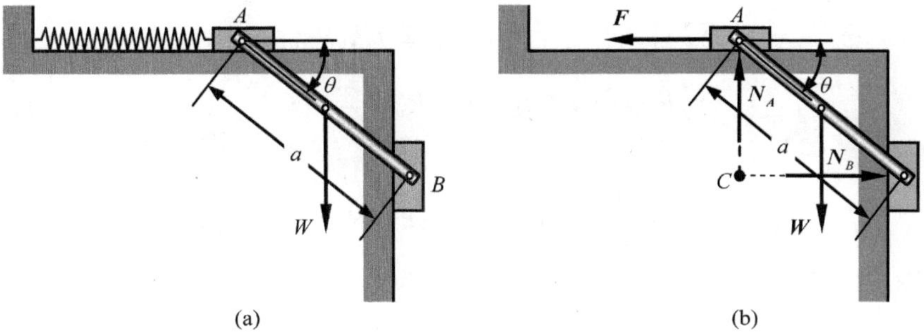

图 E4.2

解上式，得

$$\theta = 50.7639^\circ$$

例 4.3

图 E4.3(a)所示框架，试求作用于 EBG 杆 E 和 B 处的力的分量。

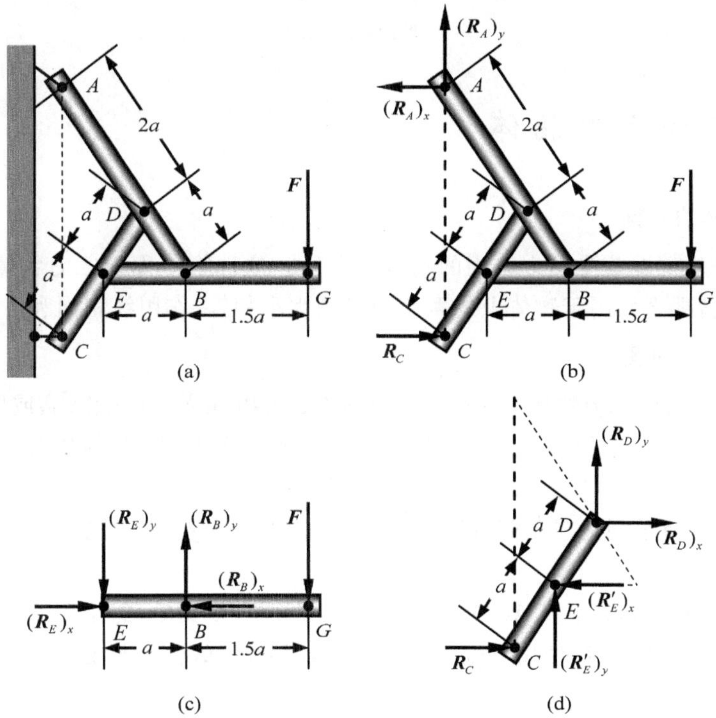

图 E4.3

解

(1) 如图 E4.3(b)，取整个框架为自由体，点 A 为矩心，得

$$\sum M_A = 0,\ R_C \cdot 2\sqrt{3}a - F \cdot 3a = 0,\text{即 } R_C = \frac{\sqrt{3}}{2}F$$

(2) 如图 E4.3(c)，取杆 EBG 为自由体，得

$$\sum M_B = 0,\ (R_E)_y \cdot a - F \cdot 1.5a = 0$$

$$\sum M_E = 0,\ (R_B)_y \cdot a - F \cdot 2.5a = 0$$

$$\sum F_x = 0,\ (R_E)_x - (R_B)_x = 0$$

解上述方程,得

$$(R_E)_y = \frac{3}{2}F, (R_B)_y = \frac{5}{2}F, (R_E)_x = (R_B)_x$$

(3) 如图 E4.3(d),取杆 CED 为自由体,点 D 为矩心,得

$$\sum M_D = 0,\ R_C \cdot \sqrt{3}a - (R'_E)_x \cdot \frac{\sqrt{3}}{2}a - (R'_E)_y \cdot \frac{1}{2}a = 0$$

式中,$(R'_E)_y = (R_E)_y = \frac{3}{2}F$ 和 $R_C = \frac{\sqrt{3}}{2}F$。解上式,得

$$(R'_E)_x = \frac{\sqrt{3}}{2}F,\text{即}(R_E)_x = \frac{\sqrt{3}}{2}F$$

利用$(R_E)_x = (R_B)_x$,得

$$(R_B)_x = \frac{\sqrt{3}}{2}F$$

4.2 二力和三力物体

1. 二力物体的平衡

一种特殊平衡情况是受两力作用的物体的平衡。这种物体称为二力物体。可以证明,如果二力物体平衡,那么作用于物体上的两个力必须大小相等、作用线相同和方向相反。

2. 三力物体的平衡

另一种平衡情况是三力物体的平衡。三力物体是指物体受三个力作用,或更一般地,物体仅在三个点受力作用。可以证明,如果三力物体平衡,那么三个力的作用线要么汇交,要么平行。

例 4.4

如图 E4.4(a)所示刚架与载荷,已知 $F=100$ N 和 $a=1$ m,试求 A 和 B 处反力。

解

如图 E4.4(b)所示,L 型杆 ACD 是二力构件,故 A 处反力$\boldsymbol{R}_A$ 的作用线必通过 A 和 D 点。因为整个结构在$\boldsymbol{R}_A$、$\boldsymbol{R}_B$ 和 $\boldsymbol{F}$ 作用下处于平衡状态,因此$\boldsymbol{R}_B$ 的作用线必经过$\boldsymbol{R}_A$ 和 $\boldsymbol{F}$ 作用线的交点。

根据整个结构平衡,得

$$\sum F_x = 0,\ -R_A\cos\theta_{R_A} + R_B\cos\theta_{R_B} = 0$$

$$\sum F_y = 0,\ -R_A\sin\theta_{R_A} + R_B\sin\theta_{R_B} - F = 0$$

利用 $\tan\theta_{R_A} = 1$、$\tan\theta_{R_B} = 2$ 和 $F=100$ N,得

$$R_A = 141.4\ \text{N},\ \theta_{R_A} = 45^\circ$$

$$R_B = 223.6\ \text{N},\ \theta_{R_B} = 63.43^\circ$$

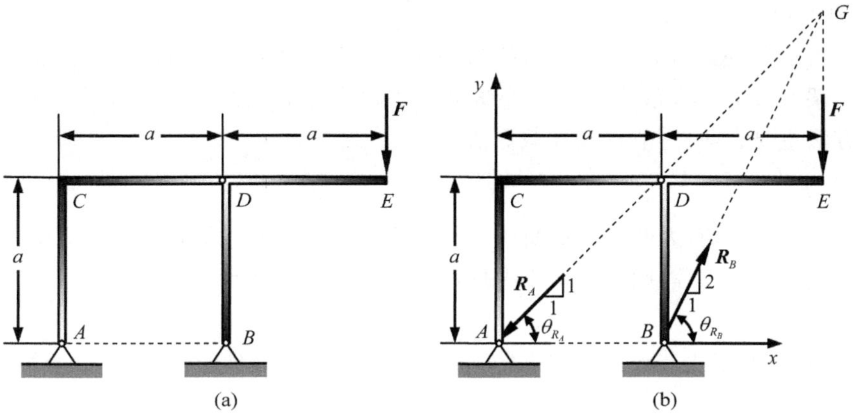

图 E4.4

4.3　平面桁架

平面桁架是由直杆通过铰接而形成的二维结构。在平面桁架中，虽然杆件实际上是通过螺栓或焊接进行连接的，但是通常都假定杆件是通过光滑销钉进行连接的。平面桁架是用于承受作用于结构平面内的载荷，然而常常假设所有载荷都是作用于平面桁架的节点，并且每根杆件的重量也平分到节点。

1. 简单桁架

桁架都是设计成能承受载荷的，因此桁架在载荷作用下必须保持稳定。图 4.1(a)所示桁架由四根杆件通过四个销钉连接而成。如果载荷作用于 C 点，则桁架是不稳定的。与此相反，如图 4.1(b)所示，由三根杆件通过三个销钉连接而成的桁架在载荷作用下是稳定的。

如图 4.2 所示，在最基本的三角形桁架 ABC 上添加杆件 BD 和 CD 即可得到较大桁架。该过程可以进行多次重复，最终的桁架是稳定的，只要每次增加两根杆件和一个节点。由一个三角形桁架通过上述方法构成的桁架称为简单桁架。

假设简单桁架的杆件总数和节点总数分别为 m 和 n，则有

$$(m-3)=2(n-3)\text{ 或 } m=2n-3 \tag{4.6}$$

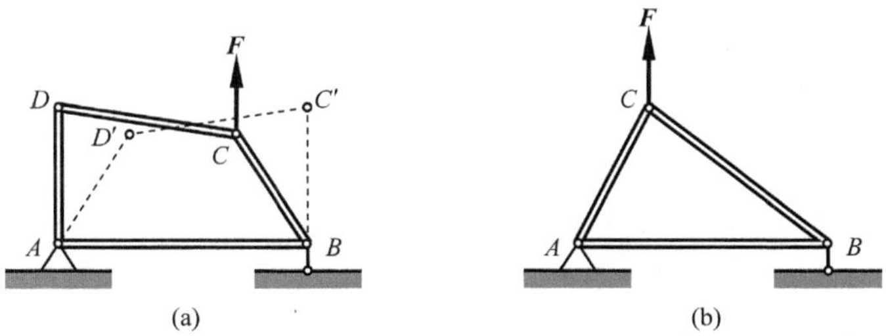

图 4.1

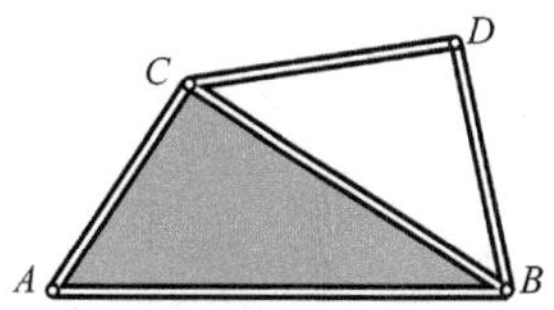

图 4.2

2. 桁架内力

桁架分析不仅需要确定外力,而且还需要确定内力。桁架可以看成是杆件和销钉的组合。因为整个桁架处于平衡状态,因此桁架的每根杆件和每个销钉也必须处于平衡状态。确定桁架内力的方法主要有两种:节点法和截面法。

3. 节点法

因为整个桁架平衡,因此桁架的每个销钉也必须平衡。通过画每个销钉的受力图,并利用每个节点的两个平衡方程,则可确定桁架的内力。

例 4.5

已知 $F=80$ N,采用节点法求图 E4.5(a)所示桁架每根杆件的内力,并说明每根杆件是受拉还是受压。

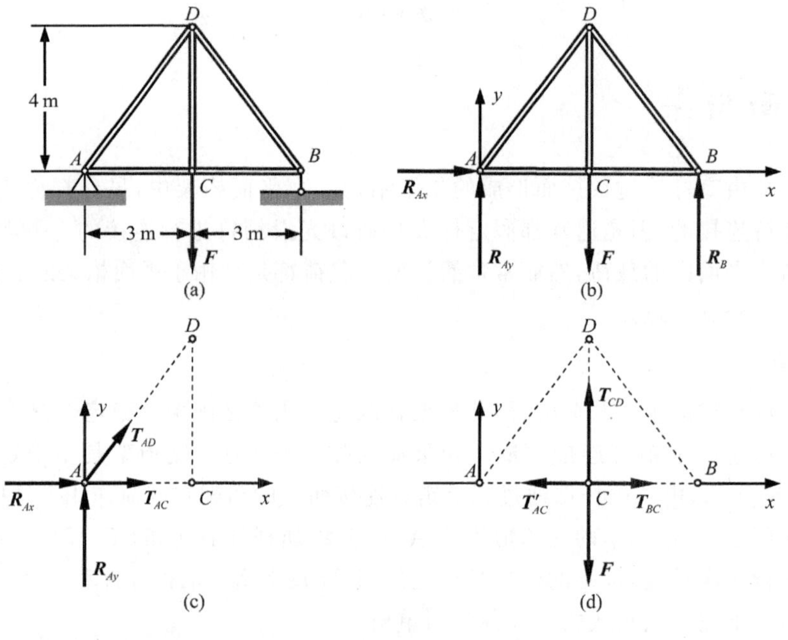

图 E4.5

解

(1) 如图 E4.5(b)所示,取整个桁架为自由体,得

$$\sum F_x = 0,\ R_{Ax} = 0$$

$$\sum M_A = 0,\ R_B \times 6 - F \times 3 = 0$$

$$\sum M_B = 0,\ -R_{Ay} \times 6 + F \times 3 = 0$$

解上述方程,得

$$R_{Ax} = 0, R_{Ay} = 40\ \text{N}, R_B = 40\ \text{N}$$

(2) 取节点 A 为自由体,如图 E4.5(c)所示,有

$$\sum F_x = 0,\ R_{Ax} + T_{AC} + T_{AD} \times \frac{3}{5} = 0$$

$$\sum F_y = 0,\ R_{Ay} + T_{AD} \times \frac{4}{5} = 0$$

求解，得

$$T_{AC} = 30\ \text{N(拉力)},T_{AD} = -50\ \text{N(压力)}$$

（3）取节点 C 为自由体，如图 E4.5(d)所示，有

$$\sum F_x = 0,\ T_{BC} - T'_{AC} = 0$$

$$\sum F_y = 0,\ T_{CD} - F = 0$$

求解，得

$$T_{BC} = 30\ \text{N(拉力)},T_{CD} = 80\ \text{N(拉力)}$$

（4）同理，取节点 B 或 D 为自由体，得

$$T_{BD} = -50\ \text{N(压力)}$$

4. 零力杆

如果杆件中的内力等于零，则该杆称为零力杆。零力杆用于增加桁架稳定性。另外，当作用于桁架的载荷发生变化时，零力杆也用于支撑载荷。

例 4.6

对图 E4.6(a)所示桁架和载荷，求零力杆。

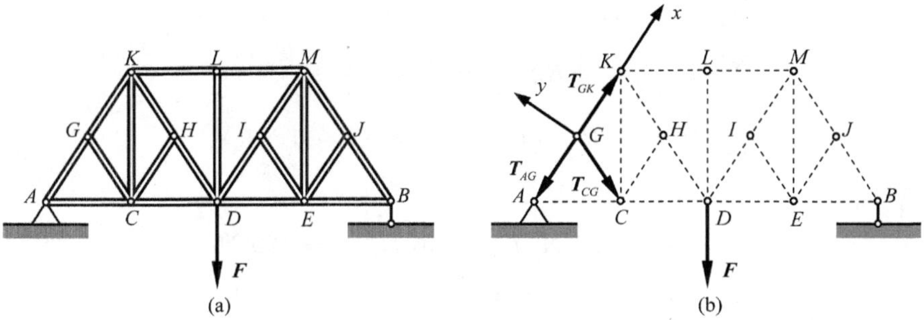

图 E4.6

解

考虑图 E4.6(b)所示节点 G，利用 $\sum F_y = 0$，得

$$T_{CG} = 0$$

同理，提供检查每个节点，得

$$T_{CH} = T_{CK} = T_{DL} = T_{EI} = T_{EJ} = T_{EM} = 0$$

5. 截面法

当需要求解桁架中所有杆件的内力时，节点法是最有效的方法。然而，如果仅需求一根杆件或很少几根杆件的内力，那么截面法将更为有效。

因为整个桁架平衡，因此桁架的任何部分也必须平衡。桁架任何部分的内力都可以通过取该部分为自由体并利用平衡方程而得到。

例 4.7

桁架加载如图 E4.7(a)所示。已知 $F_1=80$ N 和 $F_2=40$ N,试用截面法求杆件 CD、CH 和 GH 的内力。

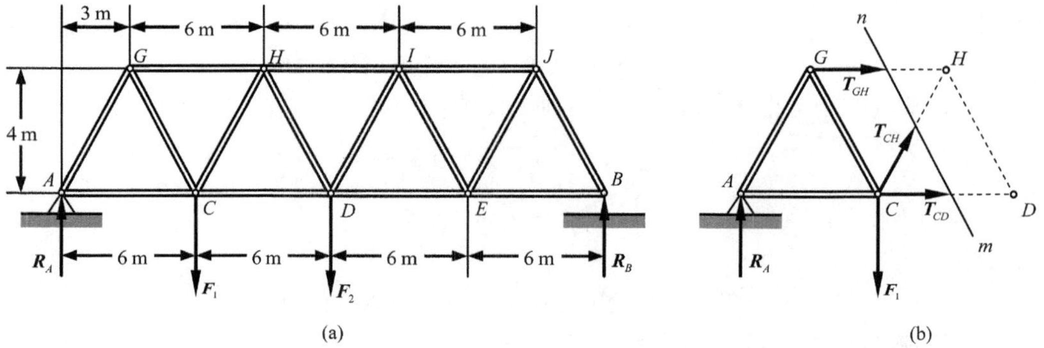

图 E4.7

解

取整个桁架为自由体,得

$$R_A=80\text{ N}, R_B=40\text{ N}$$

为了求杆件 CD、CH 和 GH 的内力,这些杆件应该被截开,如图 E4.7(b)。考虑桁架截开后左部的平衡,得

$$\sum M_H=0, -R_A\times 9+F_1\times 3+T_{CD}\times 4=0$$

$$\sum M_C=0, -R_A\times 6-T_{GH}\times 4=0$$

$$\sum F_y=0, R_A-F_1+T_{CH}\times\frac{4}{5}=0$$

解上述方程,得

$$T_{CD}=120\text{ N}, T_{CH}=0, T_{GH}=-120\text{ N}$$

4.4 三维刚体平衡

1. 平衡方程

三维刚体的平衡方程为

$$\sum F_x=0, \sum F_y=0, \sum F_z=0, \sum M_x(\boldsymbol{F})=0, \sum M_y(\boldsymbol{F})=0, \sum M_z(\boldsymbol{F})=0 \tag{4.7}$$

上述六个方程相互独立,因此可以求解不超过六个未知量。

2. 三维刚体的约束反力

作用于三维刚体的反力在光滑表面产生的单个反力和固定支撑产生的力一力偶系范围内变动。

例 4.8

重量为 150 N 的均质方板由三根垂直缆绳支撑,如图 E4.8(a)所示。试求每根缆绳的拉力。

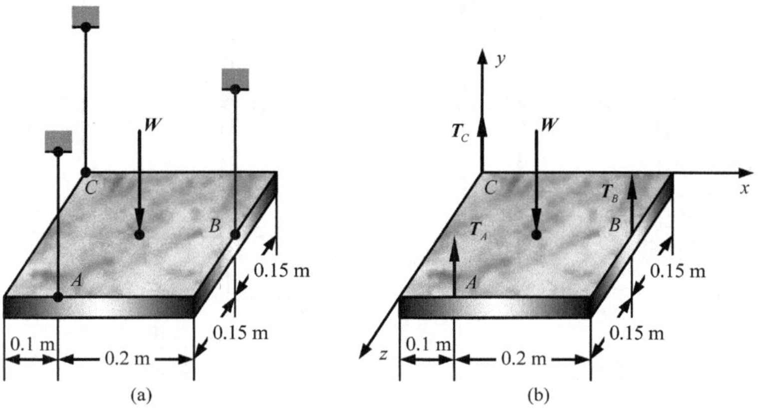

图 E4.8

解

取板为自由体，如图 E4.8(b)。根据板的平衡，得

$$\sum F_y = 0,\ T_A + T_B + T_C - W = 0$$

$$\sum M_x = 0,\ -T_A \times 0.3 - T_B \times 0.15 + W \times 0.15 = 0$$

$$\sum M_z = 0,\ T_A \times 0.1 + T_B \times 0.3 - W \times 0.15 = 0$$

求解，得

$$T_A = 45\ \text{N},\ T_B = 60\ \text{N},\ T_C = 45\ \text{N}$$

习　题

4.1 如图 P4.1 所示，T 型构件 $ABCD$ 由 C 处销钉支撑，并由穿过定滑轮 E 的缆绳 AED 连接。已知 $q=250$ N/m，不计摩擦，试求缆绳拉力和 C 处反力。

4.2 如图 P4.2 所示，T 型构件 $ABCD$ 在 A 处固定，并由缆绳 DE 连接。已知 $F=250$ N 和缆绳拉力 $T=50$ N，试求 A 处反力。

4.3 重量为 W 的均质杆 AB 连接到可沿光滑表面自由运动的物块 A 和 B 上，如图 P4.3 所示。与物块 A 相连弹簧的刚度为 k，当杆水平时弹簧未伸长。(a)假设每个物块的重量均为 W，推导当杆平衡时 W、k、a 和 θ 需要满足的方程。(b)当 $W=15$ N、$a=1$ m 和 $k=50$ N/m 时，求 θ 值。

4.4 重量为 W 的均质杆 AB 连接到可沿光滑表面自由运动的物块 A 和 B 上，如图 P4.4 所示。物块由穿过定滑轮 C 的缆绳连接。(a)不计物块重量，试用 W 和 θ 表示当杆平衡时的缆绳拉力。(b)当缆绳拉力等于 W 时，求 θ 值。

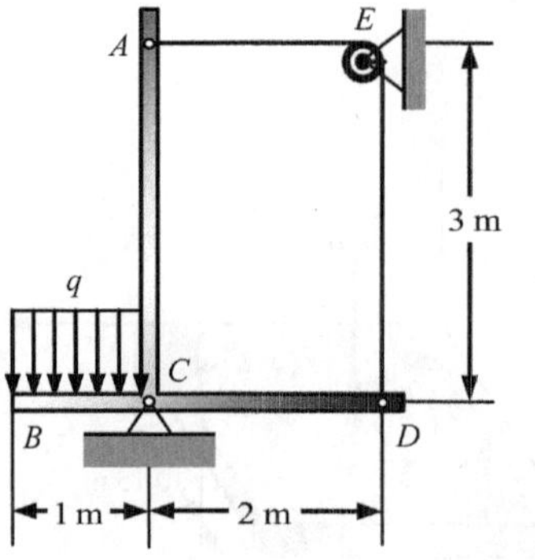

图 P4.1

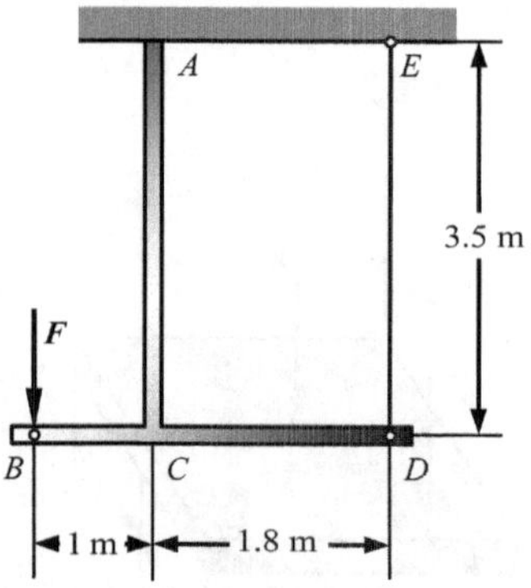

图 P4.2

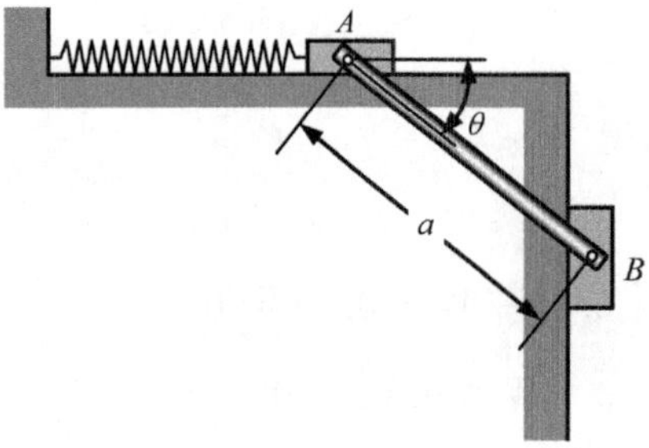

图 P4.3

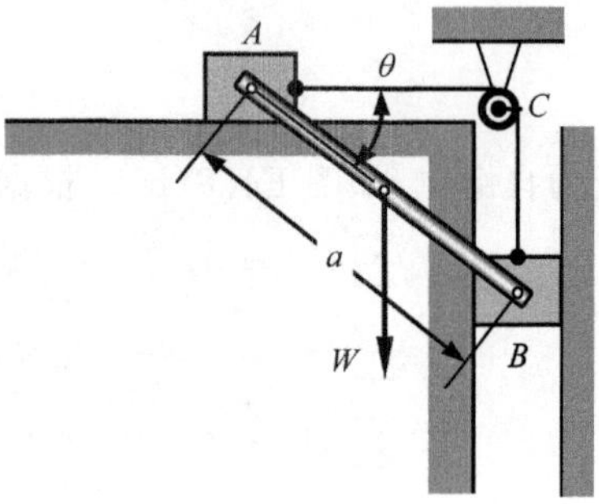

图 P4.4

4.5 如图 P4.5 所示，A 和 B 处反力的最大允许值为 450 N。不计梁重，求在保证梁安全的情况下距离 d 的取值范围。

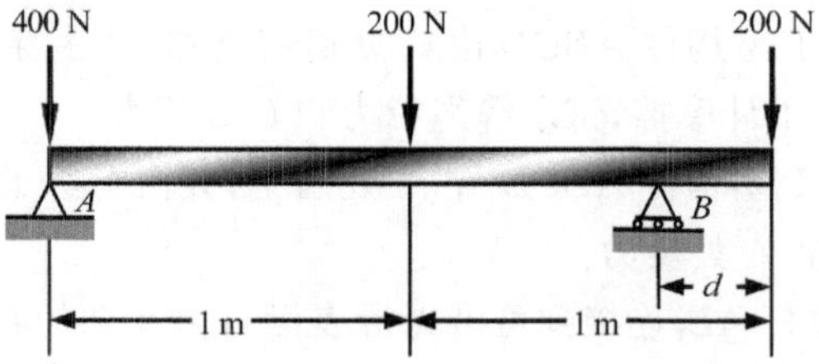

图 P4.5

4.6 如图 P4.6 所示，垂直力 $\boldsymbol{F}$ 作用于平面桁架的点 C。已知 $F=80$ N，求 A 和 H 处反力。

4.7 桁架承受图 P4.7 所示载荷。已知 $F_1=160$ N、$F_2=150$ N 和 $F_3=80$ N，求 A 和 H 处反力。

4.8 悬臂梁 AB 支撑图 4.8 所示载荷。已知 $a=2$ m 和 $q=100$ N/m，求 B 处反力。

4.9 悬臂梁 AB 支撑图 P4.9 所示载荷。已知 $l=8$ m 和 $q=80$ N/m，求 B 处反力。

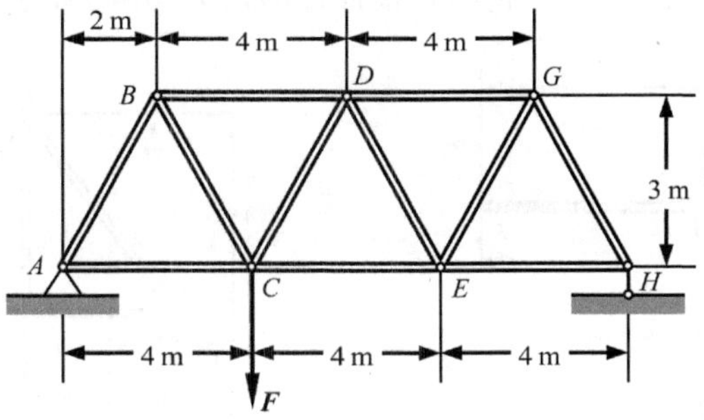

图 P4.6

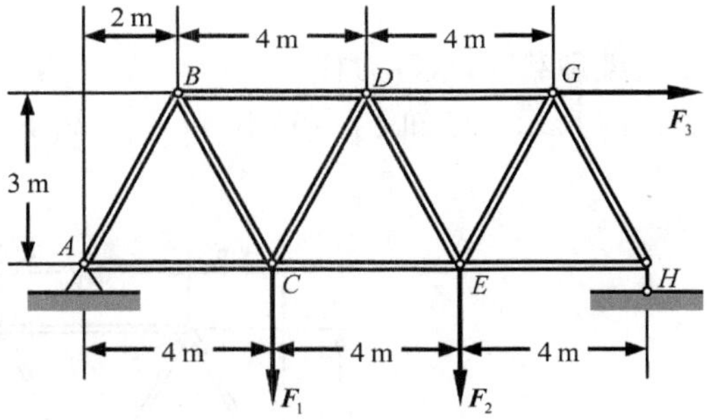

图 P4.7

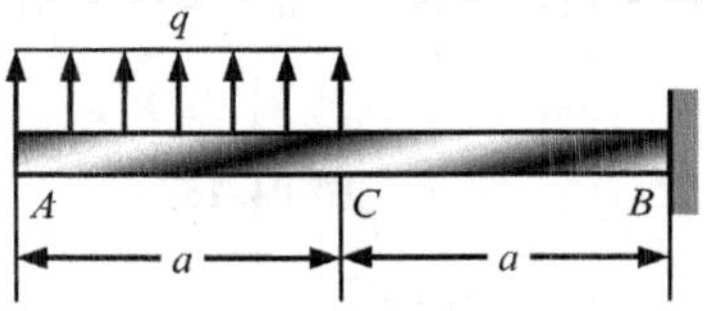

图 P4.8

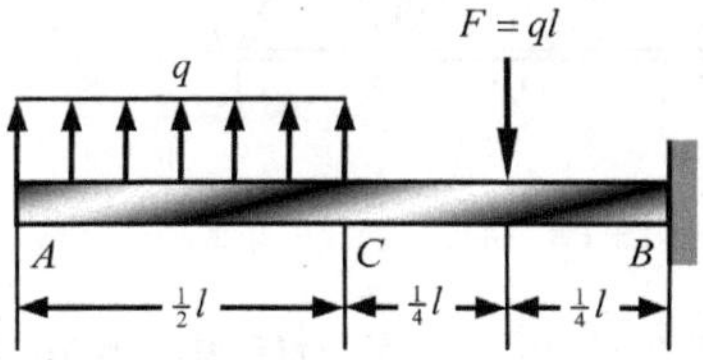

图 P4.9

4.10 刚架和载荷如图 P4.10 所示，已知 $F=200$ N 和 $a=1$ m，求 A 和 B 处反力。

4.11 刚架和载荷如图 P4.11 所示，已知 $q=100$ N/m 和 $a=1$ m，求 A 和 B 处反力。

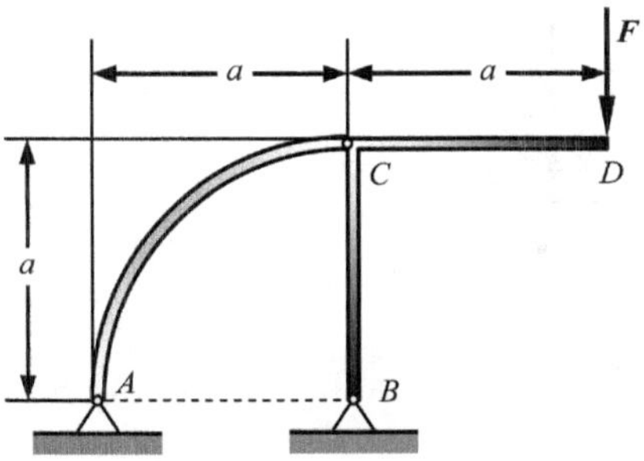

图 P4.10

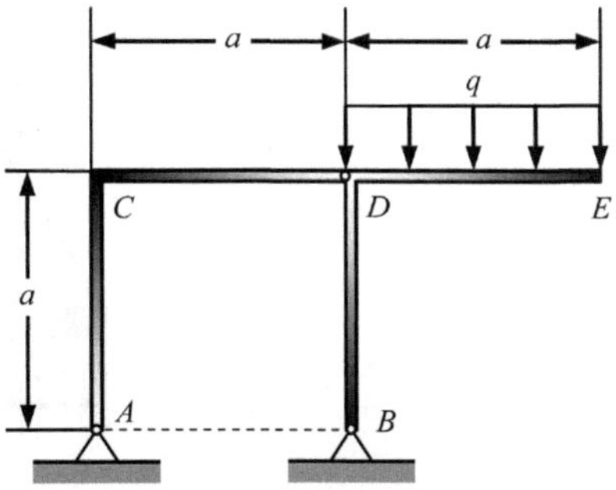

图 P4.11

4.12 如图 P4.12 所示已知 $M_e=100$ N·m 和 $a=1$ m，求 A 和 B 处反力。

4.13 用节点法求图 P4.13 所示桁架每根杆件的内力,并说明每根杆件是受拉还是受压。

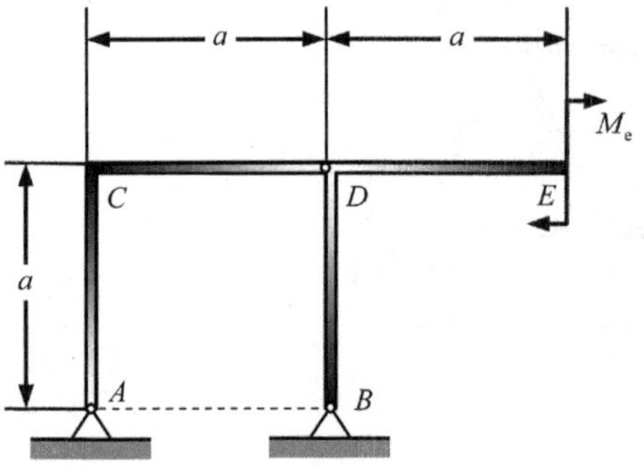

图 P4.12

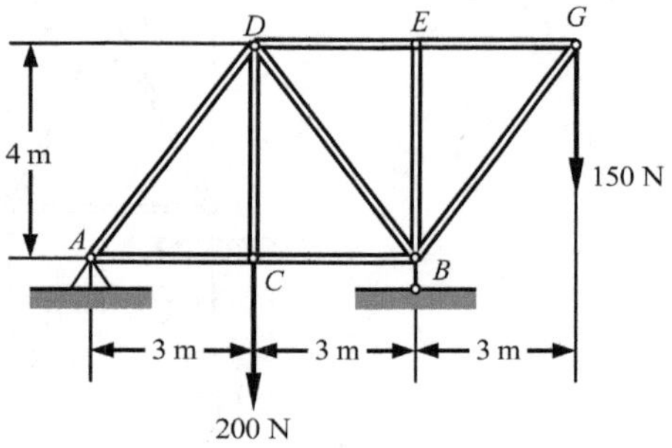

图 P4.13

4.14 对图 P4.14 所示桁架和载荷,求零力杆。

4.15 桁架加载如图 P4.15 所示。已知 $F_1=90$ N 和 $F_2=60$ N,试用截面法求杆件 CD、DG 和 GH 的内力。

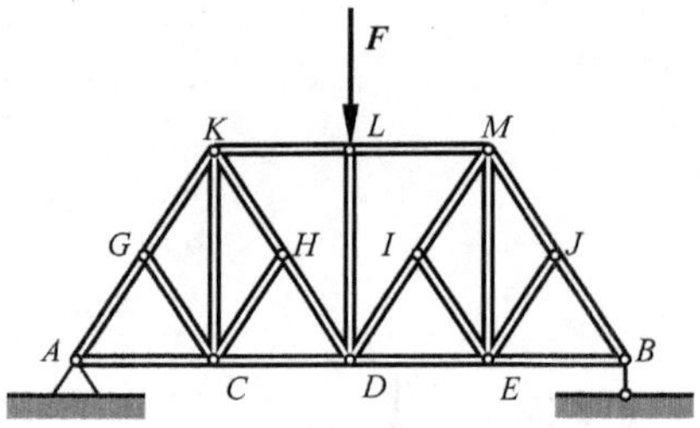

图 P4.14

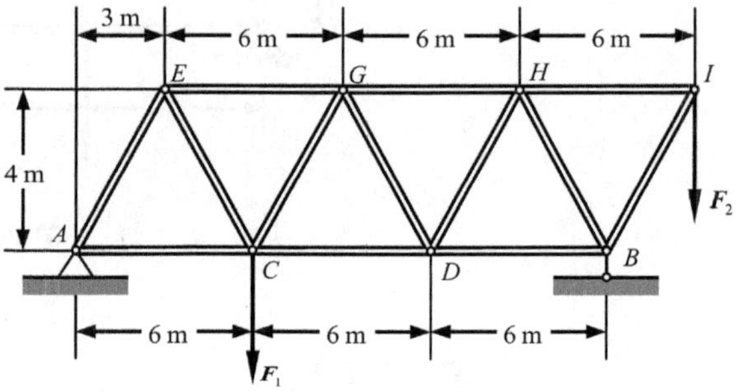

图 P4.15

4.16 重量为 250 N 的均质组合板由三根垂直缆绳支撑,如图 P4.16 所示。试求每根缆绳的拉力。

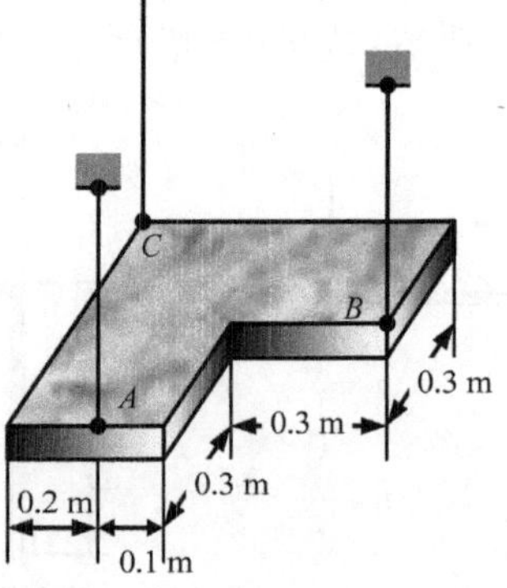

图 P4.16

第5章 摩 擦

两个相互接触的物体，如果一个物体有相对于另一个物体发生运动的趋势，则将会产生切向力(称为摩擦力)。摩擦力的最大值是有限量。

摩擦分为干摩擦和流体摩擦。干摩擦(库仑摩擦)发生在物体的无润滑液的接触面上。流体摩擦存在于通过液体隔开的物体的接触面上。流体摩擦将在流体力学中进行研究。

5.1 滑动摩擦

1. 滑动摩擦分析

滑动摩擦现象可通过如下实验进行解释：

(1) 如图5.1(a)所示，重量为 $\boldsymbol{W}$ 的物块放在粗糙水平面上。作用于物块上的力是物块重量 $\boldsymbol{W}$ 和由粗糙面施加的法向力 $\boldsymbol{N}$。这些垂直力没有沿粗糙面移动物块的趋势。

(2) 如图5.1(b)所示，假设有水平力 $\boldsymbol{F}$ 作用于物块，该力有沿粗糙面移动物块的趋势。然而，当 $\boldsymbol{F}$ 不大时，物块并不运动。这就表明，必定存在由粗糙面施加到物块的切向力与 $\boldsymbol{F}$ 平衡。该切向力称为静摩擦力 $\boldsymbol{F}_s$，可通过求解物块平衡方程而得到。

(3) 如图5.1(c)所示，如果水平力 $\boldsymbol{F}$ 增大，则静摩擦力 $\boldsymbol{F}_s$ 也相应增大，并继续与 $\boldsymbol{F}$ 平衡，直到静摩擦力 $\boldsymbol{F}_s$ 达到最大静摩擦力 $\boldsymbol{F}_{max}$。当达到最大静摩擦力，则物块将处于滑动临界状态。

(4) 如图5.1(d)所示，如果水平力 $\boldsymbol{F}$ 进一步增大，静摩擦力 $\boldsymbol{F}_s$ 不再能与 $\boldsymbol{F}$ 平衡，则物块开始滑动。物块一开始滑动，摩擦力就从最大静摩擦力 $\boldsymbol{F}_{max}$ 降为动摩擦力 $\boldsymbol{F}_k$。物块运动后，动摩擦力 $\boldsymbol{F}_k$ 近似保持常数。

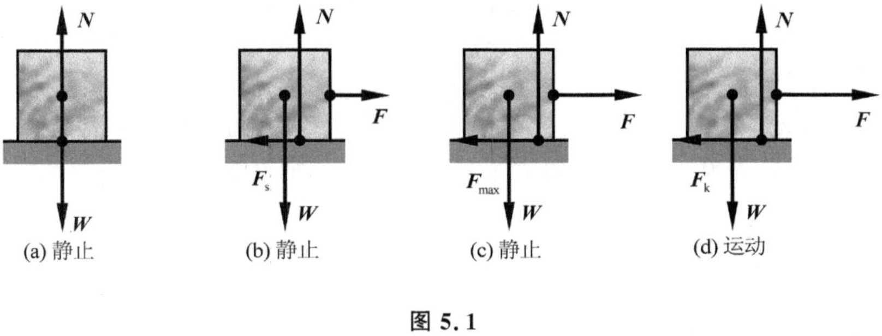

图5.1

2. 静摩擦定律

实验已经表明，最大静摩擦力值 F_{max} 与粗糙面施加于物块的法向力 N 成正比。因此，有

$$F_{max} = \mu_s N \tag{5.1}$$

式中，μ_s 为静摩擦系数。上述关系称为静摩擦定律。

3. 动摩擦定律

同样，动摩擦力值 F_k 也与作用于物块的法向力 N 成正比。因此，动摩擦定律可表示为

$$F_k = \mu_k N \tag{5.2}$$

式中,μ_k 为动摩擦系数。

摩擦系数 μ_s 和 μ_k 与接触面积无关,但与接触物体的材料特性和表面性质有关。

5.2 摩擦角

1. 摩擦角

重为 $\boldsymbol{W}$ 的物块放在粗糙水平面上,并且水平力 $\boldsymbol{F}$ 作用于物块。当静摩擦力达到最大值 F_{max},则法向力 $\boldsymbol{N}$ 与由 $\boldsymbol{N}$ 和 $\boldsymbol{F}_{max}$ 形成的合力 $\boldsymbol{R}$ 之间的夹角定义为静摩擦角 φ_s。根据图 5.2(a),有

$$\tan\varphi_s = \frac{F_{max}}{N} = \mu_s \tag{5.3}$$

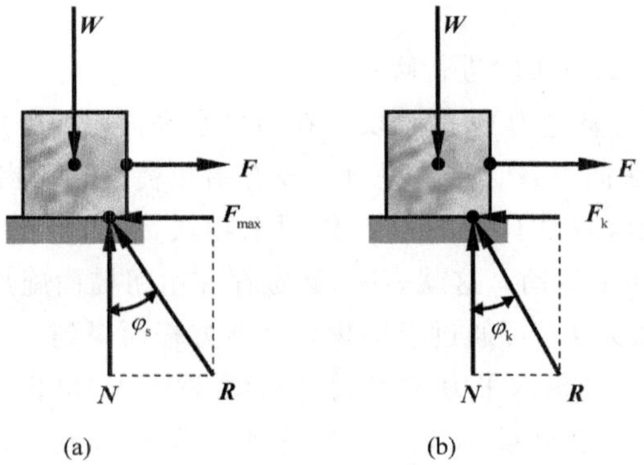

图 5.2

如果运动已经发生,则法向力 $\boldsymbol{N}$ 与合力 $\boldsymbol{R}$ 之间的夹角由静摩擦角 φ_s 降为动摩擦角 φ_k。根据图 5.2(b),动摩擦角表示为

$$\tan\varphi_k = \frac{F_k}{N} = \mu_k \tag{5.4}$$

2. 自　锁

如图 5.2(a)所示,静摩擦角 φ_s 是接触面法线与合力 $\boldsymbol{R}$ 之间的最大夹角,因此接触面法线与由 $\boldsymbol{W}$ 和 $\boldsymbol{F}$ 形成的合力 $\boldsymbol{R}'$ 之间的夹角大于 φ_s,即 $\varphi > \varphi_s$,那么物块会由于 $\boldsymbol{R}'$ 和 $\boldsymbol{R}$ 不可能共线而运动,如图 5.3(a)所示。然而,如果 φ 等于或小于 φ_s,即 $\varphi \leqslant \varphi_s$,那么物块处于静止,即物块自锁,如图 5.3(b)或 5.3(c)所示。

5.3 含有滑动摩擦的问题

含有滑动摩擦的问题都属于如下三种类型之一:

(1) 第一种类型是:所有作用力给定和静摩擦系数已知,需要判断物体是静止还是运动。

例 5.1

$F=200$ N 的力作用于质量 $m=100$ kg 的物块上,物块放在倾角 $\theta=30°$ 的斜面上,如图 E5.1(a)所示。已知物块与平面之间的摩擦系数 $\mu_s=0.3$ 和 $\mu_k=0.2$,确定物块是否平衡,并

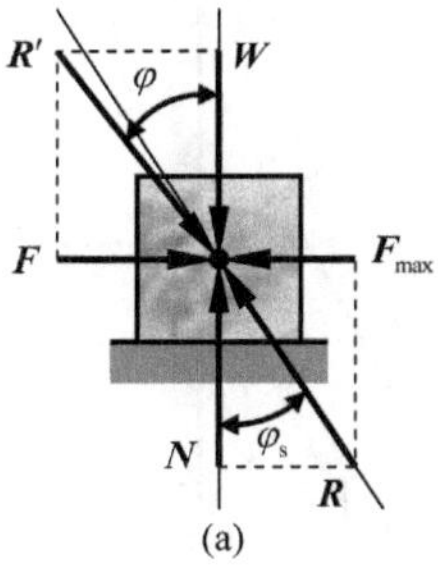

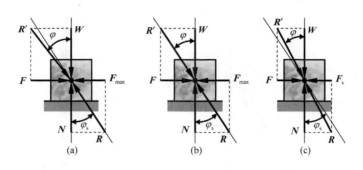

图 5.3

求摩擦力的大小和方向。

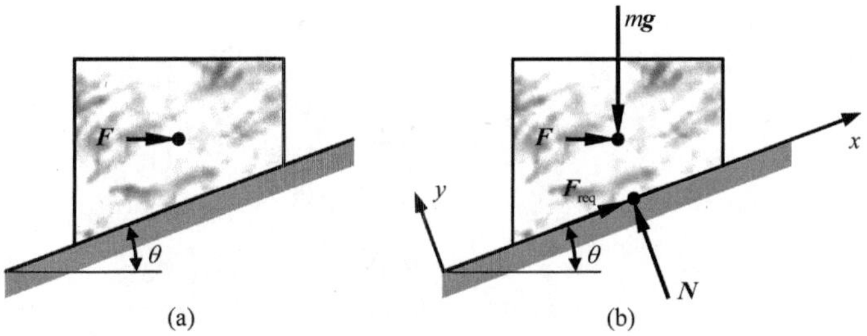

图 E5.1

解

画物块受力图，如图 E5.1(b)所示。假设物块平衡所需摩擦力为 $\boldsymbol{F}_{req}$（方向沿斜面向上），则物块平衡方程为

$$\sum F_x = 0,\ F_{req} + F\cos\theta - mg\sin\theta = 0$$

$$\sum F_y = 0,\ \mathrm{N} - F\sin\theta - mg\cos\theta = 0$$

解方程，得

$$F_{req} = 317.29\ \mathrm{N}$$

$$N = 949.57\ \mathrm{N}$$

最大静摩擦力等于

$$F_{max} = \mu_s N = 284.87\ \mathrm{N}$$

因 $F_{req} > F_{max}$，则物块不能平衡，即物块将沿斜面下滑。根据动摩擦定律，摩擦力大小为

$$F_k = \mu_k N = 189.91\ \mathrm{N}$$

方向沿斜面向上。

(2) 第二种类型是：所有作用力给定和运动处于临界情况，需要求静摩擦系数。

例 5.2

如图 E5.2(a)，长度为 l、质量为 m 的均质梯子 AB 靠在墙上。假设 A 和 B 处具有相同静摩擦系数 μ_s，求梯子在 $\theta=60°$保持平衡时 μ_s 的最小值。

解

假设 A 和 B 处都处于临界运动状态，梯子受力图如图 E5.2(b)所示。利用静摩擦定律，

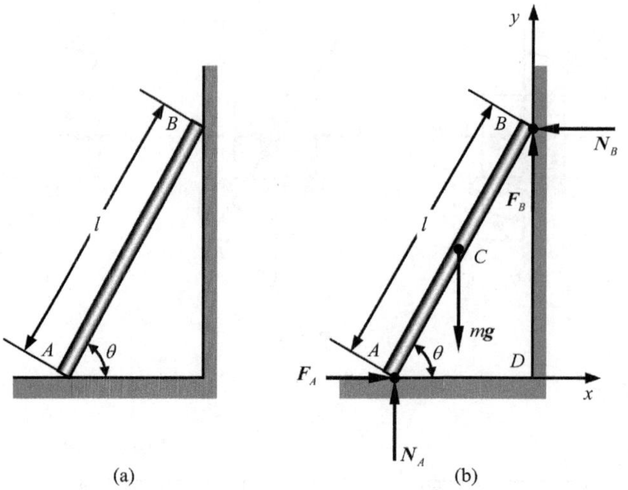

图 E5.2

补充方程为

$$F_A = \mu_s N_A$$
$$F_B = \mu_s N_B$$

考虑梯子平衡,平衡方程可写为

$$\sum F_x = 0,\ F_A - N_B = 0$$
$$\sum F_y = 0,\ F_B + N_A - mg = 0$$
$$\sum M_D = 0,\ N_B l\sin\theta - N_A l\cos\theta + \frac{1}{2}mgl\cos\theta = 0$$

求解补充和平衡方程,得

$$\mu_s^2 + 2\mu_s\tan\theta - 1 = 0$$

利用 $\theta=60°$并解方程,得

$$\mu_s = 0.27\ 或\ \mu_s = -3.73$$

物理上,正根才是合理的,因此 μ_s 的最小值等于 0.27。

(3) 第三种类型是:静摩擦系数已知和运动处于临界情况,需要求作用力的大小或方向。

例 5.3

如图 E5.3(a)所示,重量为 W 的环 B 连到弹簧 AB 并可沿杆运动。弹簧刚度 $k=1.8\ \text{kN/m}$,且当 $\theta=0$ 时弹簧未伸长。已知环与杆之间的静摩擦系数 $\mu_s=0.25$,点 A 和 B 之间的水平距离 $l=0.6\ \text{m}$,求当 $\theta=30°$环维持平衡时 W 值的范围。

解

如图 E5.3(b)所示,假设环平衡,并且静摩擦力 $\boldsymbol{F}_s$ 向上,则根据环受力图,得环平衡方程为

$$\sum F_x = 0,\ N - F_{spr}\cos\theta = 0$$
$$\sum F_y = 0,\ F_s + F_{spr}\sin\theta - W = 0$$

解方程,得

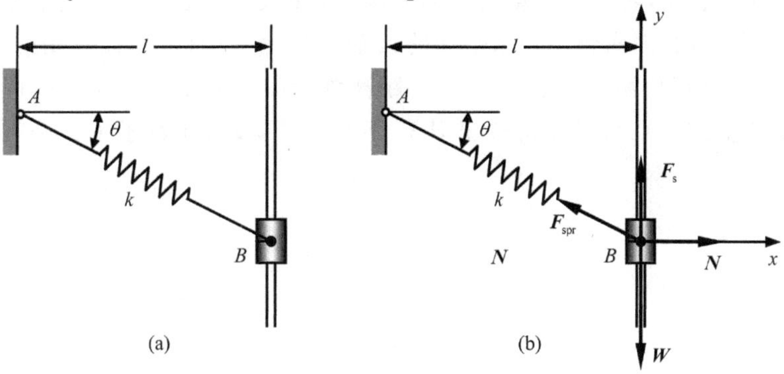

图 E5.3

$$N = F_{spr}\cos\theta$$
$$F_s = W - F_{spr}\sin\theta$$

当环平衡，则

$$|F_s| \leqslant \mu_s N$$

即

$$-\mu_s F_{spr}\cos\theta \leqslant W - F_{spr}\sin\theta \leqslant \mu_s F_{spr}\cos\theta$$

或

$$F_{spr}(\sin\theta - \mu_s\cos\theta) \leqslant W \leqslant F_{spr}(\sin\theta + \mu_s\cos\theta)$$

利用 $F_{spr}=kl(1/\cos\theta-1)$ 和 $\theta=30°$，则得

$$47.37\ \text{N} \leqslant W \leqslant 119.71\ \text{N}$$

5.4　滚动摩阻

如图 5.4(a)所示，半径为 r 的轮子沿水平面无滑动自由滚动。如果轮子和水平面都具有刚性，那么水平面施加于轮子的法向力作用于轮子的切点 A 处，即在该情况下轮子仅受两个力作用：自身重力 $\boldsymbol{W}$ 和法向反力 $\boldsymbol{N}$。不管轮子与水平面之间的摩擦系数的数值怎样，都没有摩擦力作用于轮子。因此，沿水平面无滑动自由滚动的轮子应该保持永无止境的滚动。

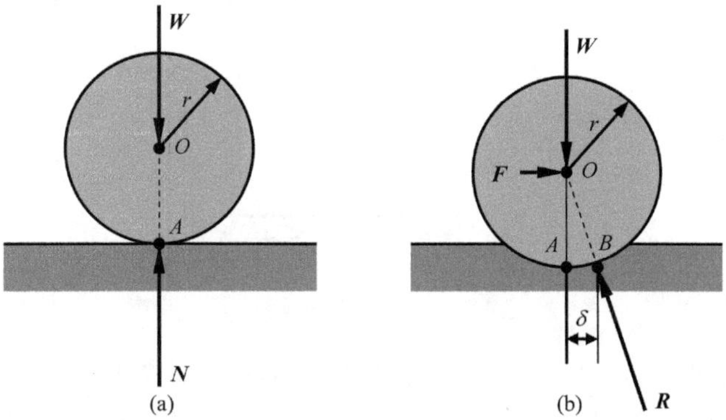

图 5.4

然而,实验证据表明,轮子的滚动将会越来越慢,并最终停止滚动。这是由于轮子和水平面不可能具有完全刚性,因而轮子与水平面的接触始终是一个区域,而不是一个点。所以,水平面对轮子的反力 $\boldsymbol{R}$ 将作用于距点 A 的水平距离为 δ 的点 B 处,如图 5.4(b)所示。为了与重力 $\boldsymbol{W}$ 对点 B 产生的矩平衡以保持轮子以不变速度滚动,则在轮子中心必须要施加水平力 $\boldsymbol{F}$。

根据 $\sum M_B = 0$, 得

$$F = \frac{\delta}{r} W \tag{5.5}$$

式中,δ 称为滚阻系数,单位通常取为 mm。滚阻系数 δ 与轮子的半径无关,但与轮子和水平面的材料特性和表面性质有关。滚阻系数 δ 的值在 0.25 mm(钢轮放在钢面上)到 125 mm(钢轮放在软面上)之间变换。

由于 $\frac{\delta}{r} \ll \mu_s$(或 μ_k),即 $F_{\text{rolling}} = \frac{\delta}{r} W \ll F_{\text{sliding}} = \mu_s W$(或 $\mu_k W$),因此使物体沿水平面滚动要比使其沿水平面滑动容易得多。

习 题

5.1 如图 P5.1 所示,力 $\boldsymbol{F}$ 作用于质量 $m=100$ kg 的物块上,物块放在倾角 $\theta=30°$ 的斜面上。已知物块与平面之间的摩擦系数 $\mu_s=0.3$ 和 $\mu_k=0.2$,确定物块是否平衡,并求当(a)$F=500$ N 和(b)$F=1200$ N 时的摩擦力的大小和方向。

5.2 如图 P5.2 所示,长度为 l、质量为 m 的均质梯子 AB 靠在墙上。假设 A 处静摩擦系数为 μ_s,B 处静摩擦系数为零,求梯子在 $\theta=60°$ 保持平衡时 μ_s 的最小值。

5.3 重量为 W 的环 B 与弹簧 AB 连接后沿杆运动,如图 P5.3 所示。弹簧刚度 $k=1.8$ kN/m,当 $\theta=0$ 时弹簧未伸长。已知环与杆之间的静摩擦系数 $\mu_s=0.25$,点 A、B 之间的水平距离 $l=0.6$ m,求当 $\theta=30°$ 环保持平衡时 W 值的范围。

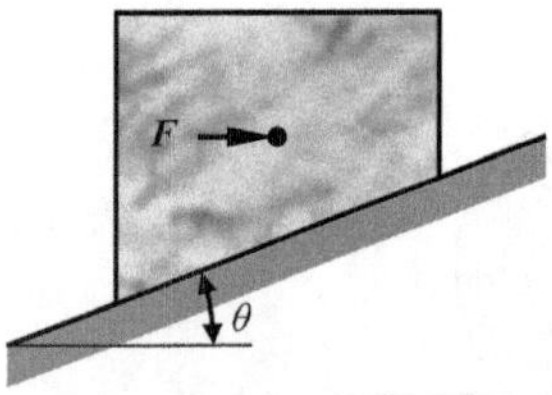

图 P5.1

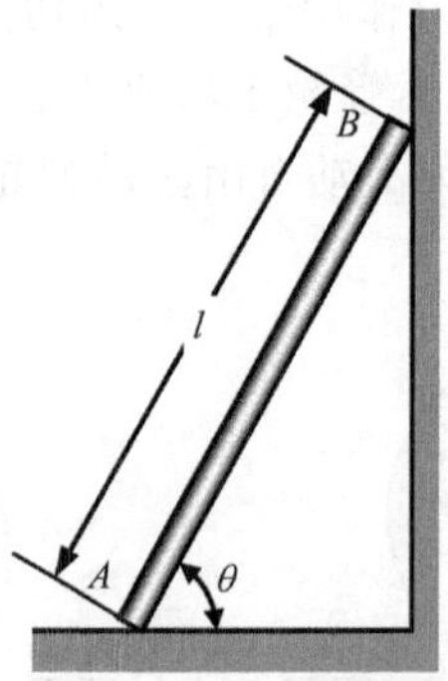

图 P5.2

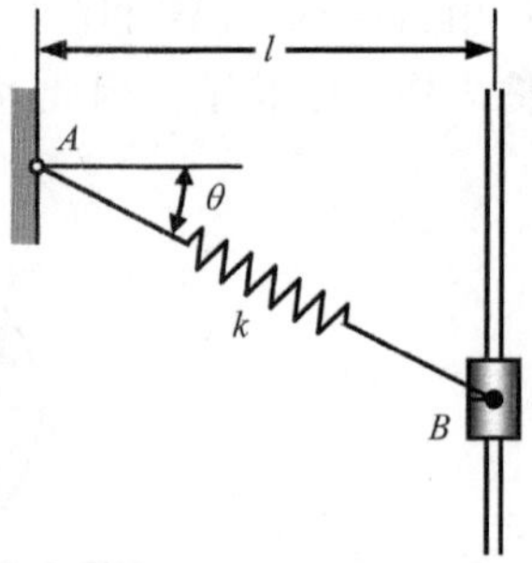

图 P5.3

5.4 如图 P5.4 所示,$F=100$ N 的力作用于质量为 m 的物块上,物块放在水平面上。已知物块与平面之间的摩擦系数 $\mu_s=0.3$ 和 $\mu_k=0.2$,仅考虑 θ 小于或等于 90°,求当(a)$m=15$ kg和(b)$m=30$ kg 物块处于向右滑动临界状态时的最小 θ 值。

5.5 如图 P5.5 所示，$F=150$ N 的力作用于质量 $m=10$ kg 的物块上。已知物块与平面之间的摩擦系数 $\mu_s=0.3$ 和 $\mu_k=0.2$，求物块保持平衡时 θ 值的范围。

5.6 质量 $m=50$ kg 的橱柜安装在小脚轮上，小脚轮能够锁定以阻止其转动，如图 P5.6 所示。地板与小脚轮之间的静摩擦系数 $\mu_s=0.3$。已知 $b=500$ mm 和 $h=650$ mm，求在下列三种情形下橱柜具有向右临界运动时力 F 的大小：(a)所有小脚轮都锁定；(b)B 处小脚轮锁定，A 处小脚轮可以自由转动；(c)A 处小脚轮锁定，B 处小脚轮可以自由转动。

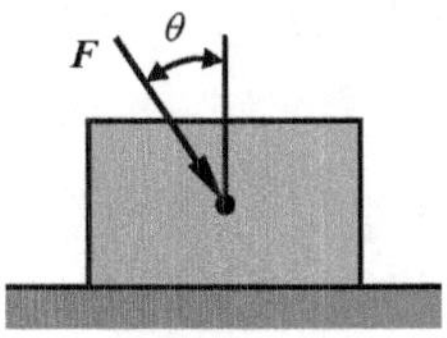

图 P5.4

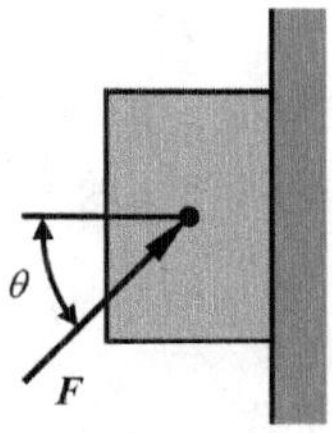

图 P5.5

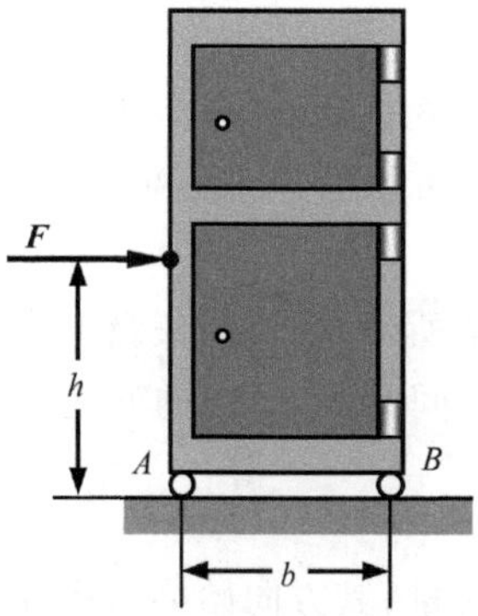

图 P5.6

第 6 章　质点运动学

沿直线运动的质点称为在做直线运动，而沿曲线运动的质点称为在做曲线运动。

6.1　质点运动的矢量表示

如图 6.1 所示，质点 P 沿曲线运动，它在时间 t 的位置可由连接 O 和 P 的位置矢量 $\boldsymbol{r}$ 表示，其中 O 为在空间选择的参考点。当质点运动时，位置矢量 $\boldsymbol{r}$ 为时间 t 的函数，即

$$\boldsymbol{r} = \boldsymbol{r}(t) \tag{6.1}$$

如图 6.2 所示，假设用矢量 $\boldsymbol{r}'$ 定义质点在时间 $t+\Delta t$ 的位置，那么矢量 $\Delta\boldsymbol{r}=\boldsymbol{r}'-\boldsymbol{r}$ 表示位置矢量 $\boldsymbol{r}$ 在时间间隔 Δt 内的变化。因此，质点在时间 t 的速度可表示为

$$\boldsymbol{v} = \lim_{\Delta t \to 0} \frac{\Delta \boldsymbol{r}}{\Delta t} = \dot{\boldsymbol{r}} \tag{6.2}$$

速度是矢量，其方向始终与运动轨迹相切，如图 6.3 所示。

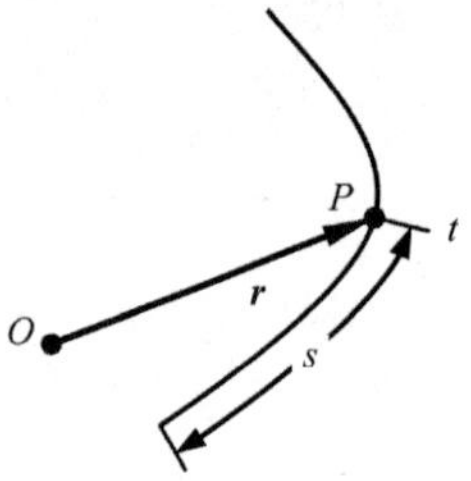

图 6.1

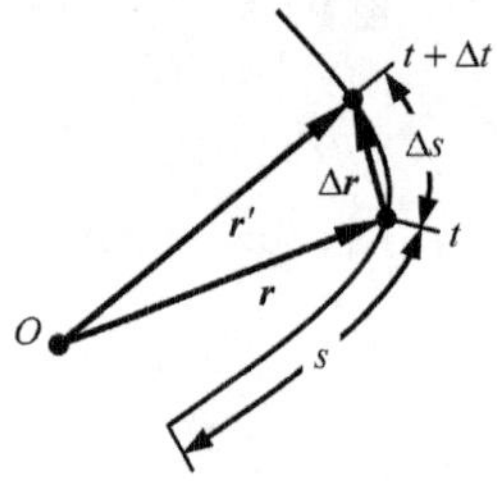

图 6.2

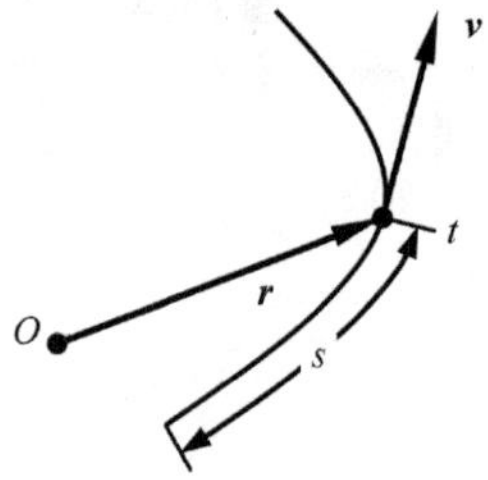

图 6.3

如图 6.4 所示，假设用矢量 $\boldsymbol{v}'$ 定义质点在时间 $t+\Delta t$ 的速度，那么矢量 $\Delta\boldsymbol{v}=\boldsymbol{v}'-\boldsymbol{v}$ 表示速度 $\boldsymbol{v}$ 在时间间隔 Δt 内的变化。因此，质点在时间 t 的加速度可表示为

$$\boldsymbol{a} = \lim_{\Delta t \to 0} \frac{\Delta \boldsymbol{v}}{\Delta t} = \dot{\boldsymbol{v}} = \ddot{\boldsymbol{r}} \tag{6.3}$$

加速度是矢量，其方向通常不与运动轨迹相切，如图 6.5 所示。

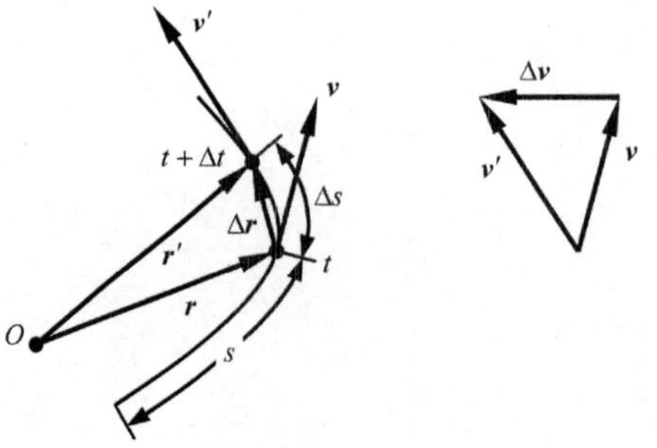

图 6.4

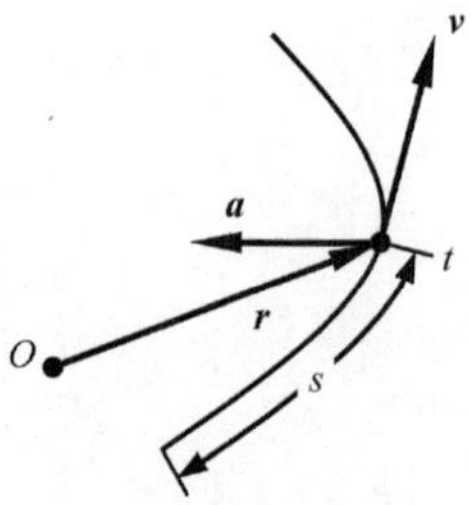

图 6.5

6.2　质点运动的直角坐标表示

建立图 6.6 所示直角坐标系 $Oxyz$，则质点 P 的位置矢量 $\boldsymbol{r}$ 可写为

$$\boldsymbol{r} = x\boldsymbol{i} + y\boldsymbol{j} + z\boldsymbol{k} \tag{6.4}$$

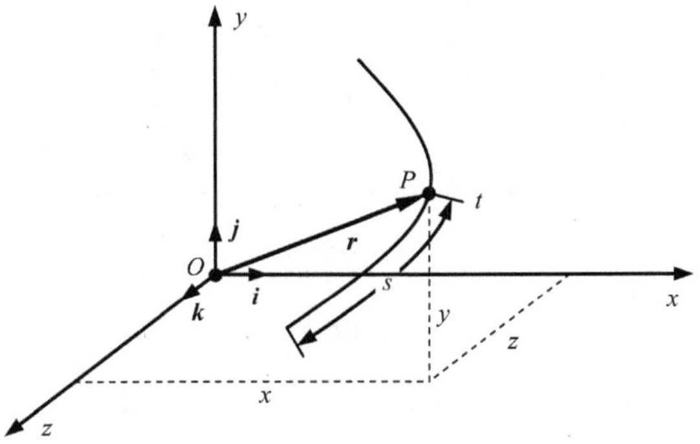

图 6.6

式中，$\boldsymbol{i}$、$\boldsymbol{j}$ 和 $\boldsymbol{k}$ 分别为沿正坐标轴的单位矢量，$x=x(t)$、$y=y(t)$ 和 $z=z(t)$ 分别为位置矢量 $\boldsymbol{r}$ 在三个坐标轴上的标量分量。

因为 $\boldsymbol{i}$、$\boldsymbol{j}$ 和 $\boldsymbol{k}$ 的大小和方向均保持不变，因此质点 P 在时间 t 的速度和加速度可分别表示为

$$\boldsymbol{v} = \dot{\boldsymbol{r}} = \dot{x}\boldsymbol{i} + \dot{y}\boldsymbol{j} + \dot{z}\boldsymbol{k} \tag{6.5}$$

$$\boldsymbol{a} = \ddot{\boldsymbol{r}} = \ddot{x}\boldsymbol{i} + \ddot{y}\boldsymbol{j} + \ddot{z}\boldsymbol{k} \tag{6.6}$$

式中，$\dot{x}$，$\dot{y}$，$\dot{z}$ 和 $\ddot{x}$，$\ddot{y}$，$\ddot{z}$ 分别表示速度 $\boldsymbol{v}$ 和加速度 a 的标量分量，即

$$v_x = \dot{x}\ v_y = \dot{y}\ v_z = \dot{z} \tag{6.7}$$

$$a_x = \ddot{x}\ a_y = \ddot{y}\ a_z = \ddot{z} \tag{6.8}$$

例 6.1

质点的运动由位置矢量 $\boldsymbol{r}=A(\cos t+t\sin t)\boldsymbol{i}+A(\sin t-t\cos t)\boldsymbol{j}$ 定义，其中 t 的单位为 s。求当位置矢量与加速度矢量分别(a)垂直和(b)平行时的 t 值。

解

根据 $\boldsymbol{r}=A(\cos t+t\sin t)\boldsymbol{i}+A(\sin t-t\cos t)\boldsymbol{j}$，有

$$\boldsymbol{v} = \dot{\boldsymbol{r}} = A(t\cos t)\boldsymbol{i} + A(t\sin t)\boldsymbol{j}$$

$$\boldsymbol{a} = \dot{\boldsymbol{v}} = A(\cos t - t\sin t)\boldsymbol{i} + A(\sin t + t\cos t)\boldsymbol{j}$$

(a) 当位置矢量与加速度矢量垂直，有

$$\boldsymbol{r} \cdot \boldsymbol{a} = 0$$

由上式得

$$t = 1\ \text{s}$$

(b) 当位置矢量与加速度矢量平行，有

$$\boldsymbol{r} \times \boldsymbol{a} = 0$$

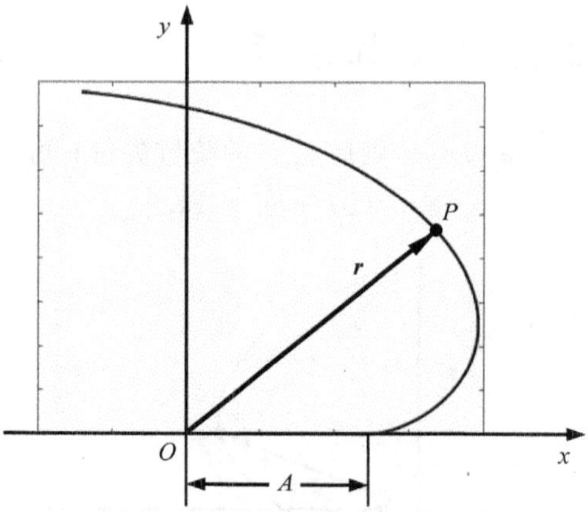

图 E6.1

由上式得

$$t = 0$$

6.3 质点运动的自然坐标表示

如图 6.7 所示,在沿曲线运动的质点 P 上建立自然坐标系,并定义三个单位矢量$\boldsymbol{e}_t$、$\boldsymbol{e}_n$ 和 $\boldsymbol{e}_b$,其中$\boldsymbol{e}_t$ 为指向运动方向的与运动轨迹相切的切向单位矢量,$\boldsymbol{e}_n$ 为指向运动轨迹曲率中心的与运动轨迹垂直的主法向单位矢量,$\boldsymbol{e}_b$ 为指向$\boldsymbol{e}_t\times\boldsymbol{e}_n$ 方向(通过右手法则确定)的与包含$\boldsymbol{e}_t$ 和 $\boldsymbol{e}_n$ 的平面垂直的副法向单位矢量,即$\boldsymbol{e}_b=\boldsymbol{e}_t\times\boldsymbol{e}_n$。包含$\boldsymbol{e}_t$ 和$\boldsymbol{e}_n$ 的平面称为密切面。

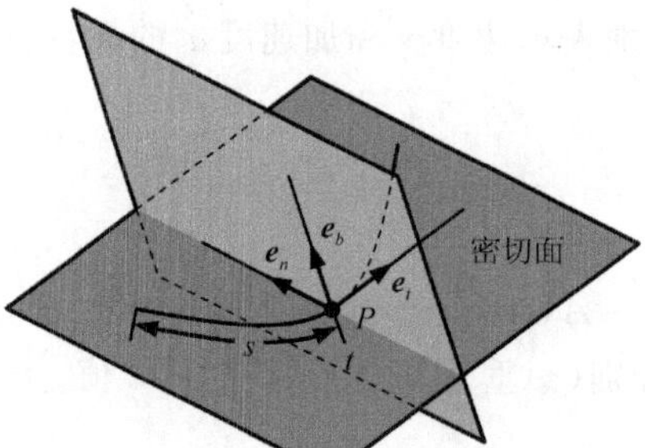

图 6.7

在自然坐标系中,质点 P 在时间 t 的速度矢量 $\boldsymbol{v}$ 可写为

$$\boldsymbol{v} = v\boldsymbol{e}_t \tag{6.9}$$

式中,$\boldsymbol{e}_t$ 为切向单位矢量。利用式(6.9),质点 P 在时间 t 的加速度矢量 $\boldsymbol{a}$ 可写为

$$\boldsymbol{a} = \dot{\boldsymbol{v}} = \dot{v}\boldsymbol{e}_t + v\dot{\boldsymbol{e}}_t \tag{6.10}$$

利用$\dot{\boldsymbol{e}}_t=\dfrac{v}{\rho}\boldsymbol{e}_n$,其中$\boldsymbol{e}_n$ 为主法向单位矢量,ρ 为运动轨迹的曲率半径,则质点 P 在时间 t 的加速度矢量 $\boldsymbol{a}$ 可重写为

$$\boldsymbol{a} = \dot{v}\boldsymbol{e}_t + \frac{v^2}{\rho}\boldsymbol{e}_n \tag{6.11}$$

式中，$\dot{v}$ 和$\frac{v^2}{\rho}$分别表示加速度 $\boldsymbol{a}$ 的切向和法向分量，即

$$a_t = \dot{v},\ a_n = \frac{v^2}{\rho} \tag{6.12}$$

式(6.12)表明，加速度的切向分量 a_t 等于质点速率的变化率，而法向分量 a_n 等于速率平方除以运动轨迹的曲率半径。如果质点速率增加，a_t 为正，$\boldsymbol{a}_t$ 与运动方向相同；如果质点速率减小，a_t 为负，$\boldsymbol{a}_t$ 与运动方向相反。然而，a_n 始终为正，并始终指向运动轨迹的曲率中心。

由此得出结论，加速度的切向分量反映质点速度的大小变化，而法向分量则反映质点速度的方向变化。

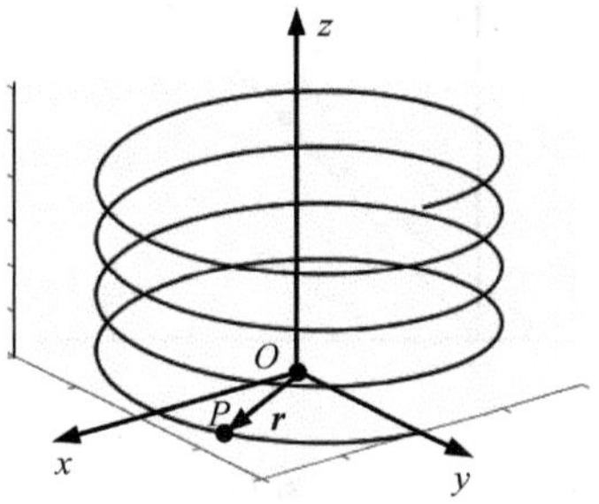

图 E6.2

例 6.2

质点的运动由位置矢量 $\boldsymbol{r}=(2\sin 4t)\boldsymbol{i}+(2\cos 4t)\boldsymbol{j}+(4t)\boldsymbol{k}$ m 定义，其中 t 的单位为 s。求质点运动轨迹的曲率半径。

解

轨迹 $\boldsymbol{r}=(2\sin 4t)\boldsymbol{i}+(2\cos 4t)\boldsymbol{j}+(4t)\boldsymbol{k}$ m，有

$$\boldsymbol{v} = \dot{\boldsymbol{r}} = (8\cos 4t)\boldsymbol{i} + (-8\sin 4t)\boldsymbol{j} + (4)\boldsymbol{k}\ \text{m/s}$$

$$\boldsymbol{a} = \dot{\boldsymbol{v}} = (-32\sin 4t)\boldsymbol{i} + (-32\cos 4t)\boldsymbol{j}\ \text{m/s}^2$$

因此，得

$$v = \sqrt{(8\cos 4t)^2 + (-8\sin 4t)^2 + (4)^2} = 4\sqrt{5}\ \text{m/s}$$

$$a = \sqrt{(-32\sin 4t)^2 + (-32\cos 4t)^2} = 32\ \text{m/s}^2$$

利用 $a^2=(a_t)^2+(a_n)^2$，其中 $a_t=\dot{v}=0$ 和 $a_n=\frac{v^2}{\rho}$，得

$$\rho = \frac{v^2}{a_n} = \frac{v^2}{a} = 2.5\ \text{m}$$

习　题

6.1 如图 P6.1 所示，质点的运动由位置矢量 $\boldsymbol{r}=(4t^2-3t)\boldsymbol{i}+t^3\boldsymbol{j}$ 定义，其中 r 和 t 的单位分别为 m 和 s。求(a)$t=0.2$ s 和(b)$t=1$ s 时质点的速度和加速度。

6.2 如图 P6.2 所示，质点 P 沿抛物线轨迹 $y=\frac{1}{20}x^2$ 运动，其中 x 和 y 的单位均为 m。当 $x=6$ m，质点速率为 6 m/s，速率增加率为 2 m/s^2。求该瞬时速度的方向以及加速度的大小和方向。

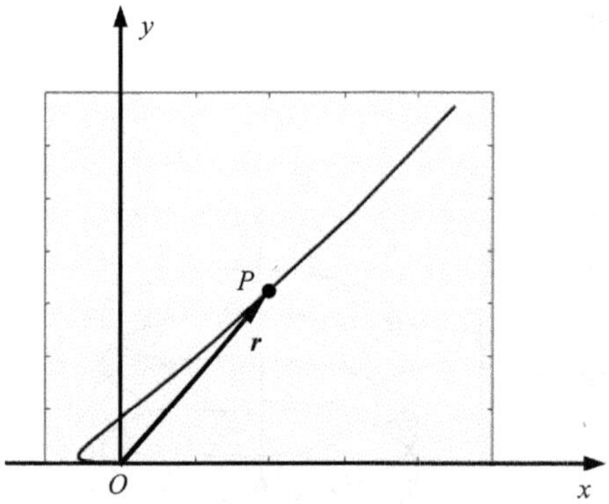

图 P6.1

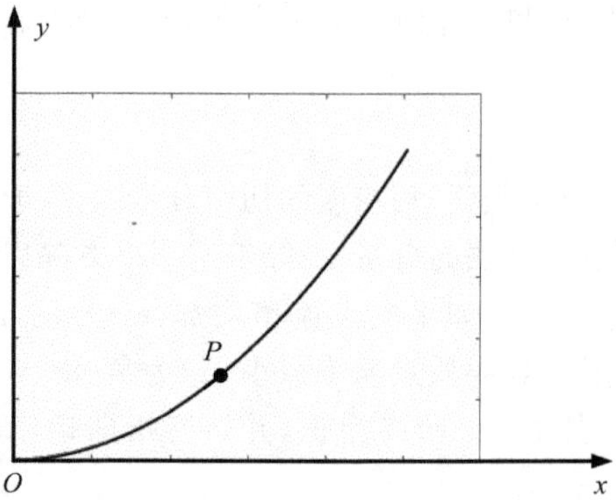

图 P6.2

6.3 如图 P6.3 所示，质点的运动由位置矢量 $\boldsymbol{r}=\frac{3}{4}[1-1/(t+1)]\boldsymbol{i}+\frac{1}{2}[\exp(-\pi t/2)\cos 2\pi t]\boldsymbol{j}$ 定义，其中 r 和 t 的单位分别为 m 和 s。求(a) $t=0.5$ s 和(b) $t=1$ s 时质点的位置、速度和加速度。

6.4 如图 P6.4 所示，质点的运动由位置矢量 $\boldsymbol{r}=(\omega t-\sin\omega t)\boldsymbol{i}+(1-\cos\omega t)\boldsymbol{j}$ m 定义，其中 ω 和 t 的单位分别为 rad/s 和 s。求在时间 t 质点的切向和法向加速度。

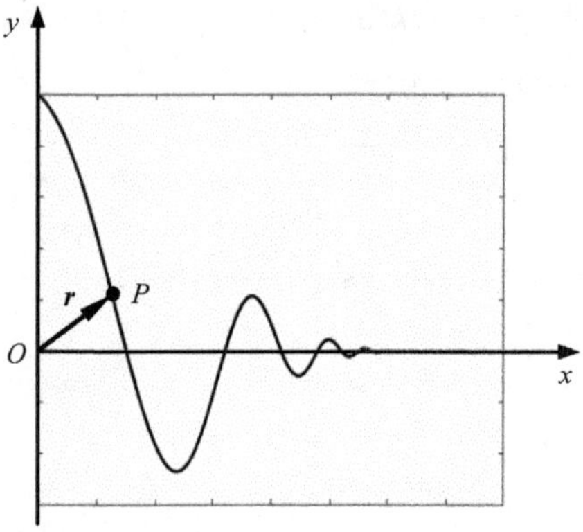

图 P6.3

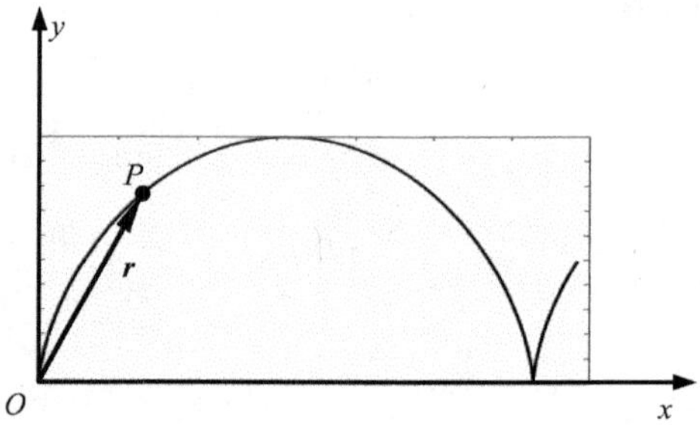

图 P6.4

第 7 章　刚体平面运动学

7.1　刚体平面运动

当刚体上所有质点都沿距某一固定平面等距离的轨迹移动，则这种运动称为平面运动。平面运动包含三种类型：

(1) 平　移

如果刚体内任何直线在刚体运动期间保持方向不变，则这种运动称为平移。当刚体平移时，刚体内所有质点都沿平行运动轨迹移动。如果运动轨迹是直线，则称为直线平移，如图 7.1(a)所示；如果运动轨迹是曲线，则称为曲线平移，如图 7.1(b)所示。

(a) 直线平移

(b) 曲线平移

图 7.1

(2) 定轴转动

当刚体绕某一固定轴旋转，则除转动轴上的点之外，形成刚体的所有质点都沿垂直于转动轴的圆形轨迹移动，如图 7.2 所示。

(3) 一般运动

物体一般运动可以看成是参考面内的平移和绕参考面垂直轴的转动的合成，如图 7.3 所示。

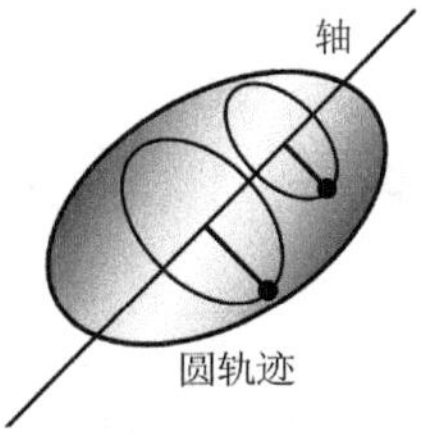

图 7.2

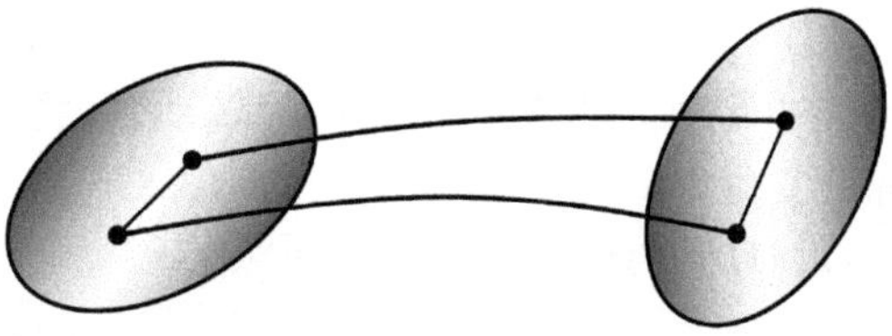

图 7.3

7.2 平　移

如图 7.4 所示,设 A 和 B 是平移(直线平移或曲线平移)刚体内任意两点,$\boldsymbol{r}_P$ 和$\boldsymbol{r}_B$ 是质点 A 和 B 相对固定参考系 $Oxyz$ 的位置矢量,$\boldsymbol{r}_{A/B}$是质点 A 相对质点 B 的位置矢量。由图 7.4 得

$$\boldsymbol{r}_A = \boldsymbol{r}_B + \boldsymbol{r}_{A/B} \tag{7.1}$$

式(7.1)对时间 t 进行微分,得

$$\boldsymbol{v}_A = \boldsymbol{v}_B + \boldsymbol{v}_{A/B} \tag{7.2}$$

式中,$\boldsymbol{v}_A = \dot{\boldsymbol{r}}_A$ 和 $\boldsymbol{v}_B = \dot{\boldsymbol{r}}_B$ 分别为质点 A 和 B 的速度,$\boldsymbol{v}_{A/B} = \dot{\boldsymbol{r}}_{A/B}$为质点 A 相对质点 B 的速度。当刚体发生平移,$\boldsymbol{r}_{A/B}$为常量,即 $\dot{\boldsymbol{r}}_{A/B}=0$,因此

$$\boldsymbol{v}_A = \boldsymbol{v}_B \tag{7.3}$$

对时间 t 微分,得

$$\boldsymbol{a}_A = \boldsymbol{a}_B \tag{7.4}$$

由此得到结论,刚体发生平移时刚体上所有质点在任意给定时刻具有相同的速度和加速度,如图 7.5 所示。

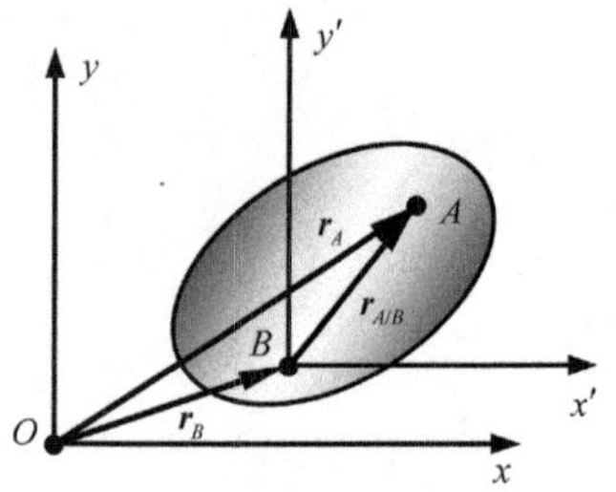

图 7.4

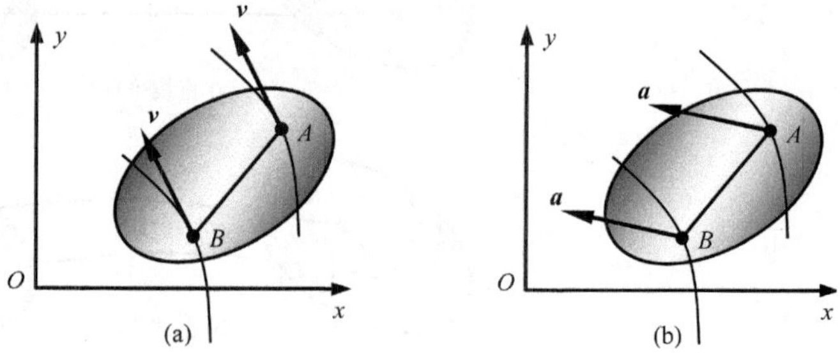

图 7.5

对于曲线平移,速度和加速度的方向不断改变,然而,对于直线平移,速度和加速度的方向保持不变。

7.3 定轴转动

如图 7.6 所示,设 A 是绕定轴 λ 旋转刚体上任意一点,$\boldsymbol{r}_A$ 是质点 A 相对转动轴 λ 上任意点 O 的位置矢量。由图 7.6 得

$$\boldsymbol{v}_A = \boldsymbol{\omega} \times \boldsymbol{r}_A \tag{7.5}$$

式中,$\boldsymbol{\omega}$ 为刚体角速度。角速度的方向沿转动轴,指向通过右手螺旋法则确定,角速度的大小等于角坐标的变化率。

式(7.5)对时间 t 微分,得

$$\boldsymbol{a}_A = \boldsymbol{\alpha} \times \boldsymbol{r}_A + \boldsymbol{\omega} \times \boldsymbol{v}_A \tag{7.6}$$

式中，$\boldsymbol{\alpha}=\dot{\boldsymbol{\omega}}$ 为刚体角加速度。对于刚体绕定轴转动，角加速度 $\boldsymbol{\alpha}$ 的方向沿转动轴，大小等于角速度变化率。利用式(7.5)，式(7.6)可写为

$$\boldsymbol{a}_A=\boldsymbol{\alpha}\times\boldsymbol{r}_A+\boldsymbol{\omega}\times(\boldsymbol{\omega}\times\boldsymbol{r}_A) \tag{7.7}$$

也可表示为

$$\boldsymbol{a}_A=\boldsymbol{a}_t+\boldsymbol{a}_n \tag{7.8}$$

式中，$\boldsymbol{a}_t=\boldsymbol{\alpha}\times\boldsymbol{r}_A$ 为加速度 $\boldsymbol{a}_A$ 的切向分量，沿轨迹切向，$\boldsymbol{a}_n=\boldsymbol{\omega}\times\boldsymbol{v}_A=\boldsymbol{\omega}\times(\boldsymbol{\omega}\times\boldsymbol{r}_A)$ 为加速度 $\boldsymbol{a}_A$ 的法向分量，指向轨迹圆心。式(7.8)表明，加速度 $\boldsymbol{a}_A$ 等于切向分量 $\boldsymbol{a}_t$ 和法向分量 $\boldsymbol{a}_n$ 的矢量和，如图 7.7 所示。

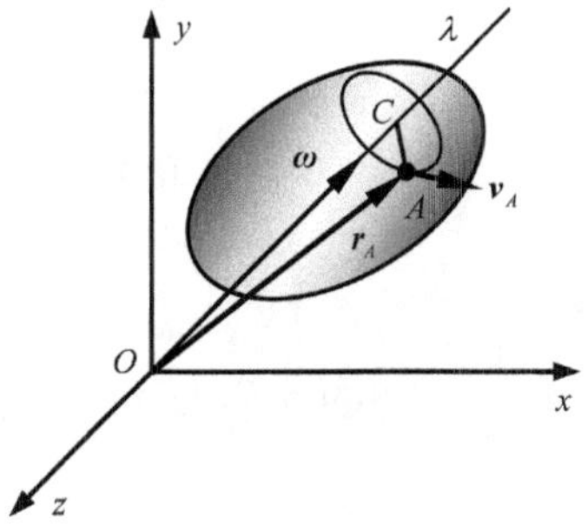

图 7.6

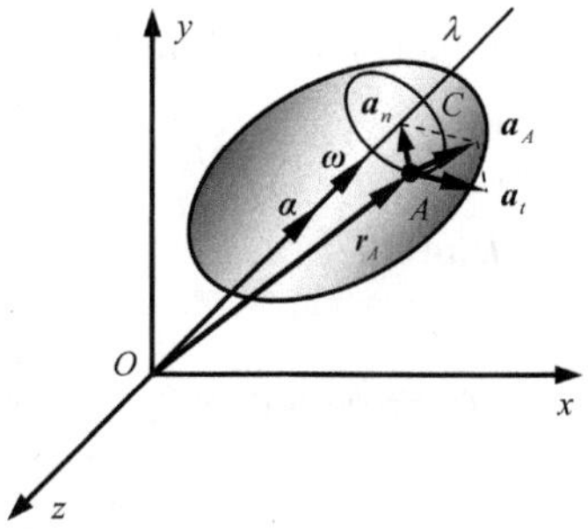

图 7.7

刚体定轴转动可简化为板的面内转动，即板面垂直刚体转动轴并绕板面与刚体转动轴的交点 O 发生面内转动，如图 7.8 所示。

对于板绕定点 O 的面内转动，角速度 $\boldsymbol{\omega}$ 可表示为

$$\boldsymbol{\omega}=\omega\boldsymbol{k} \tag{7.9}$$

式中，$\boldsymbol{k}$ 为垂直板面向外的单位矢量，ω 为角速度大小，逆时针转动为正，顺时针转动为负。

利用式(7.5)，得图 7.9 所示质点 A 的速度

$$\boldsymbol{v}=\boldsymbol{\omega}\times\boldsymbol{r}=r\omega\boldsymbol{e}_t \tag{7.10}$$

式中，$\boldsymbol{e}_t$ 为指向运动方向的切向单位矢量。因此，速度大小等于

$$v=r\omega \tag{7.11}$$

方向由 $\boldsymbol{r}$ 沿转动方向旋转 90°确定。

利用式(7.7)和(7.8)，得图 7.10 所示质点 A 的加速度

$$\boldsymbol{a}=\boldsymbol{a}_t+\boldsymbol{a}_n=\boldsymbol{\alpha}\times\boldsymbol{r}+\boldsymbol{\omega}\times(\boldsymbol{\omega}\times\boldsymbol{r})=r\alpha\boldsymbol{e}_t+r\omega^2\boldsymbol{e}_n \tag{7.12}$$

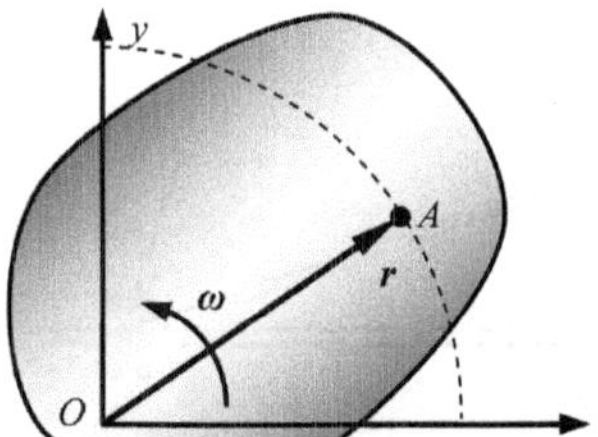

图 7.8

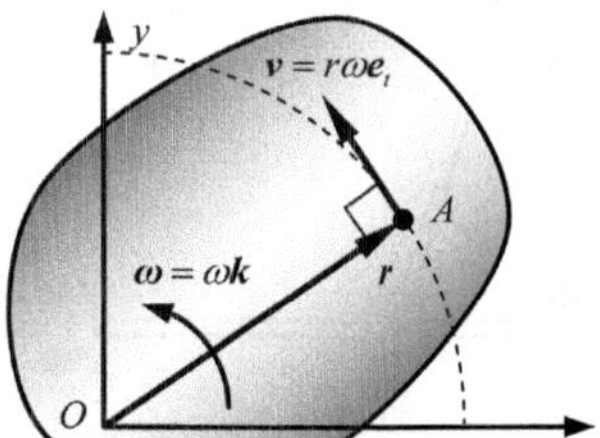

图 7.9

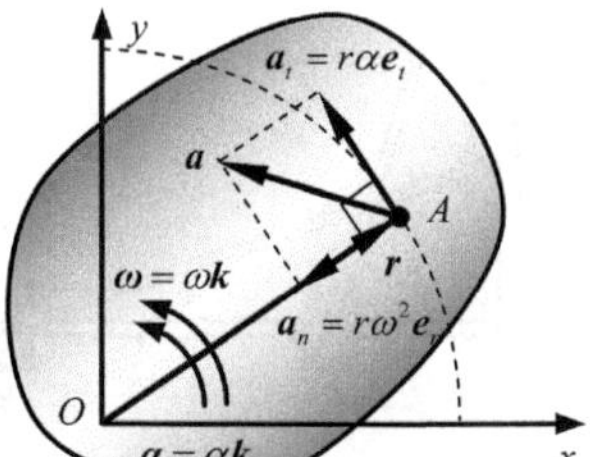

图 7.10

式中，e_n 为指向圆心的单位矢量。如果 α 为正，则 $a_t = r\alpha e_t$ 指向逆时针方向，否则指向顺时针方向。$a_n = r\omega^2 e_n$ 始终指向圆心。

例 7.1

图 E7.1 所示结构由杆 AE，CE 和矩形板 $ABCD$ 焊接而成。结构绕轴 AE 以不变角速度 $\omega = 5$ rad/s 旋转。已知从点 A 观察时转动为逆时针，求点 B 的速度和加速度。

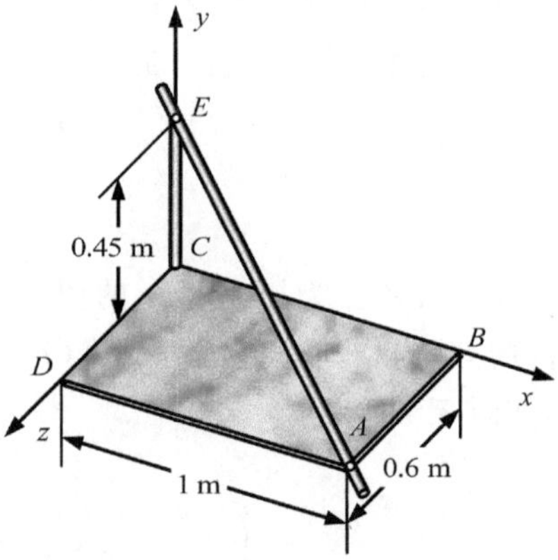

图 E7.1

解

利用 $\boldsymbol{r}_{A/E} = \boldsymbol{i} - 0.45\boldsymbol{j} + 0.6\boldsymbol{k}$ m，有

$$\boldsymbol{\omega} = \omega \frac{\boldsymbol{r}_{A/E}}{r_{A/E}} = 4\boldsymbol{i} - 1.8\boldsymbol{j} + 2.4\boldsymbol{k} \text{ rad/s}$$

利用 $\boldsymbol{r}_{B/A} = -0.6\boldsymbol{k}$ m，得

$$\boldsymbol{v}_B = \boldsymbol{\omega} \times \boldsymbol{r}_{B/A} = \begin{vmatrix} \boldsymbol{i} & \boldsymbol{j} & \boldsymbol{k} \\ 4 & -1.8 & 2.4 \\ 0 & 0 & -0.6 \end{vmatrix} = 1.08\boldsymbol{i} + 2.4\boldsymbol{j} \text{ m/s}$$

由 $\boldsymbol{\alpha} = \boldsymbol{0}$，得

$$\boldsymbol{a}_B = \boldsymbol{\alpha} \times \boldsymbol{r}_{B/A} + \boldsymbol{\omega} \times \boldsymbol{v}_B = \begin{vmatrix} \boldsymbol{i} & \boldsymbol{j} & \boldsymbol{k} \\ 4 & -1.8 & 2.4 \\ 1.08 & 2.4 & 0 \end{vmatrix} = -5.76\boldsymbol{i} + 2.59\boldsymbol{j} + 11.54\boldsymbol{k} \text{ m/s}^2$$

7.4 一般平面运动

一般平面运动既不是平移，也不是转动，但它始终可以看成为平移和转动的合成。

如图 7.11(a)所示，两端分别在水平和垂直轨道中滑动的杆 AB 做一般平面运动。杆从 A_1B_1 到 A_2B_2 的运动可以分解为随基点 B 从 A_1B_1 到 $A_1'B_2$ 的向右平移(如图 7.11(b)所示)和绕基点 B 从 $A_1'B_2$ 到 A_2B_2 的逆时针转动(如图 7.11(c)所示)。

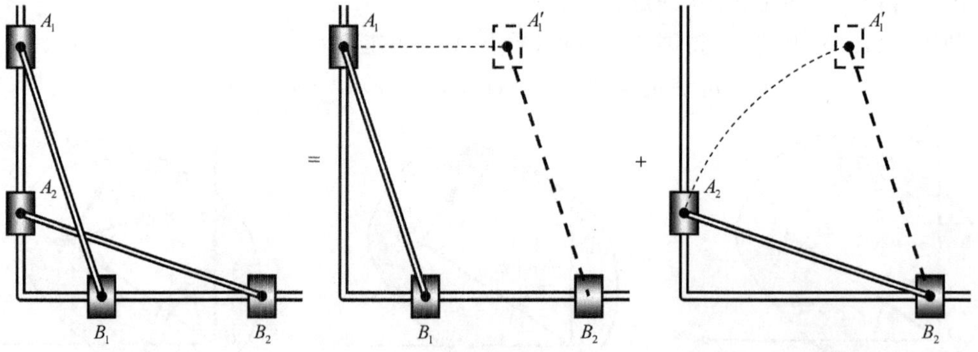

图 7.11

如图 7.12(a)所示，水平面上只滚不滑的轮子的运动也为一般平面运动。轮子从 A_1B_1 到 A_2B_2 的运动也可以分解为随基点 B 从 A_1B_1 到 $A_1'B_2$ 的向右平移(如图 7.12(b)所示)和绕基点 B 从 $A_1'B_2$ 到 A_2B_2 的顺时针转动(如图 7.12(c)所示)。

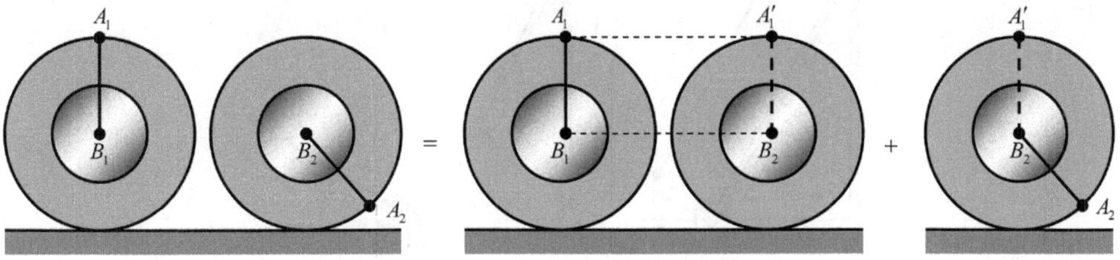

(a) 一般平面运动　　(b) 随基点平移　　(c) 绕基点转动

图 7.12

1. 基点法求一般平面运动速度

因为任何一般平面运动都可分解为随基点的平移和绕基点的转动，因此图 7.13(a)所示刚体上给定点 A 的速度 $\boldsymbol{v}_A$ 可表示为

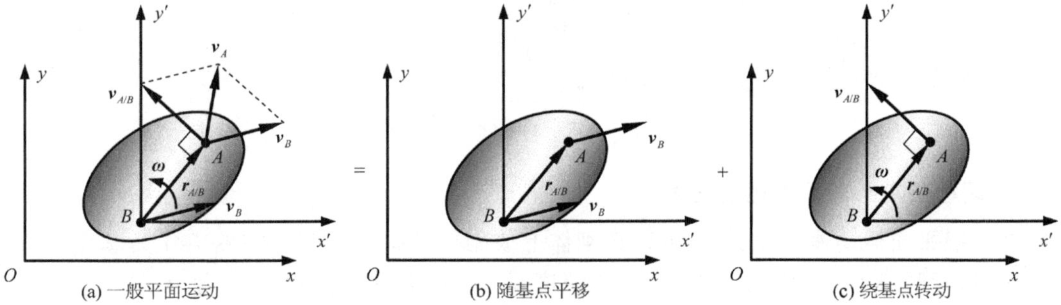

(a) 一般平面运动　　(b) 随基点平移　　(c) 绕基点转动

图 7.13

$$\boldsymbol{v}_A = \boldsymbol{v}_B + \boldsymbol{v}_{A/B} \tag{7.13}$$

式中，$\boldsymbol{v}_B$ 为基点 B 的速度，如图 7.13(b)；$\boldsymbol{v}_{A/B} = \boldsymbol{\omega} \times \boldsymbol{r}_{A/B}$ 为质点 A 相对基点 B 的速度，如图 7.13(c)。

应该注意，尽管基点可以任意选择，但是为了方便，通常选择运动已知的点作为基点，另外刚体的角速度和角加速度与基点选择无关。

例 7.2

如图 E7.2(a)所示，环 A 以不变速度 $v_A = 1$ m/s 向下移动。图示瞬时 $\theta = 30°$，求：(a)杆 AB 的角速度；(b)环 B 的速度。

解

选点 A 作为基点，则点 B 的速度为

$$\boldsymbol{v}_B = \boldsymbol{v}_A + \boldsymbol{v}_{B/A}$$

$\boldsymbol{v}_B$ 水平向右和 $\boldsymbol{v}_{B/A}$ 垂直向上，则速度平行四边形如图 E7.2(b)所示。根据平行四边形，则在该瞬时有

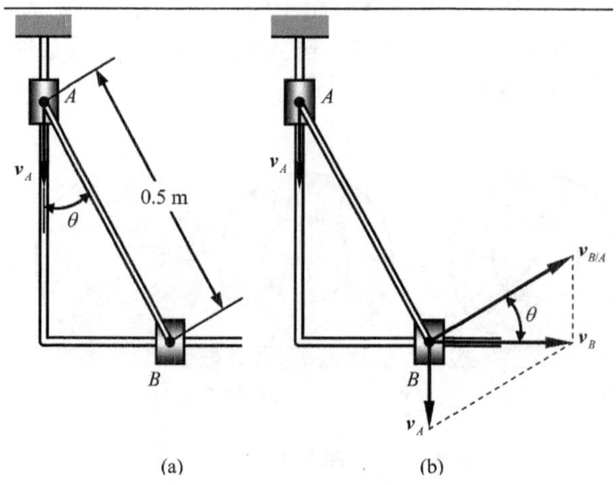

图 E7.2

$$v_B = \frac{v_A}{\tan\theta} = 1.73 \text{ m/s}$$

$$v_{B/A} = \frac{v_A}{\sin\theta} = 2 \text{ m/s}$$

利用 $v_{B/A} = AB \cdot \omega$,可得

$$\omega = \frac{v_{B/A}}{AB} = 4 \text{ rad/s (逆时针)}$$

2. 瞬心法求一般平面运动速度

在一般平面运动速度分析中,可以选刚体上或刚体外瞬时速度为零的点作为基点。假设图 7.14 所示瞬时点 C 的速度为零(C 称为速度瞬心),则点 A 的速度$\boldsymbol{v}_A$ 可写为

$$\boldsymbol{v}_A = \boldsymbol{\omega} \times \boldsymbol{r} \tag{7.14}$$

式中,$\boldsymbol{\omega}$ 和 $\boldsymbol{r}$ 分别为刚体的角速度和刚体上点 A 相对速度瞬心 C 的位置矢量。应该注意,速度瞬心的加速度通常不等于零。

在任何给定瞬时,速度瞬心可以通过下述方法进行确定:

(1) 已知刚体上点 A 速度$\boldsymbol{v}_A$ 和刚体角速度 $\boldsymbol{\omega}$,如图 7.15 所示。速度瞬心 C 在$\boldsymbol{v}_A$ 沿 $\boldsymbol{\omega}$ 旋转方向旋转 90°后的方向上,C 到 A 的距离等于 $r = \frac{v_A}{\omega}$。

(2) 已知不平行速度$\boldsymbol{v}_A$ 和$\boldsymbol{v}_B$ 的作用线,如图 7.16 所示。速度瞬心 C 位于分别通过点 A 和 B 并与$\boldsymbol{v}_A$ 和$\boldsymbol{v}_B$ 分别垂直的直线交点。

(3) 已知平行速度$\boldsymbol{v}_A$ 和$\boldsymbol{v}_B$ 的大小和方向以及 A 和 B 连线垂直于$\boldsymbol{v}_A$(或$\boldsymbol{v}_B$),如图 7.17 所示。速度瞬心 C 位于 AB 与速度$\boldsymbol{v}_A$ 和$\boldsymbol{v}_B$ 末端连线的交点。对图 7.17(a),如果在某一瞬时有 $\boldsymbol{v}_A = \boldsymbol{v}_B$,则速度瞬心在无穷远处。这种特殊情况称为瞬时平移。

速度瞬心形成的轨迹称为瞬心轨迹。在空间形成的轨迹称为空间/固定瞬心轨迹,而在物体上形成的轨迹则称为物体瞬心轨迹。例如,水平面上只滚不滑的轮子的空间瞬心轨迹为与水平面重合的水平直线,而物体瞬心轨迹为与轮缘重合的圆。

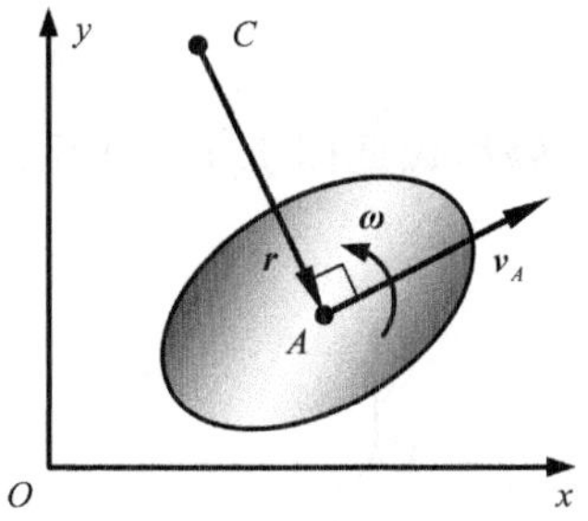

图 7.14

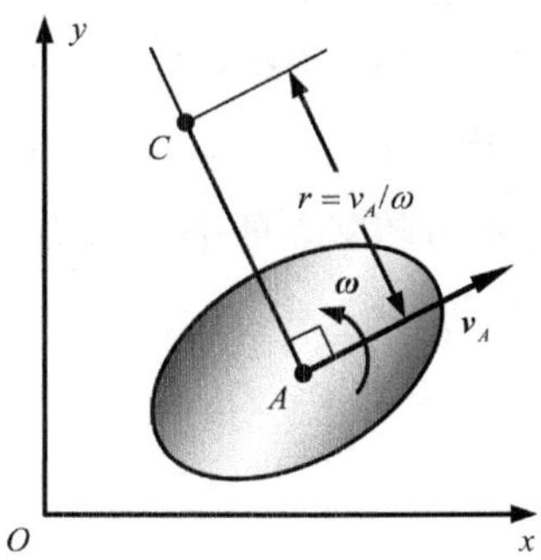

图 7.15

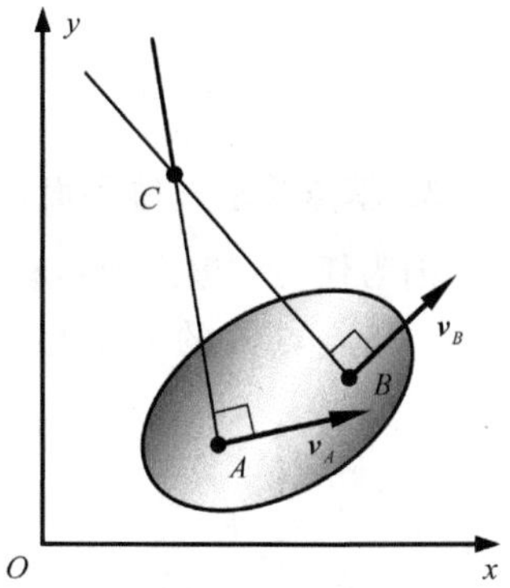

图 7.16

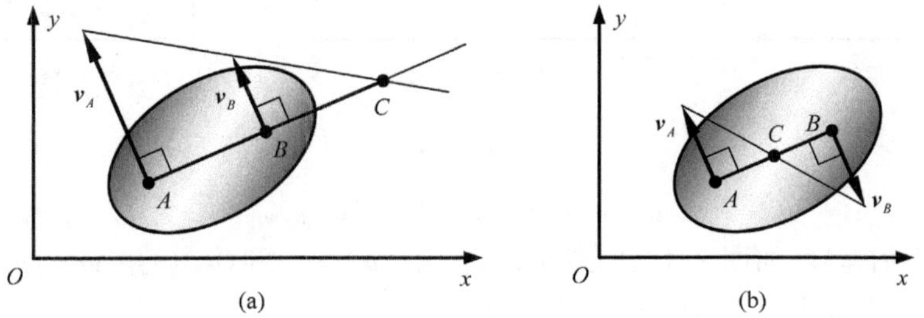

图 7.17

例 7.3

如图 E7.3(a)所示,环 A 以不变速度 $v_A=1$ m/s 向下移动。图示瞬时 $\theta=30°$,求:(a)杆 AB 的角速度;(b)环 B 的速度。

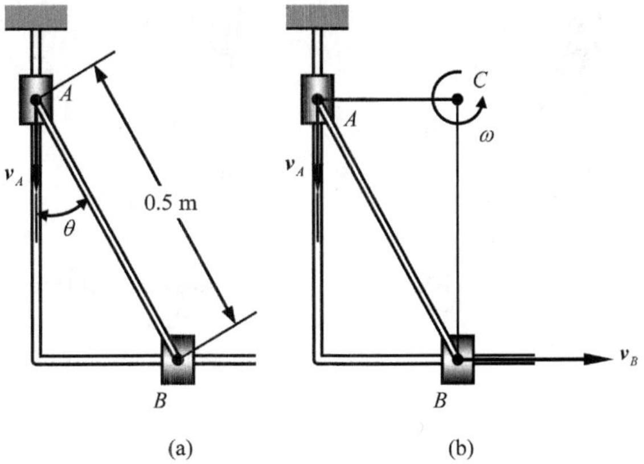

图 E7.3

解

在图示给定瞬时,速度瞬心在 C 处,则有

$$\omega = \frac{v_A}{AC} = \frac{v_A}{AB \cdot \sin\theta} = 4\ \mathrm{rad/s}$$

$$v_B = BC \cdot \omega = (AB \cdot \cos\theta)\omega = 1.73\ \mathrm{m/s}$$

3. 基点法求一般平面运动加速度

因为任何一般平面运动都可分解为随基点的平移和绕基点的转动,因此图 7.18(a)所示刚体上给定点 A 的加速度 $\boldsymbol{a}_A$ 可表示为

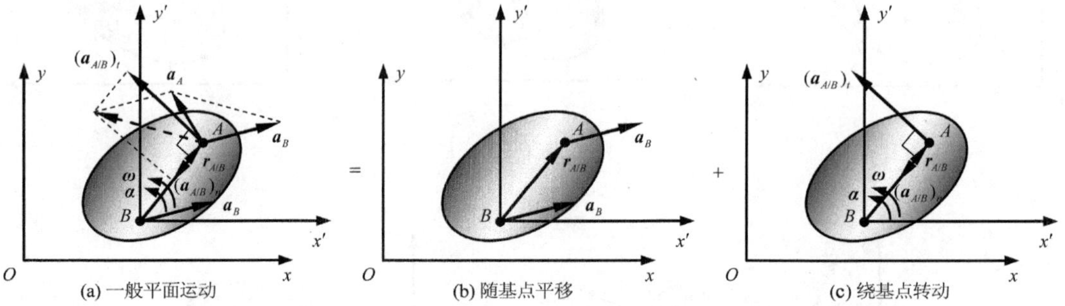

图 7.18

$$\boldsymbol{a}_A = \boldsymbol{a}_B + (\boldsymbol{a}_{A/B})_t + (\boldsymbol{a}_{A/B})_n \tag{7.15}$$

式中,$\boldsymbol{a}_B$ 为基点加速度,如图 7.18(b)所示;$(\boldsymbol{a}_{A/B})_t = \boldsymbol{\alpha} \times \boldsymbol{r}_{A/B}$ 和 $(\boldsymbol{a}_{A/B})_n = \boldsymbol{\omega} \times (\boldsymbol{\omega} \times \boldsymbol{r}_{A/B})$ 分别为质点 A 相对基点 B 的加速度的切向和法向分量。

例 7.4

如图 E7.4(a)所示,环 A 以不变速度 $v_A = 1$ m/s 向下移动。图示瞬时 $\theta = 30°$,求:(a)杆 AB 的角加速度;(b)环 B 的加速度。

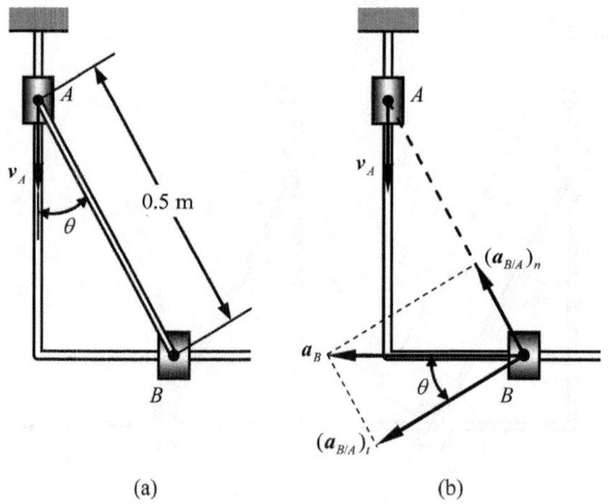

图 E7.4

解

选 A 点为基点,则 B 点加速度为

$$\boldsymbol{a}_B = \boldsymbol{a}_A + (\boldsymbol{a}_{B/A})_t + (\boldsymbol{a}_{B/A})_n$$

$\boldsymbol{a}_B$ 水平和 $\boldsymbol{a}_A$ 恒等于零,则加速度平行四边形如图 E7.4(b)所示。根据平行四边形,则在该瞬

时有

$$a_B = \frac{(a_{B/A})_n}{\sin\theta},\ (a_{B/A})_t = \frac{(a_{B/A})_n}{\tan\theta}$$

利用 $\omega=4\ \mathrm{m/s}$(参考例 7.2 或 7.3),即 $(a_{B/A})_n = AB\cdot\omega^2 = 8\ \mathrm{m/s^2}$(指向 A 点),则得

$$a_B = \frac{(a_{B/A})_n}{\sin\theta} = 16\ \mathrm{m/s^2},\ (a_{B/A})_t = \frac{(a_{B/A})_n}{\tan\theta} = 13.86\ \mathrm{m/s^2}$$

根据 $(a_{B/A})_t = AB\cdot\alpha$,有

$$\alpha = \frac{(a_{B/A})_t}{AB} = 27.72\ \mathrm{rad/s^2}\text{(顺时针)}$$

例 7.5

如图 E7.5(a)所示,杆 BD 在 B 和 D 处与连杆 AB 和 CD 相连。已知图示瞬时连杆 AB 以不变角速度 $\omega=2\ \mathrm{rad/s}$ 顺时针旋转,求(a)杆 BD 和(b)连杆 CD 的角速度和角加速度。

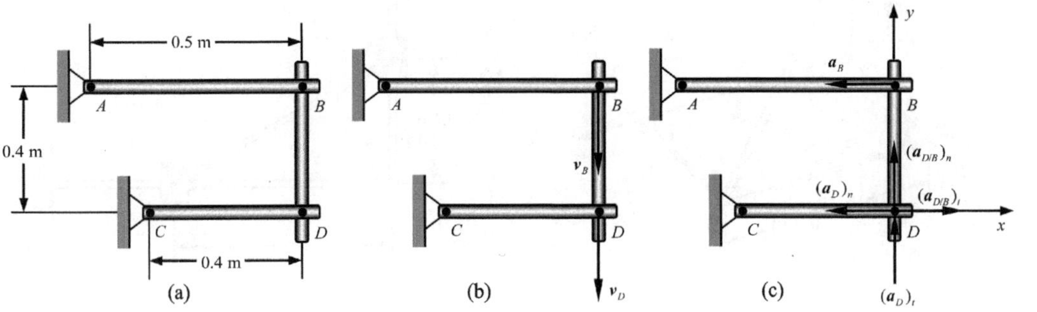

图 E7.5

解

由连杆 AB 绕 A 点顺时针旋转,则 B 点速度垂直向下,如图 E7.5(b)。同理,因连杆 CD 绕 C 点顺时针旋转,故 D 点速度也垂直向下。在图示瞬时杆 BD 在 B 和 D 处具有相同速度(包括大小和方向),因此图示瞬时杆 BD 做瞬时平移,由此得

$$v_D = v_B = AB\cdot\omega_{AB} = 1\ \mathrm{m/s},\ \omega_{BD} = 0,\ \omega_{CD} = \frac{v_D}{CD} = 2.5\ \mathrm{rad/s}\text{(顺时针)}$$

选杆 BD 上的 B 点为基点,那么 D 点加速度为

$$(\boldsymbol{a}_D)_t + (\boldsymbol{a}_D)_n = \boldsymbol{a}_B + (\boldsymbol{a}_{D/B})_t + (\boldsymbol{a}_{D/B})_n$$

如图 E7.5(c)建参考系 xy,设 $\boldsymbol{i}$ 和 $\boldsymbol{j}$ 分别为沿 x 和 y 方向的单位矢量,则有

$$(a_D)_t\boldsymbol{j} - (a_D)_n\boldsymbol{i} = -a_B\boldsymbol{i} + (a_{D/B})_t\boldsymbol{i} + (a_{D/B})_n\boldsymbol{j}$$

或

$$-(a_D)_n = -a_B + (a_{D/B})_t,\ (a_D)_t = (a_{D/B})_n$$

利用 $(a_D)_n = CD\cdot(\omega_{CD})^2 = 2.5\ \mathrm{m/s^2}$, $a_B = AB\cdot(\omega_{AB})^2 = 2\ \mathrm{m/s^2}$ 和 $(a_{D/B})_n = BD\cdot(\omega_{BD})^2 = 0$,得

$$(a_{D/B})_t = a_B - (a_D)_n = -0.5\ \mathrm{m/s^2}\text{(向左)},\ (a_D)_t = 0$$

在利用 $(a_{D/B})_t = BD\cdot\alpha_{BD}$ 和 $(a_D)_t = CD\cdot\alpha_{CD}$,有

$$\alpha_{BD} = -1.25\ \mathrm{rad/s^2}\text{(顺时针)},\ \alpha_{CD} = 0$$

习　题

7.1 图 P7.1 所示结构由杆 AE，CE 和矩形板 $ABCD$ 焊接而成。结构绕轴 AE 以角速度 $\omega=5$ rad/s 和角加速度 $\alpha=10$ rad/s^2 旋转。已知从点 A 观察时角速度和角加速度为逆时针，求点 D 的速度和加速度。

7.2 环 A 以不变速度 $v_A=0.4$ m/s 向上移动如图 P7.2 所示。在图示瞬时，求(a)杆 AB 的角速度和环 B 的速度，(b)杆 AB 的角加速度和环 B 的加速度。

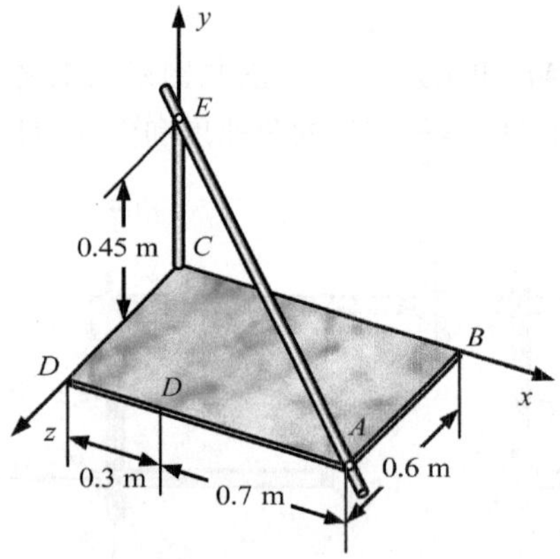

图 P7.1

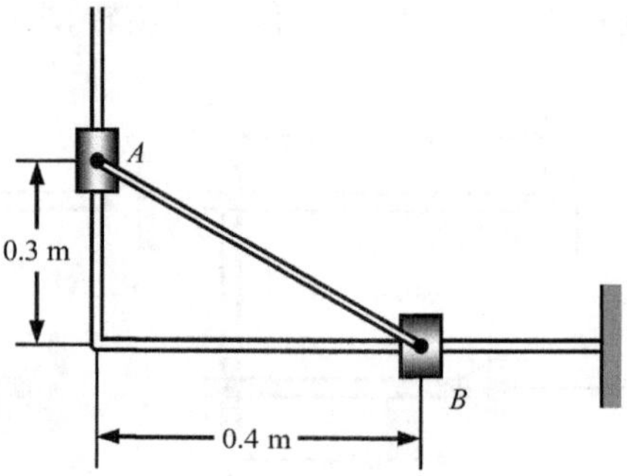

图 P7.2

7.3 杆 BDE 在 B 和 D 处与连杆 AB 和 CD 相连。已知图 P7.3 所示瞬时连杆 AB 以不变角速度 $\omega=2$ rad/s 顺时针旋转，求点 E 的速度和加速度。

7.4 如图 P7.4 所示，半径为 r 的圆柱的运动通过缆绳 AG 控制。已知 $v_G=0.3$ m/s(向上)，$a_G=0.5$ m/s^2(向上)和 $r=0.15$ m，求：(a)B 和 D 点的加速度；(b)E 点的速度和加速度。

7.5 当杆 AB 的端点 A 以不变速度 600 mm/s 向右移动时，杆 AB 将在 C 处小轮上运动。在图 P7.5 所示瞬时，求：(a)杆的角速度和端点 B 的速度；(b)杆的角加速度和端点 B 的加速度。

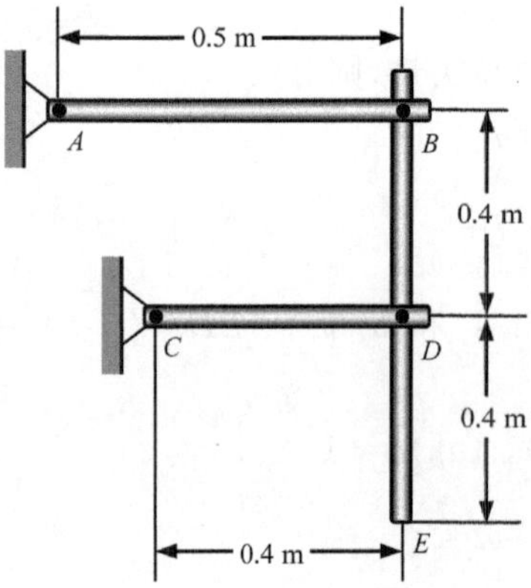

图 P7.3

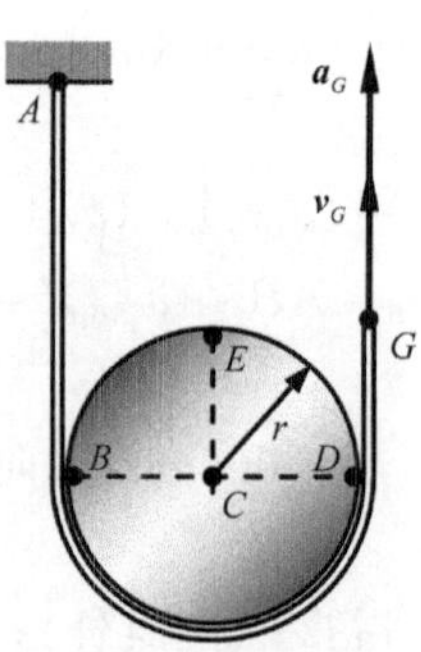

图 P7.4

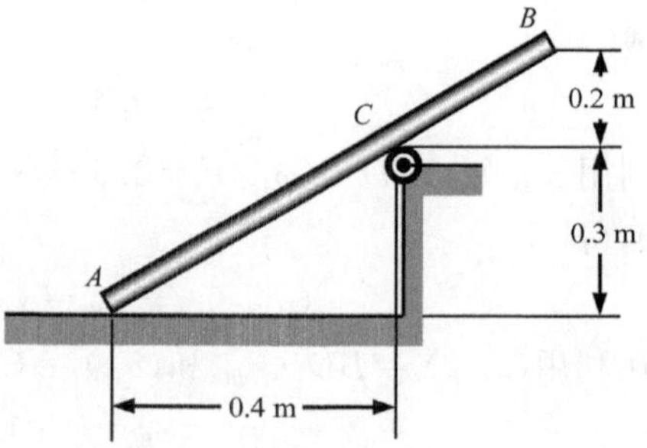

图 P7.5

第 8 章　质点合成运动

在前面的章节中已经利用单一参考系分析了质点的运动。然而，在很多情况下质点相对单一参考系的运动轨迹十分复杂，因此需要通过两个或更多参考系进行质点运动分析。例如，当飞机在飞行的时候，首先观察飞机相对地面的运动，然后通过平行四边形定律叠加螺旋桨端点相对飞机的运动，可更容易描述螺旋桨端点质点的运动。

与地球相连的参考系称为固定参考系（简称定系），而相对地球运动的参考系则称为运动参考系（简称动系）。然而，应该注意，定系的选择具有任意性。

质点相对定系的运动称为绝对运动，而质点相对动系的运动则称为相对运动。

8.1　矢量变化率

假设 $Oxyz$ 为定系，$O'x'y'z'$ 为动系，见图 8.1。

如果动系 $O'x'y'z'$ 以瞬时角速度 ω 绕轴 λ 旋转，则矢量 $\boldsymbol{V}$ 可表示为

$$\boldsymbol{V} = V'_x\boldsymbol{i}' + V'_y\boldsymbol{j}' + V'_z\boldsymbol{k}' \tag{8.1}$$

式中，V'_x，V'_y，V'_z和 $\boldsymbol{i}'$，$\boldsymbol{j}'$，$\boldsymbol{k}'$分别为矢量 $\boldsymbol{V}$ 在动系中的直角分量和动系的单位矢量。

在动系 $O'x'y'z'$ 中，$\boldsymbol{i}'$，$\boldsymbol{j}'$和 $\boldsymbol{k}'$的大小和方向均保持不变，因此有

$$\{\dot{\boldsymbol{V}}\}_{O'} = \dot{V}'_x\boldsymbol{i}' + \dot{V}'_y\boldsymbol{j}' + \dot{V}'_z\boldsymbol{k}' \tag{8.2}$$

然而，在定系 $Oxyz$ 中，$\boldsymbol{i}'$，$\boldsymbol{j}'$和 $\boldsymbol{k}'$的大小不变，但方向发生变化，因此得

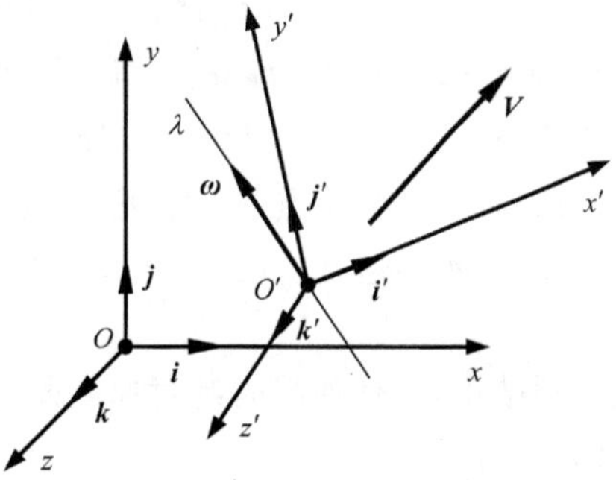

图 8.1

$$\{\dot{\boldsymbol{V}}\}_O = \dot{V}'_x\boldsymbol{i}' + \dot{V}'_y\boldsymbol{j}' + \dot{V}'_z\boldsymbol{k}' + V'_x\dot{\boldsymbol{i}}' + V'_y\dot{\boldsymbol{j}}' + V'_z\dot{\boldsymbol{k}}' \tag{8.3}$$

利用 $\dot{\boldsymbol{i}}'=\boldsymbol{\omega}\times\boldsymbol{i}'$，$\dot{\boldsymbol{j}}'=\boldsymbol{\omega}\times\boldsymbol{j}'$和 $\dot{\boldsymbol{k}}'=\boldsymbol{\omega}\times\boldsymbol{k}'$，得

$$\{\dot{\boldsymbol{V}}\}_O = \dot{V}'_x\boldsymbol{i}' + \dot{V}'_y\boldsymbol{j}' + \dot{V}'_z\boldsymbol{k}' + \boldsymbol{\omega}\times(V'_x\boldsymbol{i}' + V'_y\boldsymbol{j}' + V'_z\boldsymbol{k}') \tag{8.4}$$

再利用式(8.1)和(8.2)，得

$$\{\dot{\boldsymbol{V}}\}_O = \{\dot{\boldsymbol{V}}\}_{O'} + \boldsymbol{\omega}\times\boldsymbol{V} \tag{8.5}$$

8.2　速度合成

如图 8.2 所示，假设 $Oxyz$ 为定系，$O'x'y'z'$为动系。考虑沿空间曲线移动的质点 P，则有

$$\boldsymbol{r}_P = \boldsymbol{r}_{O'} + \boldsymbol{r}_{P/O'} \tag{8.6}$$

式中，$\boldsymbol{r}_P$ 为质点 P 的位置矢量，$\boldsymbol{r}_{O'}$ 为动系原点 O' 的位置矢量，$\boldsymbol{r}_{P/O'}$ 为质点 P 相对动系原点 O' 的位置矢量。

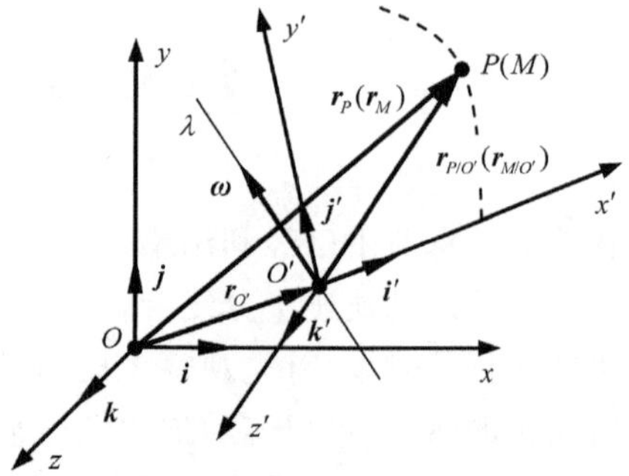

图 8.2

在定系 $Oxyz$ 中,式(8.6)对时间进行微分,得

$$\{\dot{\boldsymbol{r}}_P\}_O = \{\dot{\boldsymbol{r}}_{O'}\}_O + \{\dot{\boldsymbol{r}}_{P/O'}\}_O \tag{8.7}$$

设动系 $O'x'y'z'$ 以瞬时角速度 $\boldsymbol{\omega}$ 绕轴 λ 旋转,利用式(8.5),得

$$\{\dot{\boldsymbol{r}}_{P/O'}\}_O = \{\dot{\boldsymbol{r}}_{P/O'}\}_{O'} + \boldsymbol{\omega}\times\boldsymbol{r}_{P/O'} \tag{8.8}$$

把式(8.8)代入式(8.7),得

$$\{\dot{\boldsymbol{r}}_P\}_O = \{\dot{\boldsymbol{r}}_{O'}\}_O + \{\dot{\boldsymbol{r}}_{P/O'}\}_{O'} + \boldsymbol{\omega}\times\boldsymbol{r}_{P/O'} \tag{8.9}$$

假设牵连点(即动系上与质点 P 瞬时重合的点)由 M 表示,则有

$$\boldsymbol{r}_M = \boldsymbol{r}_{O'} + \boldsymbol{r}_{M/O'} \tag{8.10}$$

通过类似推导,得

$$\{\dot{\boldsymbol{r}}_M\}_O = \{\dot{\boldsymbol{r}}_{O'}\}_{O'} + \{\dot{\boldsymbol{r}}_{M/O'}\}_{O'} + \boldsymbol{\omega}\times\boldsymbol{r}_{M/O'} \tag{8.11}$$

式(8.9)减式(8.11),有

$$\{\dot{\boldsymbol{r}}_P\}_O - \{\dot{\boldsymbol{r}}_M\}_O = \{\dot{\boldsymbol{r}}_{P/O'} - \dot{\boldsymbol{r}}_{M/O'}\}_{O'} + \boldsymbol{\omega}\times(\boldsymbol{r}_{P/O'} - \boldsymbol{r}_{M/O'}) \tag{8.12}$$

利用 $\{\boldsymbol{r}_{P/O'} - \dot{\boldsymbol{r}}_{M/O'}\}_{O'} = \{\boldsymbol{r}_{P/M'}\}_{O'}$ 和 $\dot{\boldsymbol{r}}_{P/O'} - \dot{\boldsymbol{r}}_{M/O'} = \dot{\boldsymbol{r}}_{P/M} = 0$,可得

$$\{\dot{\boldsymbol{r}}_P\}_O - \{\dot{\boldsymbol{r}}_M\}_O = \{\dot{\boldsymbol{r}}_{P/M}\}_{O'} \tag{8.13}$$

式中,$\{\dot{\boldsymbol{r}}_P\}_O = \boldsymbol{v}_P$ 为质点 P 的速度,$\{\dot{\boldsymbol{r}}_M\}_O = \boldsymbol{v}_M$ 为牵连点 M 的速度,$\{\dot{\boldsymbol{r}}_{P/M}\}_{O'} = \boldsymbol{v}_{P/M}$ 为质点 P 相对牵连点 M 的速度。因此,式(8.13)可重写为

$$\boldsymbol{v}_P = \boldsymbol{v}_M + \boldsymbol{v}_{P/M} \tag{8.14}$$

由此得到结论,在任何瞬时绝对速度 $\boldsymbol{v}_P$ 等于牵连速度 $\boldsymbol{v}_M$ 和相对速度 $\boldsymbol{v}_{P/M}$ 的矢量和。

例 8.1

如图 E8.1(a)所示,飞机 A 和 B 在相同高度飞行,假设飞机 A 以速度 $v_A = 400$ km/h 向南飞行,飞机 B 以速度 $v_B = 500$ km/h 向东偏北 30°方向飞行,求飞机 B 相对飞机 A 的相对速度。

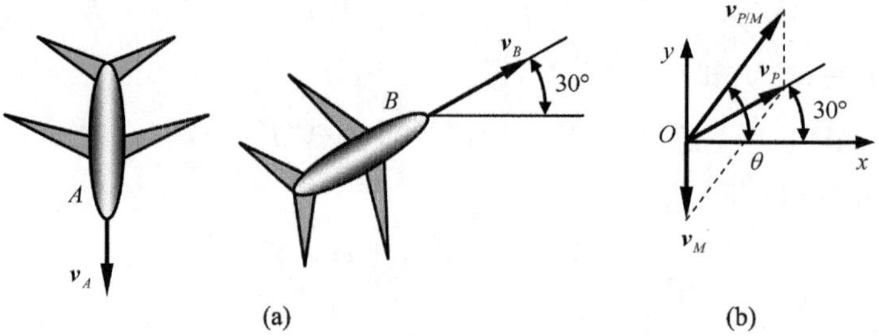

图 E8.1

解

选地球为定系,飞机 A 为动系,飞机 B 为动点,则有

$$\boldsymbol{v}_P = \boldsymbol{v}_B,\ \boldsymbol{v}_M = \boldsymbol{v}_A$$

如图 E8.1(b)所示,利用 $\boldsymbol{v}_P = \boldsymbol{v}_M + \boldsymbol{v}_{P/M}$,得

$$v_P\cos 30°\boldsymbol{i} + v_P\sin 30°\boldsymbol{j} = -v_M\boldsymbol{j} + v_{P/M}\cos\theta\boldsymbol{i} + v_{P/M}\sin\theta\boldsymbol{j}$$

式中，$\boldsymbol{i}$ 和 $\boldsymbol{j}$ 为分别对应 x 和 y 方向的单位矢量，θ 为 $v_{P/M}$ 与 x 轴的夹角。利用 $v_P=500$ km/h 和 $v_M=400$ km/h，由上式得

$$v_{P/M}=781.0\ \text{km/h},\ \theta=56.33°$$

因此，飞机 B 相对飞机 A 的相对速度大小为 781.0 km/h，方向为东偏北 56.33°。

8.3 加速度合成

如图 8.2 所示，在定系 $Oxyz$ 中由式(8.9)对时间进行微分，得

$$\{\ddot{\boldsymbol{r}}_P\}_O=\{\ddot{\boldsymbol{r}}_{O'}\}_O+\left\{\frac{\mathrm{d}}{\mathrm{d}t}\{\dot{\boldsymbol{r}}_{P/O'}\}_{O'}\right\}_O+\dot{\boldsymbol{\omega}}\times\boldsymbol{r}_{P/O'}+\boldsymbol{\omega}\times\{\dot{\boldsymbol{r}}_{P/O'}\}_O \tag{8.15}$$

利用式(8.5)，有

$$\left\{\frac{\mathrm{d}}{\mathrm{d}t}\{\dot{\boldsymbol{r}}_{P/O'}\}_{O'}\right\}_O=\{\ddot{\boldsymbol{r}}_{P/O'}\}_{O'}+\boldsymbol{\omega}\times\{\dot{\boldsymbol{r}}_{P/O'}\}_{O'} \tag{8.16}$$

和

$$\{\dot{\boldsymbol{r}}_{P/O'}\}_O=\{\dot{\boldsymbol{r}}_{P/O'}\}_{O'}+\boldsymbol{\omega}\times\boldsymbol{r}_{P/O'} \tag{8.17}$$

式(8.16)和(8.17)代入式(8.15)，得

$$\{\ddot{\boldsymbol{r}}_P\}_O=\{\ddot{\boldsymbol{r}}_{O'}\}_O+\{\ddot{\boldsymbol{r}}_{P/O'}\}_{O'}+2\boldsymbol{\omega}\times\{\dot{\boldsymbol{r}}_{P/O'}\}_{O'}+\dot{\boldsymbol{\omega}}\times\boldsymbol{r}_{P/O'}+\boldsymbol{\omega}\times(\boldsymbol{\omega}\times\boldsymbol{r}_{P/O'}) \tag{8.18}$$

对牵连点 M，通过类似推导，得

$$\{\ddot{\boldsymbol{r}}_M\}_O=\{\ddot{\boldsymbol{r}}_{O'}\}_O+\{\ddot{\boldsymbol{r}}_{M/O'}\}_{O'}+2\boldsymbol{\omega}\times\{\dot{\boldsymbol{r}}_{M/O'}\}_{O'}+\dot{\boldsymbol{\omega}}\times\boldsymbol{r}_{M/O'}+\boldsymbol{\omega}\times(\boldsymbol{\omega}\times\boldsymbol{r}_{M/O'}) \tag{8.19}$$

式(8.18)减式(8.19)，有

$$\begin{aligned}\{\ddot{\boldsymbol{r}}_P\}_O-\{\ddot{\boldsymbol{r}}_M\}_O=&\{\ddot{\boldsymbol{r}}_{P/O'}-\ddot{\boldsymbol{r}}_{M/O'}\}_{O'}+2\boldsymbol{\omega}\times\{\dot{\boldsymbol{r}}_{P/O'}-\dot{\boldsymbol{r}}_{M/O'}\}_{O'}+\\&\dot{\boldsymbol{\omega}}\times(\boldsymbol{r}_{P/O'}-\boldsymbol{r}_{M/O'})+\boldsymbol{\omega}\times[\boldsymbol{\omega}\times(\boldsymbol{r}_{P/O'}-\boldsymbol{r}_{M/O'})]\end{aligned} \tag{8.20}$$

利用 $\{\ddot{\boldsymbol{r}}_{P/O'}-\ddot{\boldsymbol{r}}_{M/O'}\}_{O'}=\{\ddot{\boldsymbol{r}}_{P/M}\}_{O'}$，$\{\dot{\boldsymbol{r}}_{P/O'}-\dot{\boldsymbol{r}}_{M/O'}\}_{O'}=\{\dot{\boldsymbol{r}}_{P/M}\}_{O'}$ 和 $\boldsymbol{r}_{P/O'}-\boldsymbol{r}_{M/O'}=\boldsymbol{r}_{P/M}=0$，得

$$\{\ddot{\boldsymbol{r}}_P\}_O-\{\ddot{\boldsymbol{r}}_M\}_O=\{\ddot{\boldsymbol{r}}_{P/M}\}_{O'}+2\boldsymbol{\omega}\times\{\dot{\boldsymbol{r}}_{P/M}\}_{O'} \tag{8.21}$$

再利用 $\{\ddot{\boldsymbol{r}}_P\}_O=\boldsymbol{a}_P$，$\{\ddot{\boldsymbol{r}}_M\}_O=\boldsymbol{a}_M$，$\{\ddot{\boldsymbol{r}}_{P/M}\}_{O'}=\boldsymbol{a}_{P/M}$ 和 $\{\dot{\boldsymbol{r}}_{P/M}\}_{O'}=\boldsymbol{v}_{P/M}$，得

$$\boldsymbol{a}_P=\boldsymbol{a}_M+\boldsymbol{a}_{P/M}+\boldsymbol{a}_C \tag{8.22}$$

式中，$\boldsymbol{a}_P$ 为质点 P 的绝对加速度，$\boldsymbol{a}_M$ 为牵连点 M 的牵连加速度，$\boldsymbol{a}_{P/M}$ 为质点 P 相对牵连点 M 的相对加速度，$\boldsymbol{a}_C=2\boldsymbol{\omega}\times\boldsymbol{v}_{P/M}$ 为科氏加速度。

由此得到结论，在任何瞬时，绝对加速度 $\boldsymbol{a}_P$ 等于牵连加速度 $\boldsymbol{a}_M$、相对加速度 $\boldsymbol{a}_{P/M}$ 和科氏加速度 $\boldsymbol{a}_C$ 的矢量和。

例 8.2

已知图 E8.2(a)所示瞬时，杆 BD 以不变角速度 $\omega_B=2$ rad/s 逆时针旋转，求：(a)杆 AD 的角速度和环 D 相对杆 BD 的相对速度；(b)杆 AD 的角加速度和环 D 相对杆 BD 的相对加速度。

解

取地球为定系、杆 BD 为动系和环 D 为动点。

(a)利用 $\boldsymbol{v}_P=\boldsymbol{v}_M+\boldsymbol{v}_{P/M}$，如图 E8.2(b)，得

$$\frac{3}{5}v_P\boldsymbol{i}+\frac{4}{5}v_P\boldsymbol{j}=v_M\boldsymbol{i}+v_{P/M}\boldsymbol{j}$$

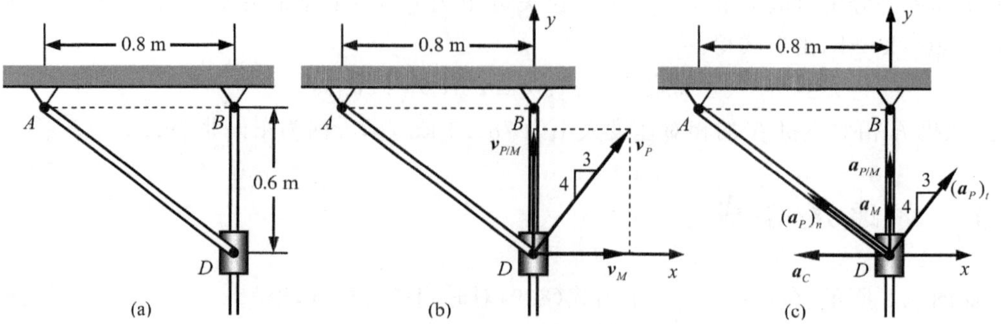

图 E8.2

式中,$\boldsymbol{i}$ 和 $\boldsymbol{j}$ 为单位矢量。解上式,并利用 $v_M=BD\cdot\omega_B=1.2\ \mathrm{m/s}$,得

$$v_P=\frac{5}{3}v_M=2\ \mathrm{m/s},\ v_{P/M}=\frac{4}{3}v_M=1.6\ \mathrm{m/s}$$

再利用 $v_P=AD\cdot\omega_A$,得

$$\omega_A=\frac{v_P}{AD}=2\ \mathrm{rad/s}$$

(b)利用 $(\boldsymbol{a}_P)_t+(\boldsymbol{a}_P)_n=\boldsymbol{a}_M+\boldsymbol{a}_{P/M}+\boldsymbol{a}_C$,如图 E8.2(c),得

$$\left[\frac{3}{5}(a_P)_t\boldsymbol{i}+\frac{4}{5}(a_P)_t\boldsymbol{j}\right]+\left[-\frac{4}{5}(a_P)_n\boldsymbol{i}+\frac{3}{5}(a_P)_n\boldsymbol{j}\right]=a_M\boldsymbol{j}+a_{P/M}\boldsymbol{j}-a_C\boldsymbol{i}$$

式中,$(a_P)_n=AD\cdot\omega_A{}^2=4\ \mathrm{m/s^2}$,$a_M=BD\cdot\omega_B{}^2=2.4\ \mathrm{m/s^2}$ 和 $a_C=2\omega_B v_{P/M}=6.4\ \mathrm{m/s^2}$。解上式,得

$$(a_P)_t=-5.33\ \mathrm{m/s^2},\ a_{P/M}=-4.27\ \mathrm{m/s^2}$$

利用 $(a_P)_t=AD\cdot\alpha_A$,得

$$\alpha_A=\frac{(a_P)_t}{AD}=-5.33\ \mathrm{rad/s^2}$$

如图 8.3 所示,当动系 $O'x'y'z'$ 相对定系 $Oxyz$ 平移,$\boldsymbol{\omega}=0$,式(8.22)可简化为

$$\boldsymbol{a}_P=\boldsymbol{a}_M+\boldsymbol{a}_{P/M} \tag{8.23}$$

由此得到结论,对平移参考系,在任何瞬时绝对加速度 $\boldsymbol{a}_P$ 等于牵连加速度 $\boldsymbol{a}_M$ 和相对加速度 $\boldsymbol{a}_{P/M}$ 的矢量和。

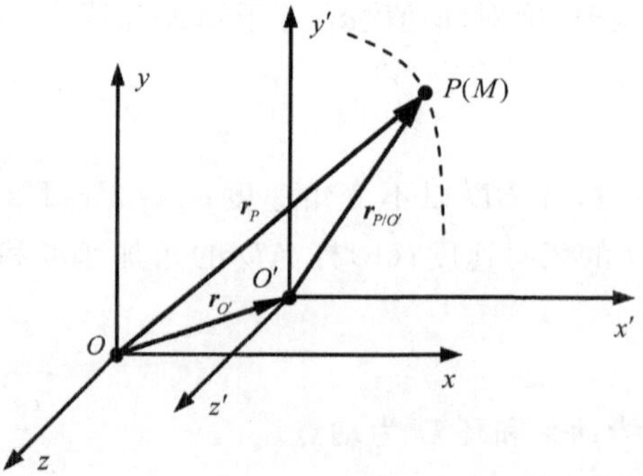

图 8.3

例 8.3

在卡车以加速度 2 m/s^2 倒车的同时，吊杆外段 D 相对卡车以加速度 1 m/s^2 缩回，如图 E8.3(a)所示。求吊杆外段 D 的加速度。

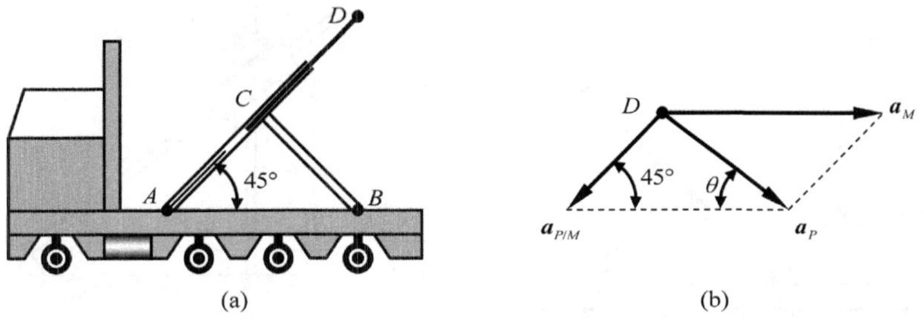

图 E8.3

解

取地球为定系、卡车为动系和吊杆外段 D 为动点。根据 $\boldsymbol{a}_P = \boldsymbol{a}_M + \boldsymbol{a}_{P/M}$ 画加速度三角形，如图 E8.3(b)，得

$$a_P = \sqrt{(a_M)^2 + (a_{P/M})^2 - 2a_M a_{P/M}\cos 45^\circ} = 1.47\ \text{m/s}^2$$

$$\theta = \arcsin\left(\frac{a_{P/M}}{a_P}\sin 45^\circ\right) = 28.75^\circ$$

因此，吊杆外段 D 的加速度大小为 1.47 m/s^2，方向为东偏南 28.75°。

习　题

8.1 如图 P8.1 所示环以不变相对速率 v 沿杆 OAB 向外滑动，同时杆 OAB 以不变角速度 $\omega_B = 2$ rad/s 逆时针转动。已知 $\theta = 0$ 时环位于 A 处，$\theta = 90^\circ$ 时环到达 B 处，求 $\theta = 30^\circ$ 时环的加速度。

8.2 已知图 P8.2 所示瞬时杆 BD 以不变角速度 $\omega_B = 2$ rad/s 逆时针旋转，求：(a)杆 AD 的角速度和环 D 相对杆 AD 的相对速度；(b)杆 AD 的角加速度和环 D 相对杆 AD 的相对加速度。

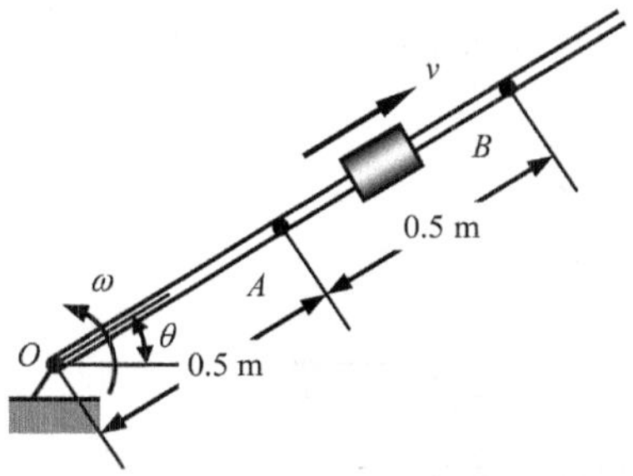

图 P8.1

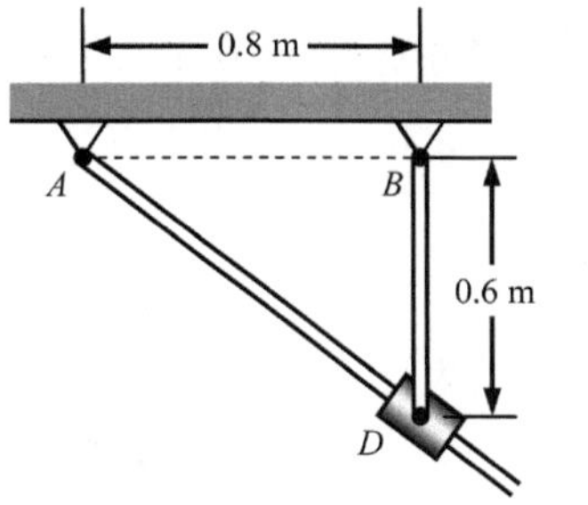

图 P8.2

8.3 如图 P8.3 所示，在卡车以加速度 3 m/s^2 前进的同时，吊杆外段 D 相对卡车以加速度 2 m/s^2 缩回。求吊杆外段 D 的加速度。

8.4 如图 P8.4 所示，环 E 沿杆 BD 滑动，并与沿垂直杆 AC 运动的环 F 相连。已知杆 BD

的角速度和角加速度分别为 $\omega=6$ rad/s 和 $\alpha=4$ rad/s^2,两者都为顺时针,求环 E 的速度和加速度。

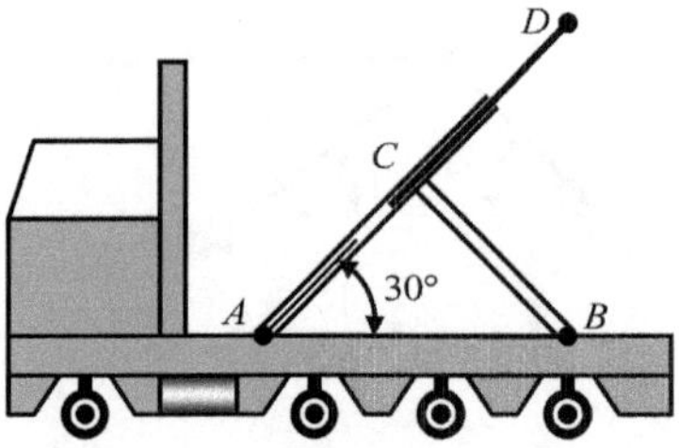

图 P8.3

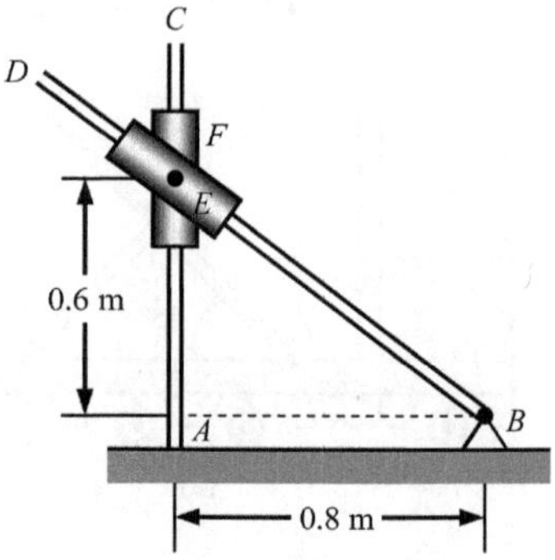

图 P8.4

8.5 沿直线水平飞行的飞机 A 在图 P8.5 所示瞬时的速度为 300 km/h,加速度为 5 m/s^2。飞机 B 在相同高度沿半径为 200 m 的圆形轨迹飞行。已知飞机 B 在图示瞬时的速度为 400 km/h,减速度为 3 m/s^2,求图示瞬时(a)B 相对 A 的速度,(b)B 相对 A 的加速度。

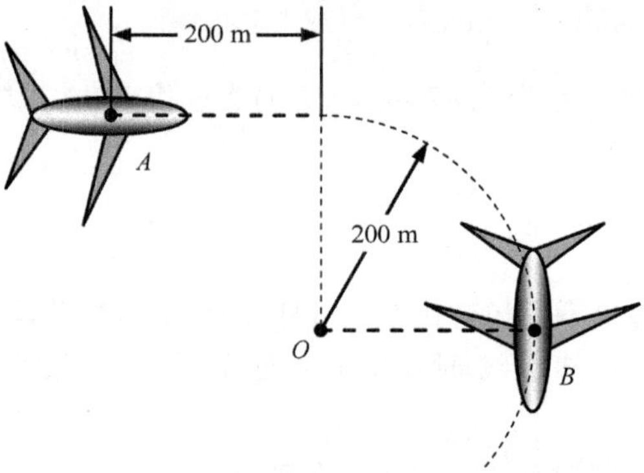

图 P8.5

8.6 如图 P8.6 所示,杆 OA 始终与半径 $r=0.4$ m 的半圆柱 C 的表面相切。已知半圆柱的速度和加速度分别为 $v=0.1$ m/s 和 $a=0.2$ m/s^2,方向都向右,求 $\theta=30°$时杆的角速度和角加速度。

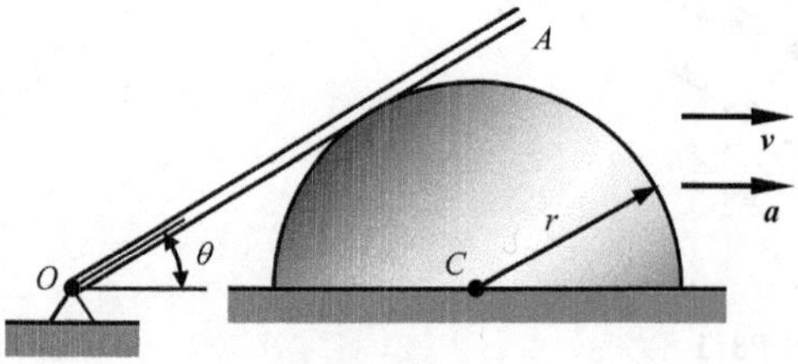

图 P8.6

第 9 章　质点动力学

牛顿第二定律表述如下：如果作用于质点上的合力不为零，则质点将具有与合力大小成正比并沿合力方向的加速度。数学上，该定律可表示为

$$m\boldsymbol{a} = \sum \boldsymbol{F} \tag{9.1}$$

式中，m 为质点的质量，$\boldsymbol{a}$ 为质点的加速度，$\sum \boldsymbol{F}$ 为作用于质点上的合力。

应该注意，牛顿第二运动定律仅在牛顿参考系或惯性参考系中成立。

9.1　质点运动方程

如图 9.1 所示，质量为 m 的质点受力 $\boldsymbol{F}_1$，$\boldsymbol{F}_2$，$\boldsymbol{F}_3$，…作用，牛顿第二定律可表示为

$$m\boldsymbol{a} = \sum \boldsymbol{F} \tag{9.2}$$

或

$$m\ddot{\boldsymbol{r}} = \sum \boldsymbol{F} \tag{9.3}$$

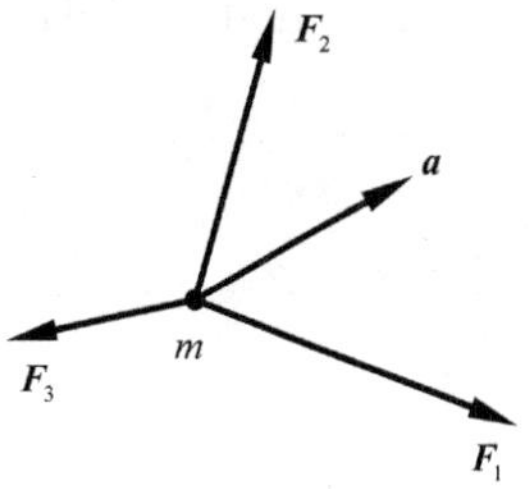

图 9.1

在质点运动问题求解中，上述矢量方程替换为由直角或自然坐标表示的等效标量方程将会给解题带来极大方便。

1. 直角坐标法

把加速度和力分解为直角分量，得

$$ma_x = \sum F_x,\ ma_y = \sum F_y,\ ma_z = \sum F_z \tag{9.4}$$

或

$$m\ddot{x} = \sum F_x,\ m\ddot{y} = \sum F_y,\ m\ddot{z} = \sum F_z \tag{9.5}$$

2. 自然坐标法

把加速度和力分解为切向分量（指向运动方向）和法向分量（指向曲率中心），得

$$ma_t = \sum F_t,\ ma_n = \sum F_n,\ 0 = \sum F_b \tag{9.6}$$

或

$$m\dot{v} = \sum F_t,\ m\frac{v^2}{\rho} = \sum F_n,\ 0 = \sum F_b \tag{9.7}$$

例 9.1

如图 E9.1(a)所示，刚度为 k 的弹簧 AB 与支撑 A 和质量为 m 的环 B 相连，弹簧原长为 l。已知环从 $x=\sqrt{3}l$ 处静止释放，不计环与水平杆之间的摩擦，求环通过中点 O 时的速度大小。

解

当环在 x 处，弹簧伸长可表示为

$$\delta = \sqrt{l^2 + x^2} - l$$

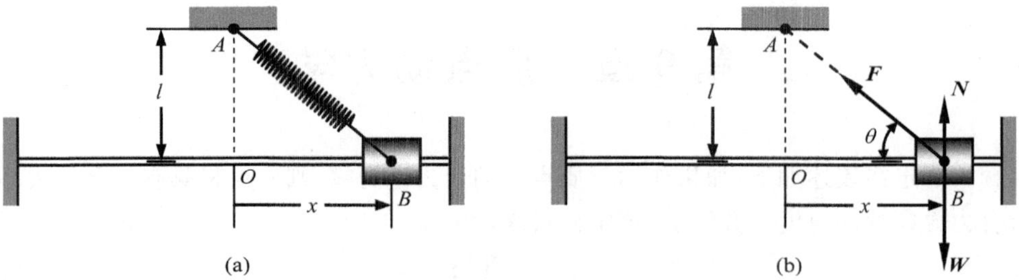

图 E9.1

相应的弹簧弹力为

$$F = k\delta = k(\sqrt{l^2 + x^2} - l)$$

由运动方程，$ma_x = \sum F_x$，得

$$ma = -F\cos\theta$$

利用 $a=\dfrac{\mathrm{d}v}{\mathrm{d}t}=\dfrac{\mathrm{d}x}{\mathrm{d}t}\dfrac{\mathrm{d}v}{\mathrm{d}x}=v\dfrac{\mathrm{d}v}{\mathrm{d}x}$和 $\cos\theta=\dfrac{x}{\sqrt{l^2+x^2}}$,得

$$v\,\mathrm{d}v = -\frac{k}{m}\left(x - \frac{lx}{\sqrt{l^2 + x^2}}\right)\mathrm{d}x$$

进行积分,得

$$\int_0^{v_O} v\,\mathrm{d}v = -\frac{k}{m}\int_{\sqrt{3}l}^{0}\left(x - \frac{lx}{\sqrt{l^2 + x^2}}\right)\mathrm{d}x$$

即

$$v_O = l\sqrt{k/m}$$

例 9.2

如图 E9.2(a),物块在 A 处的初速度 $v_A=8\ \mathrm{m/s}$。已知物块与斜面之间的动摩擦系数 $\mu_k=0.25$,$\theta=10°$,物块到达 B 处时静止,求物块移动距离和所用时间。

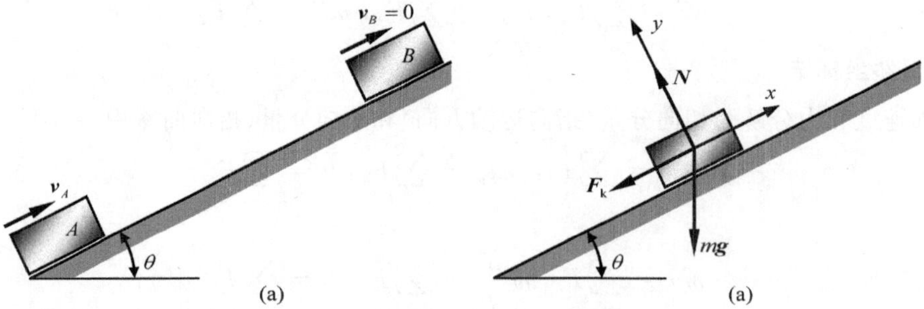

图 E9.2

解

如图 E9.2(b)所示，根据运动方程，$ma_x = \sum F_x$，得

$$ma_x = \sum F_x,\ ma = -mg\sin\theta - F_k$$

$$ma_y = \sum F_y,\ 0 = N - mg\cos\theta$$

利用 $F_k = \mu_k N$，求解上述方程，得

$$a = -(\sin\theta + \mu_k\cos\theta)g$$

(1) 利用 $a = \frac{dv}{dt} = \frac{dx}{dt}\frac{dv}{dx} = v\frac{dv}{dx}$，有

$$v\,dv = -(\sin\theta + \mu_k\cos\theta)g\,dx$$

进行积分，得

$$\int_{v_A}^{v_B} v\,dv = \int_0^{x_B} -(\sin\theta + \mu_k\cos\theta)g\,dx$$

即

$$x_B = \frac{{v_A}^2 - {v_B}^2}{2(\sin\theta + \mu_k\cos\theta)g}$$

代入 $v_A = 8\ \text{m/s}$，$v_B = 0$，$\mu_k = 0.25$，$\theta = 10°$ 和 $g = 9.81\ \text{m/s}^2$，得

$$x_B = 7.77\ \text{m}$$

(2) 利用 $a = \frac{dv}{dt}$，得

$$dv = -(\sin\theta + \mu_k\cos\theta)g\,dt$$

积分，得

$$\int_{v_A}^{v_B} dv = \int_0^{t_B} -(\sin\theta + \mu_k\cos\theta)g\,dt$$

即

$$t_B = \frac{v_A - v_B}{(\sin\theta + \mu_k\cos\theta)g}$$

代入 $v_A = 8\ \text{m/s}$，$v_B = 0$，$\mu_k = 0.25$，$\theta = 10°$ 和 $g = 9.81\ \text{m/s}^2$，得

$$t_B = 1.94\ \text{s}$$

9.2 运动质点的惯性力法

由式(9.1)表示的牛顿第二定律还可表示为

$$\sum \boldsymbol{F} - m\boldsymbol{a} = 0 \tag{9.8}$$

定义 $\boldsymbol{F}_I = -m\boldsymbol{a}$，则式(9.8)可重写为

$$\sum \boldsymbol{F} + \boldsymbol{F}_I = 0 \tag{9.9}$$

式中，$\boldsymbol{F}_I = -m\boldsymbol{a}$ 称为惯性力，其大小等于 ma，方向与加速度 $\boldsymbol{a}$ 相反。这种方法由达朗贝尔提出用于分析质点的运动，通常称为惯性力法或达朗贝尔原理。

惯性力法可以把动力学问题在形式上转换为等效的平衡问题，因而可以采用静力学方法求解动力学问题。虽然质点并不平衡，但是如果在运动质点上施加惯性力，那么即可采用静力学平衡方程研究质点的运动。

应该注意，惯性力是假想的力，实际并不存在。

9.3 运动质点的功能法

功能法可用于求解涉及力、质量和位移的动力学问题。

1. 力的功

如图 9.2 所示,质点从点 A 移动到相邻点 A'。如果 A 和 A'的位置矢量分别用 $\boldsymbol{r}$ 和 $\boldsymbol{r}'$表示,则微分 $\mathrm{d}\boldsymbol{r}=\boldsymbol{r}'-\boldsymbol{r}$ 称为质点的位移。假设力 $\boldsymbol{F}$ 作用于质点,那么力 $\boldsymbol{F}$ 在位移 $\mathrm{d}\boldsymbol{r}$ 上所做的元功定义为

$$\mathrm{d}W = \boldsymbol{F}\cdot\mathrm{d}\boldsymbol{r} \tag{9.10}$$

当质点从 A_1 移动到 A_2,则力 $\boldsymbol{F}$ 在质点发生有限位移期间的功等于

$$W_{12} = \int_{A_1}^{A_2}\boldsymbol{F}\cdot\mathrm{d}\boldsymbol{r} \tag{9.11}$$

2. 功能原理

如图 9.3 所示,质量为 m 的质点在力 $\boldsymbol{F}_1$, $\boldsymbol{F}_2$, $\boldsymbol{F}_3$, …作用下沿曲线移动,则牛顿第二定律可表示为

$$m\dot{\boldsymbol{v}} = \sum\boldsymbol{F} \tag{9.12}$$

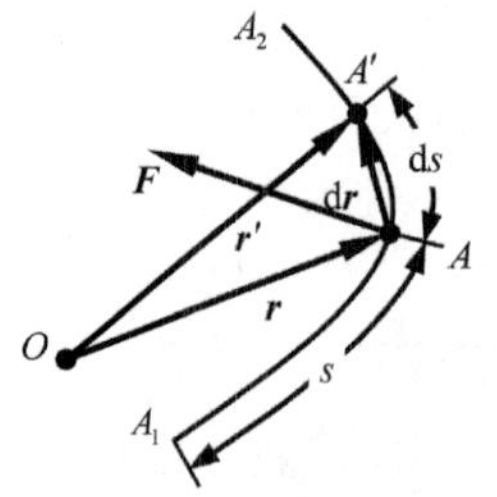

图 9.2

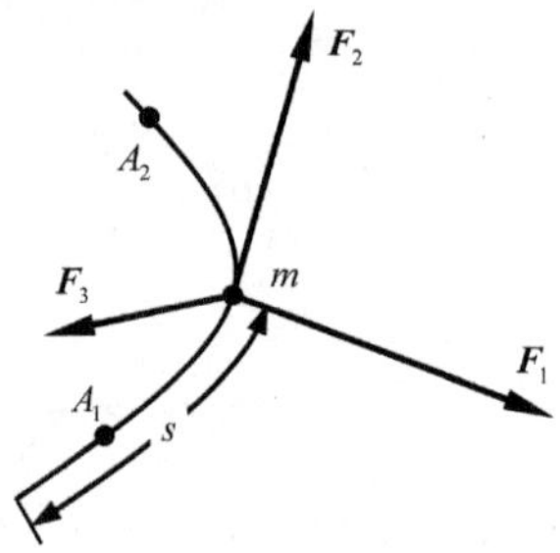

图 9.3

式(9.12)两边点乘 $\mathrm{d}\boldsymbol{r}$,得

$$m\dot{\boldsymbol{v}}\cdot\mathrm{d}\boldsymbol{r} = \sum\boldsymbol{F}\cdot\mathrm{d}\boldsymbol{r} \tag{9.13}$$

利用 $\dot{\boldsymbol{v}}\cdot\mathrm{d}\boldsymbol{r}=\boldsymbol{v}\cdot\mathrm{d}\boldsymbol{v}=\frac{1}{2}d(\boldsymbol{v}\cdot\boldsymbol{v})=\frac{1}{2}\mathrm{d}(v^2)$,得

$$\mathrm{d}T = \mathrm{d}W \tag{9.14}$$

式中,$\mathrm{d}T=\mathrm{d}\left(\frac{1}{2}mv^2\right)$和 $\mathrm{d}W = \sum\boldsymbol{F}\cdot\mathrm{d}\boldsymbol{r}$ 分别是质点动能增量和力 $\sum\boldsymbol{F}$ 在位移 $\mathrm{d}\boldsymbol{r}$ 上所做的元功。积分,得

$$T_2 - T_1 = W_{12} \tag{9.15}$$

式中,$T_1=\frac{1}{2}mv_1^2$ 和 $T_2=\frac{1}{2}mv_2^2$ 为质点分别在 A_1 和 A_2 处的动能,$W_{12} = \sum\int_{A_1}^{A_2}\boldsymbol{F}\cdot\mathrm{d}\boldsymbol{r}$ 为质点从 A_1 移动到 A_2 过程中作用在质点上的力所做的功。因此,从式(9.14)和(9.15)得到结论,当质点从一点移动到另一点时,质点动能的变化等于作用于质点上的力所做的功。

式(9.15)也可表示为

$$T_1 + W_{12} = T_2 \tag{9.16}$$

式(9.16)表明,质点在 A_2 处的动能等于质点在 A_1 处的动能加上质点从 A_1 移动到 A_2 过程中作用在质点上的力所做的功。

由式(9.14)、(9.15)或(9.16)所表示的功和能之间的关系称为功能原理。应该注意,功能原理仅在牛顿参考系成立。

例 9.3

如图 E9.3,物块在 A 处的初速度 $v_A = 8\ \mathrm{m/s}$。已知物块与斜面之间的动摩擦系数 $\mu_k = 0.25$,$\theta = 10°$,物块到达 B 处时静止,求物块移动距离。

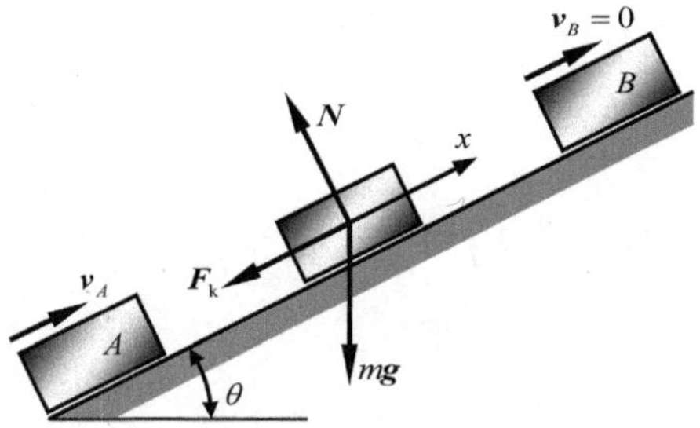

图 E9.3

解

利用功能原理,$T_A + W_{AB} = T_B$,得

$$\frac{1}{2}mv_A^2 + (-F_k - mg\sin\theta)x_B = \frac{1}{2}mv_B^2$$

式中,$F_k = \mu_k N = \mu_k mg\cos\theta$。

代入 $v_A = 8\ \mathrm{m/s}$,$v_B = 0$,$\mu_k = 0.25$,$\theta = 10°$和 $g = 9.81\ \mathrm{m/s^2}$,得

$$x_B = \frac{v_A^2 - v_B^2}{2(\sin\theta + \mu_k\cos\theta)g} = 7.77\ \mathrm{m}$$

3. 能量守恒

当质点从一点移动到另一点,如果作用于质点上的力所做的功与质点的移动轨迹无关,则作用于质点上的力称为保守力。重力和弹力是典型的保守力。

当质点从 A_1 移动到 A_2 时作用于质点上的保守力所做的功定义为质点在 A_1 处相对于 A_2 处的势能,即

$$V_1 - V_2 = W_{12} = \sum \int_{A_1}^{A_2} \boldsymbol{F} \cdot \mathrm{d}\boldsymbol{r} \tag{9.17}$$

式中,V_1 和 V_2 分别表示质点在 A_1 和 A_2 处的势能。

结合式(9.15)和(9.17),得

$$T_1 + V_1 = T_2 + V_2 \tag{9.18}$$

式(9.18)表明,当质点在保守力作用下发生运动,则质点的动能和势能之和保持不变。动能和势能之和称为质点的机械能。

例 9.4

如图 E9.4 所示,刚度为 k 的弹簧 AB 与支撑 A 和质量为 m 的环 B 相连,弹簧原长为 l。已知环从 $x=\sqrt{3}l$ 处静止释放,不计环与水平杆之间的摩擦,求环通过中点 O 时速度的大小。

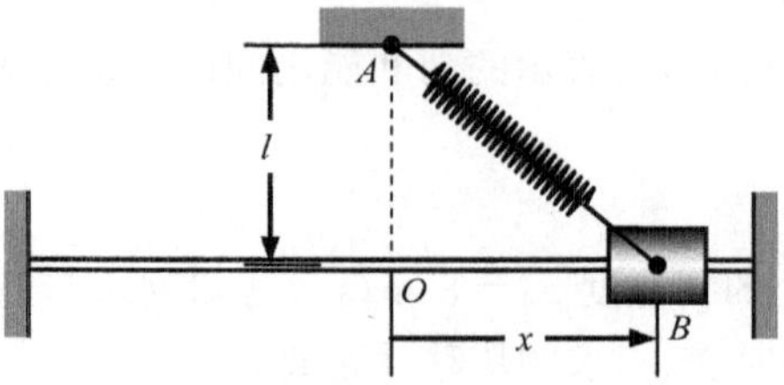

图 E9.4

解

当环在 $x=\sqrt{3}l$ 处,有

$$T_1 = 0,\ V_1 = \frac{1}{2}k\left[\sqrt{l^2+(\sqrt{3}l)^2}-l\right]^2 = \frac{1}{2}kl^2$$

当环在中点 O 处,有

$$T_2 = \frac{1}{2}mv_O^2,\ V_2 = 0$$

利用机械能守恒,$T_1+V_1=T_2+V_2$,得

$$0+\frac{1}{2}kl^2 = \frac{1}{2}mv_O^2+0$$

$$v_O = l\sqrt{k/m}$$

9.4 运动质点的冲量动量法

冲量动量法可用于求解涉及力、质量、速度和时间的动力学问题。

1. 冲量动量原理

如图 9.4 所示,质量为 m 的质点在力 $\boldsymbol{F}_1$,$\boldsymbol{F}_2$,$\boldsymbol{F}_3$,…作用下沿曲线移动,则牛顿第二定律可表示为

$$m\dot{\boldsymbol{v}} = \sum \boldsymbol{F} \tag{9.19}$$

式(9.19)两边乘 dt,得

$$\mathrm{d}\boldsymbol{L} = \mathrm{d}\boldsymbol{I} \tag{9.20}$$

式中,$\mathrm{d}\boldsymbol{L} = \mathrm{d}(m\boldsymbol{v})$ 和 $\mathrm{d}\boldsymbol{I} = \sum \boldsymbol{F}\mathrm{d}t$ 分别为质点动量的增量和作用于质点上的力 $\sum \boldsymbol{F}$ 在时间间隔 dt 内的冲量。积分,得

$$\boldsymbol{L}_2 - \boldsymbol{L}_1 = \boldsymbol{I}_{12} \tag{9.21}$$

式中,$\boldsymbol{L}_1=m\,\boldsymbol{v}_1$ 和 $\boldsymbol{L}_2=m\,\boldsymbol{v}_2$ 为质点分别在 t_1 和 t_2 时的动

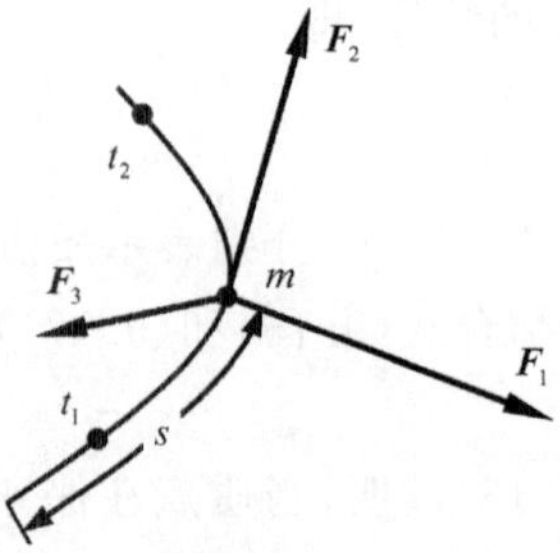

图 9.4

量，$\boldsymbol{I}_{12}=\sum\int_{t_1}^{t_2}\boldsymbol{F}\mathrm{d}t$ 为时间从 t_1 到 t_2 间隔中作用在质点上的力的冲量。因此，从式(9.20)或(9.21)得到结论，在所考虑的时间间隔内，质点动量的变化等于作用于质点上的力的冲量。式(9.21)也可表示为

$$\boldsymbol{L}_1+\boldsymbol{I}_{12}=\boldsymbol{L}_2 \tag{9.22}$$

式(9.22)表明，质点在 t_2 时的动量等于质点在 t_1 时的动量加上时间从 t_1 到 t_2 间隔中作用在质点上的力的冲量，如图 9.5 所示。

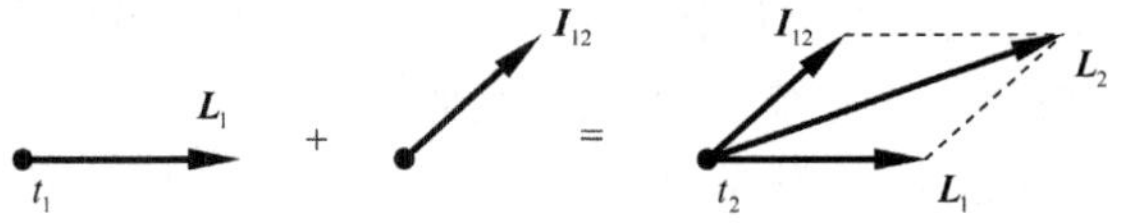

图 9.5

由式(9.20)、(9.21)或(9.22)所表示的冲量动量原理是矢量方程。当用于求解问题时，式(9.20)、(9.21)或(9.22)需要分解为直角坐标方程，即

$$\left.\begin{aligned}(\mathrm{d}L)_x=(\mathrm{d}I)_x\quad(L_2)_x-(L_1)_x=(I_{12})_x\quad\ (L_1)_x+(I_{12})_x=(L_2)_x\\(\mathrm{d}L)_y=(\mathrm{d}I)_y、(L_2)_y-(L_1)_y=(I_{12})_y\text{或}(L_1)_y+(I_{12})_y=(L_2)_y\\(\mathrm{d}L)_z=(\mathrm{d}I)_z\quad(L_2)_z-(L_1)_z=(I_{12})_z\quad\ (L_1)_z+(I_{12})_z=(L_2)_z\end{aligned}\right\} \tag{9.23}$$

例 9.5

如图 E9.5 所示，物块在 A 处的初速度 $v_A=8\ \mathrm{m/s}$。已知物块与斜面之间的动摩擦系数 $\mu_k=0.25$，$\theta=10°$，物块到达 B 处时静止，求所用时间。

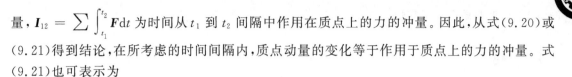

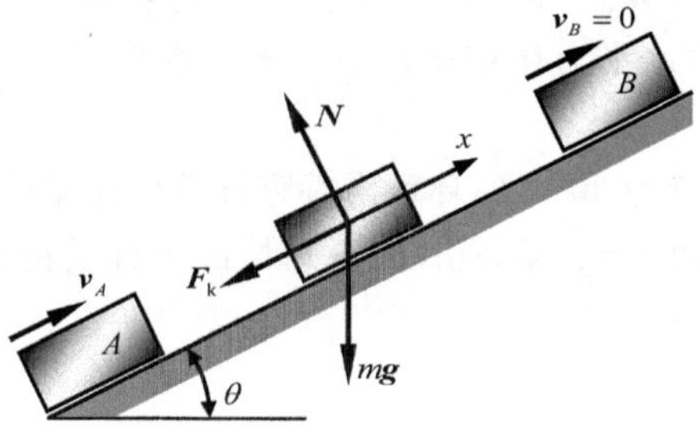

图 E9.5

解

由冲量动量原理，$mv_A+I_{AB}=mv_B$，得

$$mv_A+(-F_k-mg\sin\theta)t_B=mv_B$$

式中，$F_k=\mu_k N=\mu_k mg\cos\theta$。

代入 $v_A=8\ \mathrm{m/s}$，$v_B=0$，$\mu_k=0.25$，$\theta=10°$和 $g=9.81\ \mathrm{m/s^2}$，得

$$t_B=\frac{v_A-v_B}{(\sin\theta+\mu_k\cos\theta)g}=1.94\ \mathrm{s}$$

2. 动量守恒

如果作用于质点上的力的合力为零，则式(9.21)或(9.22)可化简为

$$\boldsymbol{L}_1=\boldsymbol{L}_2 \tag{9.24}$$

式(9.24)表明质点动量守恒。

习 题

9.1 承包人使用含有平台的电驱升降机把质量为 m 的物块从地面 A 运送到屋顶 B,升降机平台可以沿着固定在梯子上的导轨滑动,如图 P9.1 所示。升降机首先以不变加速度 a_1 从 A 处由静止开始加速运动,然后再以不变减速度 a_2 做减速运动,一直运动到梯子顶部附近的 B 处停止。已知物块与平台之间的静摩擦系数 $\mu_s=0.3$ 和梯子倾角 $\theta=60°$,求物块不在平台上发生滑动的最大允许加速度 a_1 和最大允许减速度 a_2。

9.2 为了从卡车上卸下石块,驾驶员首先倾斜卡车底板,然后从静止开始加速,如图 P9.2 所示。已知物块与底板之间的摩擦系数 $\mu_s=0.4$ 和 $\mu_k=0.3$,求物块发生滑动的最小加速度。

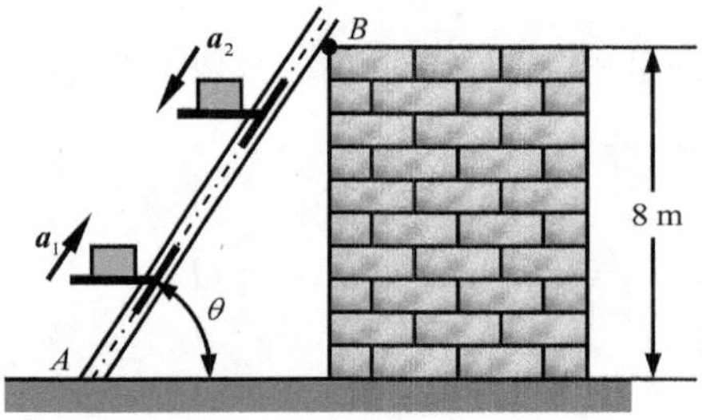

图 P9.1

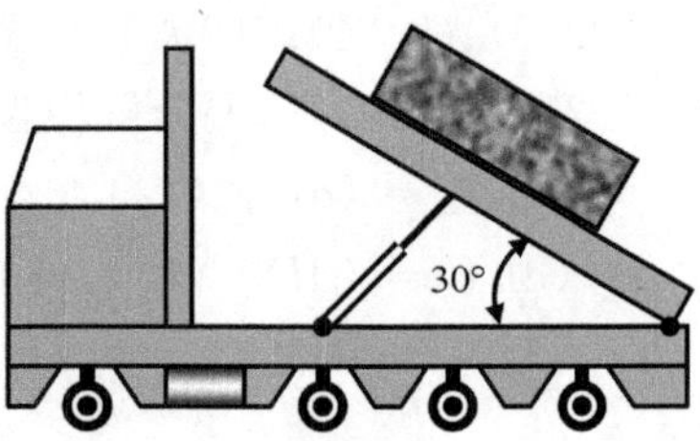

图 P9.2

9.3 如图 P9.3 所示,重量 $W_B=100$ N 的物块 B 与绳索相连后放在重量 $W_A=300$ N 的物块 A 上面,大小 $F=200$ N 的水平力作用于绳索。不计摩擦,求:(a)物块 A 的加速度;(b)物块 B 相对物块 A 的加速度。

9.4 如图 P9.4 所示,重量为 30 N 的环可沿垂直杆发生无摩擦滑动,握住环使之刚好与未伸长弹簧接触。已知弹簧刚度 $k=2$ kN/m,求下列情况下弹簧的最大挠度:(a)环慢慢释放,直到达到平衡位置;(b)环突然释放。

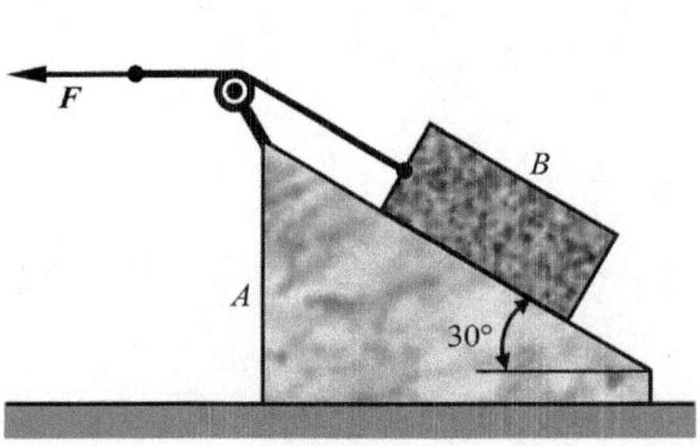

图 P9.3

图 P9.4

9.5 如图 P9.5 所示,质量为 2 kg 的环与弹簧相连,在垂直平面内沿圆杆发生无摩擦滑动。弹簧原长为0.1 m,刚度为 k。环在 A 处于静止状态时受到轻微推动而开始向右运动。已知环通过 B 处时将获得最大速度,求:(a)弹簧刚度 k;(b)环的最大速度。

9.6 物块在 A 处的初速度 $v_A=2\ \mathrm{m/s}$。已知物块与水平面之间的动摩擦系数 $\mu_k=0.2$，物块到达 B 处时静止，求物块移动距离和所用时间。

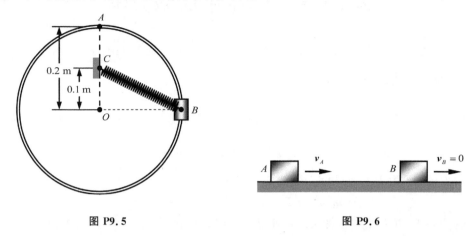

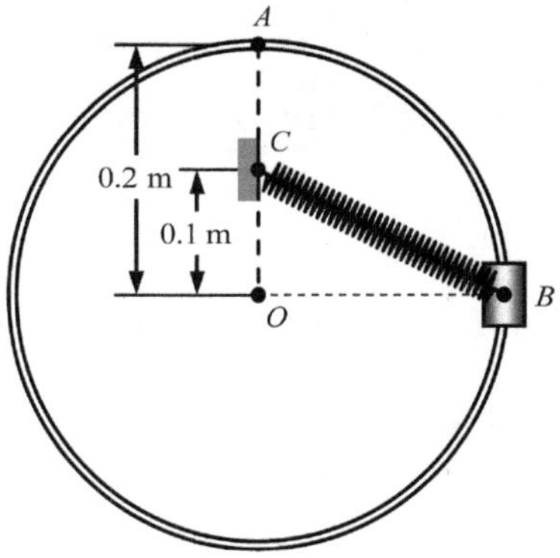

图 P9.5

图 P9.6

9.7 如图 P9.7 所示，质量为 2 kg 的球通过长度为 0.5 m 的刚性绳索与固定点 O 相连。球在 A 处静止于光滑水平面上，在垂直于 OA 的方向突然给球一个速度 $\boldsymbol{v}_0$，球开始无摩擦运动，直到球到达 B 处而使绳索拉紧。如果作用于绳索的冲量不超过 $I=5\ \mathrm{N\cdot s}$，求最大允许速度。

9.8 三级跳远是田径运动比赛项目。假设重量为 800 N 的运动员以 8 m/s 的水平速度从左侧接近起跳线，保持与地面接触的时间为 0.2 s，并以 10 m/s 的速度沿 60°角起跳如图 P9.8 所示，求地面作用于脚上的平均冲力。

9.9 如图 P9.9 所示，质量 $m=10$ kg 的物块沿倾角 $\theta=30°$ 的平面由静止开始向下滑动。已知物块与平面之间的动摩擦系数 $\mu_k=0.3$，求物块运动 $t=5$ s 后的速度。

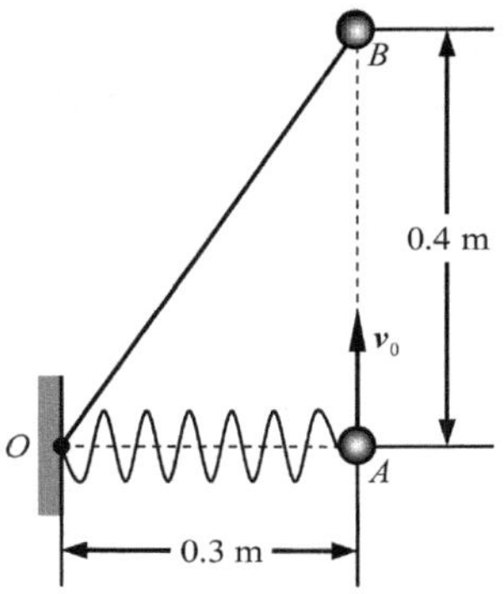

图 P9.7

图 P9.8

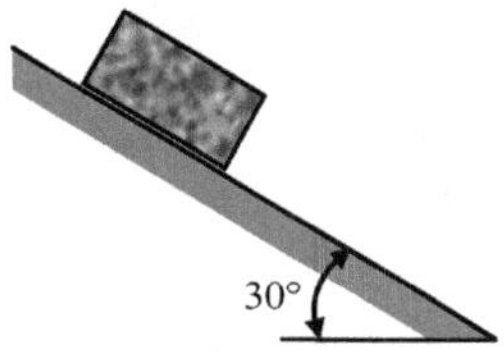

图 P9.9

第 10 章　刚体平面动力学

10.1　质点系的运动

为推导由 n 个质点构成的质点系的运动方程，可把牛顿第二定律应用于系统的每一个质点。考虑质点 P_i，质量为 m_i，外力合力为 $\boldsymbol{F}_i^{(e)}$，内力合力(即系统内所有其他质点 $\sum\limits_{j=1,j\neq i}^{n} P_j$ 施加在质点 P_i 上的力)为 $\boldsymbol{F}_i^{(i)}$。对质点 P_i，牛顿第二定律可表示为

$$m_i \boldsymbol{a}_i = \boldsymbol{F}_i^{(e)} + \boldsymbol{F}_i^{(i)} \quad (i = 1,2,\cdots,n) \tag{10.1}$$

式中，$\boldsymbol{a}_i$ 为质点 P_i 相对牛顿参考系的加速度。式(10.1)各项都对定点 O 取矩，有

$$\boldsymbol{r}_i \times m_i \boldsymbol{a}_i = \boldsymbol{r}_i \times \boldsymbol{F}_i^{(e)} + \boldsymbol{r}_i \times \boldsymbol{F}_i^{(i)} \quad (i = 1,2,\cdots,n) \tag{10.2}$$

式中，$\boldsymbol{r}_i$ 为质点 P_i 相对点 O 的位置矢量。

考虑系统中的所有质点，并利用式(10.1)和(10.2)，得

$$\sum_{i=1}^{n} m_i \boldsymbol{a}_i = \sum_{i=1}^{n} \boldsymbol{F}_i^{(e)} + \sum_{i=1}^{n} \boldsymbol{F}_i^{(i)} \tag{10.3}$$

$$\sum_{i=1}^{n} (\boldsymbol{r}_i \times m_i \boldsymbol{a}_i) = \sum_{i=1}^{n} (\boldsymbol{r}_i \times \boldsymbol{F}_i^{(e)}) + \sum_{i=1}^{n} (\boldsymbol{r}_i \times \boldsymbol{F}_i^{(i)}) \tag{10.4}$$

代入 $\sum\limits_{i=1}^{n} \boldsymbol{F}_i^{(i)} = 0$ 和 $\sum\limits_{i=1}^{n} (\boldsymbol{r}_i \times \boldsymbol{F}_i^{(i)}) = 0$，式(10.3)和(10.4)简化为

$$\sum_{i=1}^{n} m_i \boldsymbol{a}_i = \sum_{i=1}^{n} \boldsymbol{F}_i^{(e)} \tag{10.5}$$

$$\sum_{i=1}^{n} (\boldsymbol{r}_i \times m_i \boldsymbol{a}_i) = \sum_{i=1}^{n} (\boldsymbol{r}_i \times \boldsymbol{F}_i^{(e)}) \tag{10.6}$$

质点系的线动量定义为系统中各个质点的线动量之和，即

$$\boldsymbol{L} = \sum_{i=1}^{n} m_i \boldsymbol{v}_i \tag{10.7}$$

式(10.7)对时间微分，有

$$\dot{\boldsymbol{L}} = \sum_{i=1}^{n} m_i \boldsymbol{a}_i \tag{10.8}$$

质点系对定点 O 的角动量定义为系统中各个质点对相同点 O 的角动量之和，即

$$\boldsymbol{H}_O = \sum_{i=1}^{n} (\boldsymbol{r}_i \times m_i \boldsymbol{v}_i) \tag{10.9}$$

式中，$\boldsymbol{r}_i \times m_i \boldsymbol{v}_i$ 为质点 P_i 对定点 O 的角动量。

式(10.9)对时间微分，得

$$\dot{\boldsymbol{H}}_O = \sum_{i=1}^{n} (\boldsymbol{v}_i \times m_i \boldsymbol{v}_i) + \sum_{i=1}^{n} (\boldsymbol{r}_i \times m_i \boldsymbol{a}_i) \tag{10.10}$$

利用 $\boldsymbol{v}_i \times \boldsymbol{v}_i = 0$，则简化为

$$\dot{\boldsymbol{H}}_O = \sum_{i=1}^{n}(\boldsymbol{r}_i \times m_i \boldsymbol{a}_i) \tag{10.11}$$

把式(10.8)和(10.11)代入式(10.5)和(10.6),得

$$\dot{\boldsymbol{L}} = \sum_{i=1}^{n}\boldsymbol{F}_i^{(e)} \tag{10.12}$$

$$\dot{\boldsymbol{H}}_O = \sum_{i=1}^{n}(\boldsymbol{r}_i \times \boldsymbol{F}_i^{(e)}) = \sum_{i=1}^{n}\boldsymbol{M}_O(\boldsymbol{F}_i^{(e)}) \tag{10.13}$$

由此得出结论,质点系线动量的变化率和质点系对定点 O 的角动量的变化率分别等于作用于系统质点上的外力的合力和作用在系统质点上的外力对 O 的合力矩。

例 10.1

系统由质点 A、B 和 C 构成,如图 E10.1。已知质量 $m_A = m_B = m_C = 1$ kg 和速度 $\boldsymbol{v}_A = 2\boldsymbol{i} + 3\boldsymbol{j}$ m/s、$\boldsymbol{v}_B = -2\boldsymbol{i} + 3\boldsymbol{j}$ m/s 和 $\boldsymbol{v}_C = -3\boldsymbol{j} - 2\boldsymbol{k}$ m/s,求系统的线动量 $\boldsymbol{L}$ 和系统对点 O 的角动量 $\boldsymbol{H}_O$。

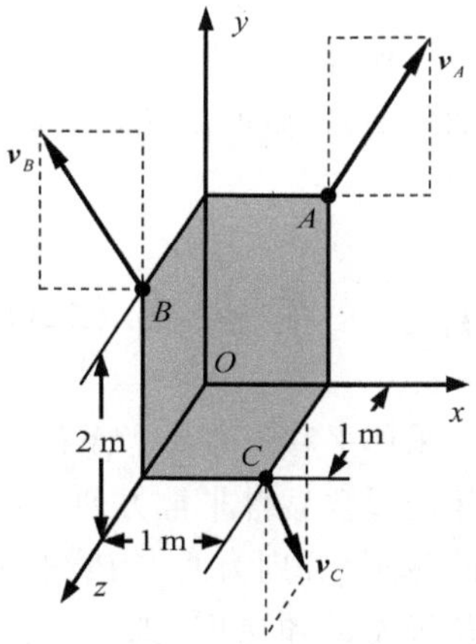

图 E10.1

解

质点 A、B 和 C 的位置矢量分别为

$\boldsymbol{r}_A = \boldsymbol{i} + 2\boldsymbol{j}$ m, $\boldsymbol{r}_B = 2\boldsymbol{j} + \boldsymbol{k}$ m, $\boldsymbol{r}_C = \boldsymbol{i} + \boldsymbol{k}$ m

因此,系统的线动量 $\boldsymbol{L}$ 和系统对点 O 的角动量 $\boldsymbol{H}_O$ 分别为

$$\begin{aligned} L &= m_A \boldsymbol{v}_A + m_B \boldsymbol{v}_B + m_C \boldsymbol{v}_C \\ &= (2\boldsymbol{i} + 3\boldsymbol{j}) + (-2\boldsymbol{i} + 3\boldsymbol{j}) + (-3\boldsymbol{j} - 2\boldsymbol{k}) \\ &= 3\boldsymbol{j} - 2\boldsymbol{k} \text{ kg} \cdot \text{m/s} \end{aligned}$$

$$\begin{aligned} \boldsymbol{H}_O &= \boldsymbol{r}_A \times m_A \boldsymbol{v}_A + \boldsymbol{r}_B \times m_B \boldsymbol{v}_B + \boldsymbol{r}_C \times m_C \boldsymbol{v}_C \\ &= \begin{vmatrix} \boldsymbol{i} & \boldsymbol{j} & \boldsymbol{k} \\ 1 & 2 & 0 \\ 2 & 3 & 0 \end{vmatrix} + \begin{vmatrix} \boldsymbol{i} & \boldsymbol{j} & \boldsymbol{k} \\ 0 & 2 & 1 \\ -2 & 3 & 0 \end{vmatrix} + \begin{vmatrix} \boldsymbol{i} & \boldsymbol{j} & \boldsymbol{k} \\ 1 & 0 & 1 \\ 0 & -3 & -2 \end{vmatrix} \\ &= 0 \end{aligned}$$

10.2　质点系质心的运动

假设 $\boldsymbol{r}_C$ 为质点系质心的位置矢量,则有

$$\boldsymbol{r}_C = \frac{\sum_{i=1}^{n} m_i \boldsymbol{r}_i}{m} \tag{10.14}$$

式中,$m = \sum_{i=1}^{n} m_i$ 为质点系总质量。

式(10.14)对时间微分两次,得

$$a_C = \frac{\sum_{i=1}^{n} m_i \boldsymbol{a}_i}{m} \tag{10.15}$$

式中,$\boldsymbol{a}_C$ 为质点系质心的加速度。

把 $\dot{\boldsymbol{L}} = \sum_{i=1}^{n} m_i \boldsymbol{a}_i = m\boldsymbol{a}_C$ 代入式(10.12),得

$$m\boldsymbol{a}_C = \sum_{i=1}^{n} \boldsymbol{F}_i^{(e)} \tag{10.16}$$

上述方程可以描述质点系的质心的运动。因此得到结论,质点系的质心的运动相当于一个质点的运动,该质点位于质点系的质心位置,并集中质点系的全部质量和作用在质点系上的所有外力。

10.3 质点系相对质心的运动

采用平移形心参考系研究质点系中质点的运动通常显得非常方便。如图 10.1 所示,设$\boldsymbol{r}'_i$ 和$\boldsymbol{v}'_i$ 分别为质点 P_i 相对动系 $O'x'y'z'$ 的位置矢量和速度,则质点系相对质心的角动量定义如下:

$$\boldsymbol{H}'_C = \sum_{i=1}^{n} (\boldsymbol{r}'_i \times m_i \boldsymbol{v}'_i) \tag{10.17}$$

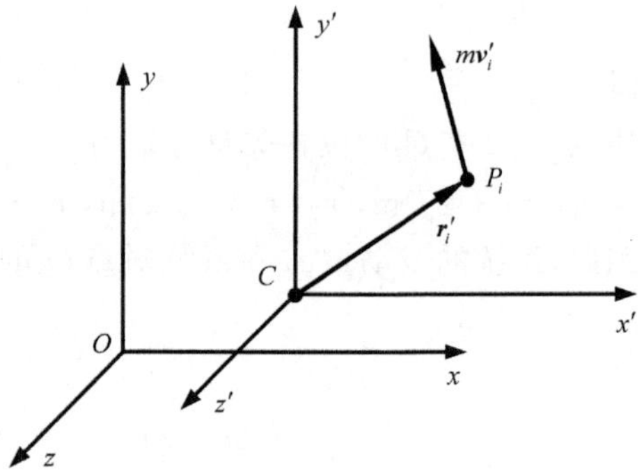

图 10.1

同理,可定义

$$\boldsymbol{H}_C = \sum_{i=1}^{n} (\boldsymbol{r}'_i \times m_i \boldsymbol{v}_i) \tag{10.18}$$

式中,$\boldsymbol{v}_i$ 为在牛顿参考系 $Oxyz$ 中观察到的绝对速度。

利用$\boldsymbol{v}_i - \boldsymbol{v}'_i = \boldsymbol{v}_C$,得

$$\boldsymbol{H}_C - \boldsymbol{H}'_C = \sum_{i=1}^{n} (\boldsymbol{r}'_i \times m_i \boldsymbol{v}_i) - \sum_{i=1}^{n} (\boldsymbol{r}'_i \times m_i \boldsymbol{v}'_i) = (\sum_{i=1}^{n} m_i \boldsymbol{r}'_i) \times \boldsymbol{v}_C \tag{10.19}$$

代入 $\sum_{i=1}^{n} m_i \boldsymbol{r}'_i = m\boldsymbol{r}'_C = 0$,得

$$\boldsymbol{H}_C = \boldsymbol{H}'_C \tag{10.20}$$

利用$\boldsymbol{r}_i - \boldsymbol{r}'_i = \boldsymbol{r}_C$,得

$$\boldsymbol{H}_O - \boldsymbol{H}_C = \sum_{i=1}^{n} (\boldsymbol{r}_i \times m_i \boldsymbol{v}_i) - \sum_{i=1}^{n} (\boldsymbol{r}'_i \times m_i \boldsymbol{v}_i) = \boldsymbol{r}_C \times (\sum_{i=1}^{n} m_i \boldsymbol{v}_i) \tag{10.21}$$

代入 $\sum_{i=1}^{n} m_i \boldsymbol{v}_i = m\boldsymbol{v}_C$,得

$$\boldsymbol{H}_O - \boldsymbol{H}_C = \boldsymbol{r}_C \times m\boldsymbol{v}_C \tag{10.22}$$

对时间微分,并利用$\boldsymbol{v}_C \times \boldsymbol{v}_C = 0$,得

$$\dot{\boldsymbol{H}}_O - \dot{\boldsymbol{H}}_C = \boldsymbol{r}_C \times m\boldsymbol{a}_C \tag{10.23}$$

再利用式(10.13)和(10.16),式(10.23)可写为

$$\dot{\boldsymbol{H}}_C = \dot{\boldsymbol{H}}_O - \boldsymbol{r}_C \times m\,\boldsymbol{a}_C = \sum_{i=1}^{n}(\boldsymbol{r}_i \times \boldsymbol{F}_i^{(e)}) - \boldsymbol{r}_C \times \sum_{i=1}^{n}\boldsymbol{F}_i^{(e)} = \sum_{i=1}^{n}[(\boldsymbol{r}_i - \boldsymbol{r}_C) \times \boldsymbol{F}_i^{(e)}] \tag{10.24}$$

利用$\boldsymbol{r}_i - \boldsymbol{r}_C = \boldsymbol{r}'_i$,有

$$\dot{\boldsymbol{H}}_C = \sum_{i=1}^{n}(\boldsymbol{r}'_i \times \boldsymbol{F}_i^{(e)}) = \sum_{i=1}^{n}\boldsymbol{M}_C(\boldsymbol{F}_i^{(e)}) \tag{10.25}$$

由此得到结论,质点系相对质心 C 的角动量的变化率等于作用于系统质点上的外力对 C 的合力矩。

10.4　平面运动刚体的运动方程

参考式(10.17),并假设刚体由 n 个质量为 m_i 的质点 P_i 组成,则得

$$\boldsymbol{H}_C = \boldsymbol{H}'_C = \sum_{i=1}^{n}(\boldsymbol{r}'_i \times m_i\,\boldsymbol{v}'_i) \tag{10.26}$$

利用$\boldsymbol{v}'_i = \boldsymbol{\omega} \times \boldsymbol{r}'_i$,其中 $\boldsymbol{\omega}$ 为刚体角速度,则有

$$\boldsymbol{H}_C = \sum_{i=1}^{n}[\boldsymbol{r}'_i \times m_i(\boldsymbol{\omega} \times \boldsymbol{r}'_i)] = \left\{\sum_{i=1}^{n}[(r'_i)^2 m_i]\right\}\boldsymbol{\omega} \tag{10.27}$$

当 $n \to \infty$,即 $m_i \to 0$,则式(10.27)可重写为

$$\boldsymbol{H}_C = (\int r'^2 \mathrm{d}m)\omega \tag{10.28}$$

定义 $I_C = \int r'^2 \mathrm{d}m$ (即刚体相对质心的转动惯量),则得

$$\boldsymbol{H}_C = I_C\boldsymbol{\omega} \tag{10.29}$$

对时间微分,得

$$\dot{\boldsymbol{H}}_C = I_C\boldsymbol{\alpha} \tag{10.30}$$

参考式(10.16)和(10.25),假设刚体具有对称面,受对称面内的平面力 $\boldsymbol{F}_1,\boldsymbol{F}_2,\boldsymbol{F}_3,\cdots$作用而发生对称面内的平面运动,则运动方程可表示为

$$m\boldsymbol{a}_C = \sum \boldsymbol{F},\ I_C\alpha = \sum M_C(\boldsymbol{F}) \tag{10.31}$$

把上述方程分解为直角分量,得

$$m(a_C)_x = \sum F_x,\ m(a_C)_y = \sum F_y,\ I_C\alpha = \sum M_C(\boldsymbol{F}) \tag{10.32}$$

例 10.2

质量为 40 kg 的均质薄板放在卡车上,A 端由光滑垂直面支撑,B 端静止放在粗糙水平面上,如图 E10.2 所示。已知卡车减速度为 2 m/s^2,l=2 m 和 θ=60°,求:(a)A 和B 处反力;(b)B 处最小静摩擦系数。

解

取 AB 为自由体,画受力图,如图 E10.2(b)所示,则有

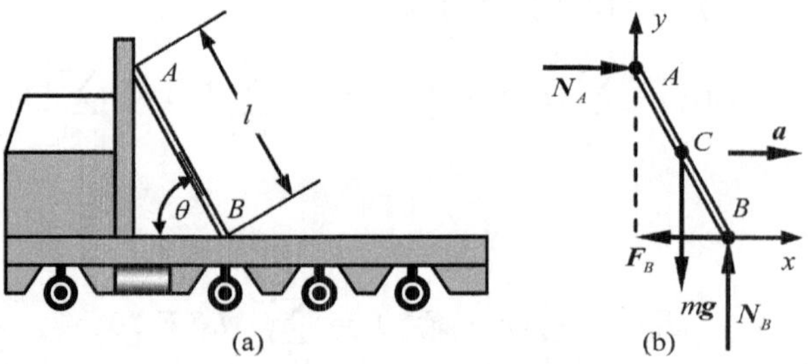

图 E10.2

$$m(a_C)_x = \sum F_x,\ ma = N_A - F_B$$

$$m(a_C)_y = \sum F_y,\ 0 = N_B - mg$$

$$I_C\alpha = \sum M_C(\boldsymbol{F}),\ 0 = -N_A\left(\frac{1}{2}l\sin\theta\right) + N_B\left(\frac{1}{2}l\cos\theta\right) - F_B\left(\frac{1}{2}l\sin\theta\right)$$

解方程,并利用 $g=9.81\ \mathrm{m/s^2}$,得

$$N_A = 153.3\ \mathrm{N},\ N_B = 392.4\ \mathrm{N},\ F_B = 73.27\ \mathrm{N}$$

利用 $F_B \leqslant \mu_s N_B$,有

$$\mu_s \geqslant 0.1867$$

例 10.3

长度为 l、质量为 m 的均质杆 AB 支撑如图 E10.3(a)所示。如果 B 处绳子突然断裂,求:(a)杆 AB 的角加速度;(b)支座 A 处的反力。

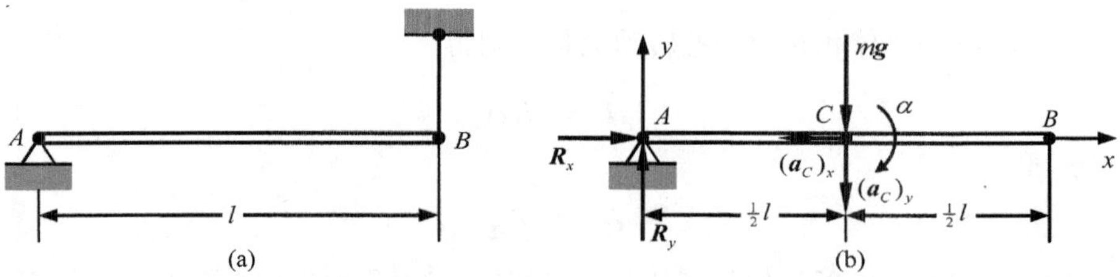

图 E10.3

解

取 AB 为自由体,画受力图,如图 E10.3(b)所示,则有

$$m(a_C)_x = \sum F_x,\quad -m(a_C)_x = R_x$$

$$m(a_C)_y = \sum F_y,\quad -m(a_C)_y = R_y - mg$$

$$I_C\alpha = \sum M_C(\boldsymbol{F}),\ -I_C\alpha = -R_y\left(\frac{1}{2}l\right)$$

式中,$(a_C)_x = AC\cdot\omega^2 = 0$,$(a_C)_y = AC\cdot\alpha = \frac{1}{2}l\alpha$,$I_C = \frac{1}{12}ml^2$。

解上述方程,得

$$R_x = 0,\ R_y = \frac{1}{4}mg,\ \alpha = \frac{3g}{2l}$$

10.5　平面运动刚体的惯性力法

惯性力法可用于分析质点的运动，同样该方法也可用于分析运动刚体。对运动刚体，不仅需要把惯性力(大小与 $m\boldsymbol{a}_C$ 相等，方向与 $m\boldsymbol{a}_C$ 相反)加到刚体质心，而且需要把惯性力偶(大小与 $I_C\boldsymbol{\alpha}$ 相等，方向与 $I_C\boldsymbol{\alpha}$ 相反)加到刚体。因此，刚体的运动方程可表示为

$$\sum \boldsymbol{F} + \boldsymbol{F}_{\mathrm{I}} = 0,\ \sum \boldsymbol{M}_C(\boldsymbol{F}) + \boldsymbol{M}_{\mathrm{I}C} = 0 \tag{10.33}$$

式中，$\boldsymbol{F}_{\mathrm{I}} = -m\boldsymbol{a}_C$ 和 $\boldsymbol{M}_{\mathrm{I}C} = -\boldsymbol{I}_C\boldsymbol{\alpha}$ 分别称为惯性力和惯性力偶。

虽然所考虑的刚体并不平衡，但是平衡方程可用于分析刚体的运动，只需在基于达朗贝尔原理的惯性力法中把惯性力和惯性力偶加到运动刚体。

对具有对称面的刚体，当受对称面内的平面力 $\boldsymbol{F}_1, \boldsymbol{F}_2, \boldsymbol{F}_3, \cdots$ 作用而发生对称面内的平面运动时，刚体的运动方程可表示为

$$\sum \boldsymbol{F} + \boldsymbol{F}_{\mathrm{I}} = 0,\ \sum \boldsymbol{M}_C(\boldsymbol{F}) + M_{\mathrm{I}C} = 0 \tag{10.34}$$

把上述方程分解为直角分量，得

$$\sum F_x + (F_{\mathrm{I}})_x = 0,\ \sum F_y + (F_{\mathrm{I}})_y = 0,\ \sum M_C(\boldsymbol{F}) + M_{\mathrm{I}C} = 0 \tag{10.35}$$

式中，$(F_{\mathrm{I}})_x = m(a_C)_x$、$(F_{\mathrm{I}})_y = m(a_C)_y$ 和 $M_{\mathrm{I}C} = I_C\alpha$。

惯性力法的优点是可以把刚体动力学问题转化为等效的平衡问题，因而可以对任何轴取矩，而不限于形心轴。因此，上述方程也可表示为

$$\sum F_x + (F_{\mathrm{I}})_x = 0,\ \sum F_y + (F_{\mathrm{I}})_y = 0,\ \sum M_A(F) + M_A(F_{\mathrm{I}}) + M_{\mathrm{I}C} = 0 \tag{10.36}$$

式中，A 为平面运动刚体上的任意一点。

上述方程的两种替代形式分别为

$$\sum F_x + (\boldsymbol{F}_{\mathrm{I}})_x = 0,\ \sum M_A(\boldsymbol{F}) + M_A(\boldsymbol{F}_{\mathrm{I}}) + M_{\mathrm{I}C} = 0,\ \sum M_B(\boldsymbol{F}) + M_B(\boldsymbol{F}_{\mathrm{I}}) + M_{\mathrm{I}C} = 0$$

(其中 A 和 B 连线不能垂直 x 轴)　(10.37)

$$\sum M_A(\boldsymbol{F}) + M_A(\boldsymbol{F}_{\mathrm{I}}) + M_{\mathrm{I}C} = 0$$

$$\sum M_B(\boldsymbol{F}) + M_B(\boldsymbol{F}_{\mathrm{I}}) + M_{\mathrm{I}C} = 0$$

$$\sum M_C(\boldsymbol{F}) + M_C(\boldsymbol{F}_{\mathrm{I}}) + M_{\mathrm{I}C} = 0$$

(其中 A、B 和 C 三点不能共线)　(10.38)

例 10.4

长度为 l、质量为 m 的均质杆 AB 支撑如图 E10.4(a)所示。如果 B 处绳子突然断裂，求：(a)杆 AB 的角加速度；(b)支座 A 处的反力。

解

取 AB 为自由体，画受力图，如图 E10.4(b)，则有

$$\sum F_x + (F_{\mathrm{I}})_x = 0,\ R_x + m(a_C)_x = 0$$

$$\sum F_y + (F_{\mathrm{I}})_y = 0,\ R_y - mg + m(a_C)_y = 0$$

$$\sum M_A(\boldsymbol{F}) + M_A(\boldsymbol{F}_{\mathrm{I}}) + M_{\mathrm{I}C} = 0,\ -mg\left(\frac{1}{2}l\right) + m(a_C)_y\left(\frac{1}{2}l\right) + I_C\alpha = 0$$

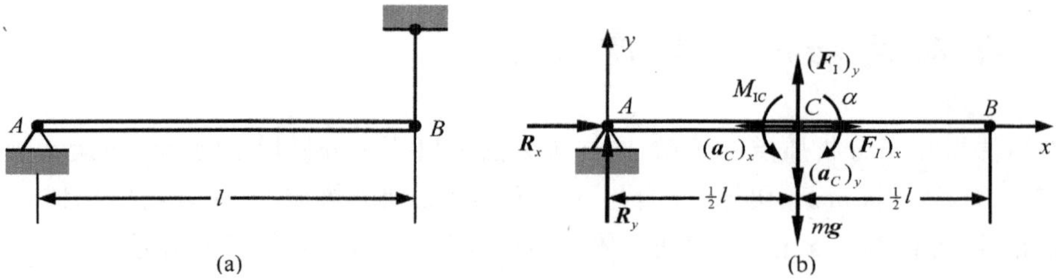

图 E10.4

利用$(a_C)_x=AC\cdot\omega^2=0$、$(a_C)_y=AC\cdot\alpha=\dfrac{1}{2}l\alpha$ 和 $I_C=\dfrac{1}{12}ml^2$，解方程得

$$R_x=0,\ R_y=\frac{1}{4}mg,\ \alpha=\frac{3g}{2l}$$

10.6 平面运动刚体的功能法

1. 作用于刚体上的力和力偶的功

作用于刚体上的外力 $\boldsymbol{F}$ 在力的作用点从 A_1 移到 A_2 期间所做的功可表示为

$$W_{12}=\int_{A_1}^{A_2}\boldsymbol{F}\cdot\mathrm{d}\boldsymbol{r} \tag{10.39}$$

同理，作用于刚体上的外力偶矩 $\boldsymbol{M}$ 在力偶从 θ_1 转到 θ_2 期间所做的功可表示为

$$W_{12}=\int_{\theta_1}^{\theta_2}M\mathrm{d}\theta \tag{10.40}$$

2. 平面运动刚体的动能

刚体的动能定义为刚体中所有质点的动能之和，因而有

$$T=\frac{1}{2}\sum_{i=1}^{n}m_iv_i^2 \tag{10.41}$$

当计算刚体动能时，可以分别考虑刚体质心的运动和刚体相对平移形心参考系的运动。利用$\boldsymbol{v}_i=\boldsymbol{v}_C+\boldsymbol{v}'_i$，则式(10.41)可表示为

$$T=\frac{1}{2}\sum_{i=1}^{n}m_i(\boldsymbol{v}_i\cdot\boldsymbol{v}_i)=\frac{1}{2}\sum_{i=1}^{n}m_iv_C^2+\boldsymbol{v}_C\cdot\left(\sum_{i=1}^{n}m_i\,\boldsymbol{v}'_i\right)+\frac{1}{2}\sum_{i=1}^{n}m_iv'^2_i \tag{10.42}$$

利用 $\sum\limits_{i=1}^{n}m_i\,\boldsymbol{v}'_i=m\,\boldsymbol{v}'_C=0$，有

$$T=\frac{1}{2}mv_C^2+\frac{1}{2}\sum_{i=1}^{n}m_iv'^2_i \tag{10.43}$$

代入 $v'_i=r'_i\omega$，得

$$T=\frac{1}{2}mv_C{}^2+\frac{1}{2}\left\{\sum_{i=1}^{n}[m_i\,(r'_i)^2]\right\}\omega^2=\frac{1}{2}mv_C{}^2+\frac{1}{2}I_C\omega^2 \tag{10.44}$$

式中，$I_C=\sum\limits_{i=1}^{n}[m_i\,(r'_i)^2]$ 为刚体转动惯量。

式(10.44)表明，平面运动刚体的动能等于刚体随质心的平移动能和刚体相对平移形心参考系的转动动能之和。

3. 平面运动刚体的功能原理

质点的功能原理可以应用于刚体上的每一个质点。相加所有质点的动能，并考虑所有力的功，即可得到平面运动刚体的功能原理

$$dT = dW \tag{10.45}$$

式中，dT 和 dW 是刚体动能的增量和作用于刚体上的所有力所做的元功。积分，得

$$T_2 - T_1 = W_{12} \text{ 或 } T_1 + W_{12} = T_2 \tag{10.46}$$

式中，T_1 和 T_2 为刚体的初动能和末动能，W_{12} 为刚体上所有力的功。

4. 能量守恒

如果作用于刚体上的所有力都是保守力，则功能原理可替换为

$$T_1 + V_1 = T_2 + V_2 \tag{10.47}$$

式中，V 为刚体的势能。该式即为刚体的机械能守恒原理。

例 10.5

长度为 l、质量为 m 的均质杆可绕点 O 转动，如图 E10.5(a)所示。杆从水平位置由静止状态开始释放而发生自由摆动。求：(a)当杆通过垂直位置时杆具有最大角速度的距离 d；(b)相应的角速度值和 O 处反力值。

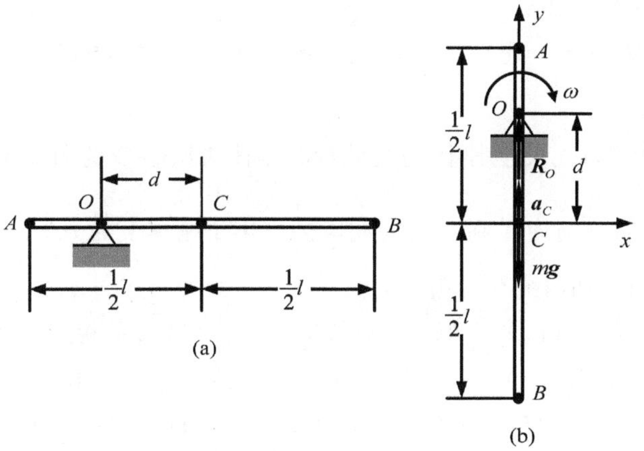

图 E10.5

解

在位置 1，如图 E10.5(a)所示，有

$$T_1 = V_1 = 0$$

在位置 2，如图 E10.5(b)所示，有

$$T_2 = \frac{1}{2}I_O\omega^2 = \frac{1}{2}(I_C + md^2)\omega^2 = \frac{1}{2}\left(\frac{1}{12}ml^2 + md^2\right)\omega^2, V_2 = -mgd$$

利用 $T_1 + V_1 = T_2 + V_2$，得

$$\omega = \sqrt{\frac{2g}{d + \frac{1}{12}l^2/d}}$$

显然,当 $d=\frac{1}{12}l^2/d$,即 $d=\frac{\sqrt{3}}{6}l$ 时,角速度 ω 将有最大值,即

$$\omega_{\max}=\sqrt{\frac{2g}{2\sqrt{d(\frac{1}{12}l^2/d)}}}=\sqrt[4]{12}\sqrt{\frac{g}{l}}=1.86\sqrt{\frac{g}{l}}$$

利用在位置 2 处,$ma_C=\sum F_y=R_O-mg$ 和 $a_C=d(\omega_{max})^2=g$,如图 E10.5(b),有

$$R_O=2mg$$

10.7 平面运动刚体的冲量动量法

1. 平面运动刚体的冲量动量原理

考虑质量为 m 的刚体在平面力 $\boldsymbol{F}_1$, $\boldsymbol{F}_2$, $\boldsymbol{F}_3$, …作用下做平面运动,从 t_1 到 t_2 对式(10.12)和(10.25)进行积分,有

$$\mathrm{d}\boldsymbol{L}=\mathrm{d}\boldsymbol{I},\ \mathrm{d}\boldsymbol{H}_C=\mathrm{d}\boldsymbol{G}_C \tag{10.48}$$

式中,$\mathrm{d}\boldsymbol{L}=\mathrm{d}(m\boldsymbol{v}_C)$ 和 $\mathrm{d}\boldsymbol{H}_C=\mathrm{d}(\boldsymbol{I}_C\boldsymbol{\omega})$ 分别为刚体的线动量增量和刚体绕质心 C 的角动量增量,$\mathrm{d}\boldsymbol{I}=\sum\boldsymbol{F}\mathrm{d}t$ 和 $\mathrm{d}\boldsymbol{G}_C=\sum\boldsymbol{M}_C(\boldsymbol{F})\mathrm{d}t$ 分别为作用于刚体上力在时间间隔 $\mathrm{d}t$ 内的线冲量和绕质心 C 的角冲量。从 t_1 到 t_2 积分,

$$\boldsymbol{L}_2-\boldsymbol{L}_1=\boldsymbol{I}_{12},\ (\boldsymbol{H}_C)_2-(\boldsymbol{H}_C)_1=(\boldsymbol{G}_C)_{12},\ \text{or}\ \boldsymbol{L}_1+\boldsymbol{I}_{12}=\boldsymbol{L}_2,\ (\boldsymbol{H}_C)_1+(\boldsymbol{G}_C)_{12}=(\boldsymbol{H}_C)_2 \tag{10.49}$$

式中,$\boldsymbol{L}=m\boldsymbol{v}_C$ 和 $\boldsymbol{H}_C=I_C\boldsymbol{\omega}$ 分别为刚体的线动量和刚体对质心 C 的角动量,$\boldsymbol{I}_{12}=\sum\int_{t_1}^{t_2}\boldsymbol{F}\mathrm{d}t$ 和 $\boldsymbol{G}_{12}=\sum\int_{t_1}^{t_2}\boldsymbol{M}_C(\boldsymbol{F})\mathrm{d}t$ 分别为 t_1 到 t_2 时间间隔内作用于刚体上的力的线冲量和力对质心 C 的角冲量。分解式(10.48)和(10.49),得

$$\left.\begin{array}{lll}(\mathrm{d}L)_x=(\mathrm{d}I)_x & (L_2)_x-(L_1)_x=(I_{12})_x & (L_1)_x+(\mathrm{I}_{12})_x=(L_2)_x\\ (\mathrm{d}L)_y=(\mathrm{d}I)_y、 & (L_2)_y-(L_1)_y=(I_{12})_y & \text{或}(L_1)_y+(\mathrm{I}_{12})_y=(L_2)_y\\ \mathrm{d}H_C=\mathrm{d}G_C & (H_C)_2-(H_C)_1=(G_C)_{12} & (H_C)_1+(G_C)_{12}=(H_C)_2\end{array}\right\} \tag{10.50}$$

例 10.6

半径为 r、重量为 W 的均质圆柱,具有逆时针初始角速度 ω_0,放在由地板和墙壁形成的墙角,如图 E10.6(a)所示。设 A 和 B 处的动摩擦系数均为 μ_k,试求圆柱停止转动时所需时间。

解

取圆柱为自由体,画受力图,如图 E10.6(b)所示,则有

$$(L_1)_x+(I_{12})_x=(L_2)_x,\quad 0+(N_A-F_B)t=0$$
$$(L_1)_y+(I_{12})_y=(L_2)_y,\quad 0+(N_B+F_A-W)t=0$$
$$(H_C)_1+(G_C)_{12}=(H_C)_2,\quad I_C\omega_0-(F_A+F_B)rt=0$$

代入 $F_A=\mu_k N_A$、$F_B=\mu_k N_B$ 和 $I_C=\frac{1}{2}mr^2$,解方程,得

$$t=\frac{(1+\mu_k^2)I_C\omega_0}{\mu_k(1+\mu_k)Wr}=\frac{(1+\mu_k^2)r\omega_0}{2\mu_k(1+\mu_k)g}$$

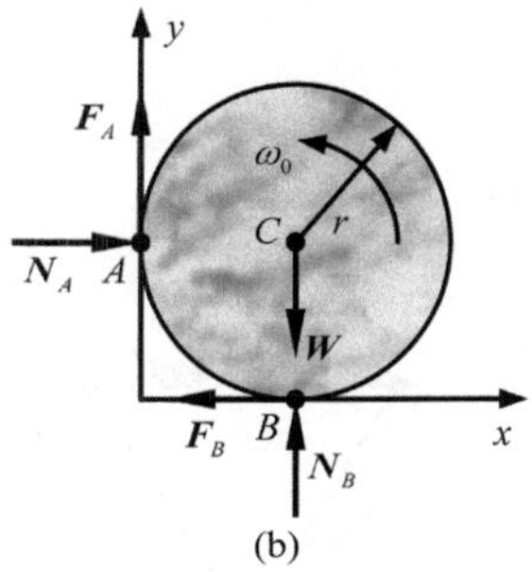

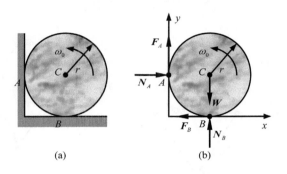

图 E10.6

式中,g 为重力加速度。

2. 动量守恒

如果没有外力作用于刚体,则式(10.49)化简为

$$\boldsymbol{L}_1 = \boldsymbol{L}_2,\ (\boldsymbol{H}_C)_1 = (\boldsymbol{H}_C)_2 \tag{10.51}$$

该式表明,如果没有外力作用于刚体,则刚体的线动量和刚体对质心 C 的角动量守恒。

习 题

10.1 图 P10.1 所示,系统由质点 A 和 B 组成。已知质量 $m_A=1$ kg 和 $m_B=2$ kg,速度 $\boldsymbol{v}_A=2\boldsymbol{i}+3\boldsymbol{j}$ m/s和$\boldsymbol{v}_B=-2\boldsymbol{i}+3\boldsymbol{j}$ m/s,求系统对 O 的角动量$\boldsymbol{H}_O$ 和系统对质心 C 的角动量$\boldsymbol{H}_C$。

10.2 如图 P10.2 所示,质量 $m=20$ kg 的均质橱柜安装在小脚轮上可以在地板上自由移动。已知 $F=100$ N、$b=500$ mm 和 $h_C=800$ mm,求:(a)橱柜加速度;(b)保证橱柜不发生翻倒的 h 值范围。

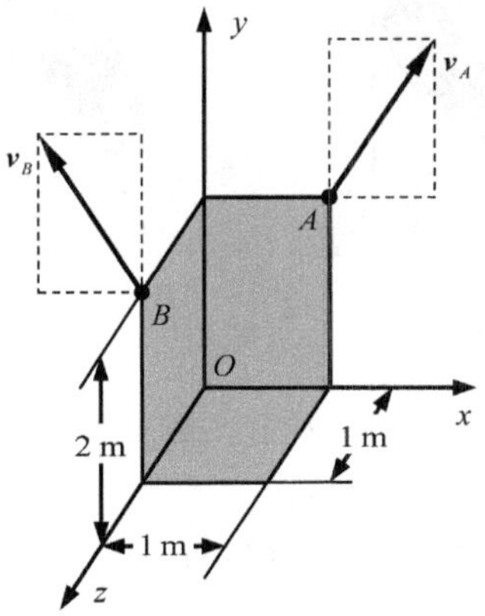

图 P10.1

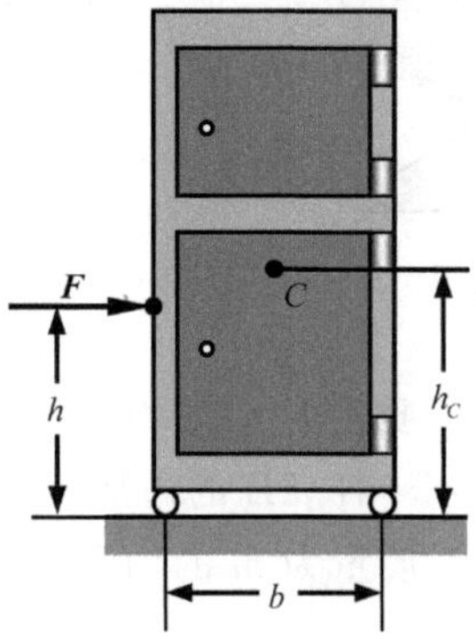

图 P10.2

10.3 质量为 2 kg 的均质细杆 AB 由长度相同的三根绳子支撑在图 P10.3 所示位置。求绳子 BE 突然剪断后的瞬间:(a)杆 AB 的加速度;(b)每一绳子的拉力。

10.4 质量为 5 kg 的均质圆板由长度相同的三根绳子悬挂在图 P10.4 所示位置。已知 CA 和 CB 分别水平和垂直,求绳子 BF 突然剪断后的瞬间:(a)板的加速度;(b)每一绳子的拉力。

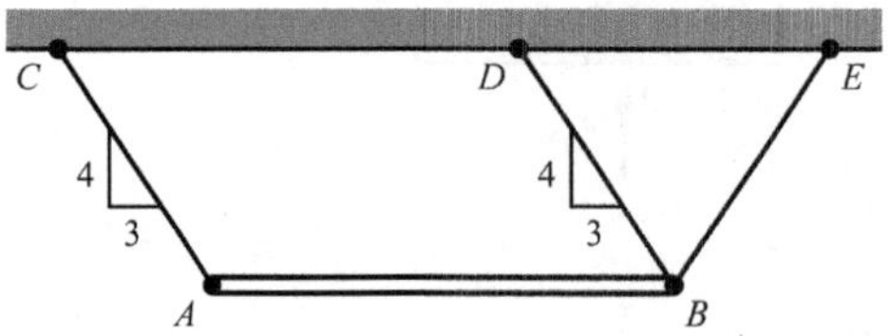

图 P10.3

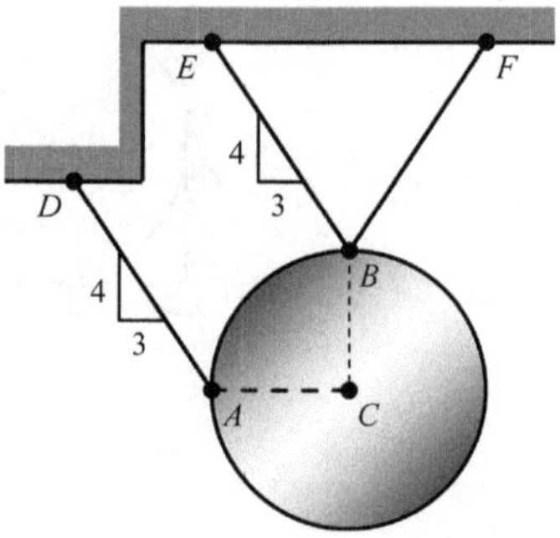

图 P10.4

10.5 如图 P10.5 所示,质量 $m=8$ kg、半径 $r=0.15$ m 的均质圆柱沿倾角 $\theta=30°$的斜面向下只滚不滑。求摩擦力和质心加速度。

10.6 图 P10.6 所示,飞轮的半径为 500 mm,质量为 150 kg,回转半径为 400 mm。质量为 20 kg 的物块 A 与绕在飞轮上的绳子相连,系统由静止释放。不计摩擦,求:(a)物块 A 的加速度;(b)物块 A 移动 2 m 后的速度。

10.7 半径为 0.4 m 的鼓轮与半径为 0.3 m 的圆盘相连,如图 P10.7 所示。圆盘和鼓轮具有组合质量 5 kg 和组合回转半径 0.25 m,通过两根绳子悬挂在图示位置。已知 $T_A=50$ N 和 $T_B=30$ N,求绳上点 A 和 B 的加速度。

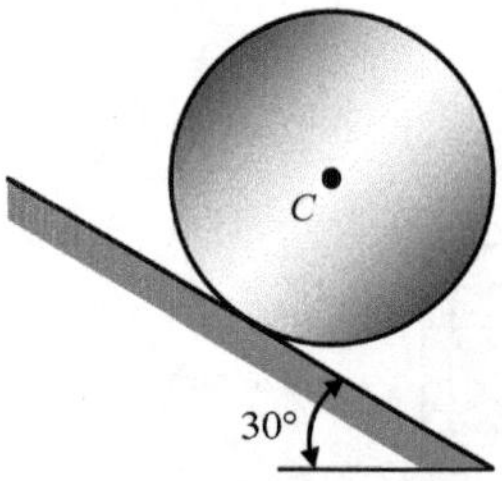

图 P10.5

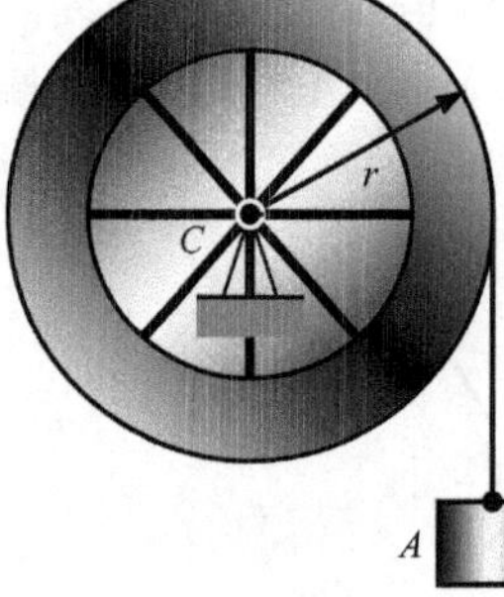

图 P10.6

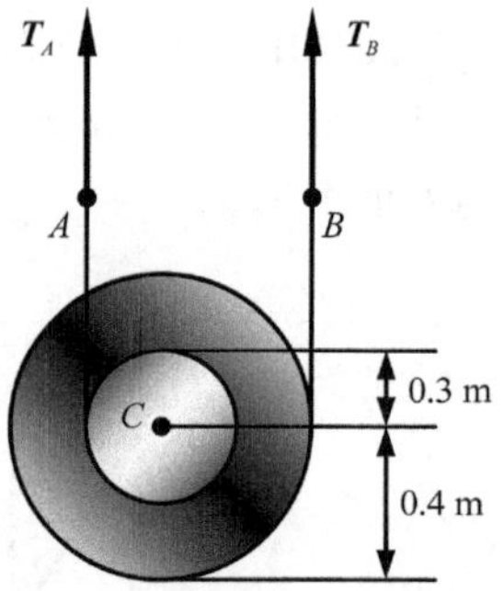

图 P10.7

10.8 长度为 l、质量为 m 的均质杆 AB 通过图 P10.8 所示弹簧支撑。如果 B 处弹簧突然断裂,求断裂瞬时:(a)杆 AB 的角加速度;(b)A 点的加速度。

10.9 长度为 l、质量为 m 的均质杆可绕点 O 转动,如图 P10.9 所示。杆从水平位置由静止状态开始释放而发生自由摆动。已知 $d=\frac{3}{8}l$,求当杆旋转 90°时杆的角速度和 O 处反力。

10.10 如图 P10.10 所示,半径为 r、重量为 W 的均质圆柱,具有逆时针初始角速度 ω_0,放在由粗糙地板和光滑墙壁形成的墙角。设圆柱和地板之间的动摩擦系数均为 μ_k,试求圆柱停止转动时所需时间。

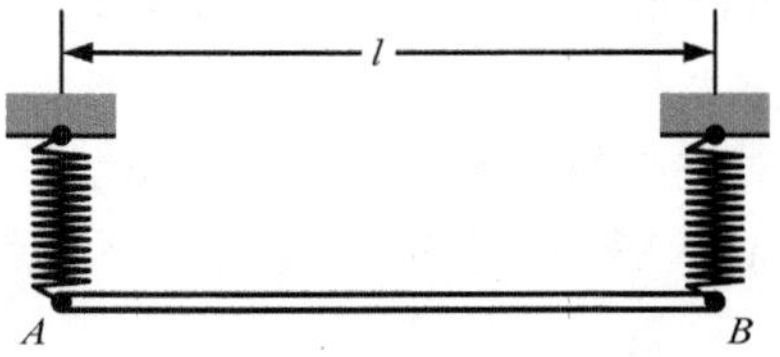

图 P10.8

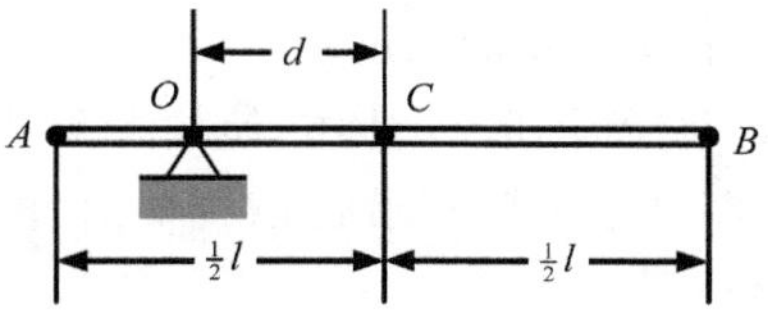

图 P10.9

10.11 质量 $m=25$ kg、半径 $r=0.2$ m 的均质圆柱滚轮，初始静止，受 $F=120$ N 的力作用，如图 P10.11 所示。已知滚轮只滚不滑，求：(a)滚轮运动 1.5 m 时的质心速度；(b)不发生滑动所需要的摩擦力。

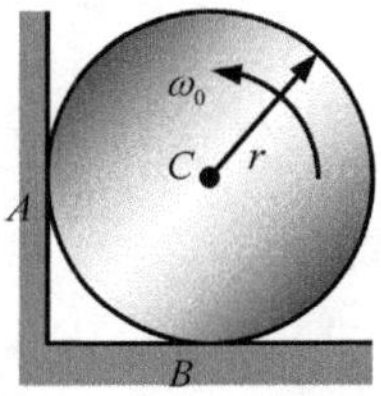

图 P10.10

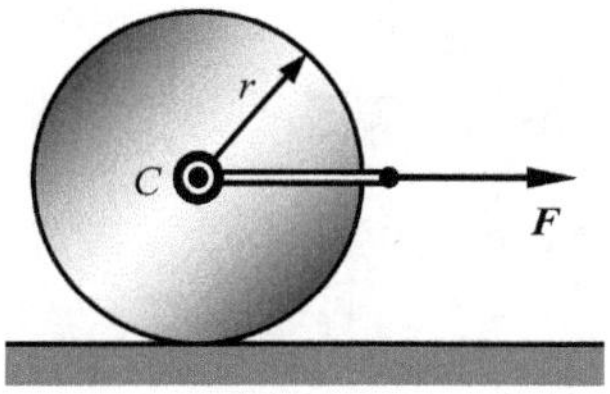

图 P10.11

10.12 如图 P10.12 所示，绳子绕在半径为 r、质量为 m 的均质圆柱上。已知圆柱由静止释放，求圆柱向下运动距离 s 后的圆柱质心的速度。

10.13 如图 P10.13 所示，质量为 10 kg 的支架，受力 $F=30$ N 作用，由两个只滚不滑的均质圆盘支撑。每个圆盘的质量 $m=5$ kg、半径 $r=0.1$ m。已知系统初始静止，求支架运动 0.5 m后的速度。

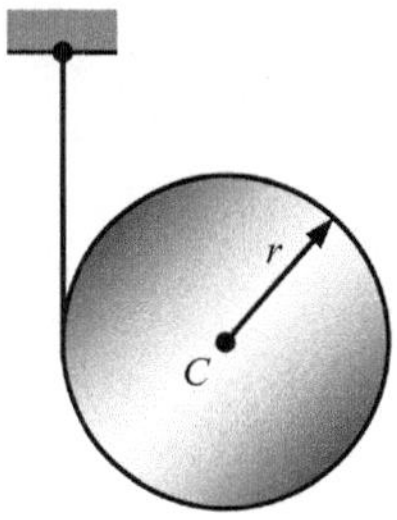

图 P10.12

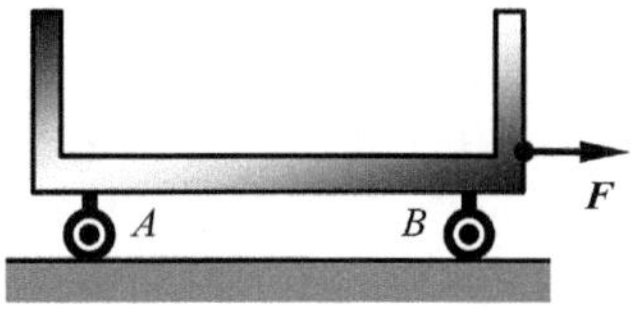

图 P10.13

10.14 如图 P10.14 所示，半径为 r、质量为 m 的均质圆柱，初始线速度为零，逆时针初始角速度为 ω_0，放在水平地板上。设圆柱和地板之间的动摩擦系数均为 μ_k，试求：(a)圆柱开始只滚不滑的时间；(b)圆柱开始只滚不滑时的线速度和角速度。

10.15 物块 A 和 B 通过绕在圆盘 O 表面的绳子 AB 连接，如图 P10.15 所示，。假设绳子和圆盘之间没有相对滑动，圆盘和轴承之间没有摩擦。已知图示瞬时，物块 B 以 0.4 m/s 的速度向下移动，弹簧压缩量为 0.2 m，求物块 B 下降 0.5 m 后的速度。

10.16 物块 A 和 B 通过绕在圆盘 O 表面的绳子 AB 连接，如图 P10.16 所示。假设绳子和圆盘之间没有相对滑动,圆盘和轴承之间没有摩擦。如果物块 A 在 6 s 内速度从 4 m/s 变为 8 m/s,求物块 B 的质量 m_B。

10.17 如图 P10.17 所示,质量为 75 kg 轮子相对质心的回转半径为 0.9 m。不计定滑轮质量和摩擦。已知轮子最初以逆时针角速度 10 rad/s 滚动,求轮子到达以顺时针角速度 6 rad/s滚动所需时间。

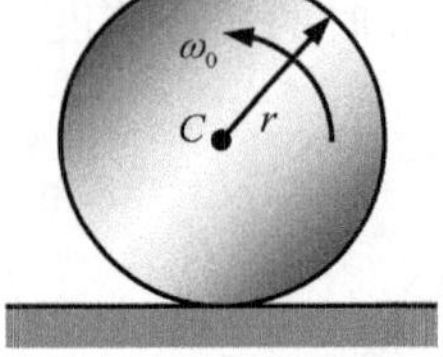

图 P10.14

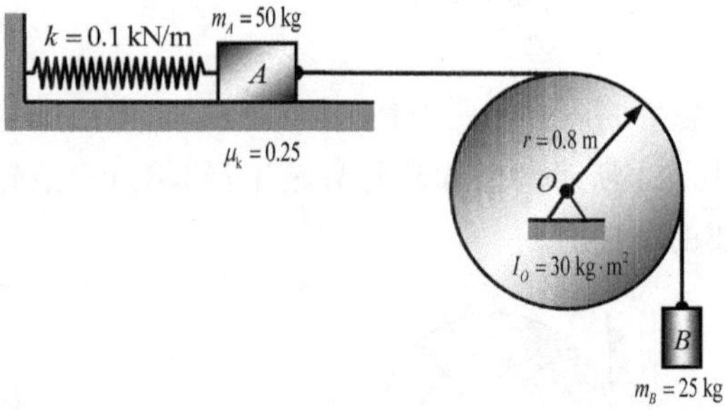

图 P10.15

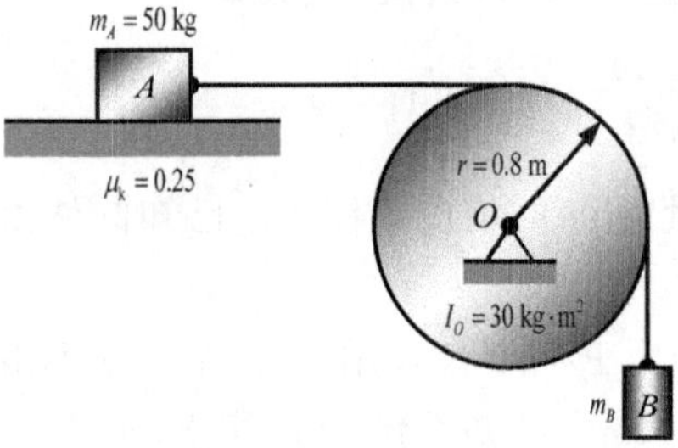

图 P10.16

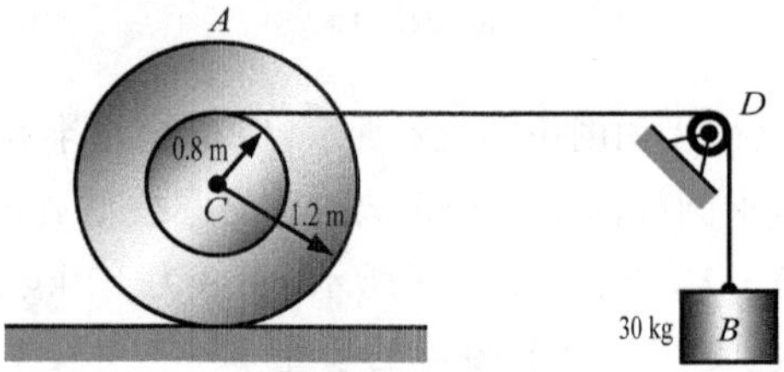

图 P10.17

第 11 章　机械振动

机械振动定义为物体的周期运动，即物体在弹力或重力恢复力作用下围绕平衡位置的往复运动。

当物体偏离其平衡位置时机械振动将会产生，因为物体在恢复力作用下具有回到平衡位置的趋势。然而，通常物体在到达平衡位置时都具有一定的速度，从而会使物体越过平衡位置。这种过程可以无限重复，从而物体能够保持围绕平衡位置的往复运动。

当运动仅由恢复力来维持，这种振动称为自由振动。当外部周期力施加于物体，这种振动称为受迫振动。当摩擦效应可以忽略不计时，这种振动称为无阻尼振动。然而，所有振动实际上都是有阻尼振动，因为摩擦力总会存在。

11.1　无阻尼自由振动

1. 质点无阻尼自由振动

质量为 m 的物体与刚度为 k 的弹簧相连。当物体位于如图 11.1(a)所示的静平衡位置时，作用于物体的力包括重力和弹力。弹力大小为 $k\delta_{st}$，其中 δ_{st} 为平衡位置所对应的弹簧伸长。因此，有 $mg=k\delta_{st}$。

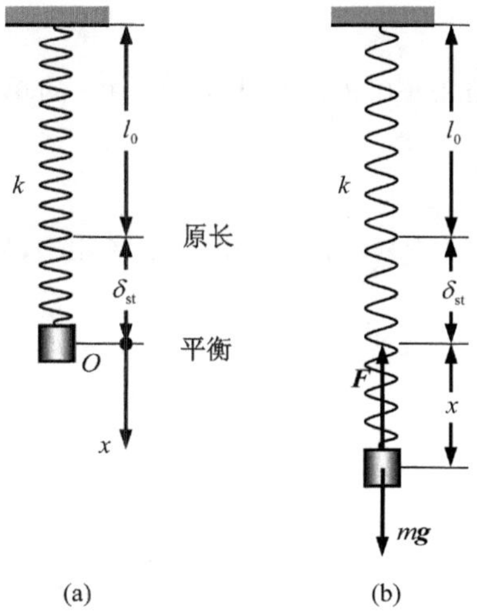

图 11.1

假设物体偏离平衡位置 O 任意位移 x，如图 11.1(b)所示所示，那么作用于物体的力有重力和大小为 $F=k(\delta_{st}+x)$的弹力。利用 $ma_x = \sum F_x$，可得

$$m\ddot{x} = mg - k(\delta_{st} + x) \tag{11.1}$$

利用 $mg=k\delta_{st}$，式(11.1)可重写为

$$m\ddot{x} + kx = 0 \tag{11.2}$$

该式是齐次二阶微分方程。由式(11.2)定义的运动称为简谐振动。式(11.2)的通解可表示为

$$x = C_1\cos\omega_n t + C_2\sin\omega_n t \tag{11.3}$$

式中,$\omega_n=\sqrt{k/m}$称为振动的固有频率或圆频率,单位为 rad/s;C_1 和 C_2 是常数,由运动的初始条件确定。式(11.2)的通解也可写为更紧凑形式,即

$$x = A\sin(\omega_n t+\alpha) \tag{11.4}$$

式中,A 称为振幅,即物体离开平衡位置的最大位移,α 称为初始相位。

式(11.4)对时间微分,速度和加速度可分别表示为

$$\begin{aligned} v &= \dot{x} = \omega_n A\cos(\omega_n t+\alpha) \\ a &= \ddot{x} = -\omega_n^2 A\sin(\omega_n t+\alpha) \end{aligned} \tag{11.5}$$

速度和加速度的最大值分别为

$$\begin{aligned} v_{\max} &= \omega_n A \\ a_{\max} &= \omega_n^2 A \end{aligned} \tag{11.6}$$

物体完成完整运动循环所需要的时间间隔称为振动周期,由 T_n 表示。由式(11.4)可以看出,相位增加 2π 弧度则完成一个完整循环。因此,周期可表示为

$$T_n = \frac{2\pi}{\omega_n} \tag{11.7}$$

单位时间内完成的循环次数称为振动频率,由 f_n 表示。频率可写为

$$f_n = \frac{\omega_n}{2\pi} \tag{11.8}$$

在国际单位制中,频率单位是 Hz。

例 11.1

质点运动由方程 $x=0.4\sin 2t+0.3\cos 2t$ 表示,其中 x 的单位为 m、t 的单位为 s。求(a)振幅、(b)周期、(c)初始相位。

解

利用三角恒等式,即 $\sin(A+B)=\sin A\cos B+\cos A\sin B$,则方程 $x=0.4\sin 2t+0.3\cos 2t$ 可重写为

$$x = 0.5\sin[2t+\arctan(3/4)]$$

(a) 振幅

$$A = 0.5\ \text{m}$$

(b) 周期

$$T_n = \frac{2\pi}{\omega_n} = \frac{2\pi}{2} = 3.14\ \text{s}$$

(c) 初始相位

$$\alpha = \arctan(3/4) = 0.75\ \text{rad} = 36.9^\circ$$

例 11.2

握住质量为 5 kg 的物块使所连接的图示垂直弹簧不发生变形。已知弹簧刚度为 3 kN/m,物块由静止突然释放,求(a)运动的振幅和频率,(b)物块的最大速度和最大加速度。

解

(a)振幅

$$A=\delta_{st}=\frac{mg}{k}=16.4\ \text{mm}$$

频率

$$f_n=\frac{\omega_n}{2\pi}=\frac{1}{2\pi}\sqrt{\frac{k}{m}}=3.90\ \text{Hz}$$

(b) 最大速度

$$v_{max}=\omega_n A=\sqrt{\frac{k}{m}}\frac{mg}{k}=g\sqrt{\frac{m}{k}}=400\ \text{mm/s}$$

最大加速度

$$a_{max}=\omega_n^2 A=\frac{k}{m}\frac{mg}{k}=g=9.81\ \text{m/s}^2$$

例 11.3

质量为 2 kg 的物块 A 静止放置于质量为 8 kg 的平板 B 的顶部，平板 B 与刚度为 $k=400$ N/m的未伸长弹簧相连，如图 E11.3 所示。平板 B 向右缓慢移动 65 mm 后从静止释放。假设物块 A 在平板上不发生滑动，求(a)运动的振幅和周期，(b)静摩擦系数的最小值。

图 E11.2

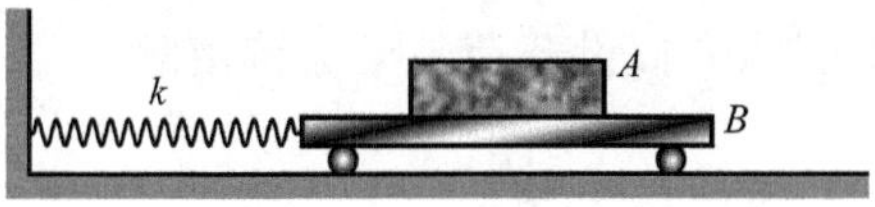

图 E11.3

解

(a) 振幅

$$A=65\ \text{mm}$$

周期

$$T_n=\frac{2\pi}{\omega_n}=2\pi\sqrt{\frac{m_A+m_B}{k}}=0.993\ \text{s}$$

(b) 利用 $a_{max}=\omega_n^2 A=\dfrac{k}{m_A+m_B}A=2.60\ \text{m/s}^2$，得

$$\mu_s=\frac{F_s}{N_A}=\frac{m_A a_{max}}{m_A g}=\frac{a_{max}}{g}=0.265$$

例 11.4

图 E11.4 所示质量为 m 的物块在垂直方向发生振动。已知弹簧刚度分别为 k_1 和 k_2，求振动周期。

解

(a) 弹簧并联。物块位移 δ 时，弹力大小分别为 $F_1=k_1\delta$ 和 $F_2=k_2\delta$，则有

$$F=F_1+F_2=k_1\delta+k_2\delta=(k_1+k_2)\delta$$

因此，等效弹簧刚度 k_{eq} 可表示为

$$k_{eq}=\frac{F}{\delta}=k_1+k_2$$

利用 $T_n = 2\pi/\omega_n$ 和 $\omega_n = \sqrt{k_{eq}/m}$,那么振动频率等于

$$T_n = \frac{2\pi}{\omega_n} = 2\pi\sqrt{\frac{m}{k_{eq}}} = 2\pi\sqrt{\frac{m}{k_1 + k_2}}$$

(b) 弹簧串联。物块受力 F 时,弹簧变形分别为 $\delta_1 = F/k_1$ 和 $F_2 = k_2\delta$,则有

$$\delta = \delta_1 + \delta_2 = \frac{F}{k_1} + \frac{F}{k_2} = F\left(\frac{1}{k_1} + \frac{1}{k_2}\right)$$

因此,等效弹簧刚度 k_{eq} 等于

$$k_{eq} = \frac{F}{\delta} = \frac{k_1 k_2}{k_1 + k_2}$$

及振动频率可表示为

$$T_n = \frac{2\pi}{\omega_n} = 2\pi\sqrt{\frac{m}{k_{eq}}} = 2\pi\sqrt{\frac{m(k_1 + k_2)}{k_1 k_2}}$$

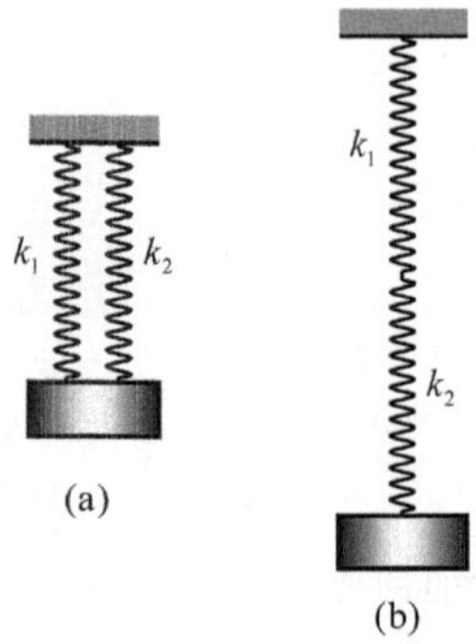

图 E11.4

例 11.5

根据材料力学,对等截面简支梁,中点作用静载荷 F 后所产生的中点挠度为 $w_C = Fl^3/48EI$,其中 l 为梁长、E 为弹性模量、I 为横截面惯性矩。已知 $l = 5$ m、$E = 200$ GPa、$I = 2370$ cm^4,求图 E11.5(a)梁的等效弹簧刚度,图(b)250 kg 的物块放置在梁的中点时的振动频率。忽略梁的质量,并假设载荷与梁保持接触。

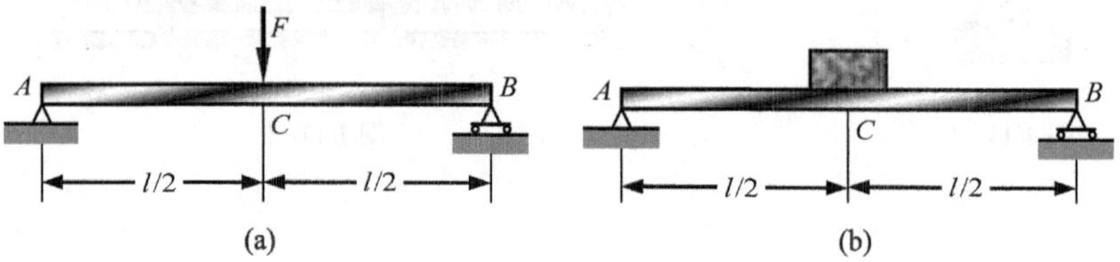

图 E11.5

解

(a) 梁的等效弹簧刚度等于

$$k_{eq} = \frac{48EI}{l^3} = 1820 \text{ kN/m}$$

(b) 振动频率为

$$f_n = \frac{\omega_n}{2\pi} = \frac{1}{2\pi}\sqrt{\frac{k_{eq}}{m}} = 13.58 \text{ Hz}$$

2. 刚体无阻尼自由振动

刚体的振动分析与质点类似。需要选择合适的变量(如线位移 x 或角位移 φ)用以确定物体位置,并写出与该变量及其时间二阶导数有关的方程。如果所得方程具有式(11.2)的形式,即

$$\ddot{x} + \omega_n^2 x = 0 \text{ 或 } \ddot{\varphi} + \omega_n^2 \varphi = 0 \tag{11.9}$$

则所考虑的振动即为简谐振动。上述方法可用于分析严格由简谐振动表示的振动,也可用于分析近似由简谐振动表示的微幅振动。

例如,如图 11.2(a)所示质量为 m、半径为 r 的均质半圆板悬挂于其水平直径中点 O,求微

幅振动固有频率。

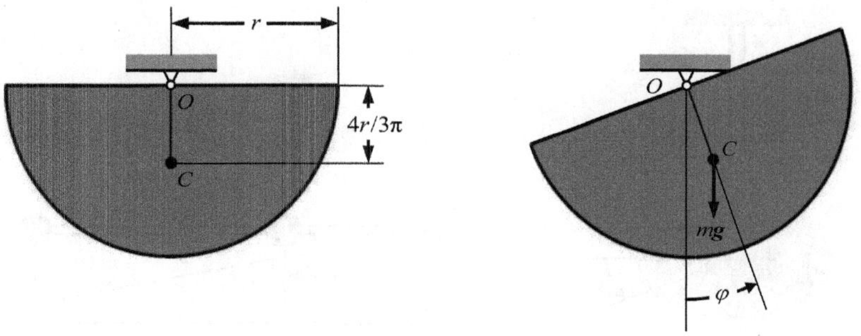

(a)　　　(b)

图 11.2

考虑板处于由 φ 定义的任意位置，其中 φ 是线 OC 与垂直方向的夹角，如图 11.2(b)所示。利用 $I_O\alpha = \sum M_O$，有

$$\left(\frac{1}{2}mr^2\right)\ddot{\varphi} = -(mg)\left(\frac{4r}{3\pi}\sin\varphi\right) \tag{11.10}$$

即

$$\ddot{\varphi} + \frac{8g}{3\pi r}\sin\varphi = 0 \tag{11.11}$$

对微幅振动，$\sin\varphi$ 由 φ(弧度)替换，得

$$\ddot{\varphi} + \omega_n\varphi = 0 \tag{11.12}$$

式中，ω_n 为固有频率，可表示为

$$\omega_n = \sqrt{\frac{8g}{3\pi r}} \tag{11.13}$$

例 11.6

质量为 m、半径为 r 的均质圆盘在中点 O 处与杆件 AO 连接(见图 E11.6)。假设杆件的扭转弹簧刚度为 k_t，求圆盘绕轴 AO 的振动周期。

解

假设圆盘自平衡位置转动任意角度 φ，那么杆件作用于圆盘的力偶矩大小为 $k_t\varphi$。根据 $I_O\alpha = \sum M_O$，有

$$I_O\ddot{\varphi} = -k_t\varphi$$

式中，$I_O = \frac{1}{2}mr^2$ 是圆盘对轴 AO 的转动惯量(或质量惯性矩)。上述方程也可重写为

$$\ddot{\varphi} + \omega_n^2\varphi = 0$$

式中，$\omega_n = \sqrt{k_t/I_O}$。利用 $T_n = 2\pi/\omega_n$，圆盘绕轴 AO 的振动周期可表示为

$$T_n = 2\pi\sqrt{\frac{I_O}{k_t}} = \pi\sqrt{\frac{2mr^2}{k_t}}$$

例 11.7

如图 E11.7 所示，质量均为 m 的小球 B 和 C 与杆件 AB 连接，杆件由销钉 A 和刚度为 k

的弹簧CD支撑。不计杆件质量,并已知杆件水平时系统平衡,求系统微幅振动频率。

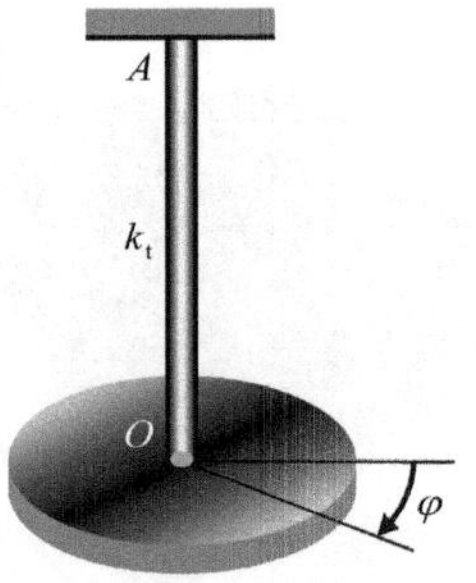

图 E11.6

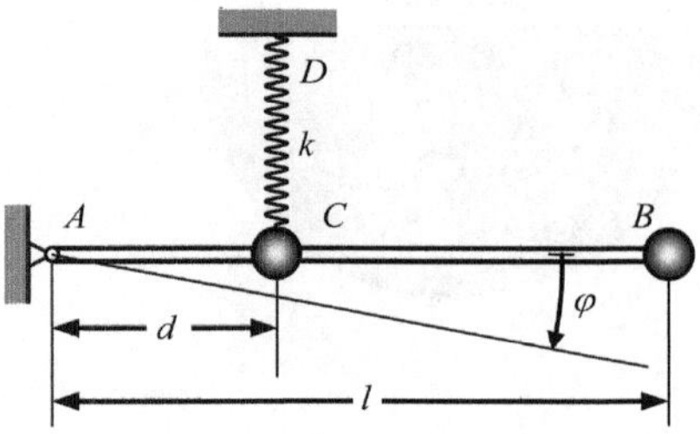

图 E11.7

解

假设杆件自平衡位置转动任意角度φ,那么弹簧作用于杆件的力偶矩大小为$kd^2\varphi$。根据$I_A\alpha = \sum M_A$,有

$$I_A\ddot{\varphi} = -kd^2\varphi$$

式中,$I_A=m(d^2+l^2)$是两球对点A的转动惯量。上述方程可重写为

$$\ddot{\varphi}+\omega_n^2\varphi = 0$$

式中,$\omega_n=\sqrt{kd^2/I_A}$。利用$f_n=\omega_n/2\pi$,得系统微幅振动频率

$$f_n = \frac{1}{2\pi}\sqrt{\frac{kd^2}{I_A}} = \frac{1}{2\pi}\sqrt{\frac{kd^2}{m(d^2+l^2)}}$$

3. 能量法求固有频率

物体的简谐振动由施加于物体的重力或/和弹力恢复力所引起。由于这些恢复力为保守力,因此就有可能利用能量守恒获取振动周期和频率。

例如,如图11.3(a)所示质量为m、半径为r的均质半圆板悬挂于其水平直径中点O,求微幅振动固有频率。

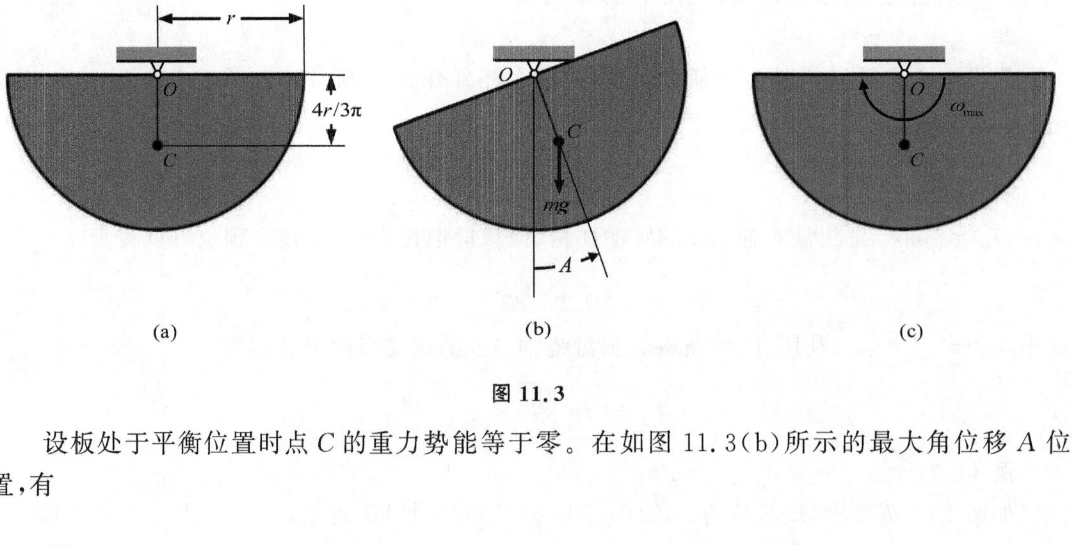

图 11.3

设板处于平衡位置时点C的重力势能等于零。在如图11.3(b)所示的最大角位移A位置,有

$$T_1=0,\ V_1=(mg)\frac{4r}{3\pi}(1-\cos A) \tag{11.14}$$

在如图 11.3(c)所示的最大角速度 $\omega_{\max}$ 位置，有

$$T_2=\frac{1}{2}I_O\omega_{\max}^2,\ V_2=0 \tag{11.15}$$

式中，$I_O=\frac{1}{2}mr^2$。利用机械能守恒，$T_1+V_1=T_2+V_2$，得

$$(mg)\frac{4r}{3\pi}(1-\cos A)=\frac{1}{2}\left(\frac{1}{2}mr^2\right)\omega_{\max}^2 \tag{11.16}$$

利用 $\cos A\approx 1-A^2/2$（微幅振动）和 $\omega_{\max}=\omega_n A$，有

$$(mg)\frac{4r}{3\pi}\frac{A^2}{2}=\frac{1}{2}\left(\frac{1}{2}mr^2\right)(\omega_n A)^2 \tag{11.17}$$

即微幅振动固有频率可表示为

$$\omega_n=\sqrt{\frac{8g}{3\pi r}} \tag{11.18}$$

例 11.8

如图 E11.8 所示，半径为 r、质量为 m 的均质圆盘沿柱面纯滚动（即只滚不滑），圆盘中心与长为 $2a$、质量不计的杆件 ABC 相连。杆件在点 A 与刚度为 k 的弹簧相连，并在垂直平面内绕点 B 自由转动。已知杆件处于垂直位置时弹簧为原长，求在 A 施加微小位移并释放后系统振动的固有频率。

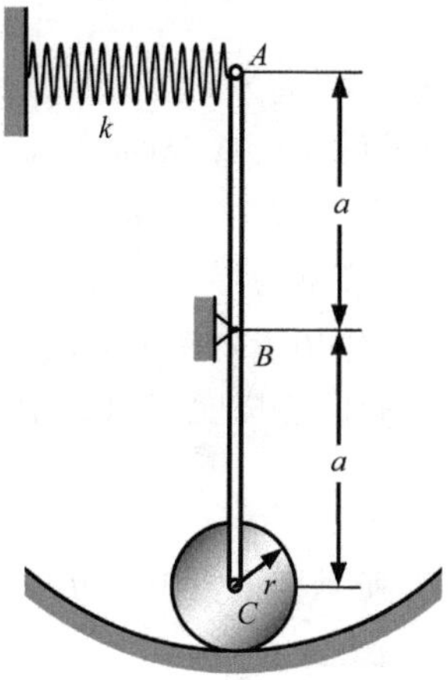

图 E11.8

解

设杆件处于垂直位置时点 C 的重力势能等于零。在杆件最大角位移 A 位置，有

$$T_1=0,\ V_1=\frac{1}{2}k(aA)^2+(mg)a(1-\cos A)$$

在杆件最大角速度 $\omega_{\max}$ 位置，有

$$T_2=\frac{1}{2}\left(\frac{1}{2}mr^2+mr^2\right)\left(\frac{a\omega_{\max}}{r}\right)^2,\ V_2=0$$

根据 $T_1+V_1=T_2+V_2$，得

$$\frac{1}{2}k(aA)^2+(mg)a(1-\cos A)=\frac{1}{2}\left(\frac{1}{2}mr^2+mr^2\right)\left(\frac{a\omega_{\max}}{r}\right)^2$$

利用 $\cos A\approx 1-A^2/2$ 和 $\omega_{\max}=\omega_n A$，有

$$\frac{1}{2}k(aA)^2+(mg)a\frac{A^2}{2}=\frac{1}{2}\left(\frac{1}{2}mr^2+mr^2\right)\left(\frac{a\omega_n A}{r}\right)^2$$

即

$$\omega_n=\sqrt{\frac{2}{3}\left(\frac{k}{m}+\frac{g}{a}\right)}$$

11.2 无阻尼受迫振动

质量为 m 的物体与刚度为 k 的弹簧相连,承受大小为 $F_f=H\sin\omega_f t$ 的周期力作用,其中 H 为施加于物体的周期力的幅值,ω_f 为周期力的圆频率,如图 11.4 所示。

假设物体位于任意位置 x,其中 x 以平衡位置 O 为原点,那么作用于物体的力有大小为 mg 的重力、大小为 $F=k(\delta_{st}+x)$ 的弹力和大小为 $F_f=H\sin\omega_f t$ 的周期力。利用 $ma_x=\sum F_x$,有

$$m\ddot{x}=mg-k(\delta_{st}+x)+H\sin\omega_f t \tag{11.19}$$

利用 $mg=k\delta_{st}$,得

$$m\ddot{x}+kx=H\sin\omega_f t \tag{11.20}$$

该式是非齐次二阶微分方程,其通解可以表示为特解 x_p 加补解 x_c(即对应齐次方程的通解)。因此式(11.20)的通解可表示为

$$x=x_c+x_p=A\sin(\omega_n t+\alpha)+B\sin\omega_f t \tag{11.21}$$

式中

$$B=\frac{H/k}{1-\lambda^2} \tag{11.22}$$

式中,$\lambda=\omega_f/\omega_n$ 为周期力圆频率与无阻尼自由振动固有频率之比。式(11.21)描述了两个叠加振动。补解 $x_c=A\sin(\omega_n t+\alpha)$ 表示无阻尼自由振动。固有频率 ω_n 仅与弹簧刚度 k 和物体质量 m 有关,而常数 A 和 α 则由初始条件确定。无阻尼自由振动是瞬态振动,由于摩擦力作用,它将很快衰减为零。特解 $x_p=B\sin\omega_f t$ 表示由周期力引起并维持的稳态振动。稳态振动的圆频率等于作用于物体上的周期力的圆频率,振幅 B 与频率比 λ 有关。稳态振动的振幅 B 与由力 H 引起的静挠度 H/k 之比称为放大因子,可表示为

$$\beta=\frac{B}{H/k}=\frac{1}{1-\lambda^2} \tag{11.23}$$

图 11.5 所示为放大因子 β 与频率比 λ 之间的关系曲线。

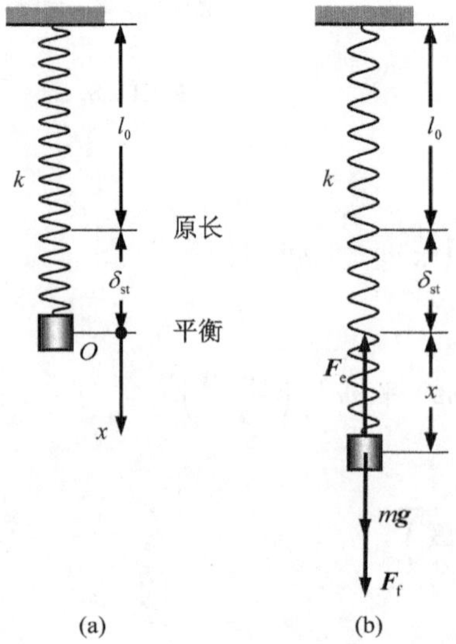

图 11.4

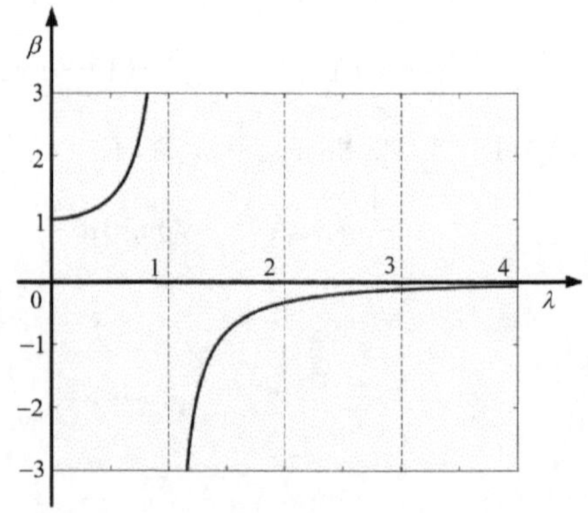

图 11.5

从图 11.5 可以看出，当 $\lambda=1$ 时，受迫振动的振幅将变得无穷大或极大，这种现象称为共振。只要稳态振动的圆频率不要太靠近无阻尼自由振动的固有频率，共振即可避免。还可看出，当 $\lambda<1$ 时 β 为正，而当 $\lambda>1$ 时 β 为负。第一种情形中受迫振动为同相，而第二种情形中受迫振动为 180°反相。

例 11.9

质量为 m 的小球 C 与不计质量的杆件 AB 连接，杆件由 A 处销钉和 D 处刚度为 k 的弹簧支撑，如图 E11.9 所示。系统可在垂直平面内运动，当杆件水平时系统平衡。杆件在 B 处受大小为 $F_f=H\sin\omega_f t$ 的周期力作用，其中 H 是周期力的幅值。已知 $\omega_f=\sqrt{k/m}$，求系统的稳态振动。

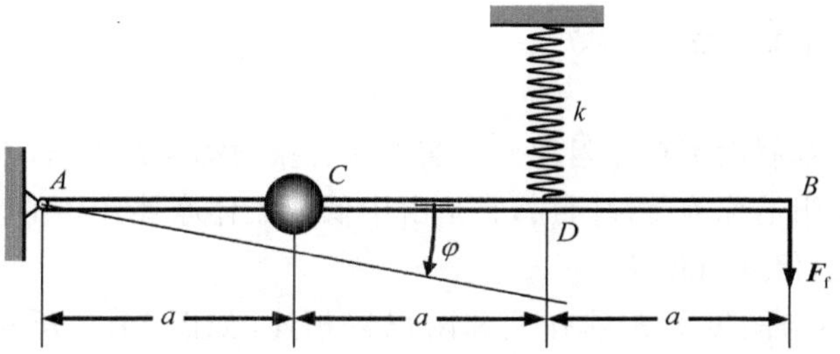

图 E11.9

解

假设杆自平衡位置旋转任意角度 φ，则有

$$(ma^2)\ddot{\varphi}+k\,(2a)^2\varphi=H(3a)\sin\omega_f t$$

利用式(11.22)和 $\omega_n=\sqrt{k\,(2a)^2/ma^2}=2\sqrt{k/m}$，稳态振动的角振幅 B 可表示为

$$B=\frac{H(3a)/k\,(2a)^2}{1-(\omega_f/\omega_n)^2}=\frac{H(3a)/k\,(2a)^2}{1-(1/2)^2}=\frac{H}{ka}$$

因此，系统的稳态振动方程为

$$\varphi_p=B\sin\omega_f t=\frac{H}{ka}\sin\sqrt{\frac{k}{m}}t$$

例 11.10

质量为 m 的小球 C 与不计质量的杆件 AB 连接，杆件由 A 处销钉支撑，并与运动支撑 D 通过刚度为 k 的弹簧相连，如图 E11.10 所示。系统可在垂直平面内运动，当杆件水平时系统平衡。已知支撑 D 的垂直位移为 $\delta_f=\Delta\sin\omega_f t$，其中 $\omega_f=\sqrt{k/m}$，Δ 是周期位移的幅值，求系统的稳态振动。

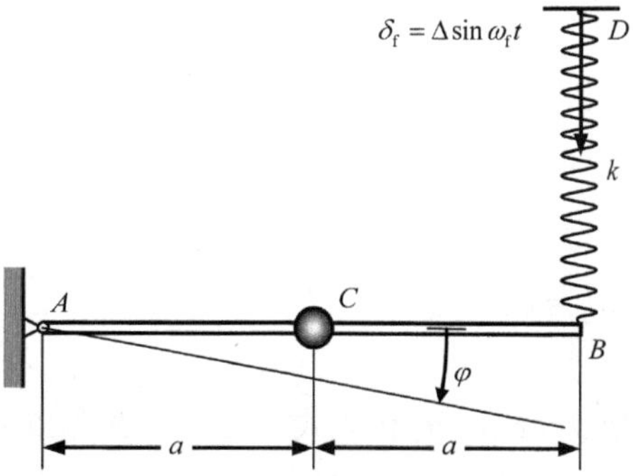

图 E11.10

解

假设杆自平衡位置旋转任意角度 φ。根据 $I_A\alpha=\sum M_A$，得

$$(ma^2)\ddot{\varphi} = -k[(2a)\varphi - \Delta\sin\omega_f t](2a)$$

即

$$ma\ddot{\varphi} + 4ka\varphi = 2k\Delta\sin\omega_f t$$

利用式(11.22)和 $\omega_n = \sqrt{4ka/ma} = 2\sqrt{k/m}$,稳态振动的角振幅$(\varphi_p)_{max}$即 B 等于

$$B = \frac{2k\Delta/4ka}{1-(\omega_f/\omega_n)^2} = \frac{2k\Delta/4ka}{1-(1/2)^2} = \frac{2\Delta}{3a}$$

因此,系统的稳态振动方程为

$$\varphi_p = B\sin\omega_f t = \frac{2\Delta}{3a}\sin\sqrt{\frac{k}{m}}t$$

11.3 有阻尼自由振动

前面考虑的振动都假设不存在阻尼。事实上所有振动都存在阻尼,因为摩擦始终存在。

一类特别重要的阻尼称为粘性阻尼,它是由中低速流体摩擦产生的。粘性阻尼摩擦与运动物体的速度成正比,但方向相反。

质量为 m 的物体与刚度为 k 的弹簧和阻尼器相连,如图 11.6 所示。摩擦力大小等于 $F_d = c\dot{x}$,其中常数 c 称为粘性阻尼系数,单位为 $N \cdot s/m$。物体运动的微分方程可表示为

$$m\ddot{x} = mg - k(\delta_{st} + x) - c\dot{x} \tag{11.24}$$

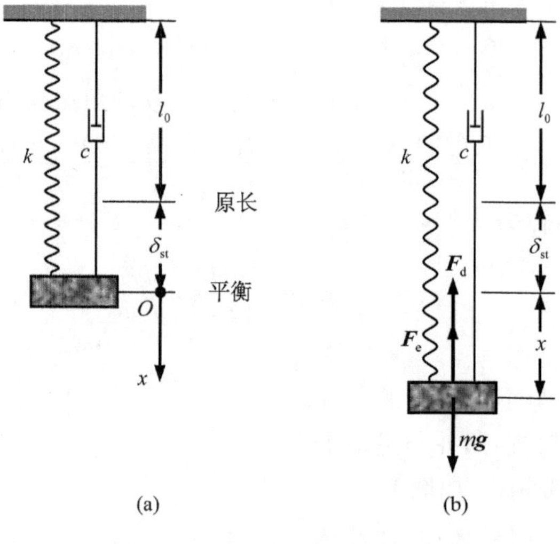

图 11.6

利用 $mg = k\delta_{st}$,得

$$m\ddot{x} + c\dot{x} + kx = 0 \tag{11.25}$$

把 $x = \exp(rt)$ 代入式(11.25),并除以 $\exp(rt)$,可得特征方程

$$mr^2 + cr + k = 0 \tag{11.26}$$

以及方程的根

$$r_{1,2}=-\frac{c}{2m}\pm\sqrt{\left(\frac{c}{2m}\right)^2-\frac{k}{m}} \tag{11.27}$$

把式(11.27)中根号等于零所对应的阻尼系数 c 值定义为临界阻尼系数 c_{cr}，则有

$$c_{cr}=2\sqrt{mk} \tag{11.28}$$

定义 $\eta=c/c_{cr}$ 为阻尼因子，并利用 $\omega_n=\sqrt{k/m}$，则式(11.27)可重写为

$$r_{1,2}=(-\eta\pm\sqrt{\eta^2-1})\omega_n \tag{11.29}$$

根据阻尼因子的数值，对三类不同的阻尼简述如下：

1. 过阻尼($\boldsymbol{\eta}>1$)

式(11.26)的根 r_1 和 r_2 是互不相同的实数，此时式(11.25)的通解可表示为

$$x=\exp(-\eta\omega_n t)[C_1\exp(\sqrt{\eta^2-1}\omega_n t)+C_2\exp(-\sqrt{\eta^2-1}\omega_n t)] \tag{11.30}$$

式中，C_1 和 C_2 是常数。该解对应非振动，因 r_1 和 r_2 均为负值，当 t 增加时 x 接近于零，即物体回到平衡位置。

2. 临界阻尼($\boldsymbol{\eta}=1$)

式(11.26)有重根 $r=-\omega_n$，此时式(11.25)的通解可表示为

$$x=\exp(-\omega_n t)(C_1+C_2 t) \tag{11.31}$$

该解也对应非振动。在临界阻尼情况下，物体以最短时间回到平衡位置。

3. 欠阻尼($\boldsymbol{\eta}<1$)

式(11.26)的根是一对共轭复数，此时式(11.25)的通解可表示为

$$x=\exp(-\eta\omega_n t)(C_1\cos\omega_d t+C_2\sin\omega_d t)=A\exp(-\eta\omega_n t)\sin(\omega_d t+\alpha) \tag{11.32}$$

式中，A 和 α 是常数，由初始条件确定；ω_d 是有阻尼自由振动的圆频率，可表示为

$$\omega_d=\omega_n\sqrt{1-\eta^2} \tag{11.33}$$

由式(11.32)定义的运动具有振动特性，振动的振幅随着每个振动周期在逐渐减小，这是因为运动被限制在两个根指数曲线的范围之内，如图 11.7 所示。由于 $\lambda<1$，因此有阻尼自由振动的周期 $T_d=2\pi/\omega_d$ 大于相应无阻尼自由振动的周期 $T_n=2\pi/\omega_n$。

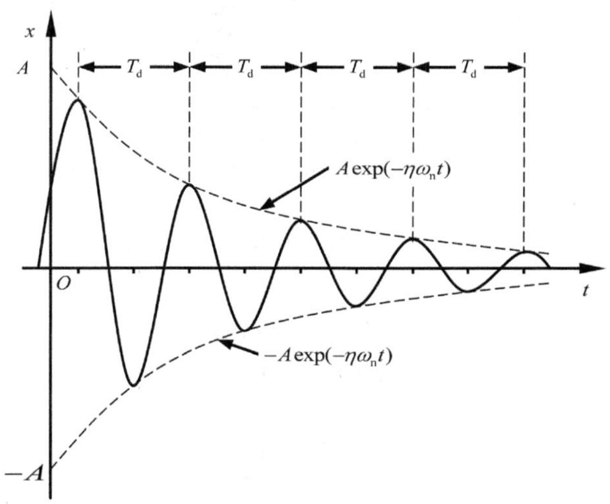

图 11.7

例 11.11

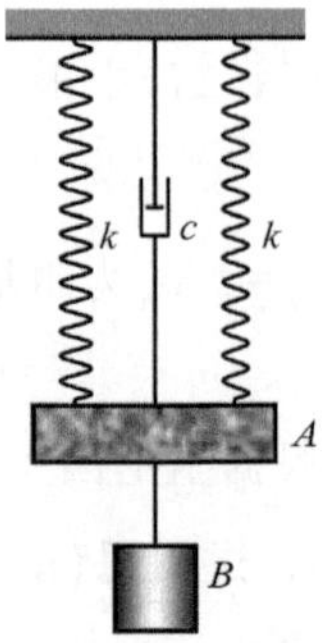

图 E11.11

如图 E11.11 所示，重量为 10 N 的物块 B 通过绳索与质量为 2.5 kg 的物块 A 相连，而物块 A 则悬挂于两根刚度均为 $k=125$ N/m 的弹簧和一根阻尼系数为 $c=30$ N·s/m 的阻尼器上。已知当连接 A 和 B 的绳索被剪断时系统处于静止状态，求物块 A 的运动。

解

根据初始条件，初振幅和初相位分别为

$$A=\frac{m_B g}{k_{\rm eq}}=40\ \text{mm},\quad \alpha=\frac{\pi}{2}$$

利用 $k_{\rm eq}=2k$，临界阻尼系数等于

$$c_{\rm cr}=2\sqrt{m_A k_{\rm eq}}=50\ \text{N}\cdot\text{s/m}$$

即阻尼因子为

$$\eta=\frac{c}{c_{\rm cr}}=0.6$$

因 $\eta<1$，故为欠阻尼振动。利用 $\omega_{\rm n}=\sqrt{k_{\rm eq}/m_A}=10$ rad/s，得有阻尼自由振动的圆频率

$$\omega_{\rm d}=\omega_{\rm n}\sqrt{1-\eta^2}=8\ \text{rad/s}$$

因此，物块 A 的运动可表示为

$$x=A\exp(-\eta\omega_{\rm n}t)\sin(\omega_{\rm d}t+\alpha)=40\exp(-6t)\sin\left(8t+\frac{\pi}{2}\right)$$

式中，x 和 t 以 mm 和 s 为单位。

例 11.12

质量为 m 的小球 C 与不计质量的杆件 AB 连接，杆件由 A 处销钉和 D 处刚度为 k 的弹簧支撑，并在 B 处与阻尼系数为 c 的阻尼器相连，如图 E11.12 所示。系统可在垂直平面内运动，当杆件水平时系统平衡。试通过 m、k 和 c 表示微幅振动的(a)运动微分方程和(b)临界阻尼系数 $c_{\rm cr}$。

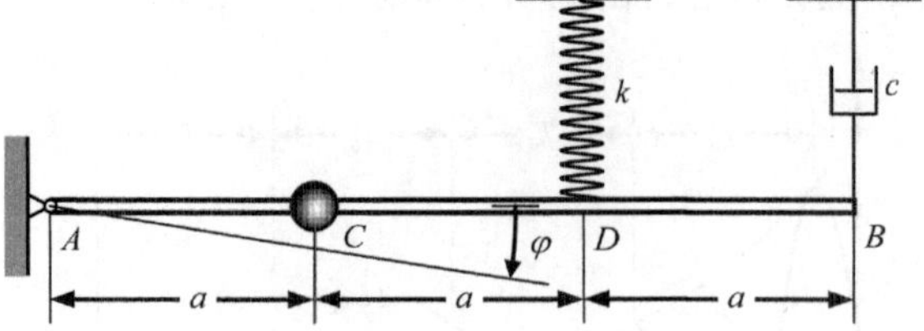

图 E11.12

解

假设杆自平衡位置旋转任意角度 φ，得运动微分方程

$$(ma^2)\ddot{\varphi}+c\,(3a)^2\dot{\varphi}+k\,(2a)^2\varphi=0$$

即

$$\ddot{\varphi}+\frac{9c}{m}\dot{\varphi}+\frac{4k}{m}\varphi=0$$

上式也可重写为

$$\frac{m}{9}\ddot{\varphi}+c\dot{\varphi}+\frac{4k}{9}\varphi=0$$

比较该式与式(11.25),并利用式(11.28),得临界阻尼系数

$$c_{\mathrm{cr}}=2\sqrt{\frac{m}{9}\frac{4k}{9}}=\frac{4}{9}\sqrt{mk}$$

11.4　有阻尼受迫振动

质量为 m 的物体与刚度为 k 的弹簧和阻尼系数为 c 的阻尼器相连,受大小为 $F_{\mathrm{f}}=H\sin\omega_{\mathrm{f}}t$ 的周期力作用,如图 11.8 所示。物体的运动方程可表示为

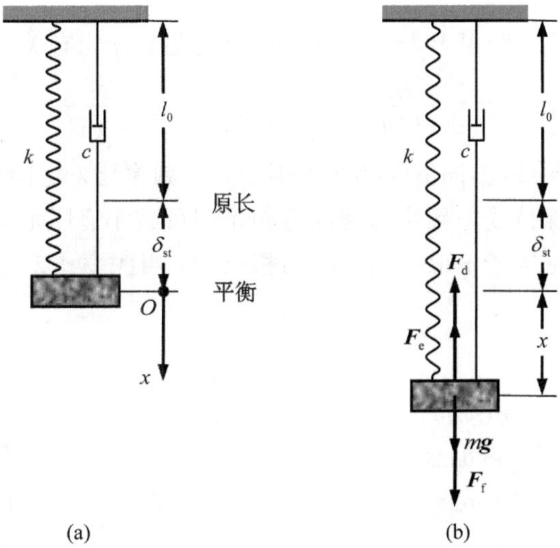

图 11.8

$$m\ddot{x}=mg-k(\delta_{\mathrm{st}}+x)-c\dot{x}+H\sin\omega_{\mathrm{f}}t \tag{11.34}$$

利用 $mg=k\delta_{\mathrm{st}}$,有

$$m\ddot{x}+c\dot{x}+kx=H\sin\omega_{\mathrm{f}}t \tag{11.35}$$

该式的通解可表示为特解 x_p 加补解 x_c,即

$$x=x_{\mathrm{c}}+x_{\mathrm{p}} \tag{11.36}$$

式中,x_{c} 可写为

$$x_{\mathrm{c}}=\begin{cases}\exp(-\eta\omega_{\mathrm{n}}t)[C_1\exp(\sqrt{\eta^2-1}\omega_{\mathrm{n}}t)+C_2\exp(-\sqrt{\eta^2-1}\omega_{\mathrm{n}}t)] & (\eta>1)\\ \exp(-\omega_{\mathrm{n}}t)(C_1+C_2t) & (\eta=1)\\ \exp(-\eta\omega_{\mathrm{n}}t)(C_1\cos\omega_{\mathrm{d}}t+C_2\sin\omega_{\mathrm{d}}t) & (\eta<1)\end{cases} \tag{11.37}$$

补解 x_{c} 表示瞬态振动,并最终会由于摩擦而衰减为零。因此,我们的兴趣主要集中在与稳态振动对应的特解上。特解 x_{p} 可表示为

$$x_{\mathrm{p}}=B\sin(\omega_{\mathrm{f}}t-\varepsilon) \tag{11.38}$$

式中,B 和 ε 分别是稳态振动的振幅和周期力与有阻尼系统稳态振动之间的相位差。用 x_{p} 代替 x 代入式(11.35),得

$$-m\omega_f^2 B\sin(\omega_f t-\varepsilon)+c\omega_f B\cos(\omega_f t-\varepsilon)+kB\sin(\omega_f t-\varepsilon)=H\sin\omega_f t \qquad (11.39)$$

利用 $\sin\omega_f t=\sin[(\omega_f t-\varepsilon)+\varepsilon]=\sin(\omega_f t-\varepsilon)\cos\varepsilon+\cos(\omega_f t-\varepsilon)\sin\varepsilon$,式(11.39)可重写为

$$(-m\omega_f^2 B+kB-F_{max}\cos\varepsilon)\sin(\omega_f t-\varepsilon)+(c\omega_f B-H\sin\varepsilon)\cos(\omega_f t-\varepsilon)=0 \qquad (11.40)$$

即有

$$\left.\begin{aligned}-m\omega_f^2 B+kB-H\cos\varepsilon=0\\ c\omega_f B-H\sin\varepsilon=0\end{aligned}\right\} \qquad (11.41)$$

联解方程(11.41)求出 B 和 ε,得

$$B=\frac{F_{max}/k}{\sqrt{(1-\lambda^2)^2+(2\eta\lambda)^2}} \qquad (11.42)$$

$$\varepsilon=\arctan\frac{2\eta\lambda}{1-\lambda^2} \qquad (11.43)$$

式中,$\eta=c/c_{cr}$是阻尼因子。振幅 B 与频率比 λ 和阻尼因子 η 均有关。放大因子 β 可表示为

$$\beta=\frac{B}{H/m}=\frac{1}{\sqrt{(1-\lambda^2)^2+(2\eta\lambda)^2}} \qquad (11.44)$$

图 11.9 所示为阻尼因子 η 取不同值时的放大因子 β 与频率比 λ 之间的关系曲线。观察发现,选择较大的阻尼因子或保持受迫频率远离固有频率即可减小有阻尼受迫振动的振幅。

图 11.10 所示为阻尼因子 η 取不同值时的相位差 ε 与频率比 λ 之间的关系曲线。

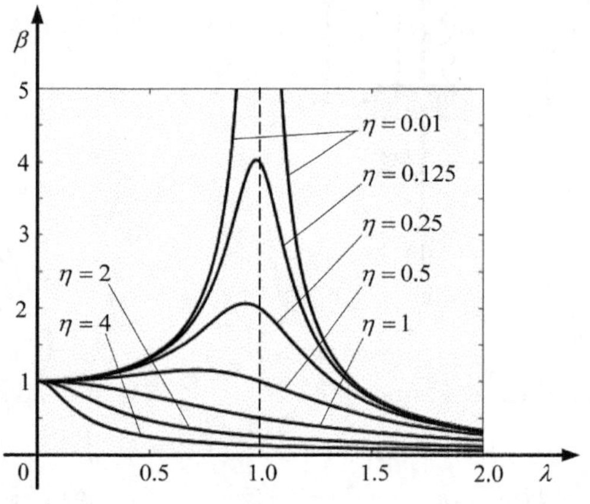

图 11.9

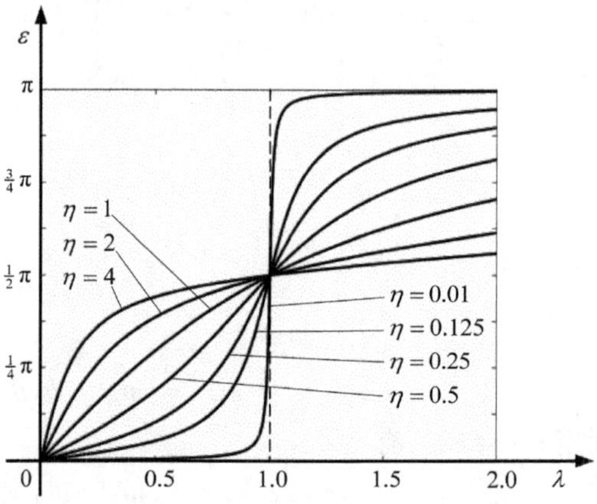

图 11.10

例 11.13

如图 E11.13 所示,质量为 5 kg 的机器部件由两根刚度 k 均为 45 N/m的弹簧支撑。幅值为 1.5 N、频率为 2.5 Hz 的周期力作用于该部件。已知阻尼系数为 20 N·s/m,求该部件做稳态振动的振幅。

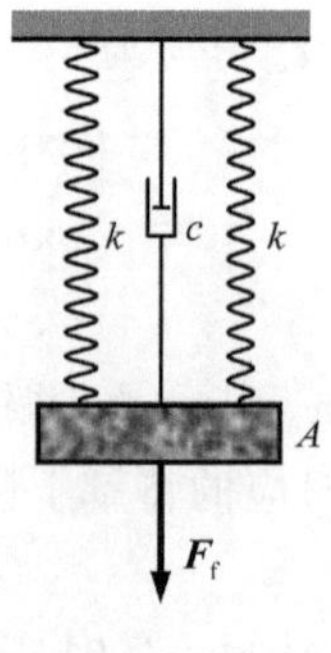

图 E11.13

解

利用 $k_{eq}=2k=90$ N/m 和 $f_f=2.5$ Hz,有

$$\omega_n=\sqrt{k_{eq}/m}=4.243\ \text{rad/s},\quad \omega_f=2\pi f_f=15.71\ \text{rad/s}$$

即

$$\lambda=\omega_f/\omega_n=3.703$$

因 $c_{cr}=2\sqrt{mk_{eq}}=42.43\ \text{N}\cdot\text{s/m}$,则有

$$\eta = c/c_{cr} = 0.4714$$

根据这些数据,计算得到的稳态振动的振幅为

$$B = \frac{H/k_{eq}}{\sqrt{(1-\lambda^2)^2+(2\eta\lambda)^2}} = 1.264\ \text{mm}$$

习　题

11.1 质量为 15 kg 的物块与弹簧相连,并可沿着如图 P11.1 所示的滑槽无摩擦滑动。当物块受到铁锤敲击时处于平衡位置,铁锤传递给物块的初始速度为 2 m/s。已知 $k=80$ kN/m,求(a)运动的周期和频率,(b)运动的振幅和物块的最大加速度。

11.2 质量为 1.4 kg 的物块由刚度为 $k=400$ N/m 的可伸缩弹簧支撑,如图 P11.2 所示。当物块下方受到铁锤敲击时处于平衡位置,铁锤传递给物块的向上速度为 2.5 m/s。求(a)物块向上运动 60 mm 时的时间、速度和加速度,(b)物块被铁锤敲击后 0.90 s 时的位置、速度和加速度。

11.3 质量为 2.5 kg 的环放在如图 P11.3 所示的弹簧上,但不与弹簧连接。观察发现,若环被压下 150 mm 或 150 mm 以上后释放,则环离开弹簧。求(a)弹簧的刚度,(b)环被物块压下 150 mm 后释放 0.2 s 时的位置、速度和加速度。

11.4 质量为 10 kg 的环在如图 P11.4 所示位置由静止释放,并沿着垂直杆件无摩擦滑动到撞击刚度为 $k=981$ N/m 的弹簧,且迫使弹簧压缩。当环的速度减小到零时,环改变运动方向回到初始位置,然后循环不断重复。求环的运动周期(注:这是周期运动,但不是简谐运动)

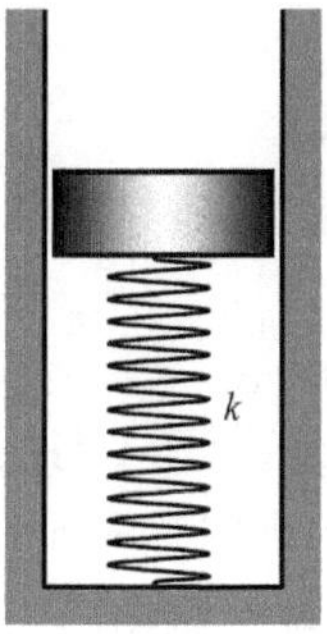

图 P11.1

图 P11.2

图 P11.3

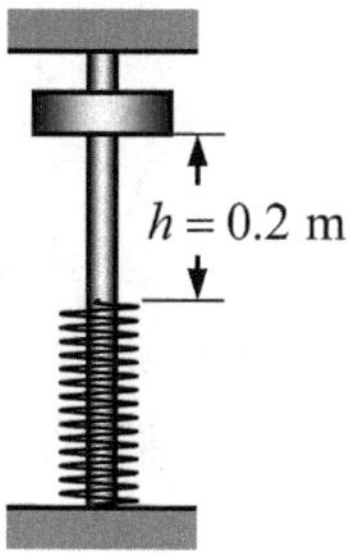

图 P11.4

11.5 根据材料力学,当静载荷 F 作用于均质杆件的 B 端(A 端固定,见图 P11.5)时,杆件长度将增加 $\delta=Fl/EA$,其中 l 为杆件的原长,E 为材料的弹性模量,A 为横截面面积。已知 $l=250$ mm,$E=200$ GPa,杆件直径 $d=10$ mm,忽略杆件重量,求(a)杆件的等效弹簧刚度,(b)质量 $m=5$ kg 的物块连接到同一杆件的 B 端做垂直振动的频率。

11.6 质量为 5 kg 的均质圆柱在水平面上只滚不滑,并在 C 处通过销钉与质量为 3 kg 的水平杆件 AB 相连,水平杆件再与刚度均为 $k=4$ kN/m 的两根弹簧相连,如图 P11.6 所示。已知杆件自平衡位置向右移动 20 mm 后释放,求系统的振动周期。

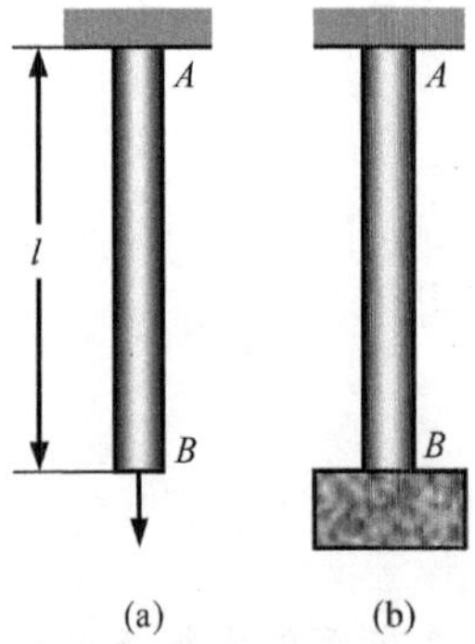

图 P11.5

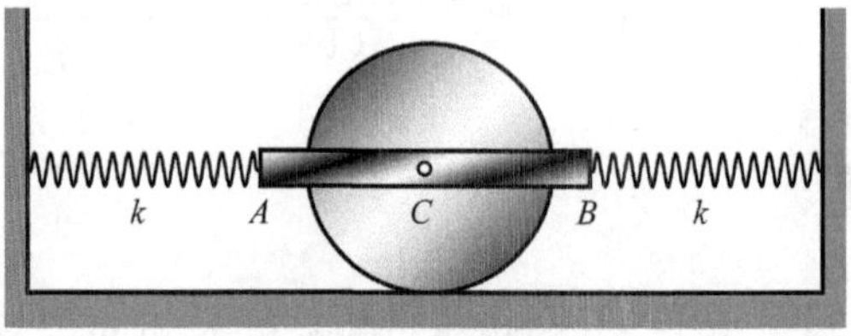

图 P11.6

11.7 如图 P11.7 所示,均质杆件 AB 在垂直平面内绕水平轴 O 转动,O 位于杆件质心 C 上方 d 处。求在微幅振动条件下当振动频率取最大值时的 d 值。

11.8 如图 P11.8 所示,求质量为 m、边长为 $2a$ 的均质方板做微幅振动的周期,设方板边缘中点 O 为悬挂点。

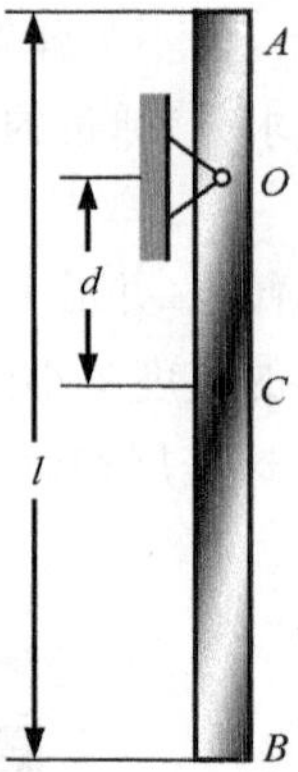

图 P11.7

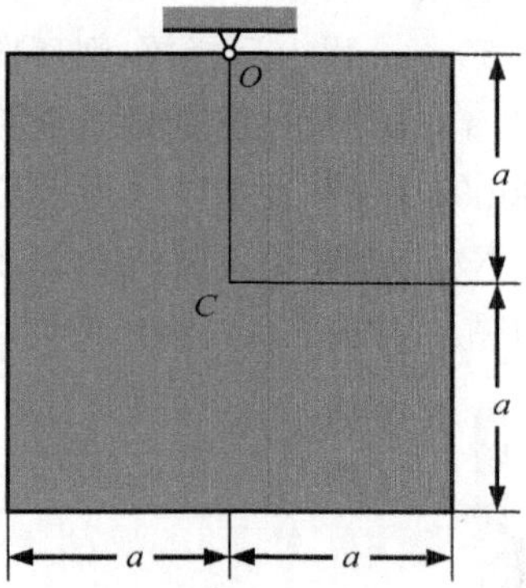

图 P11.8

11.9 质量为 m、长度为 $2a$ 的均质杆件 AB 与不计质量的两环相连。环 A 与刚度为 k 的弹簧相连,并沿水平杆件自由滑动,而环 B 则沿垂直杆件自由滑动,如图 P11.9 所示。已知当杆件 AB 处于垂直位置时系统平衡,求当环 A 偏离平衡位置释放后系统的振动周期。

11.10 质量为 m、长度为 a 的三根相同均质杆件通过如图 P11.10 所示销钉连接后在垂直平面内运动。已知杆件 BC 偏离平衡位置后释放,求系统的振动周期。

11.11 质量为 20 kg 的物块与刚度为 $k=8$ kN/m 的弹簧连接后沿如图 P11.11 所示垂直滑槽做无摩擦运动。大小为 $F_f=H\sin\omega_f t$ 的周期力作用于物块,其中 $\omega_f=10$ rad/s。已知运动的振幅等于 10 mm,求 H。

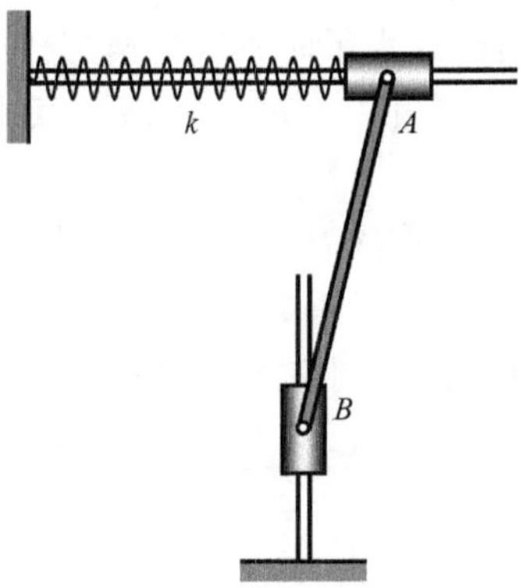

图 P11.9

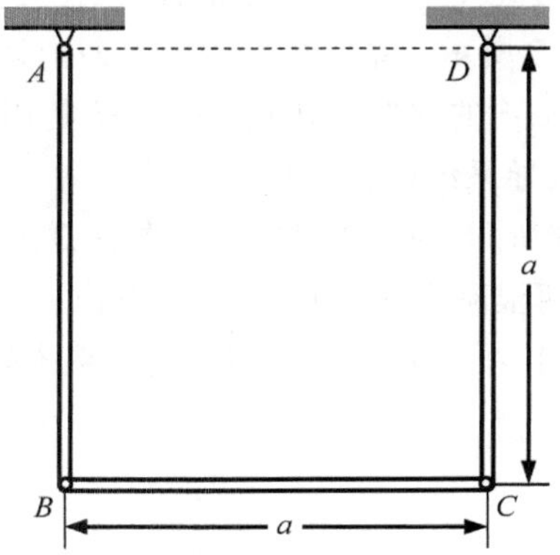

图 P11.10

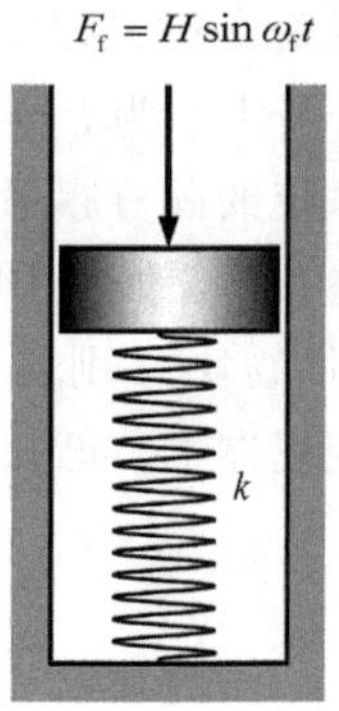

图 P11.11

11.12 质量为 5 kg 的环 C 与刚度为 500 N/m 的弹簧连接后沿光滑水平杆件 AB 滑动，如图 P11.12 所示。大小为 $F_f = H\sin\omega_f t$ 的周期力作用于环，其中 $H=15$ N。求在如下两种条件下环运动的振幅：(a)$\omega_f = 5$ rad/s，(b)$\omega_f = 10$ rad/s。

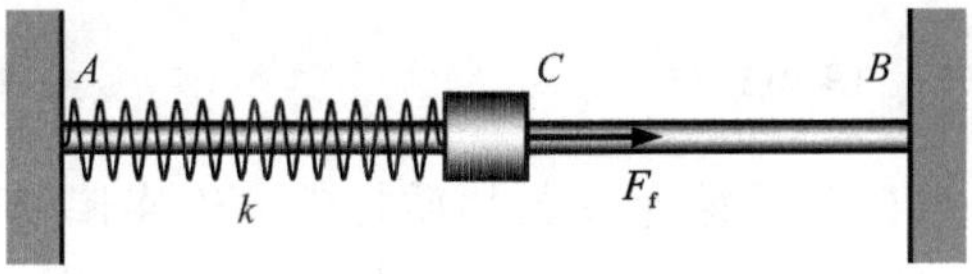

图 P11.12

11.13 质量为 5 kg 的环 C 与刚度为 k 的弹簧连接后沿光滑水平杆件 AB 滑动，如图 P11.13 所示。大小为 $F_f = H\sin\omega_f t$ 的周期力作用于环，其中 $H=15$ N 和 $\omega_f = 5$ rad/s。求在如下两种情况下的弹簧刚度 k 的值：已知环运动具有 100 mm 的振幅，相位与作用力(a)同相，(b)反相。

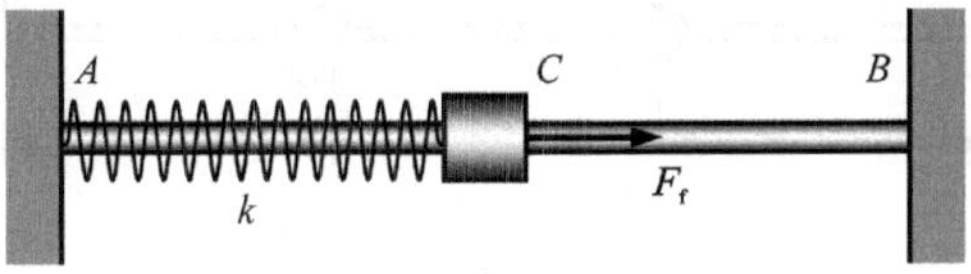

图 P11.13

11.14 质量为 m 的环 C 与刚度为 k 的弹簧连接后沿光滑水平杆件 AB 滑动，如图 P11.14 所示。大小为 $F_f = H\sin\omega_f t$ 的周期力作用于环。求当振动幅值超过由常力 H 引起的静挠度的 3 倍时的 ω_f 值的范围。

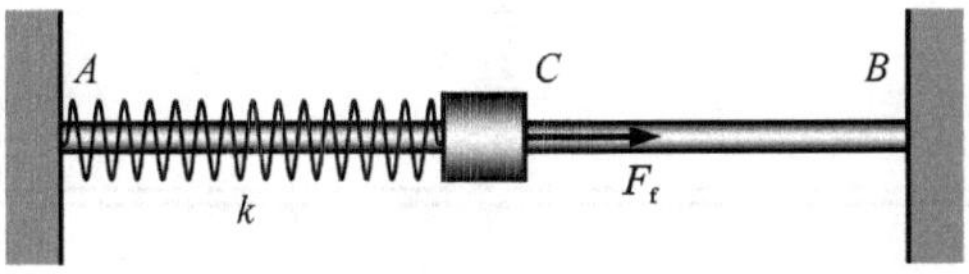

图 P11.14

11.15 质量为 m 的小球 C 与不计质量的杆件 AB 连接,杆件由 A 处销钉和 B 处刚度为 k 的弹簧支撑,如图 P11.15 所示。系统可在垂直平面内运动,当杆件水平时系统平衡。杆件在 B 处受大小为 $F_f=H\sin\omega_f t$ 的周期力作用,其中 H 是周期力的幅值。已知 $\omega_f=2\omega_n$,其中 ω_n 是系统的固有频率,试推导系统的稳态振动方程。

11.16 如图 P11.16 所示,重量为 10 N 的物块 B 通过绳索与质量为 2.5 kg 的物块 A 相连,而物块 A 则悬挂于两根刚度均为 $k=125$ N/m 的弹簧和一根阻尼系数为 $c=65$ N·s/m 的阻尼器上。已知当连接 A 和 B 的绳索被剪断时系统处于静止状态,求物块 A 的运动。

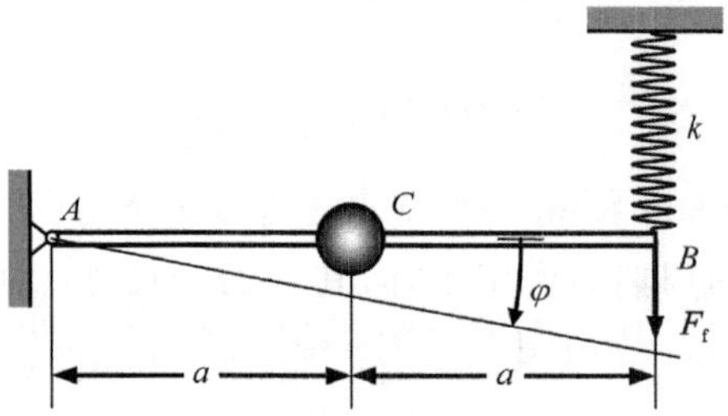

图 P11.15

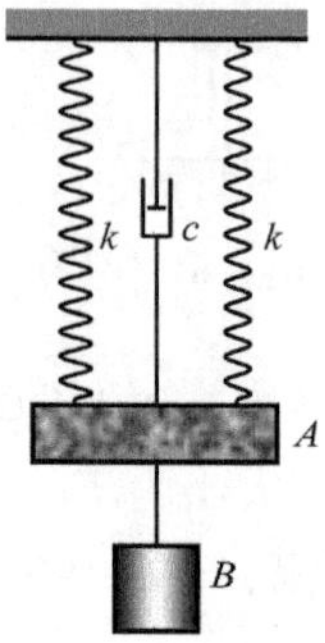

图 P11.16

11.17 质量为 m 的小球 C 与质量为 m 的均质杆件 AB 连接,杆件由 A 处销钉和 D 处刚度为 k 的弹簧支撑,并在 B 处与阻尼系数为 c 的阻尼器相连,如图 P11.17 所示。系统可在垂直平面内运动,当杆件水平时系统平衡。试通过 m、k 和 c 表示微幅振动的(a)运动微分方程和(b)临界阻尼系数 c_{cr}。

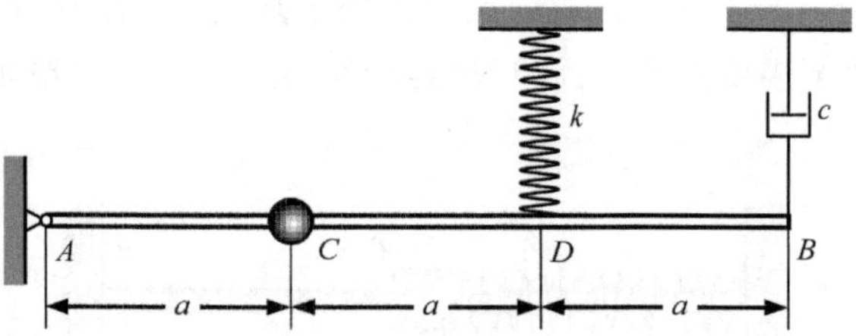

图 P11.17

11.18 如图 P11.8 所示,质量为 m 的均质杆件 AB 由 A 处销钉和 C 处刚度为 k 的弹簧支撑,并在 B 处与阻尼系数为 c 的阻尼器相连。系统可在垂直平面内运动,当杆件水平时系统平衡。试通过 m、k 和 c 表示微幅振动的(a)运动微分方程和(b)临界阻尼系数 c_{cr}。

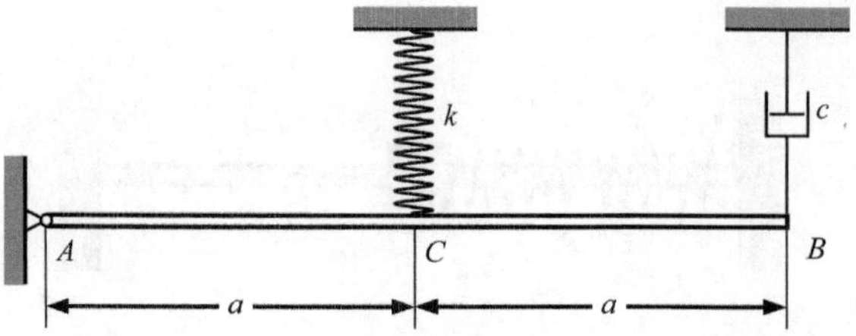

图 P11.18

11.19 如图 P11.9 所示，质量为 5 kg 的机器部件由两根刚度均为 k 的弹簧支撑。幅值为 1.5 N、频率为 2.5 Hz 的周期力作用于该部件。已知阻尼系数为 20 N·s/m，如果稳态振动的振幅等于 1.25 mm，求每根弹簧的刚度。

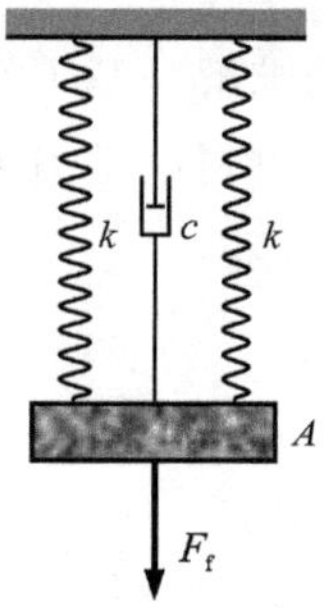

图 P11.19

第 12 章　虚功原理

在前面的静力学中，物体平衡问题的求解是基于作用于物体上的外力满足平衡条件，即首先写出平衡方程，然后再求解未知量。本章将讨论另外一种方法，这种方法是基于虚功原理，对求解一定类型的平衡问题将显得更为有效。

12.1　约束与虚功

1. 约　束

约束是指限制物体运动的几何或运动学条件。假设质点在 t 时刻的位置矢量和速度分别为 $\boldsymbol{r}$ 和 $\dot{\boldsymbol{r}}$，则关系式

$$f(\boldsymbol{r},\dot{\boldsymbol{r}},t)\leqslant 0,\ f(\boldsymbol{r},\dot{\boldsymbol{r}},t)=0,\ \text{或}\ f(\boldsymbol{r},\dot{\boldsymbol{r}},t)\geqslant 0 \tag{12.1}$$

即为约束的数学表达。

根据如下准则，约束可分为：

(1) 几何和运动学约束：速度不出现的约束称为几何约束（或有限约束），如 $f(\boldsymbol{r})=0$；而与速度有关的约束则称为运动学约束（或微分约束），如 $f(\boldsymbol{r},\dot{\boldsymbol{r}})=0$。能够表示为有限形式的微分约束称为可积分约束，如 $\dot{x}-r\dot{\varphi}=0$。

(2) 定常和非定常约束：如果时间并不显含于约束方程之中，则称为定常约束（或稳定约束），如 $f(\boldsymbol{r})=0$。如果约束与时间有关，则称为非定常约束（或非稳定约束），如 $f(\boldsymbol{r},t)=0$。

(3) 双侧和单侧约束：由等式表示的约束称为双侧约束，如 $f(\boldsymbol{r})=0$；而由不等式表示的约束则称为单侧约束，如 $f(\boldsymbol{r})\leqslant 0$。

几何约束和可积分约束统称为完整约束。不可积分约束和单侧约束则统称为非完整约束。

一个约束可能会满足上述多个准则。例如，由 $f(\boldsymbol{r},t)=0$ 表示的约束即为双侧、非定常、几何约束，而由 $f(\boldsymbol{r},\dot{\boldsymbol{r}})=0$ 表示的约束则为双侧、定常、运动学约束。

2. 自由度

自由度是指完全确定给定系统位形所需的最小变量数，即给定系统的所有质点的位置。例如，沿直线或曲线运动的质点有 1 个自由度，二维刚体有 3 个自由度（即确定刚体上指定点位置的 2 个平移自由度，确定刚体方向的 1 个转动自由度），三维刚体有 6 个自由度（即确定刚体质心的 3 个平移自由度，确定刚体方向的 3 个转动自由度）。

系统的自由度与约束数紧密关联。自由质点（即仅受外加力作用的质点）有 3 个自由度。如果质点的坐标通过关系 $f(\boldsymbol{r},t)=0$ 联系，则质点的自由度降低到 2。同样，存在 2 个约束 $f_1(\boldsymbol{r},t)=0$ 和 $f_2(\boldsymbol{r},t)=0$，则暗示质点的位置可以通过单个参数确定，即质点具有单自由度。通常，每个几何、双侧约束作用于系统时都会降低 1 个自由度。

3. 虚　功

物体的虚位移定义为与施加于物体上的约束相协调的假想的无穷小位移。

虚功定义为力或力偶经历虚位移所做的功。力 $\boldsymbol{F}$ 经历虚位移 $\delta\boldsymbol{r}$，所做的虚功可表示为

$$\delta W = \boldsymbol{F}\cdot\delta\boldsymbol{r} \tag{12.2}$$

大小为 M 的力偶经历虚位移 $\delta\theta$，所做的虚功可表示为

$$\delta W = M\delta\theta \tag{12.3}$$

4. 理想约束

如果所有约束力经历任意虚位移所做的虚功都等于零，则这类约束称为理想约束。

12.2 虚功原理

考虑由 n 个质点构成的系统。假设 $\delta\boldsymbol{r}_i$ 是质点 i 的与约束相协调的虚位移，那么作用于质点 i 上的力所做的虚功可表示为

$$\delta W_i = (\boldsymbol{F}_i + \boldsymbol{R}_i)\cdot\delta\boldsymbol{r}_i \quad (i = 1,2,\cdots,n) \tag{12.4}$$

式中，$\boldsymbol{F}_i$ 是作用于质点 i 上的主动力的合力，$\boldsymbol{R}_i$ 是作用于相同质点上的约束力的合力。如果质点系处于静平衡，即 $\boldsymbol{F}_i+\boldsymbol{R}_i=\boldsymbol{0}(i=1,2,\cdots,n)$，那么式(12.4)可重写为

$$\delta W_i = (\boldsymbol{F}_i + \boldsymbol{R}_i)\cdot\delta\boldsymbol{r}_i = 0 \quad (i = 1,2,\cdots,n) \tag{12.5}$$

对 i 求和，则总虚功等于

$$\delta W = \sum_{i=1}^{n}\delta W_i = \sum_{i=1}^{n}(\boldsymbol{F}_i + \boldsymbol{R}_i)\cdot\delta\boldsymbol{r}_i = 0 \tag{12.6}$$

对理想约束系统，有 $\sum\limits_{i=1}^{n}\boldsymbol{R}_i\cdot\delta\boldsymbol{r}_i = 0$，则

$$\delta W = \sum_{i=1}^{n}\boldsymbol{F}_i\cdot\delta\boldsymbol{r}_i = 0 \tag{12.7}$$

式(12.7)表示虚功原理：理想约束系统平衡的充分必要条件是主动力在满足约束条件下的任意虚位移所做的虚功等于零。

例 12.1

刚度 k=800 N/m 的原长弹簧与点 I 和 J 处的销钉连接，如图 E12.1 所示。B 处销钉连接到构件 BCD，并可沿着固定平板上的垂直滑槽自由滑动。求当 F=135 N 的水平力向右作用于点 G 时弹簧受力和点 H 的水平位移。

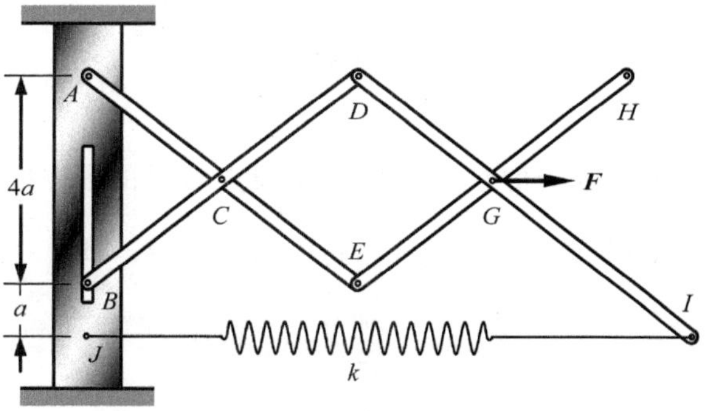

图 E12.1

解

利用 $x_G=3x_C$，$x_H=4x_C$，$x_I=4.5x_C$，有

$$\delta x_G=3\delta x_C,\ \delta x_H=4\delta x_C,\ \delta x_I=4.5\delta x_C$$

根据虚功原理，$\delta W=0$，有

$$F\delta x_G-F_{\text{spr}}\delta x_I=0$$

即

$$F_{\text{spr}}=\frac{\delta x_G}{\delta x_I}F=\frac{3\delta x_C}{4.5\delta x_C}F=90\ \text{N}$$

利用 $F_{\text{spr}}=k\delta x_I$，得

$$x_I=\frac{F_{\text{spr}}}{k}=112.5\ \text{mm}$$

利用 $\delta x_H=4\delta x_C$ 和 $\delta x_I=4.5\delta x_C$，得

$$\delta x_H=\frac{4}{4.5}\delta x_I=100\ \text{m}$$

例 12.2

结构在点 G 处受力 $\boldsymbol{F}$ 作用，如图 E12.2(a)所示。忽略结构重量，已知 $AB=AC=BC=CD=CE=DG=EG=a$，求点 B 处水平约束力的大小。

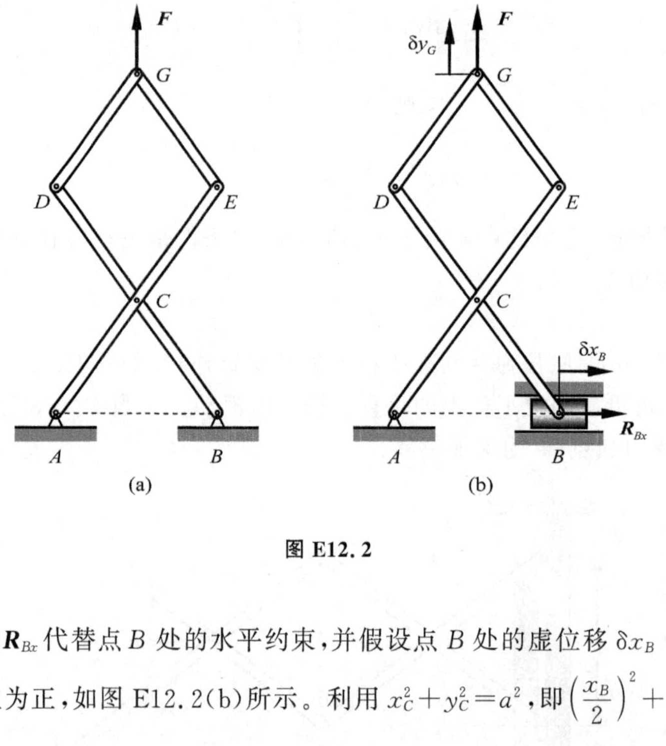

图 E12.2

解

用水平约束力$\boldsymbol{R}_{Bx}$代替点 B 处的水平约束，并假设点 B 处的虚位移 δx_B 向右为正，点 G 处的虚位移 δy_G 向上为正，如图 E12.2(b)所示。利用 $x_C^2+y_C^2=a^2$，即 $\left(\frac{x_B}{2}\right)^2+\left(\frac{y_G}{3}\right)^2=a^2$，有

$$x_B\delta x_B+\frac{4}{9}y_G\delta y_G=0$$

利用 $AB=AC=BC$，即 $y_C=\sqrt{3}x_C$，或 $y_G=\frac{3\sqrt{3}}{2}x_B$，有

$$\delta x_B = -\frac{2\sqrt{3}}{3}\delta y_G$$

根据虚功原理，$\delta W=0$，有

$$R_{Bx}\delta x_B + F\delta y_G = 0$$

即

$$R_{Bx} = -\frac{\delta y_G}{\delta x_B}F = \frac{\sqrt{3}}{2}F$$

例 12.3

重量为 W 的均质杆件 AB 与沿光滑表面自由滑动的物块 A 和 B 相连，如图 E12.3(a)所示。与物块 A 相连弹簧的刚度为 k，当杆件处于水平位置时弹簧为原长。忽略物块的重量，试推导当杆件平衡时 W、k、a 和 θ 必须满足的方程。

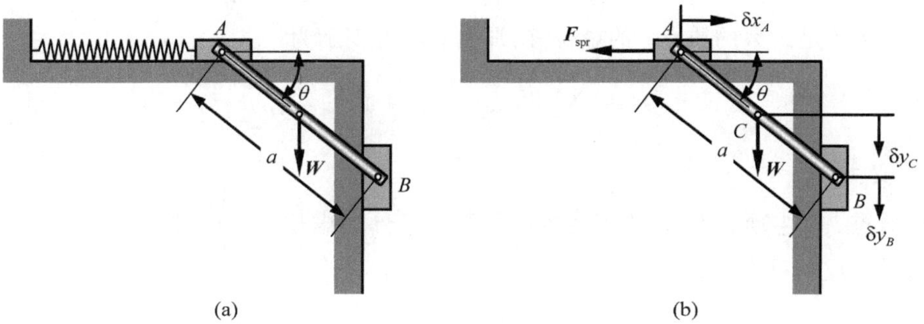

图 E12.3

解

由 δx_A 表示点 A 处的虚位移，δy_B 表示点 B 处的虚位移，δy_C 表示点 C 处的虚位移，有

$$\frac{\delta x_A}{a\sin\theta} = \frac{\delta y_B}{a\cos\theta} = \frac{2\delta y_C}{a\cos\theta}$$

即

$$\delta x_A = 2\delta y_C \tan\theta$$

根据虚功原理，$\delta W=0$，有

$$-F_{spr}\delta x_A + W\delta y_C = 0 \quad 或 \quad 2F_{spr}\tan\theta = W$$

利用 $F_{spr}=ka(1-\cos\theta)$，有

$$\tan\theta - \sin\theta = \frac{W}{2ka}$$

例 12.3 表明，与传统平衡方程方法相比，虚功方法的优越之处在于通过利用虚功原理可以消除所有未知反力。然而，应该注意的是，虚功方法的诱人之处在很大程度上取决于需要求解的给定问题中的各种虚位移之间的简单几何关系。当没有简单关系存在时，通常回到平衡方程方法是最可取的。

12.3 广义坐标和广义力

1. 广义坐标

确定给定系统位形的独立变量称为广义坐标。通常广义坐标可选为线坐标和角坐标。如果物体做直线运动,则线坐标可选为广义坐标;如果物体做定轴转动,则可选择角坐标作为广义坐标;对于平面运动刚体,广义坐标应该选为线坐标和角坐标的集合。

有时采用多于自由度的坐标比较合适,那么这些坐标之间一定通过某些约束而发生关联。例如,考虑由 n 个质点构成的系统,假设系统的位形通过广义坐标 $q_1,q_2,\cdots,q_s$ 进行表征,其中 s 是系统的自由度,那么质点 i 的位置矢量可表示为

$$\boldsymbol{r}_i=\boldsymbol{r}_i(q_1,q_2,\cdots,q_s,t)\quad(i=1,2,\cdots,n)\tag{12.8}$$

假设 $\delta\boldsymbol{r}_i$ 表示满足约束的质点 i 的虚位移,那么 $\delta\boldsymbol{r}_i$ 可表示为

$$\delta\boldsymbol{r}_i=\sum_{k=1}^{s}\frac{\partial\boldsymbol{r}_i}{\partial q_k}\delta q_k\quad(i=1,2,\cdots,n)\tag{12.9}$$

2. 广义力

假设质点 i 在主动力 $\boldsymbol{F}_i$ 作用下发生满足约束条件的虚位移 $\delta\boldsymbol{r}_i$,那么由作用于系统上的主动力所做的虚功可表示为

$$\delta W=\sum_{i=1}^{n}\boldsymbol{F}_i\cdot\delta\boldsymbol{r}_i\tag{12.10}$$

把式(12.9)代入式(12.10),有

$$\delta W=\sum_{i=1}^{n}\boldsymbol{F}_i\cdot\left(\sum_{k=1}^{s}\frac{\partial\boldsymbol{r}_i}{\partial q_k}\delta q_k\right)=\sum_{k=1}^{s}\left[\sum_{i=1}^{n}\left(\boldsymbol{F}_i\cdot\frac{\partial\boldsymbol{r}_i}{\partial q_k}\right)\right]\delta q_k=\sum_{k=1}^{s}Q_k\delta q_k\tag{12.11}$$

式中

$$Q_k=\sum_{i=1}^{n}\left(\boldsymbol{F}_i\cdot\frac{\partial\boldsymbol{r}_i}{\partial q_k}\right)\quad(k=1,2,\cdots,s)\tag{12.12}$$

称为与广义坐标 q_k 对应的广义力。由于 $Q_k\delta q_k$ 具有功的量纲,因此,如果 q_k 是线坐标,那么 Q_k 具有力的量纲;如果 q_k 是角坐标,那么 Q_k 具有矩的量纲。有两种方法计算 Q_k:一种方法是从广义力定义计算 Q_k;另一种方法是从虚功 $Q_k\delta q_k$ 计算 Q_k:$Q_k\delta q_k$ 是当 q_k(其余广义坐标保持不变)具有任意虚位移 δq_k 时,作用于系统上的主动力所做的虚功,即 $Q_k=\delta W/\delta q_k\,(k=1,2,\cdots,s)$。

对保守力系,广义力 Q_k 可表示为

$$Q_k=\sum_{i=1}^{n}\left(\boldsymbol{F}_i\cdot\frac{\partial\boldsymbol{r}_i}{\partial q_k}\right)=-\frac{\partial V}{\partial q_k}\quad(k=1,2,\cdots,s)\tag{12.13}$$

式中,$V=V(q_1,q_2,\cdots,q_s)$是系统的势能函数。

12.4 平衡条件的广义坐标表示

根据虚功原理,当系统处于静平衡时,有

$$\delta W=\sum_{i=1}^{n}\boldsymbol{F}_i\cdot\delta\boldsymbol{r}_i=\sum_{k=1}^{s}Q_k\delta q_k=0 \tag{12.14}$$

因 $q_1,q_2,\cdots,q_s$ 是独立变量，$\delta q_1,\delta q_2,\cdots,\delta q_s$ 可取任意值，故上述方程可简化为

$$Q_k=0\quad(k=1,2,\cdots,s) \tag{12.15}$$

式(12.15)表示广义坐标的虚功原理，即理想约束系统静平衡的充分必要条件是作用于系统的广义力等于零。

习　题

12.1 刚度 $k=500$ N/m 的原长弹簧与点 I 和 J 处的销钉连接，如图 P12.1 所示。销钉 B 销接到构件 BCD 上，并可沿着固定平板上的垂直滑槽自由滑动。求当 $F=100$ N 的水平力向右作用于点 H 时弹簧的受力和点 B 的反作用力。

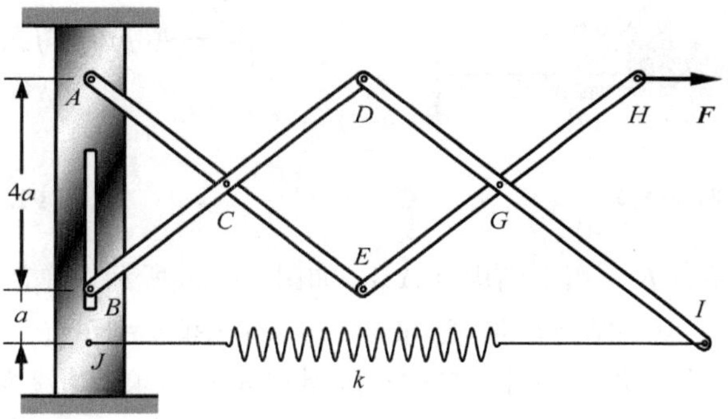

图 P12.1

12.2 图 P12.2 所示机构受力 $\boldsymbol{F}$ 作用。忽略机构重量，已知 $AC=BC=CD=CE=DG=EG=a$ 和 $AB=b$，试推导当机构平衡时所需力 $\boldsymbol{R}$ 的大小。

12.3 销钉 B 与构件 ABC 连接，并能沿着固定平板上的滑槽滑动，如图 P12.3 所示。忽略摩擦，已知 $AB=BC=CD=a$ 和 $BD=b$，当作用于 A 的力 $\boldsymbol{F}$ 分别(a)垂直向下和(b)水平向左时，试推导维持平衡所需力偶的大小 M。

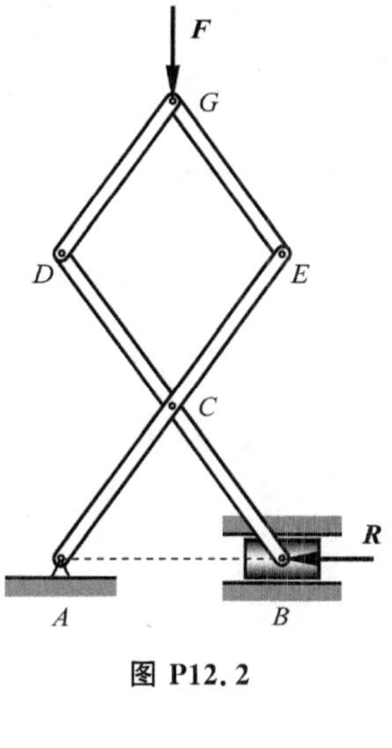

图 P12.2

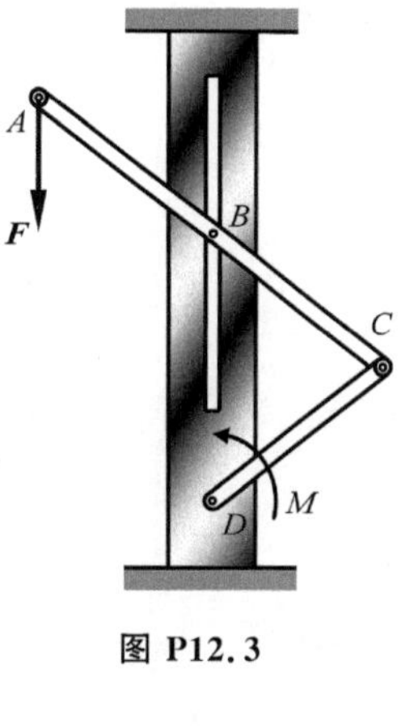

图 P12.3

12.4 长为 l 的细长杆 AB 与环 B 连接,并静止放置于半径为 r 的半圆柱面上,如图 P12.4 所示。忽略摩擦,已知结构受力 $\boldsymbol{F}_A$ 和 $\boldsymbol{F}_B$ 作用,试推导机构平衡时的 θ 表达式。

12.5 两杆 ABC 和 CDE 通过销钉 C 和弹簧 AE 连接,如图 P12.5 所示。弹簧刚度为 k,当 $\theta=30°$ 时弹簧为原长。已知 $AB=BC=CD=DE=a$,试推导当系统在如图所示载荷作用下处于平衡状态时的 F、θ、l 和 k 必须满足的方程。

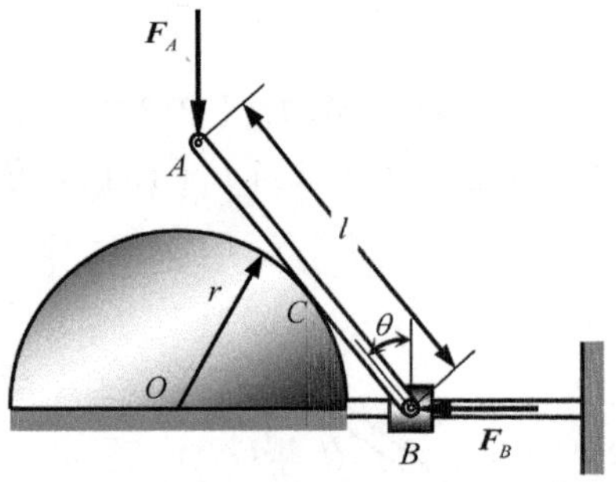

图 P12.4

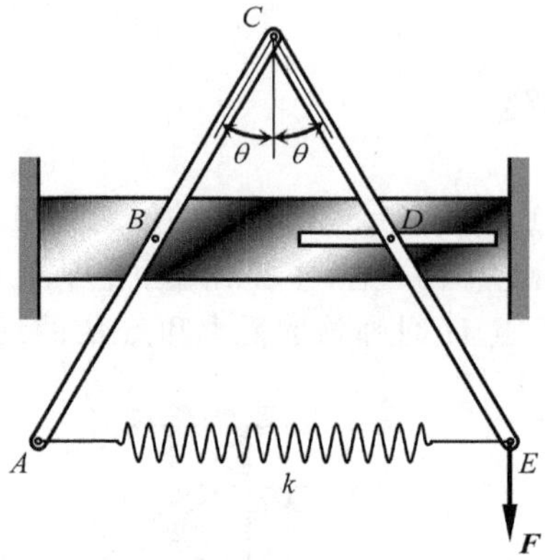

图 P12.5

12.6 150 N 的水平力 $\boldsymbol{F}$ 作用于机构点 A 上,如图 P12.6 所示。弹簧刚度为 $k=1.5$ kN/m,当 $\theta=0°$ 时弹簧为原长。忽略结构质量,已知 $a=250$ mm 和 $r=150$ mm,求平衡时的 θ 值。

12.7 如图 P12.7 所示,如果构件 AC 的长度增加 6 mm,求节点 I 的垂直位移。

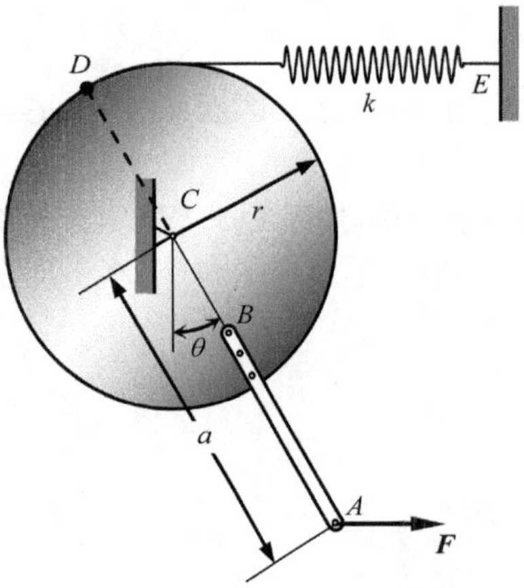

图 P12.6

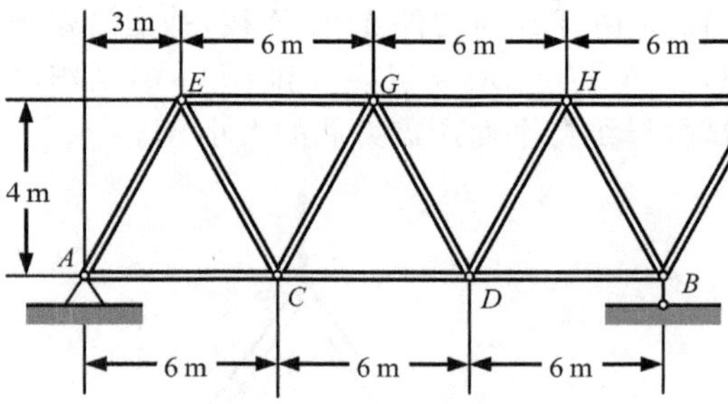

图 P12.7

第 13 章　拉格朗日方程

13.1　拉格朗日方程

考虑由 n 个质点组成的质点系，受完整约束作用。假设系统位形可通过广义坐标 q_1，q_2，…，q_s 描述，其中 s 是系统的自由度，那么质点 i 的位置矢量可表示为

$$\boldsymbol{r}_i = \boldsymbol{r}_i(q_1, q_2, \cdots, q_s, t) \quad (i = 1,2,\cdots,n) \tag{13.1}$$

对时间求导，有

$$\dot{\boldsymbol{r}}_i = \sum_{k=1}^{s} \frac{\partial \boldsymbol{r}_i}{\partial q_k}\dot{q}_k + \frac{\partial \boldsymbol{r}_i}{\partial t} \quad (i = 1,2,\cdots,n) \tag{13.2}$$

根据式(13.2)，得

$$\frac{\partial \dot{\boldsymbol{r}}_i}{\partial \dot{q}_k} = \frac{\partial \boldsymbol{r}_i}{\partial q_k} \quad (i = 1,2,\cdots,n;\ k = 1,2,\cdots,s) \tag{13.3}$$

点乘 $\dot{\boldsymbol{r}}_i$ 并对时间求导，有

$$\frac{\mathrm{d}}{\mathrm{d}t}\left(\dot{\boldsymbol{r}}_i \cdot \frac{\partial \dot{\boldsymbol{r}}_i}{\partial \dot{q}_k}\right) = \frac{\mathrm{d}}{\mathrm{d}t}\left(\dot{\boldsymbol{r}}_i \cdot \frac{\partial \boldsymbol{r}_i}{\partial q_k}\right) = \ddot{\boldsymbol{r}}_i \cdot \frac{\partial \boldsymbol{r}_i}{\partial q_k} + \dot{\boldsymbol{r}}_i \cdot \frac{\partial \dot{\boldsymbol{r}}_i}{\partial q_k} \quad (i = 1,2,\cdots,n;\ k = 1,2,\cdots,s) \tag{13.4}$$

即

$$\frac{\mathrm{d}}{\mathrm{d}t}\frac{\partial}{\partial \dot{q}_k}\left(\frac{1}{2}m_i\dot{\boldsymbol{r}}_i \cdot \dot{\boldsymbol{r}}_i\right) = (m_i\ddot{\boldsymbol{r}}_i) \cdot \frac{\partial \boldsymbol{r}_i}{\partial q_k} + \frac{\partial}{\partial q_k}\left(\frac{1}{2}m_i\dot{\boldsymbol{r}}_i \cdot \dot{\boldsymbol{r}}_i\right) \quad (i = 1,2,\cdots,n;\ k = 1,2,\cdots,s) \tag{13.5}$$

式中，m_i 是质点 i 的质量。对 i 求和，得

$$\frac{\mathrm{d}}{\mathrm{d}t}\frac{\partial}{\partial \dot{q}_k}\left(\sum_{i=1}^{n}\frac{1}{2}m_i\dot{\boldsymbol{r}}_i \cdot \dot{\boldsymbol{r}}_i\right) = \sum_{i=1}^{n}\left[(m_i\ddot{\boldsymbol{r}}_i) \cdot \frac{\partial \boldsymbol{r}_i}{\partial q_k}\right] + \frac{\partial}{\partial q_k}\left(\sum_{i=1}^{n}\frac{1}{2}m_i\dot{\boldsymbol{r}}_i \cdot \dot{\boldsymbol{r}}_i\right) \quad (k = 1,2,\cdots,s) \tag{13.6}$$

即

$$\frac{\mathrm{d}}{\mathrm{d}t}\frac{\partial T}{\partial \dot{q}_k} = \sum_{i=1}^{n}\left(\boldsymbol{F}_i \cdot \frac{\partial \boldsymbol{r}_i}{\partial q_k}\right) + \frac{\partial T}{\partial q_k} \quad (k = 1,2,\cdots,s) \tag{13.7}$$

式中，$T = \sum_{i=1}^{n}\frac{1}{2}m_i\dot{\boldsymbol{r}}_i \cdot \dot{\boldsymbol{r}}_i$ 是系统动能，$\boldsymbol{F}_i = m_i\ddot{\boldsymbol{r}}_i$ 是作用于质点 i 上的力。利用广义力的定义，方程(13.7)可重写为

$$\frac{\mathrm{d}}{\mathrm{d}t}\frac{\partial T}{\partial \dot{q}_k} - \frac{\partial T}{\partial q_k} = Q_k \quad (k = 1,2,\cdots,s) \tag{13.8}$$

式中，$Q_k = \sum_{i=1}^{n}\left(\boldsymbol{F}_i \cdot \frac{\partial \boldsymbol{r}_i}{\partial q_k}\right)$ 是对应广义坐标 q_k 的广义力。方程(13.8)称为拉格朗日方程。

对保守系统，$Q_k = -\frac{\partial V}{\partial q_k}$，则拉格朗日方程可表示为

$$\frac{\mathrm{d}}{\mathrm{d}t}\frac{\partial T}{\partial \dot{q}_k}-\frac{\partial T}{\partial q_k}=-\frac{\partial V}{\partial q_k}\quad (k=1,2,\cdots,s) \tag{13.9}$$

定义

$$L=T-V \tag{13.10}$$

式中,L 称为拉格朗日函数。利用$\frac{\partial V}{\partial \dot{q}_k}=0$,保守系统的拉格朗日方程可重写为

$$\frac{\mathrm{d}}{\mathrm{d}t}\frac{\partial L}{\partial \dot{q}_k}-\frac{\partial L}{\partial q_k}=0\quad (k=1,2,\cdots,s) \tag{13.11}$$

如果部分广义力不是保守力,例如 Q'_k,其余部分广义力是保守力,则可从势函数 V 求导得到,即

$$Q_k=Q'_k-\frac{\partial V}{\partial q_k}\quad (k=1,2,\cdots,s) \tag{13.12}$$

那么可得拉格朗日方程的最一般表达式

$$\frac{\mathrm{d}}{\mathrm{d}t}\frac{\partial L}{\partial \dot{q}_k}-\frac{\partial L}{\partial q_k}=Q'_k\quad (k=1,2,\cdots,s) \tag{13.13}$$

例 13.1

质量-弹簧系统由质量分别为 m_1 和 m_2 的物块和刚度分别为 k_1 和 k_2 的弹簧组成,如图 E13.1 所示。用拉格朗日方程法求系统运动微分方程。

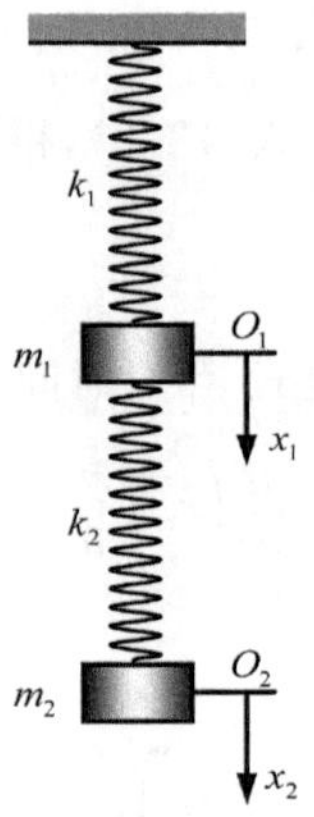

图 E13.1

解

系统有 2 个自由度,并受保守力作用。如果 x_1 和 x_2 选为广义坐标,原点分别在物块静平衡位置,向下为正,那么在任意位置系统的动能和势能可分别表示为

$$T=\frac{1}{2}m_1\dot{x}_1{}^2+\frac{1}{2}m_2\dot{x}_2{}^2,$$

$$V=\frac{1}{2}k_1(\delta_1+x_1)^2+\frac{1}{2}k_2(\delta_2+x_2-x_1)^2-m_1gx_1-m_2gx_2$$

式中,$\delta_1=\frac{(m_1+m_2)g}{k_1}$和 $\delta_2=\frac{m_2g}{k_2}$。因此,可得拉格朗日函数

$$L=T-V=\frac{1}{2}m_1\dot{x}_1{}^2+\frac{1}{2}m_2\dot{x}_2{}^2-\frac{1}{2}k_1(\delta_1+x_1)^2$$
$$-\frac{1}{2}k_2(\delta_2+x_2-x_1)^2+m_1gx_1+m_2gx_2$$

对物块 m_1,有

$$\frac{\partial L}{\partial \dot{x}_1}=m_1\dot{x}_1,\ \frac{\partial L}{\partial x_1}=-k_1(\delta_1+x_1)+k_2(\delta_2+x_2-x_1)+m_1g=-(k_1+k_2)x_1+k_2x_2$$

上式代入拉格朗日方程,得

$$m_1\ddot{x}_1+(k_1+k_2)x_1-k_2x_2=0$$

同理,有

$$\frac{\partial L}{\partial \dot{x}_2}=m_2\dot{x}_2,\ \frac{\partial L}{\partial x_2}=-k_2(\delta_2+x_2-x_1)+m_2g=k_2x_1-k_2x_2$$

及物块 m_2 的运动微分方程

$$m_2\ddot{x}_2 - k_2 x_1 + k_2 x_2 = 0$$

例 13.2

具有 2 个自由度的系统由 2 个物块和 3 个滑轮组成，如图 E13.2 所示。中间滑轮垂直自由运动，质量为 m。不计绳索重量，物块 m_1 和 m_2 分别悬挂于定滑轮的左侧和右侧。不计滑轮摩擦，绳索在滑轮上可自由滑动。用拉格朗日方程法求系统运动的微分方程。

解

本题是具有 2 个自由度的保守系。选 x_1 和 x_2 为广义坐标，原点位于参考面，向上为正，那么系统处于任意位置时的动能和势能分别为

$$T = \frac{1}{2}m_1\dot{x}_1^{\ 2} + \frac{1}{2}m_2\dot{x}_2^{\ 2} + \frac{1}{2}m\left[\frac{1}{2}(\dot{x}_1 + \dot{x}_2)\right]^2,$$

$$V = m_1 g x_1 + m_2 g x_2 + mg\left[h - \frac{1}{2}(x_1 + x_2)\right]$$

式中，h 为常数。拉格朗日函数为

$$L = T - V = \frac{1}{2}m_1\dot{x}_1^2 + \frac{1}{2}m_2\dot{x}_2^2 + \frac{1}{2}m\left[\frac{1}{2}(\dot{x}_1 + \dot{x}_2)\right]^2$$

$$- m_1 g x_1 - m_2 g x_2 - mg\left[h - \frac{1}{2}(x_1 + x_2)\right]$$

对物块 m_1，有

$$\frac{\partial L}{\partial \dot{x}_1} = m_1\dot{x}_1 + \frac{1}{4}m(\dot{x}_1 + \dot{x}_2),\ \frac{\partial L}{\partial x_1} = -m_1 g + \frac{1}{2}mg$$

上式代入拉格朗日方程，得

$$\left(m_1 + \frac{1}{4}m\right)\ddot{x}_1 + \frac{1}{4}m\ddot{x}_2 + \left(m_1 - \frac{1}{2}m\right)g = 0$$

同理，有物块 m_2 的运动微分方程

$$\frac{1}{4}m\ddot{x}_1 + \left(m_2 + \frac{1}{4}m\right)\ddot{x}_2 + \left(m_2 - \frac{1}{2}m\right)g = 0$$

例 13.3

质量为 m_1 的物块沿光滑斜面滑动，质量为 m_2 的斜面沿光滑水平面滑动，如图 E13.3 所示。求物块的水平加速度和斜面的加速度。

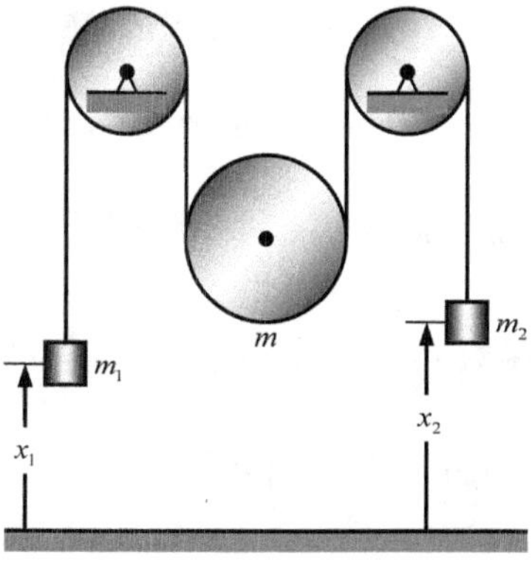

图 E13.2

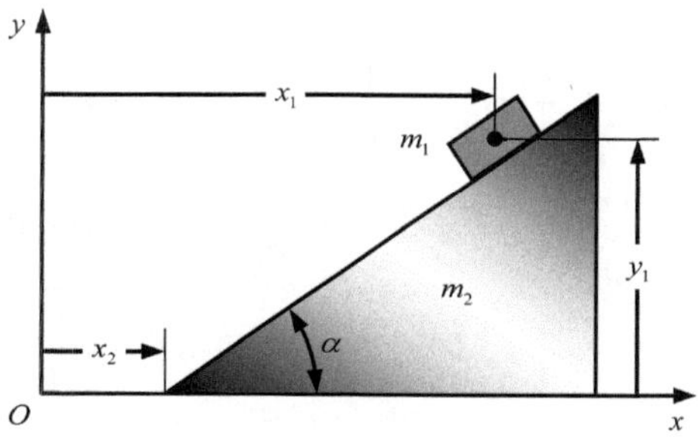

图 E13.3

解

选 x_1 和 x_2 为广义坐标，则有 $y_1=\tan\alpha(x_1-x_2)$ 和 $\dot{y}_1=\tan\alpha(\dot{x}_1-\dot{x}_2)$。因此，系统的动能和势能可分别表示为

$$T=\frac{1}{2}m_1[\dot{x}_1^{\ 2}+\tan^2\alpha\,(\dot{x}_1-\dot{x}_2)^2]+\frac{1}{2}m_2\dot{x}_2^{\ 2},\ V=m_1g\tan\alpha(x_1-x_2)+m_2gh$$

式中，h 为常数。相应的拉格朗日函数等于

$$L=T-V=T=\frac{1}{2}m_1[\dot{x}_1^{\ 2}+\tan^2\alpha\,(\dot{x}_1-\dot{x}_2)^2]+\frac{1}{2}m_2\dot{x}_2^{\ 2}-m_1g\tan\alpha(x_1-x_2)-m_2gh$$

对质量为 m_1 的物块，有

$$\frac{\partial L}{\partial\dot{x}_1}=m_1[\dot{x}_1+\tan^2\alpha(\dot{x}_1-\dot{x}_2)],\ \frac{\partial L}{\partial x_1}=-m_1g\tan\alpha$$

把上式代入拉格朗日方程，得

$$m_1[\ddot{x}_1+\tan^2\alpha(\ddot{x}_1-\ddot{x}_2)]+m_1g\tan\alpha=0$$

同理，对质量为 m_2 的斜面，有

$$\frac{\partial L}{\partial\dot{x}_2}=-m_1\tan^2\alpha(\dot{x}_1-\dot{x}_2)+m_2\dot{x}_2,\ \frac{\partial L}{\partial x_2}=m_1g\tan\alpha$$

上式代入拉格朗日方程，得

$$-m_1\tan^2\alpha(\ddot{x}_1-\ddot{x}_2)+m_2\ddot{x}_2-m_1g\tan\alpha=0$$

联立求解方程组，得

$$\ddot{x}_1=-\frac{m_2g\sin\alpha\cos\alpha}{m_1\sin^2\alpha+m_2},\ \ddot{x}_2=\frac{m_1g\sin\alpha\cos\alpha}{m_1\sin^2\alpha+m_2}$$

例 13.4

具有 2 个自由度的有阻尼受迫振动系统由质量为 m_1 和 m_2 的物块、刚度为 k_1 和 k_2 的弹簧、阻尼系数为 c_1 和 c_2 的阻尼器组成，如图 E13.4 所示。系统受大小为 $H_1\sin\omega_1t$ 和 $H_2\sin\omega_2t$ 的周期力作用，其中 H_1 和 H_2 为力的幅值，ω_1 和 ω_2 为力的圆频率。求系统运动的微分方程。

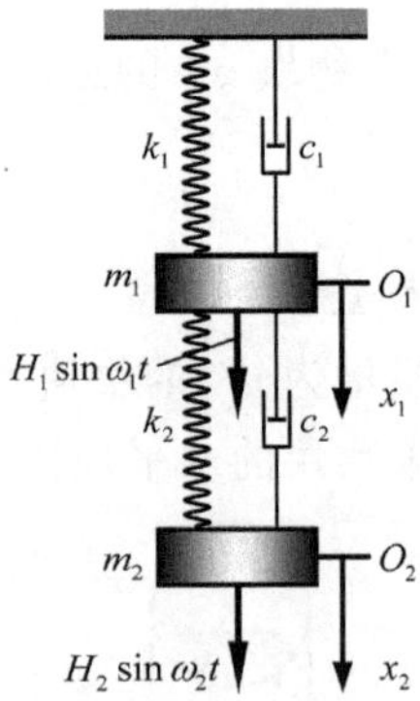

图 E13.4

解

选 x_1 和 x_2 作为广义坐标，原点分别在物块的静平衡位置，则系统处于任意位置时所具有的动能和势能分别为

$$T=\frac{1}{2}m_1\dot{x}_1^{\ 2}+\frac{1}{2}m_2\dot{x}_2^{\ 2},$$

$$V=\frac{1}{2}k_1\,(\delta_1+x_1)^2+\frac{1}{2}k_2\,(\delta_2+x_2-x_1)^2-m_1gx_1-m_2gx_2$$

式中，$\delta_1=\dfrac{(m_1+m_2)g}{k_1}$ 和 $\delta_2=\dfrac{m_2g}{k_2}$。因此，可得拉格朗日函数

$$L=T-V=\frac{1}{2}m_1\dot{x}_1^{\ 2}+\frac{1}{2}m_2\dot{x}_2^{\ 2}-\frac{1}{2}k_1\,(\delta_1+x_1)^2$$
$$-\frac{1}{2}k_2\,(\delta_2+x_2-x_1)^2+m_1gx_1+m_2gx_2$$

对物块 m_1，有

$$\frac{\partial L}{\partial \dot{x}_1}=m_1\dot{x}_1,\ \frac{\partial L}{\partial x_1}=-k_1(\delta_1+x_1)+k_2(\delta_2+x_2-x_1)+m_1 g=-(k_1+k_2)x_1+k_2x_2$$

利用拉格朗日方程，$\frac{\mathrm{d}}{\mathrm{d}t}\frac{\partial L}{\partial \dot{x}_1}-\frac{\partial L}{\partial x_1}=Q'_1$，有

$$m_1\ddot{x}_1+(k_1+k_2)x_1-k_2x_2=Q'_1$$

式中，$Q'_1=H_1\sin\omega_1 t-c_1\dot{x}_1+c_2(\dot{x}_2-\dot{x}_1)$。因此，物块 m_1 的运动微分方程可重写为

$$m_1\ddot{x}_1+(c_1+c_2)\dot{x}_1-c_2\dot{x}_2+(k_1+k_2)x_1-k_2x_2=H_1\sin\omega_1 t$$

同理，有

$$\frac{\partial L}{\partial \dot{x}_2}=m_2\dot{x}_2,\ \frac{\partial L}{\partial x_2}=-k_2(\delta_2+x_2-x_1)+m_2 g=k_2x_1-k_2x_2$$

以及物块 m_2 的微分方程

$$m_2\ddot{x}_2-k_2x_1+k_2x_2=Q'_2$$

式中，$Q'_2=H_2\sin\omega_2 t-c_2(\dot{x}_2-\dot{x}_1)$。把 Q'_2 代入上式，得物块 m_2 的运动微分方程

$$m_2\ddot{x}_2-c_2\dot{x}_1+c_2\dot{x}_2-k_2x_1+k_2x_2=H_2\sin\omega_2 t$$

13.2　拉格朗日方程的初积分

1. 能量积分

考虑由 n 个质点组成的保守系统，受定常约束作用。假设系统位形由广义坐标 $q_1,q_2,\cdots,q_s$ 描述，其中 s 是系统自由度，那么质点 i 的位置矢量可表示为

$$\boldsymbol{r}_i=\boldsymbol{r}_i(q_1,q_2,\cdots,q_s)\quad(i=1,2,\cdots,n)\tag{13.14}$$

对时间求导，有

$$\dot{\boldsymbol{r}}_i=\sum_{k=1}^{s}\frac{\partial \boldsymbol{r}_i}{\partial q_k}\dot{q}_k\quad(i=1,2,\cdots,n)\tag{13.15}$$

利用式(13.15)，系统动能可表示为

$$T=\sum_{i=1}^{n}\frac{1}{2}m_i\dot{\boldsymbol{r}}_i\cdot\dot{\boldsymbol{r}}_i=\sum_{i=1}^{n}\frac{1}{2}m_i\left(\sum_{k=1}^{s}\frac{\partial \boldsymbol{r}_i}{\partial q_k}\dot{q}_k\right)\cdot\left(\sum_{l=1}^{s}\frac{\partial \boldsymbol{r}_i}{\partial q_l}\dot{q}_l\right)=\sum_{k,l=1}^{s}\frac{1}{2}m_{kl}\dot{q}_k\dot{q}_l\tag{13.16}$$

式中

$$m_{kl}=\sum_{i=1}^{n}m_i\frac{\partial \boldsymbol{r}_i}{\partial q_k}\cdot\frac{\partial \boldsymbol{r}_i}{\partial q_l}\quad(k,l=1,2,\cdots,s)\tag{13.17}$$

称为广义质量。式(13.16)对广义速度 $\dot{q}_k$ 求导，有

$$\frac{\partial T}{\partial \dot{q}_k}=\frac{\partial}{\partial \dot{q}_k}\sum_{k,l=1}^{s}\frac{1}{2}m_{kl}\dot{q}_k\dot{q}_l=2\sum_{l=1}^{s}\frac{1}{2}m_{kl}\dot{q}_l\quad(k=1,2,\cdots,s)\tag{13.18}$$

乘 $\dot{q}_k$，并对 k 求和，有

$$\sum_{k=1}^{s}\frac{\partial T}{\partial \dot{q}_k}\dot{q}_k=2\sum_{k,l=1}^{s}\frac{1}{2}m_{kl}\dot{q}_k\dot{q}_l=2T\tag{13.19}$$

利用$\partial T/\partial \dot{q}_k=\partial L/\partial \dot{q}_k$，式(13.19)重写为

$$2T = \sum_{k=1}^{s} \frac{\partial L}{\partial \dot{q}_k}\dot{q}_k \tag{13.20}$$

式(13.20)对时间求导,有

$$2\dot{T} = \frac{\mathrm{d}}{\mathrm{d}t}\sum_{k=1}^{s} \frac{\partial L}{\partial \dot{q}_k}\dot{q}_k = \sum_{k=1}^{s}\left[\frac{\mathrm{d}}{\mathrm{d}t}\left(\frac{\partial L}{\partial \dot{q}_k}\right)\dot{q}_k + \frac{\partial L}{\partial \dot{q}_k}\ddot{q}_k\right] = \sum_{k=1}^{s}\left[\frac{\partial L}{\partial q_k}\dot{q}_k + \frac{\partial L}{\partial \dot{q}_k}\ddot{q}_k\right] = \dot{L} \tag{13.21}$$

即

$$2\dot{T} - \dot{L} = 0 \tag{13.22}$$

积分,得

$$2T - L = T + V = 常数 \tag{13.23}$$

式(13.23)表明,保守系统受定常约束作用,系统的动能和势能之和保持不变。

2. 循环积分

如果由 n 个质点组成的保守系统由广义坐标 $q_1,q_2,\cdots,q_s$ 描述,其中 s 是系统自由度,那么广义动量定义为

$$p_k = \frac{\partial L}{\partial \dot{q}_k} \quad (k = 1,2,\cdots,s) \tag{13.24}$$

利用保守系拉格朗日方程,有

$$\dot{p}_k = \frac{\partial L}{\partial q_k} \quad (k = 1,2,\cdots,s) \tag{13.25}$$

假设广义坐标之一,例如 q_λ,并不显含于拉格朗日函数 L 中,则有

$$\dot{p}_\lambda = \frac{\partial L}{\partial q_\lambda} = 0 \tag{13.26}$$

式中,q_λ 称为可遗坐标或循环坐标,积分式(13.26),得

$$p_\lambda = 常数 \tag{13.27}$$

式(13.27)表明,对应可遗坐标或循环坐标的广义动量保持不变。

习　题

13.1 质量-弹簧系统由质量分别为 m_1 和 m_2 的物块和刚度分别为 k_1 和 k_2 的弹簧组成,如图 P13.1 所示。用拉格朗日方程法求系统运动的微分方程。

13.2 如图 P13.2 所示系统由质量分别为 m_1 和 m_2 的物块组成,物块由长为 l 的轻质不可伸长绳索连接,绳索跨过质量为 m、半径为 r 的均质滑轮。绳索与滑轮之间没有相对运动。用拉格朗日方程法求系统运动的微分方程。

13.3 质量为 m_1 的小球与长为 l 的绳索相连后悬挂于质量为 m_2、半径为 r 的均质圆盘的边界,如图 P13.3 所示。假设圆盘可绕 O 自由转动,求系统运动的微分方程。

13.4 如图 P13.4 所示,质量为 m_1 的物块沿光滑斜面滑动,质量为 m_2 的斜面沿光滑水平面滑动。求物块相对斜面的相对加速度和斜面加速度。

13.5 质量为 m 的物体悬挂于刚度为 k 的弹簧和阻尼系数为 c 的阻尼器上,如图 P13.5 所示,受大小为 $H\sin\omega_{\mathrm{f}}t$ 的周期力作用,其中 H 和 ω_{f} 分别为周期力的幅值和圆频率。求系统运动的微分方程。

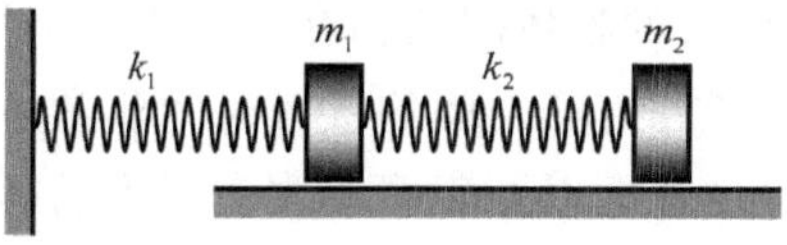

图 P13.1

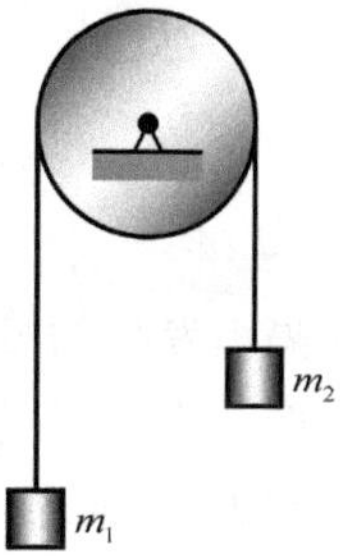

图 P13.2

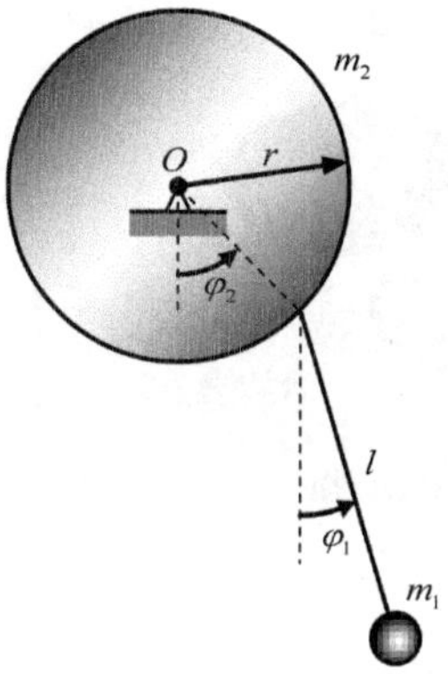

图 P13.3

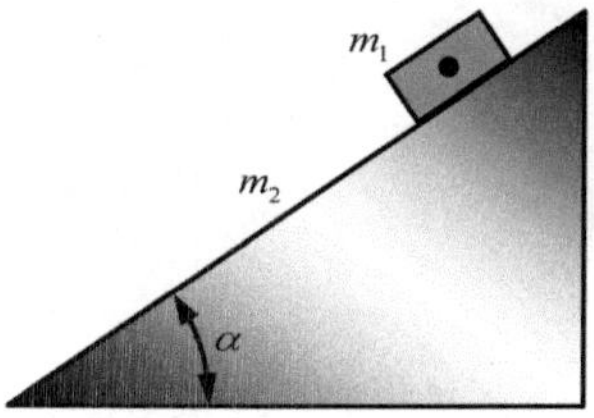

图 P13.4

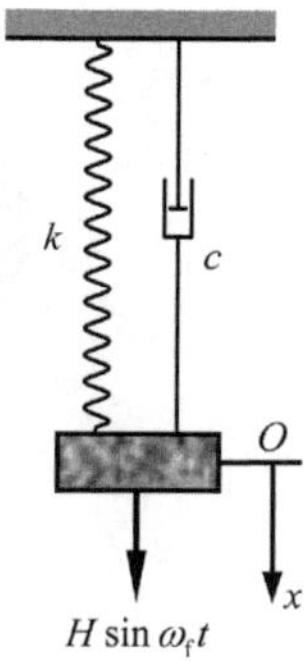

图 P13.5

13.6 具有 2 个自由度的有阻尼自由振动系统由质量为 m_1 和 m_2 的物块、刚度为 k_1 和 k_2 的弹簧、阻尼系数为 c_1 和 c_2 的阻尼器组成，如图 P13.6 所示。求系统运动的微分方程。

13.7 具有 2 个自由度的无阻尼受迫振动系统由质量均为 m 的物块、刚度均为 k 的弹簧组成，如图 P13.7 所示。系统受大小均为 $H\sin\omega_f t$ 的周期力作用，其中 H 和 ω_f 分别为周期力的幅值和圆频率。求系统运动的微分方程。

13.8 质量为 m_1 的环可沿水平梁滑动，并与刚度为 k 的弹簧相连，如图 P13.8 所示。长为 l、质量为 m_2 的均质杆与环相连，并可在包含梁的垂直平面内摆动。不计摩擦，求系统运动的拉格朗日方程。

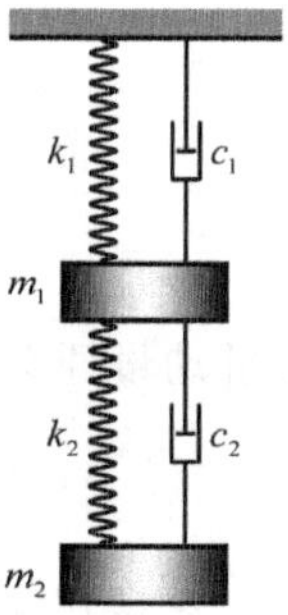

图 P13.6

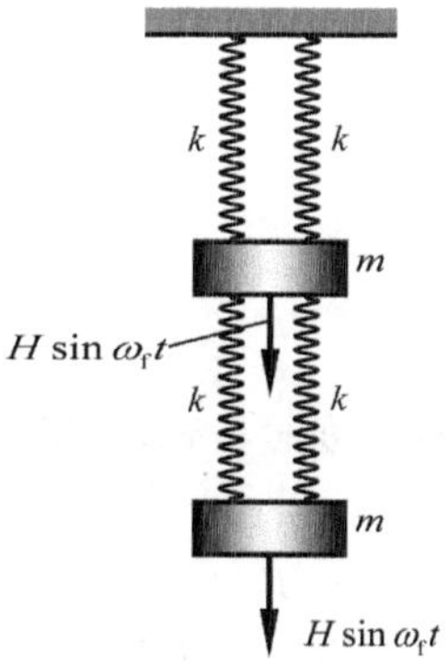

图 P13.7

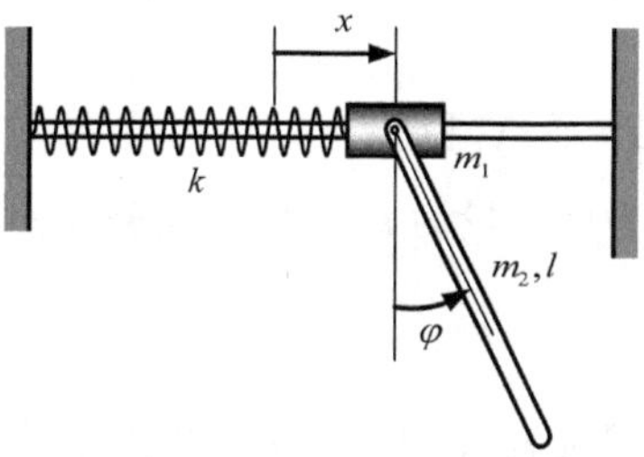

图 P13.8

第 14 章　碰　撞

在极短时间间隔内发生的具有极大作用力的物体间的相互作用称为碰撞。锤子敲击钉子即为碰撞的典型实例。当钉子受到敲击，锤子和钉子之间的接触时间非常短。然而，锤子施加在钉子上的力却非常大，因而所引起的冲量足够大以至于能够改变钉子的运动。

碰撞期间接触面的公法线称为碰撞线。如果两物体的质心在碰撞期间位于公法线上，则碰撞称为对心碰撞，如图 14.1(a)所示；否则碰撞称为偏心碰撞，如图 14.1(b)所示。

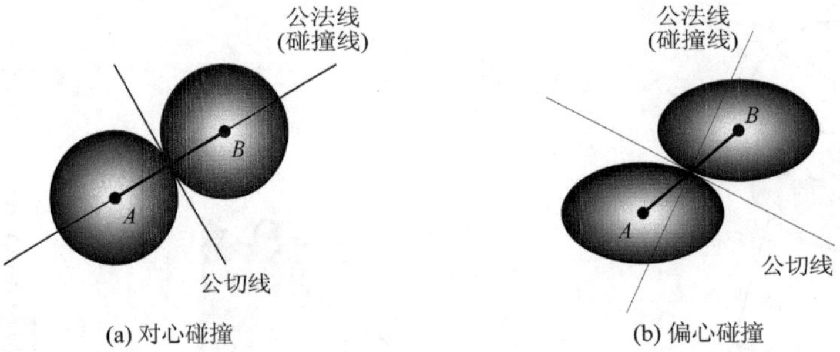

图 14.1

如果两物体的速度沿公法线，则碰撞称为正碰撞，如图 14.2 所示。如果一个或两个物体不沿公法线运动，则碰撞称为斜碰撞，如图 14.3 所示。

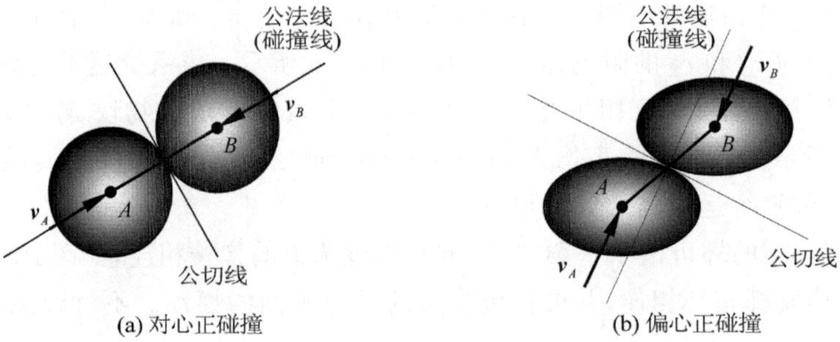

图 14.2

14.1　用于碰撞的基本原理

1. 质点碰撞

考虑质量为 m 的质点在碰撞前后分别具有速度 $\boldsymbol{v}$ 和 $\boldsymbol{v}'$，则碰撞期间的冲量动量原理可表示为

$$m\boldsymbol{v}' - m\boldsymbol{v} = \boldsymbol{I} \tag{14.1}$$

式中，$\boldsymbol{I}$ 是碰撞期间作用于质点上的冲力冲量。冲力定义为作用于质点上的力，它足够大以至于在非常短的时间间隔之内能够引起质点动量的突然改变。碰撞期间，非冲力在碰撞期间可

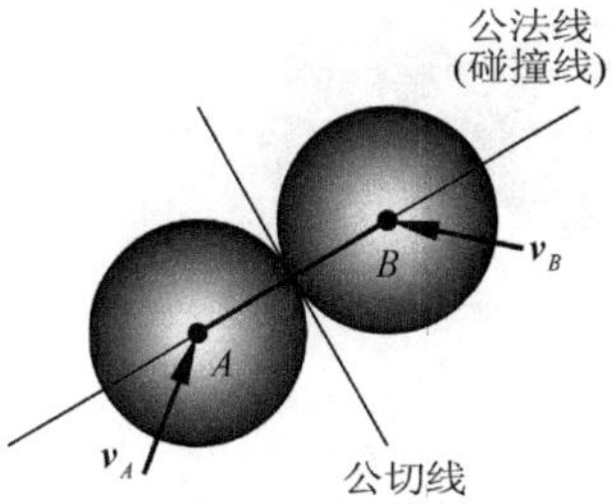

(a) 对心斜碰撞

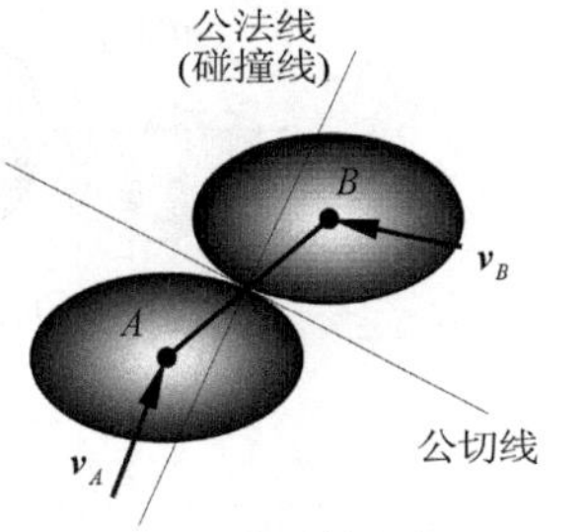

(b) 偏心斜碰撞

图 14.3

以忽略，因为由非冲力产生的冲量通常非常小。重力和弹力在碰撞期间可作为非冲力。

2. 平面运动刚体碰撞

对平面运动刚体，碰撞期间的冲量动量原理可写为

$$m\boldsymbol{v}'_C - m\boldsymbol{v}_C = \boldsymbol{I},\ I_C\omega' - I_C\omega = \boldsymbol{M}_C(\boldsymbol{I}) \tag{14.2}$$

式中，m 为刚体质量，$\boldsymbol{v}_C$ 和$\boldsymbol{v}'_C$为刚体在碰撞前后的质心速度，$\boldsymbol{I}$ 为碰撞期间作用于刚体上的冲力冲量，I_C 为刚体对质心的转动惯量，ω 和 ω'分别为碰撞前后刚体的角速度，$M_C(\boldsymbol{I})$为碰撞期间作用于刚体上的冲力对质心 C 的角动量。碰撞期间刚体质心位置的改变可忽略，因为在非常短的时间间隔之内位置改变非常小。

3. 质点系碰撞

对由 n 个质点构成的系统，碰撞期间的冲量动量原理可表示为

$$\boldsymbol{L}' - \boldsymbol{L} = \sum_{i=1}^{n}\boldsymbol{I}_i^{(e)},\ \boldsymbol{H}'_O - \boldsymbol{H}_O = \sum_{i=1}^{n}\boldsymbol{M}_O(\boldsymbol{I}_i^{(e)}) \tag{14.3}$$

式中，$\boldsymbol{L} = \sum_{i=1}^{n} m_i\boldsymbol{v}_i$ 和$\boldsymbol{L}' = \sum_{i=1}^{n} m_i\boldsymbol{v}'_i$分别为碰撞前后系统的动量，$\sum_{i=1}^{n}\boldsymbol{I}_i^{(e)}$ 为碰撞期间作用于系统上的外冲力冲量，$\boldsymbol{H}_O = \sum_{i=1}^{n}\boldsymbol{r}_i \times m_i\boldsymbol{v}_i$ 和$\boldsymbol{H}'_O = \sum_{i=1}^{n}\boldsymbol{r}_i \times m_i\boldsymbol{v}'_i$分别为碰撞前后系统对定点$O$ 的角动量，$\sum_{i=1}^{n}\boldsymbol{M}_O(\boldsymbol{I}_i^{(e)}) = \sum_{i=1}^{n}\boldsymbol{r}_i \times \boldsymbol{I}_i^{(e)}$ 为碰撞期间作用于系统上的外冲力对定点O 的角冲量。碰撞期间系统中每个质点的位移已经被忽略，因为时间间隔如此之短以至于每个质点的位移非常小。

14.2 恢复系数

1. 对心碰撞

考虑对心正碰撞，如图 14.4(a)所示。质量为 m_A 和 m_B 的两个质点在碰撞前分别具有速度$\boldsymbol{v}_A$ 和$\boldsymbol{v}_B$，如果 $v_A > v_B$，那么质点 A 最终将会与质点 B 发生碰撞。假设碰撞后两个质点具有如图 14.4(b)所示的速度$\boldsymbol{v}'_A$ 和$\boldsymbol{v}'_B$，那么恢复系数可表示为

$$e = \frac{v'_B - v'_A}{v_A - v_B} \tag{14.4}$$

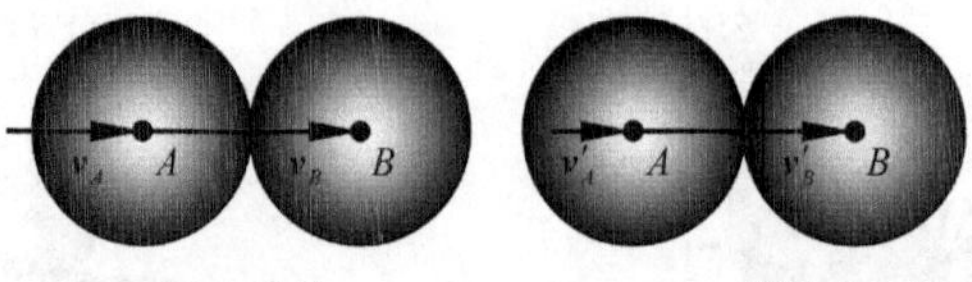

(a) 碰撞前　　(b) 碰撞后

图 14.4

式中，$v_A - v_B$ 表示碰撞前两个质点接近的相对速度，$v'_B - v'_A$ 表示碰撞后两个质点分离的相对速度。依据恢复系数值，碰撞可分类如下：

(1) 完全弹性碰撞($e=1$)

对完全弹性碰撞，有 $v'_B - v'_A = v_A - v_B$，即碰撞前后两个质点的相对速度相等。在完全弹性碰撞中，两个质点的总能量和总动量均守恒。

(2) 完全塑性碰撞($e=0$)

对完全塑性碰撞，有 $v'_B = v'_A$，即两个质点碰撞后不再分离。在完全塑性碰撞中总动量守恒，但总能量不再守恒。

(3) 弹塑性碰撞($0<e<1$)

对弹塑性碰撞，应该注意质点的总能量不守恒，但总动量仍然守恒。

对图 14.5 所示对心斜碰撞，假设两个质点在碰撞前具有速度$\boldsymbol{v}_A$ 和$\boldsymbol{v}_B$，在碰撞后具有速度 v'_A和 v'_B，那么恢复系数为

$$e = \frac{(v'_B)_n - (v'_A)_n}{(v_A)_n - (v_B)_n} \tag{14.5}$$

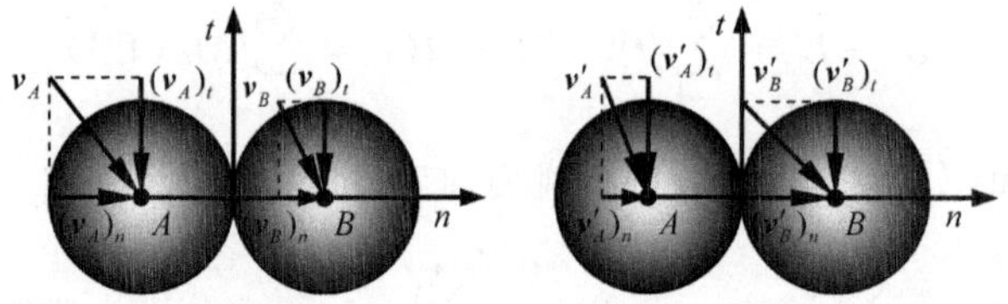

(a) 碰撞前　　(b) 碰撞后

图 14.5

式中，$(v_A)_n - (v_B)_n$ 和$(v'_B)_n - (v'_A)_n$ 分别表示两个质点碰撞前、后相对速度的法向分量。

例 14.1

如图 E14.1 所示，具有相同尺寸的两个均质球，质量分别为 $m_A = 0.5$ kg 和 $m_B = 0.3$ kg，在光滑水平面上分别以大小为 $v_A = 10$ m/s 和 $v_B = 5$ m/s 的速度滑动。已知恢复系数 $e = 0.6$，求碰撞后每个球的速度。

解

沿 t 轴和 n 轴分解每个球在碰撞前的速度，有

$$(v_A)_t = 6\ \text{m/s},\ (v_A)_n = 8\ \text{m/s},\ (v_B)_t = 4\ \text{m/s},\ (v_B)_n = 3\ \text{m/s}$$

冲量动量原理应用于每个球，并注意碰撞期间作用于每个球的冲力冲量沿 n 轴，则有

$$m_A[(v'_A)_t \boldsymbol{e}_t + (v'_A)_n \boldsymbol{e}_n] - m_A[(v_A)_t \boldsymbol{e}_t + (v_A)_n \boldsymbol{e}_n] = -I \boldsymbol{e}_n$$
$$m_B[(v'_B)_t \boldsymbol{e}_t + (v'_B)_n \boldsymbol{e}_n] - m_B[(v_B)_t \boldsymbol{e}_t + (v_B)_n \boldsymbol{e}_n] = I \boldsymbol{e}_n$$

式中，$\boldsymbol{e}_t$ 和 $\boldsymbol{e}_n$ 分别为沿 t 轴和 n 轴的单位矢量，I 为一个球作用于另一个球上的冲力冲量值。

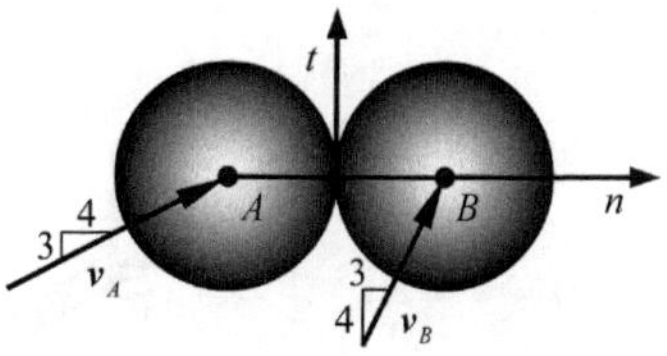

图 E14.1

因为作用于每个球上的冲力冲量的 t 分量等于零，因此碰撞期间每个球的速度的 t 分量分别保持不变，即有

$$(v'_A)_t = (v_A)_t = 6 \text{ m/s},\ (v'_B)_t = (v_B)_t = 4 \text{ m/s}$$

两个球作为一个系统考虑，尽管作用于每个球上的冲力冲量的 n 分量不等于零，但是作用于系统上的冲力冲量的 n 分量仍然等于零，因此有

$$m_A[(v'_A)_n - (\boldsymbol{v}_A)_n] + m_B[(v'_B)_n - (\boldsymbol{v}_B)_n] = 0$$

利用 $(v'_B)_n - (v'_A)_n = e[(v_A)_n - (v_B)_n]$，得

$$(v'_A)_n = (v_A)_n - \frac{(1+e)m_B}{m_A + m_B}[(v_A)_n - (v_B)_n] = 5 \text{ m/s}$$

$$(v'_B)_n = (v_B)_n + \frac{(1+e)m_A}{m_A + m_B}[(v_A)_n - (v_B)_n] = 8 \text{ m/s}$$

利用上述计算结果，则碰撞后每个球的速度分别为

$$\boldsymbol{v}'_A = (v'_A)_t \boldsymbol{e}_t + (v'_A)_n \boldsymbol{e}_n = 6\,\boldsymbol{e}_t + 5\,\boldsymbol{e}_n \text{ m/s}$$
$$\boldsymbol{v}'_B = (v'_B)_t \boldsymbol{e}_t + (v'_B)_n \boldsymbol{e}_n = 4\,\boldsymbol{e}_t + 8\,\boldsymbol{e}_n \text{ m/s}$$

2. 偏心碰撞

考虑图 14.6 所示两个物体之间发生的偏心碰撞。假设碰撞前接触点 A 和 B 的速度分别为 $\boldsymbol{v}_A$ 和 $\boldsymbol{v}_B$，并假设碰撞后的速度分别为 $\boldsymbol{v}'_A$ 和 $\boldsymbol{v}'_B$，则恢复系数可定义为

$$e = \frac{(v'_B)_n - (v'_A)_n}{(v_A)_n - (v_B)_n} \tag{14.6}$$

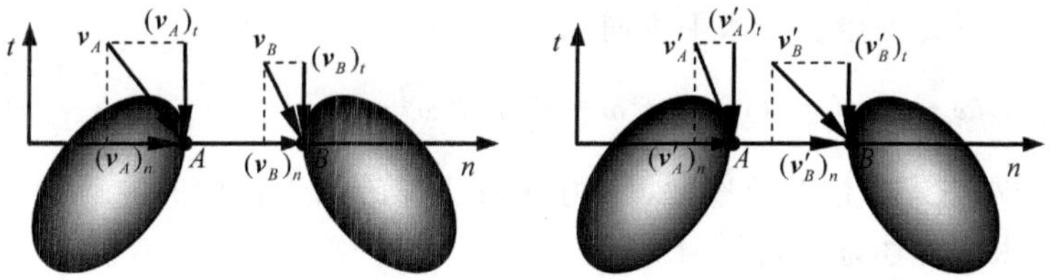

(a) 碰撞前　　(b) 碰撞后

图 14.6

式中，$(v_A)_n - (v_B)_n$ 和 $(v'_B)_n - (v'_A)_n$ 分别为碰撞前、后沿物体接触面公法线方向的相对速度分量。

例 14.2

如图 E14.2(a)所示质量为 m、长度为 l、与水平方向成 θ 角的均质细长杆 AB 以垂直速度 $\boldsymbol{v}$ 和零角速度撞击光滑表面。假设为完全弹性碰撞，试推导杆碰撞后的角速度、碰撞期间表面施加于杆的冲力冲量。

解

建立如图 E14.2(b)所示坐标系，对杆应用冲量动量原理，并注意杆的质心速度垂直，碰撞期间作用于杆的冲力冲量也垂直，因此有

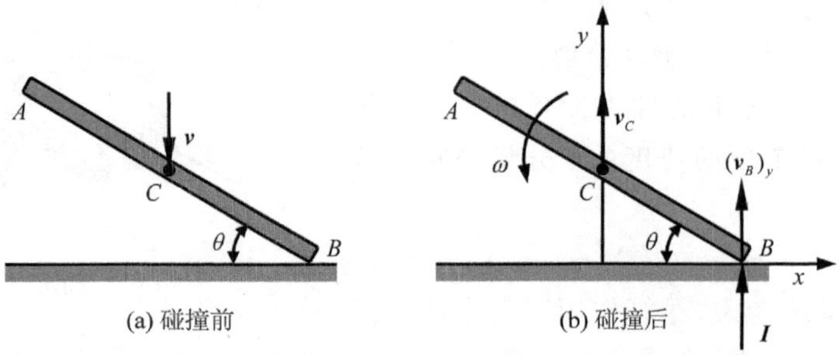

图 E14.2

$$mv_C - m(-v) = I,\ I_C\omega - 0 = I\left(\frac{1}{2}l\cos\theta\right)$$

式中,I_C 为杆对质心的转动惯量,I 为表面作用于杆的冲力冲量值。利用 $I_C=\frac{1}{12}ml^2$、$(v_B)_y=ev=v$、$v_C=(v_B)_y-\omega\left(\frac{1}{2}l\right)\cos\theta=v-\frac{1}{2}\omega l\cos\theta$,并求解上述方程,得

$$\omega = \frac{12v\cos\theta}{(1+3\cos^2\theta)l},\ I = \frac{2mv}{1+3\cos^2\theta}$$

例 14.3

质量为 m_1 的子弹以大小为 v 的水平速度射击进入质量为 m_2、长度为 l 的均质细长杆中,如图 E14.3 所示。已知杆最初静止,求:(a)子弹嵌入后杆的角速度,(b)A 处冲击反力的冲量,(c)A 处冲击反力为零时的距离 a。

解

对由子弹和杆件构成的系统应用冲量动量原理,并注意碰撞期间作用于杆件的冲击反力冲量位于水平方向,有

$$m_1a\omega + \frac{1}{2}m_2l\omega - m_1v = I_A,\ m_1a^2\omega + \frac{1}{3}m_2l^2\omega - m_1av = 0$$

式中,ω 为子弹刚刚嵌入后杆的角速度,I_A 为 A 处支撑施加于杆件的冲击反力的冲量。求解上述方程,得

$$\omega = \frac{3m_1av}{3m_1a^2+m_2l^2},\ I_A = \frac{(3a-2l)lm_1m_2v}{6m_1a^2+2m_2l^2}$$

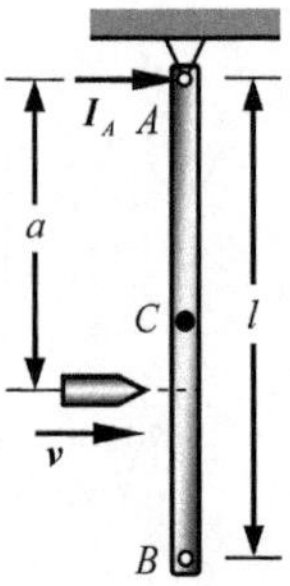

图 E14.3

由上述第 2 个表达式可以看出,如果 A 处的冲击反力为零,则距离 a 等于

$$a = \frac{2}{3}l$$

杆件上距支点为 $a=2l/3$ 的点称为撞击中心。撞击中心指,当垂直碰撞发生在撞击中心时,不会在支点引起冲力。

习　题

14.1 质量为 1 kg 的球 A 以速度 $\boldsymbol{v}_A$ 运动,受到质量为 2 kg、速度大小为 5 m/s 的球 B 撞

击，如图 P14.1 所示。已知碰撞后球 B 的速度为零，恢复系数为 0.5，求：球 A 分别在(a)碰撞前和(b)碰撞后的速度。

14.2 四个等质量的球通过等长度的绳索悬挂于天花板上，绳索之间的间隔略大于球的直径，如图 P14.2 所示。球 A 被拉回并释放后，三次碰撞将会在相邻球 A、B、C 和 D 之间依次发生。设 e 为球之间的恢复系数，$\boldsymbol{v}$ 为球 A 在撞击球 B 前的速度，求：(a)第一次碰撞后 A 和 B 的速度，(b)第二次碰撞后 B 和 C 的速度，(c)第三次碰撞后 C 和 D 的速度。

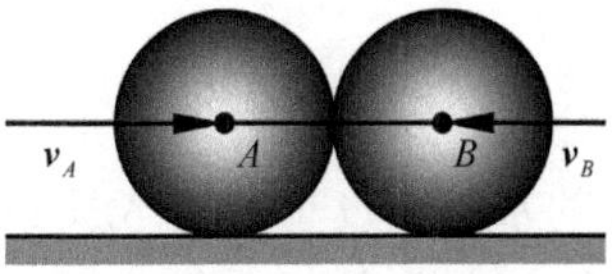

图 P14.1

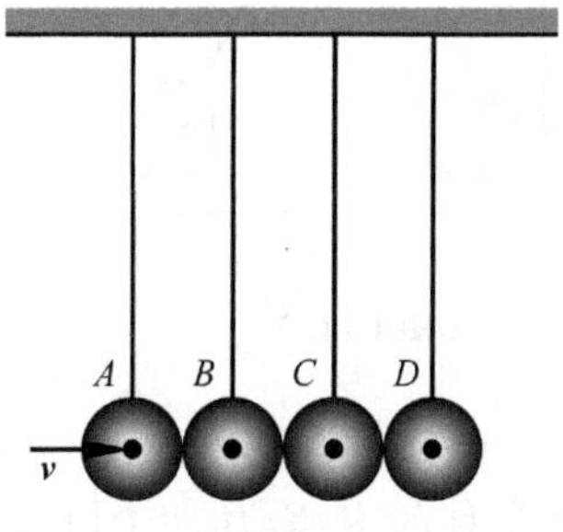

图 P14.2

14.3 质量为 m_A 的球 A 以平行于地面的速度 v 运动，撞击质量为 m_B 的物块斜面，如图 P14.3 所示。物块可在地面上自由滑动，初始静止。已知球与物块之间的恢复系数为 e，求碰撞后物块的速度。

14.4 质量为 2 kg 的物块 A 以大小为 $v=1$ m/s 的速度撞击质量为 1 kg 的物块 B，物块 B 初始静止，并通过绳索悬挂于点 O，如图 P14.4 所示。已知物块之间的恢复系数 $e=0.8$ 以及物块与水平面之间的动摩擦系数 $\mu_k=0.5$，求碰撞后：(a)物块 B 所能达到的最大高度，(b)物块 A 的移动距离。

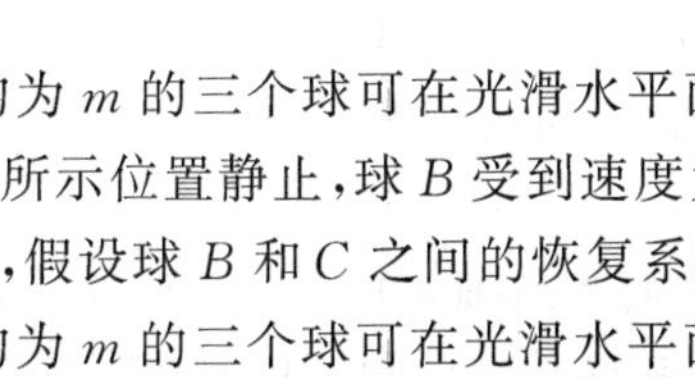

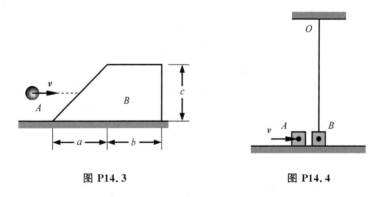

图 P14.3

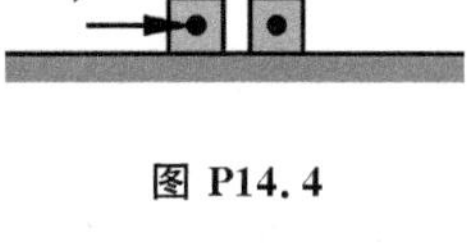

图 P14.4

14.5 质量均为 m 的三个球可在光滑水平面上滑动。球 A 和 B 通过不可伸长绳索连接，并在如图 P14.5 所示位置静止，球 B 受到速度为 $\boldsymbol{v}$ 的球 C 撞击。已知球 B 受到球 C 撞击时绳索处于拉紧状态，假设球 B 和 C 之间的恢复系数 $e=1$，求碰撞后每个球的速度。

14.6 质量均为 m 的三个球可在光滑水平面上滑动。球 A 和 B 通过不可伸长绳索连接，并在如图 P14.6 所示位置静止，球 B 受到速度为 $\boldsymbol{v}$ 的球 C 撞击。已知球 B 受到球 C 撞击时绳索处于松弛状态，假设球 B 和 C 之间的恢复系数等于 e，求绳索拉紧后每个球的速度。

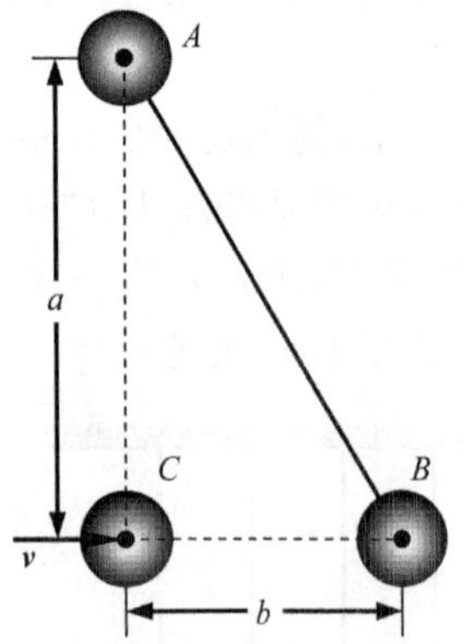

图 P14.5

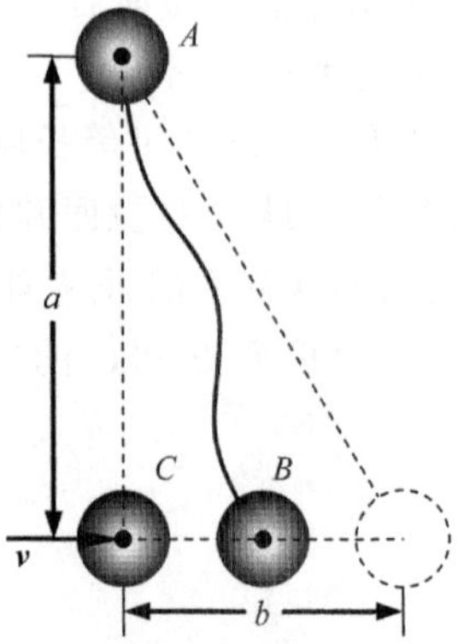

图 P14.6

14.7 质量 $m_1=10$ g 的子弹以大小 $v=500$ m/s 的水平速度射击进入质量 $m_2=5$ kg、长度 $l=1$ m 的均质细长杆的下端，如图 P14.7 所示。已知杆最初静止，求：(a)子弹嵌入后杆的角速度，(b)A 处施加于杆的冲量。

14.8 质量为 m、长度为 l 的均质细长杆 AB 以大小为 v 的垂直速度和零角速度撞击光滑刚性支撑 D，如图 P14.8 所示。已知 $a=l/5$ 和 $e=0$，求：(a)碰撞后杆的角速度和质心速度，(b)D 处作用于杆的冲量。

14.9 质量为 m、长度为 l 的均质细长杆 AB 以逆时针角速度 ω 和零质心速度撞击光滑刚性支撑 D，如图 P14.9 所示。已知 $a=l/5$ 和 $e=1$，求：(a)碰撞后杆的角速度和质心速度，(b)D 处作用于杆的冲量。

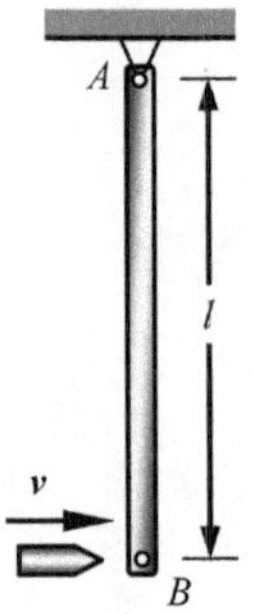

图 P14.7

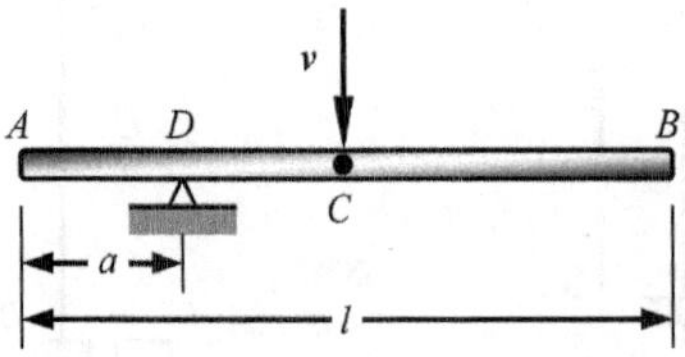

图 P14.8

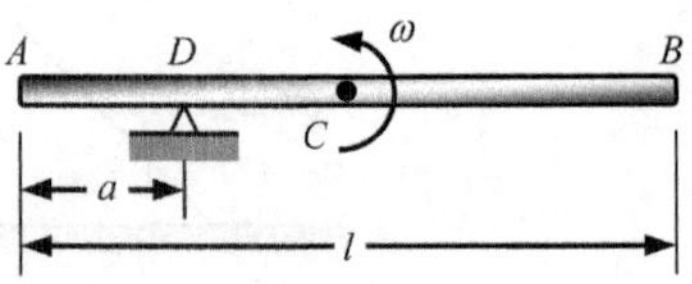

图 P14.9

14.10 如图 P14.10 所示质量为 m、长度为 l、与水平方向成 θ 角的均质细长杆 AB 以垂直速度 $\boldsymbol{v}$ 和零角速度撞击光滑表面。假设杆与表面之间的恢复系数为 e，试推导杆碰撞后的角速度。

14.11 如图 P14.11 所示质量为 m、长度为 l 的均质细长杆 AB 以速度 $\boldsymbol{v}$ 下落，其 B 端撞击倾角为 θ 的光滑斜面。假设碰撞为完全弹性，求碰撞后杆的角速度和质心速度。

14.12 如图 P14.12 所示质量为 m_1、长度为 l 的均质细长杆 AB 从水平位置静止释放，向下摆动到垂直位置时撞击质量为 m_2 的物块 C，物块 C 静止放置于光滑表面。假设杆与物块之间的恢复系数为 e，求碰撞后物块的速度。

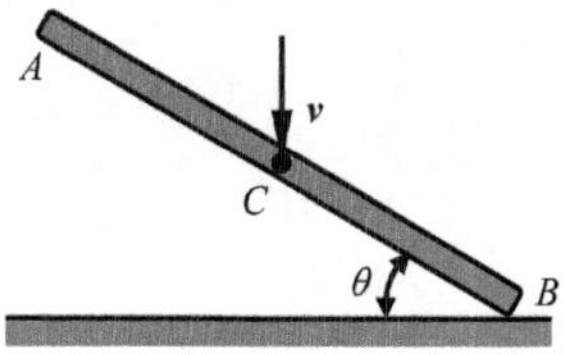

图 P14.10

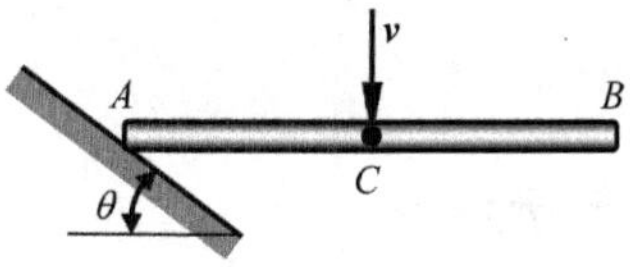

图 P14.11

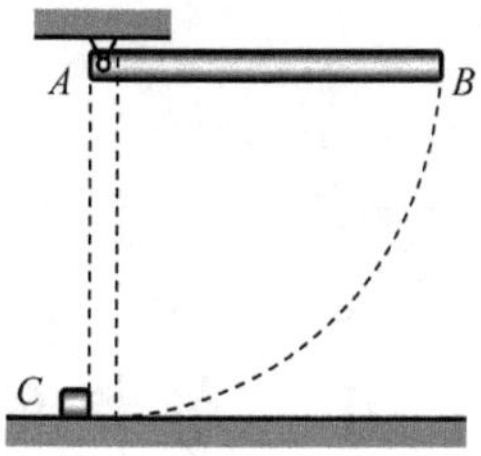

图 P14.12

附录Ⅰ　重心与形心

Ⅰ.1　薄板的重心与形心

1. 薄板的重心

考虑如图Ⅰ.1所示薄板，其重心定义为

$$\bar{x}=\frac{\int x\mathrm{d}W}{W},\ \bar{y}=\frac{\int y\mathrm{d}W}{W} \tag{Ⅰ.1}$$

式中，$W=\int \mathrm{d}W$ 为薄板的重量。

2. 薄板的形心

如果薄板具有均匀密度 ρ 和均匀厚度 t，则薄板重心 G 将与其形心 C 重合，如图Ⅰ.2所示。利用 $W=\rho gAt$，则薄板的形心可表示为

$$\bar{x}=\frac{\int x\mathrm{d}A}{A},\ \bar{y}=\frac{\int y\mathrm{d}A}{A} \tag{Ⅰ.2}$$

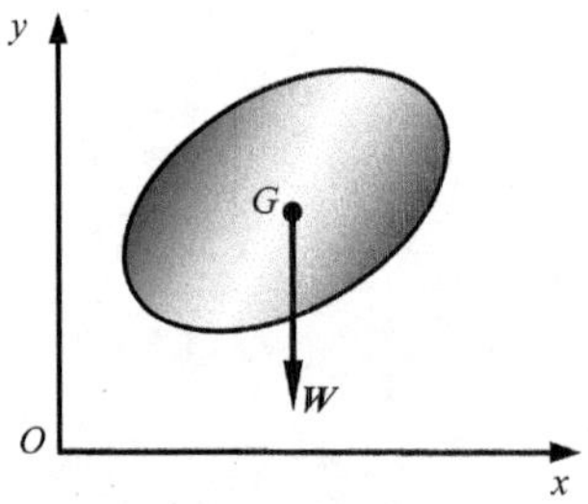

图Ⅰ.1

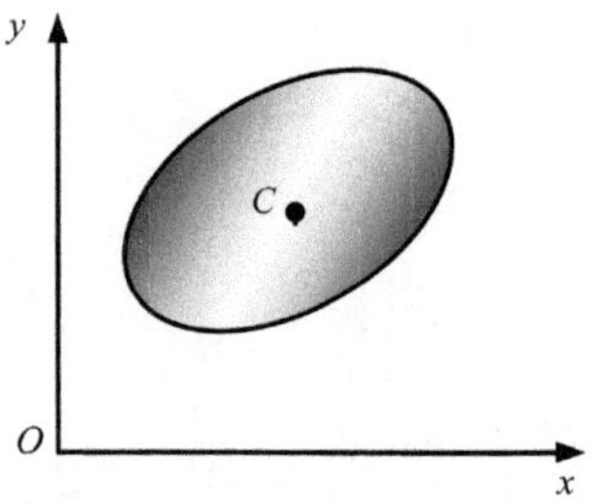

图Ⅰ.2

式中，$A=\int \mathrm{d}A$ 为薄板的面积。

Ⅰ.2　组合薄板的重心与形心

1. 组合薄板的重心

考虑如图Ⅰ.3所示组合薄板，则组合薄板的重心可表示为

$$\bar{X}=\frac{\sum \bar{x}_i W_i}{W},\ \bar{Y}=\frac{\sum \bar{y}_i W_i}{W} \tag{Ⅰ.3}$$

式中，$W=\sum W_i$ 为组合薄板的重量。

2. 组合薄板的形心

如果组合薄板具有均匀密度 ρ 和均匀厚度 t，则组合薄板的重心 G 将与其形心 C 重合，如图Ⅰ.4 所示。利用 $W_i=\rho g A_i t$，则组合薄板的形心可表示为

$$\bar{X}=\frac{\sum \bar{x}_i A_i}{A},\ \bar{Y}=\frac{\sum \bar{y}_i A_i}{A} \tag{Ⅰ.4}$$

式中，$A=\sum A_i$ 为组合薄板的面积。

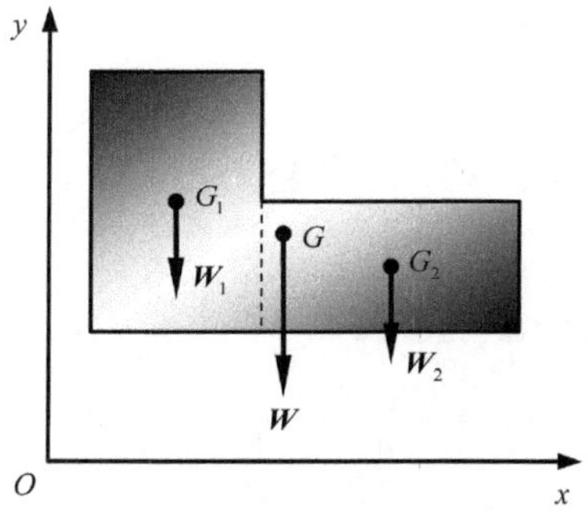

图Ⅰ.3

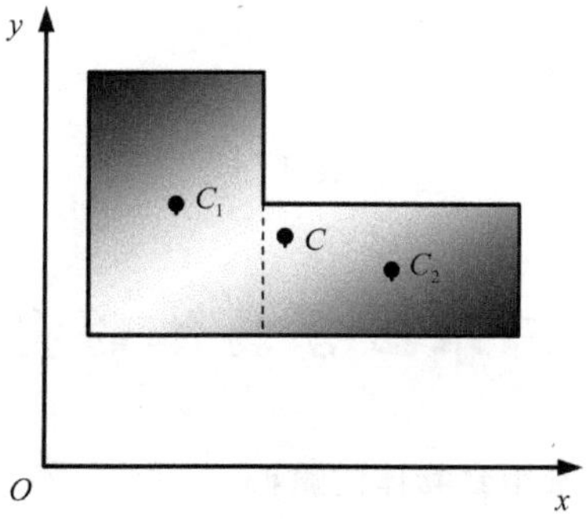

图Ⅰ.4

Ⅰ.3　三维物体的重心与形心

1. 三维物体的重心

考虑如图Ⅰ.5 所示三维物体，其重心定义为

$$\bar{x}=\frac{\int x\mathrm{d}W}{W},\ \bar{y}=\frac{\int y\mathrm{d}W}{W},\ \bar{z}=\frac{\int z\mathrm{d}W}{W} \tag{Ⅰ.5}$$

式中，$W=\int dW$ 为三维物体的重量。

2. 三维物体的形心

如果三维物体具有均匀密度 ρ，则三维物体的重心 G 将与其形心 C 重合，如图Ⅰ.6 所示。利用 $W=\rho g V$，则三维物体的形心可表示为

$$\bar{x}=\frac{\int x dV}{V},\ \bar{y}=\frac{\int y dV}{V},\ \bar{z}=\frac{\int z dV}{V} \tag{Ⅰ.6}$$

式中，$V=\int dV$ 为三维物体的体积。

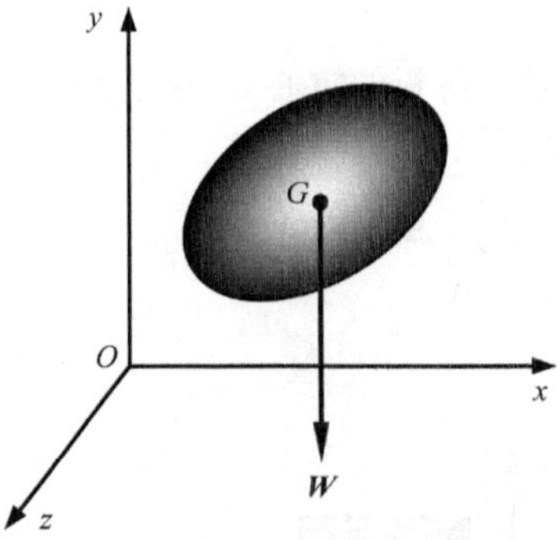

图Ⅰ.5

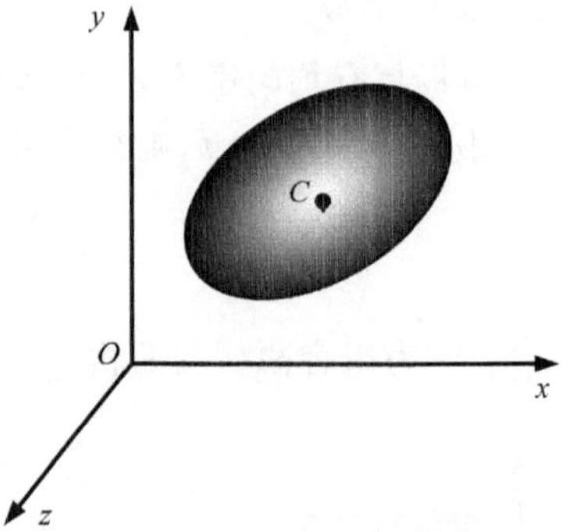

图Ⅰ.6

Ⅰ.4 三维组合物体的重心与形心

1. 三维组合物体的重心

考虑如图Ⅰ.7所示三维组合物体,则三维组合物体的重心定义为

$$\bar{X}=\frac{\sum \bar{x}_i W_i}{W},\ \bar{Y}=\frac{\sum \bar{y}_i W_i}{W},\ \bar{Z}=\frac{\sum \bar{z}_i W_i}{W} \tag{Ⅰ.7}$$

式中,$W=\sum W_i$ 为三维组合物体的重量。

2. 三维组合物体的形心

如果三维组合物体具有均匀密度 ρ,则三维组合物体的重心 G 将与其形心 C 重合,如图Ⅰ.8所示。利用 $W_i=\rho g V_i$,则三维组合物体的形心可表示为

$$\bar{X}=\frac{\sum \bar{x}_i V_i}{V},\ \bar{Y}=\frac{\sum \bar{y}_i V_i}{V},\ \bar{Z}=\frac{\sum \bar{z}_i V_i}{V} \tag{Ⅰ.8}$$

式中,$V=\sum V_i$ 为三维组合物体的体积。

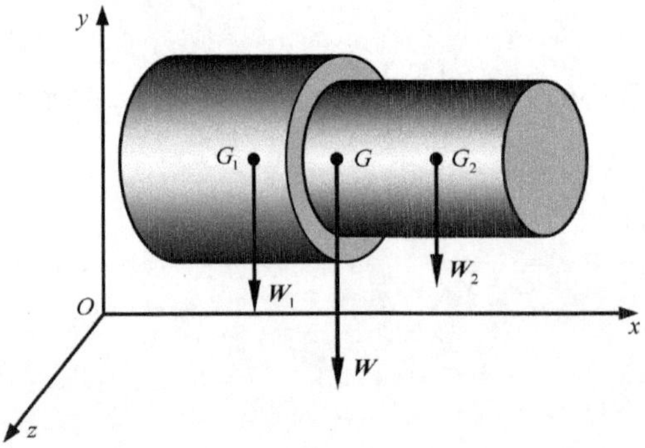

图Ⅰ.7

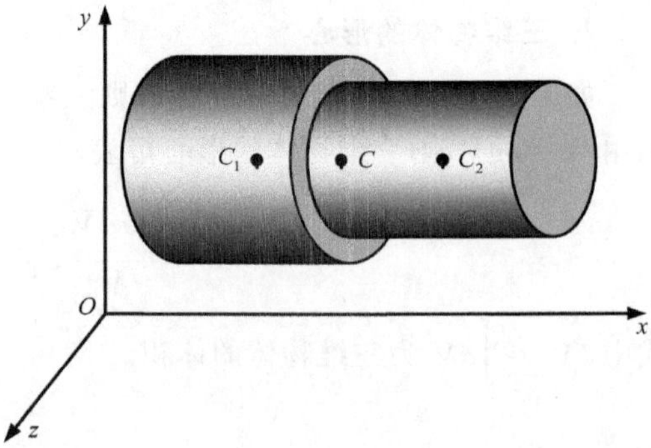

图Ⅰ.8

附录Ⅱ　转动惯量

Ⅱ.1　转动惯量与回转半径

考虑如图Ⅱ.1所示质量为 m 的三维物体,则物体对 x、y 和 z 轴的转动惯量分别定义为

$$I_x = \int (y^2 + z^2) dm\,,\ I_y = \int (z^2 + x^2) dm\,,\ I_z = \int (x^2 + y^2) \mathrm{d}m \tag{Ⅱ.1}$$

在国际单位制中,转动惯量 I_x、I_y 和 I_z 的单位为 kg·m²。

物体对 x、y 和 z 轴的回转半径分别定义为

$$i_x = \sqrt{\frac{I_x}{m}},\ i_y = \sqrt{\frac{I_y}{m}},\ i_z = \sqrt{\frac{I_z}{m}} \tag{Ⅱ.2}$$

在国际单位制中,回转半径 i_x、i_y 和 i_z 的单位为 m。

Ⅱ.2　平行移轴定理

考虑如图Ⅱ.2所示质量为 m 的三维物体,物体对 x 和 x_C 轴的转动惯量分别表示为

$$I_x = \int (y^2 + z^2) dm\,,\ I_{x_C} = \int ({y_C}^2 + {z_C}^2) \mathrm{d}m \tag{Ⅱ.3}$$

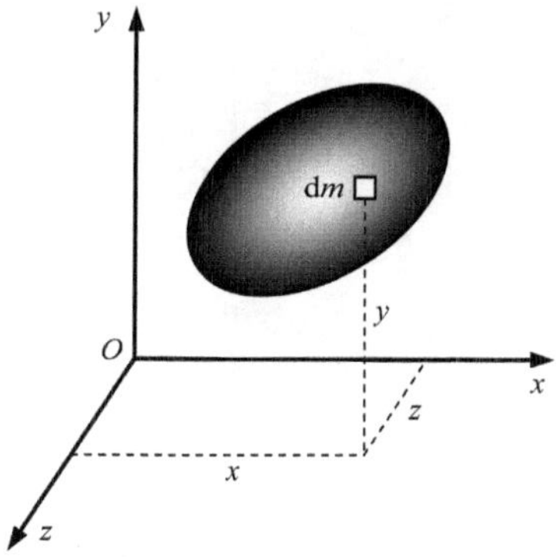

图Ⅱ.1

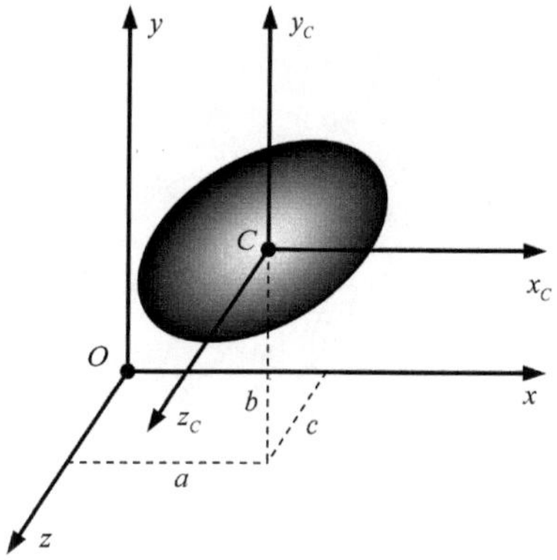

图Ⅱ.2

代入 $y=y_C+b$ 和 $z=z_C+c$,得

$$I_x = \int (y^2 + z^2) \mathrm{d}m = \int (y_C^2 + z_C^2) \mathrm{d}m + 2\int (by_C + cz_C) \mathrm{d}m + (b^2 + c^2)\int \mathrm{d}m \tag{Ⅱ.4}$$

利用 $\int (by_C + cz_C) \mathrm{d}m = 0$,方程(Ⅱ.4)可简化为

$$I_x = I_{x_C} + m(b^2 + c^2) \tag{Ⅱ.5}$$

同理,得

$$I_y = I_{y_C} + m(c^2 + a^2),\ I_z = I_{z_C} + m(a^2 + b^2) \tag{Ⅱ.6}$$

方程(Ⅱ.5)和方程(Ⅱ.6)所表示的关系称为平行移轴定理。当已知物体对形心轴的转动惯量时,上述关系常用于计算物体对任意轴的转动惯量。

参考文献

[1] 哈尔滨工业大学理论力学教研室. 理论力学(I). 7 版. 北京：高等教育出版社，2009.
[2] 哈尔滨工业大学理论力学教研室. 理论力学(II). 7 版. 北京：高等教育出版社，2009.
[3] 范钦珊，陈建平，等. 理论力学. 2 版. 北京：高等教育出版社，2010.
[4] 王开福. 工程力学. 北京：科学出版社，2012.